OBSERVATIONAL TESTS OF THE STELLAR EVOLUTION THEORY

INTERNATIONAL ASTRONOMICAL UNION
UNION ASTRONOMIQUE INTERNATIONALE

SYMPOSIUM No. 105
HELD IN GENEVA, SWITZERLAND, SEPTEMBER, 12–16, 1983

OBSERVATIONAL TESTS OF THE STELLAR EVOLUTION THEORY

EDITED BY

ANDRÉ MAEDER
Geneva Observatory, Sauverny, Switzerland

and

ALVIO RENZINI
Astronomy Department, University of Bologna, Bologna, Italy

D. REIDEL PUBLISHING COMPANY

A MEMBER OF THE KLUWER  ACADEMIC PUBLISHERS GROUP

DORDRECHT / BOSTON / LANCASTER

Library of Congress Cataloging in Publication Data

Main entry under title:

Observational tests of the stellar evolution theory.

At head of title: International Astronomical Union, Union astronomique internationale.
Includes bibliographies and index.
1. Stars–Evolution–Congresses. I. Maeder, André, 1942-
II. Renzini, Alvio. III. International Astronomical Union.
QB806.O27   1984   523.8   84-6978
ISBN 90–277–1774–5
ISBN 90–277–1775–3 (pbk.)

*Published on behalf of
the International Astronomical Union
by
D. Reidel Publishing Company, P.O. Box 17, 3300 AA Dordrecht, Holland*

*Sold and distributed in the U.S.A. and Canada
by Kluwer Academic Publishers,
190 Old Derby Street, Hingham, MA 02043, U.S.A.*

*In all other countries, sold and distributed
by Kluwer Academic Publishers Group,
P.O. Box 322, 3300 AH Dordrecht, Holland*

Printed in The Netherlands

TABLE OF CONTENTS

I. EVOLUTION OF LOW AND INTERMEDIATE MASS STARS: OBSERVATIONS AND MODELS

SOLAR CONSTRAINTS

OPEN AND POPULOUS CLUSTERS, RED GIANTS

PLANETARY NUCLEI AND OB SUBDWARFS

II. EVOLUTION OF MASSIVE STARS: EVOLUTION AND MODELS

III. BINARITY, PULSATION, ROTATION AND MIXING

BINARY EVOLUTION

EVOLUTION AND PULSATION

ROTATION AND MIXING

IV. RELATION WITH CHEMICAL EVOLUTION OF GALAXIES AND COSMOLOGY

PREFACE

*"If simple perfect laws uniquely rule the universe,
should not pure thought be capable of uncovering this
perfect set of laws without having to lean on the
crutches of tediously assembled observations? True,
the laws to be discovered may be perfect, but the
human brain is not. Left on its own, it is prone to
stray, as many past examples sadly prove. In fact,
we have missed few chances to err until new data
freshly gleaned from nature set us right again for
the next steps. Thus pillars rather than crutches
are the observations on which we base our theories;
and for the theory of stellar evolution these pillars
must be there before we can get far on the right track."*

These words written by Martin Schwarzschild in his famous book entitled "Structure and Evolution of the Stars"(1958) remind us how necessary and fruitful is the interplay of stellar evolution theory and observations. Clearly, observations are the great censor by their possibility of confirming or contradicting theoretical constructions. In addition, they have a driving role: new and sometimes unexpected facts may give rise to progressive ideas and stimulate further theoretical developments. In turn, theory, in its major role of sifting out and placing the facts in a logical sequence based on physical laws, must also be predictive and indicate new and pertinent observations to be undertaken. The aim of this Symposium is to favour these interactions, which means comparing the numerous recent results in ground-based and space observations of stars in our Galaxy and in external galaxies with the considerable developments of evolutionary models. The period of evolution considered here covers the zero-age sequence to the pre-supernova stage, or pre-white dwarf stage, according to the initial mass considered; the cases of star formation and compact stars have already been treated in other IAU symposia.

From the middle of the sixties the appearance of the second generation of computers prompted several groups to start massive calculations of stellar evolutionary sequences. Progress in this field has been steady over the past 20 years, and it is now currently feasible (at least in principle, but often also in practice) to compute in just one "run" the whole evolutionary history of a model star, from its pre-main sequence stage, to its final fate as either a white dwarf or a supernova explosion.

These models exhibit a close "resemblance" to real stars: for example, they also become red giants, expose at some stage nuclearly processed materials, produce planetary nebulae, or end their career with

violent hydrodynamical events. In this respect the theory is highly
successful. But a closer, finely quantitative comparison with the observations may often reveal subtle differences between model and real stars.
A first question to raise when such a disagreement is perceived should
be: are the "observations" correct? Indeed, quite often the procedure of
distilling physically interesting quantities (e.g. temperatures, luminosities, abundances, etc) from the brute data (magnitudes, spectra, etc)
is highly indirect, involving other theoretical tools and assumptions,
which, in turn, may well be at least as imperfect as the current theory
of stellar structure and evolution. Anyway, when the existence of some
serious disagreement is unambiguously established, the reliability of
the models is called into question, and cures and refinements have to be
found.

The astrophysical interest in such a cross-check between theory and
observations is at least twofold: first, we would like to fully understand every detail of how stars evolve and die, and, second, we are very
interested in applying the predictions of the stellar evolution theory
as a useful tool in a large variety of astrophysical fields (like age
determinations and their implications for galactic dynamics and cosmology,
stellar statistics, spectral and chemical evolution of galaxies, etc).
This symposium on "The Observational Tests of the Stellar Evolution Theory"
was conceived and organized in precisely this spirit, with its main goal
being a collective assessment of the reliability of current evolutionary
models. Obviously, the conclusions on this issue cannot be but very complex and/or rather problematic: they just form the matter of these proceedings!

The Symposium was organized about a series of review lectures
followed by contributed talks, and about a particularly exciting and numerous poster session. Some 140 participants from 25 countries attended
the meeting which was lively and successful. The local organization was
undertaken by the staff of Geneva Observatory and the Astronomical Institute of the University of Lausanne who we would like to thank for their
help. In particular, we express our gratitude to Mrs M.-L. Zeier for her
efficiency and administrative ability during the preparation and the
meeting and for her help in editing the proceedings. We are grateful to
Mr. J.-F. Bopp for his valuable technical assistance in the organization.
We are deeply indebted to the following organizations and institutions who
gave generous financial support:

- The International Astronomical Union
- The Swiss National Science Foundation
- The Swiss Academy of Sciences
- The Faculty of Sciences of Geneva University
- The Swiss Bank Corporation

André MAEDER, Alvio RENZINI
Editors of the Symposium

LIST OF PARTICIPANTS

ACKER, A., Observatoire de Strasbourg, France
AIAD, A.K. Cairo University, Egypt
ALCAINO, G., Instituto Isaac Newton, Santiago, Chile
ANDERSEN, J., Copenhagen University Observatory, Denmark
ANDERSEN, T.B., Astronomisk Institut, Aarhus, Denmark
AUDOUZE, J., Institut d'Astrophysique, Paris, France
AUVERGNE, M., Observatoire de Nice, France
BAGLIN, A., Observatoire de Nice, France
BARBARO, G., Istituto di Astronomia, Padova, Italy
BARLAI, K., Konkoly Observatory, Budapest, Hungary
BARTHOLDI, P., Observatoire de Genève, Switzerland
BECKER, S.A., Los Alamos National Laboratory, Los Alamos, U.S.A.
BENZ, W., Observatoire de Genève, Switzerland
BERTELLI, G., Istituto di Astronomia, Padova, Italy
BERTHOMIEU, G., Observatoire de Nice, France
BIENAYME, O., Observatoire de Besançon, France
BLANCO, V., Cerro Tololo Observatory, La Serena, Chile
BLECHA, A., Observatoire de Genève, Switzerland
BÖHM-VITENSE, E., University of Washington, Seattle, U.S.A.
BOUVIER, P., Observatoire de Genève, Switzerland
BRUNISH, W.M., Los Alamos National Laboratory, Los Alamos, U.S.A.
BURKI, G., Observatoire de Genève, Switzerland
CACCIARI, C., Dipartimento di Astronomia, Bologna, Italy
CANNON, R.D., Royal Observatory, Edinburgh, Scotland
CASSE, M., Centre d'Etudes Nucléaires de Saclay, Gif-sur-Yvette, France
CAYREL DE STROBEL, G., Observatoire de Paris-Meudon, France
CAYREL, R., Observatoire de Paris, France
CHMIELEWSKI, Y., Observatoire de Genève, Switzerland
CHRISTIAN, C.A., CFH Telescope Corporation, Kamuela, U.S.A.
CLAUSEN, J.V., Copenhagen University Observatory, Denmark
CLEMENTINI, G., Osservatorio Astronomico Universitario, Bologna, Italy
CONTI, P., University of Colorado, Boulder, U.S.A.
CORBALLY, C.,S.J., Specola Vaticana, Vatican City State
COX, A.N., Los Alamos National Laboratory, Los Alamos, U.S.A.
CRAMER, N., Observatoire de Genève, Switzerland
CRAWFORD, D., Kitt Peak National Observatory, Tucson, U.S.A.
CROWE, R., University of Toronto, Canada
DAVIDSON, K., University of Minnesota, Minneapolis, U.S.A.
DEMARQUE, P., Yale University Observatory, New Haven, U.S.A.
DESPAIN, K., Los Alamos National Laboratory, Los Alamos, U.S.A.
DOOM, C., Astrophysical Institute, Brussels, Belgium
DUBOIS, P., Observatoire de Strasbourg, France
EDVARDSSON, B., Astronomiska Observatoriet, Uppsala, Sweden
ENGELS, D., Sternwarte Universität Bonn, West-Germany
ERIKSSON, K., Astronomiska Observatoriet, Uppsala, Sweden

FEITZINGER, J., Ruhr-Universität Bochum, West-Germany
FERNIE, J.D., David Dunlap Observatory, Richmond Hill, Canada
FLOWER, P., Clemson University, Clemson, U.S.A.
FROGEL, J.A., Cerro Tololo Observatory, La Serena, Chile
GIMENEZ, A., Departamento de Astrofisica, Madrid, Spain
GOLAY, M., Observatoire de Genève, Switzerland
GOY, G., Observatoire de Genève, Switzerland
GREGGIO, L., Dipartimento di Astronomia, Bologna, Italy
GRENON, M., Observatoire de Genève, Switzerland
de GREVE, J.P., Astrophysical Institute, Brussels, Belgium
de GROOT, M., Armagh Observatory, Northern Ireland
GURM, H.S., Punjabi University, Patiala, India
GUTHRIE, B., Royal Observatory, Edinburgh, Scotland
HABETS, G., Astronomical Institute A. Pannekoek, Amsterdam, Netherlands
HAMANN, W.R., Universität Kiel, Inst. Theor. Phys., West-Germany
HAUCK, B., Institut d'Astronomie, Lausanne, Switzerland
HEBER, U., Universität Kiel, Inst. Theor. Phys., West-Germany
HESSER, J.E., Dominion Astrophysical Observatory, Victoria, Canada
HUMPHREYS, R.M., University of Minnesota, Minneapolis, U.S.A.
IBEN, I., University of Illinois, Urbana, U.S.A.
ISOBE, S., Tokyo Astronomical Observatory, Japan
ITOH, N., Sophia University, Tokyo, Japan
JAKOBSEN, A.M., Kitt Peak National Observatory, Tucson, U.S.A.
JANES, K., Boston University, Boston, U.S.A.
JEFFERY, S., University Observatory, St Andrews, Scotland
KUDRITZKI, R.P., Inst. Astronomie u. Astrophys. München, West-Germany
LAMBERT, D.L., University of Texas, Austin, U.S.A.
LATTANZIO, J., Monash University, Clayton, Australia
LAW, W.Y., Max-Planck-Institut, Garching bei München, West-Germany
LLOYD EVANS, T., South African Astronomical Observatory, South Africa
LODEN, L.O., Astronomiska Observatoriet, Uppsala, Sweden
de LOORE, C., Astrophysical Institute, Brussels, Belgium
LUB, J., Sterrewacht Leiden, Huygens Laboratorium, Netherlands
LYNAS-GRAY, A.E., University College London, Great-Britain
MAEDER, A., Observatoire de Genève, Switzerland
MARTINET, L., Observatoire de Genève, Switzerland
MAYOR, M., Observatoire de Genève, Switzerland
MCCARTHY, M.F., Specola Vaticana, Vatican City State
MCCLURE, R.D., Dominion Astrophysical Observatory, Victoria, Canada
MERMILLIOD, J.C., Institut d'Astronomie, Lausanne, Switzerland
MESTEL, L., University of Sussex, Astronomy Centre, Great-Britain
MEYLAN, G., Institut d'Astronomie, Lausanne, Switzerland
MITALAS, R., University of Western Ontario, London, Canada
MOULD, J., Astronomy Dept, Caltech, Pasadena, U.S.A.
MULLER, E.A., Observatoire de Genève, Switzerland
NICOLET, B., Observatoire de Genève, Switzerland
NISSEN, P.E., Astronomisk Institut, Aarhus, Denmark
NORDSTROEM, B., Copenhagen University Observatory, Denmark

OLSEN, E.H., Copenhagen University Observatory, Denmark
PENNY, A.J., Royal Greenwich Observatory, Great-Britain
PHILIP, A.G.D., Van Vleck Obs. and Union College, Schenectady, U.S.A.
PIGATTO, L., Osservatorio Astronomico, Padova, Italy
POPOVA, M., Bulgarian Academy of Sciences, Sofia, Bulgaria
PROVOST, J., Observatoire de Nice, France
RENZINI, A., Dipartimento di Astronomia, Bologna, Italy
RICHER, H.B., University of British Columbia, Vancouver, Canada
ROOD, R., University Station, Charlottesville, U.S.A.
ROXBURGH, I.W., Queen Mary College, London, Great-Britain
RUFENER, F., Observatoire de Genève, Switzerland
SCHATZMAN, E., Observatoire de Nice, France
SCHILD, H., Observatoire de Genève, Switzerland
SCHMUTZ, W., Institut für Astronomie, ETH-Zentrum Zürich, Switzerland
SCHOENBERNER, D., Universität Kiel, Inst. Theor. Physik, West-Germany
MICELA-SCIORTINO, G., Osservatorio Astronomico Palermo, Italy
SCIORTINO, S., Osservatorio Astronomico Palermo, Italy
SIMON, K.P., Universitäts-Sternwarte München, West-Germany
SMEYERS, P., Astronomisch Instituut Leuven, Belgium
SPRUIT, H., Max-Planck-Institut, Garching bei München, West-Germany
STAGG, C., University of Toronto, Canada
STELLINGWERF, R.F., Mission Research Corporation, Albuquerque, U.S.A.
SZABADOS, L., Konkoly Observatory, Budapest, Hungary
TAPIA, M., Universidad Nacional Autonoma de Mexico, Mexico
TASSOUL, M., Université de Montréal, Canada,
TASSOUL, J.-L., Université de Montréal, Canada
TAYLER, R.J., University of Sussex, Astronomy Centre, Great-Britain
TORNAMBE, A., Istituto di Astrofisica Spaziale, Roma, Italy
TOSI, M., Osservatorio Astronomico Universitario, Bologna, Italy
TREFZGER, C., Astronomisches Institut Basel, Switzerland
VANBEVEREN, D., Vrije Universiteit Brussel, Belgium
VANDENBERG, D.A., University of Victoria, Canada
VAN DER BORGHT, R., Monash University, Clayton, Australia
VAN DER HUCHT, K.A., Sterrekundig Instituut te Utrecht, Netherlands
VAUCLAIR, G., Observatoire de Toulouse, France
WAELKENS, C., Astronomisch Instituut Leuven, Belgium
WALBORN, N., Goddard Space Flight Center, Greenbelt, U.S.A.
WALKER, M., Lick Observatory, Santa Cruz, U.S.A.
WEIDEMANN, V., Universität Kiel, Inst. Theor. Physik, West-Germany
WESTERLUND, B., Astronomiska Observatoriet, Uppsala, Sweden
WILLIS, A., University College, London, Great-Britain
WILLSON, L.A., Iowa State University, Ames, U.S.A.
WING, R.F., Ohio State University, Columbus, U.S.A.
WOOD, P.R., Mount Stromlo Observatory, Australia
ZAHN, J.-P., Observatoire du Pic-du-Midi, France

OPENING OF THE SYMPOSIUM

E. SCHATZMAN
CHAIRMAN OF THE SCIENTIFIC ORGANIZING COMMITTEE

The place of stellar physics in Astrophysics is not the same as it were 50 years ago. Stellar evolution theory has been one of the preferred places of numerical calculations, and as a consequence, standard evolution models, from the homogeneous pre-main sequence star to the final stages, white dwarfs, neutron stars, supernovae, has become sort of archetype of numerical analysis and computation, the essence of perfection.

However, standard models turned out to be unable to solve a number of contradictions with the observations, surface abundances of Lithium, Carbon isotopes, mass of variable stars, Russell gap, etc., ... A number of new physical processes needed to be added to the usual variables: mass loss, rate of mixing and transfer of angular momentum, mass exchange in binaries; the theory of the convective zone has improved to give a better idea of overshooting, the problems of meridional circulation have been studied extensively, the theory of stability of rotating stars has made new steps ... On the other hand, high resolution spectroscopy has brought new results on surface abundance with large consequences. Altogether new constraints to stellar evolution models have been added one to the other.

The present meeting is probably the first one of a new species. We enter the epoch of high precision astrophysics. Within a few years, high precision parallaxes, new spectroscopic data with the space telescope will bring new constraints, not speaking of stellar sismology which must be carried as soon as possible, will put new constraints on stellar models, and stellar evolution scenarios. We can easily predict the time where stellar models will have no undeterminacy, and in fact will definitely oblige us to achieve great improvements in physical theories.

I am quite convinced that every one of us will leave Geneva with new ideas and new research plans.

This important gathering, a turning point in stellar astrophysics, would have been impossible without the devotion to the projects of A. Maeder. Let us thank him heartily!

I take this opportunity to present my gratitude to the Swiss Organizing Committee and to the Swiss institutions:
* University of Geneva,
* Swiss National Science Foundation,
* Société Helvétique des Sciences Naturelles,
for their generous assistance, in addition to the I.A.U. support.

The staff of Geneva Observatory has carried a considerable amount of work to prepare the meeting: all our thanks to the staff members.

I. EVOLUTION OF LOW AND INTERMEDIATE MASS STARS
OBSERVATIONS AND MODELS

FUNDAMENTAL PROBLEMS AND BASIC TESTS OF STELLAR EVOLUTION THEORY - THE CASE OF CARBON STARS[†]

Icko Iben, Jr.
University of Illinois at Champaign-Urbana

Abstract

Carbon stars are thought to be in the asymptotic giant branch (AGB) phase of evolution, alternately burning hydrogen and helium in shells above an electron-degenerate carbon-oxygen (CO) core. The excess of carbon relative to oxygen at the surfaces of these stars is thought to be due to convective dredge-up which occurs following a thermal pulse. During a thermal pulse, carbon and neutron-rich isotopes are made in a convective helium-burning zone. In model stars of large CO core mass, the source of neutrons for producing the neutron-rich isotopes is the $^{22}Ne(\alpha,n)^{25}Mg$ reaction and the isotopes are produced in the solar system s-process distribution. In models of small core mass, the $^{13}C(\alpha,n)^{16}O$ reaction is thought to be responsible for the release of neutrons, and the resultant distribution of neutron-rich isotopes is expected to vary considerably from one star to the next, with the distribution in isolated instances possibly resembling the solar system distribution of r-process isotopes. After the dredge-up phase following each pulse, the ^{13}C is made by the reactions $^{12}C(p,\gamma)^{13}N(\beta^+\nu)^{13}C$ in a zone of large ^{12}C abundance and small ^1H abundance that has been established by semiconvective mixing during the dredge-up phase. There is qualitative accord between the properties of carbon stars in the Magellanic Clouds and properties of model stars, but considerably more theoretical work is required before a quantitative match is achieved.

The observed paucity of AGB stars more luminous than $M_{BOL} \sim -6$ is interpreted to mean that the AGB lifetime of a star more luminous than this is at least a factor of ten smaller than the AGB lifetime of stars less luminous than this, or, at most 10^5 yr. Since, with current estimates of the $^{22}Ne(\alpha,n)^{25}Mg$ reaction rate R_{22}, only AGB model stars more luminous than $M_{BOL} \sim -6$ can produce s-process

---

[†]Supported in part by the NSF grant AST 81-15325.

A. Maeder and A. Renzini (eds.), Observational Tests of the Stellar Evolution Theory, 3–19.
© *1984 by the IAU.*

isotopes in the solar system distribution, it is inferred that either (1) the current estimates of R_{22} are too small by one to two orders of magnitude, allowing less luminous AGB stars to contribute, (2) the solar system distribution is not equivalent to the average Galactic distribution, being rather the consequence of a unique injection into the protosolar nebula of matter from a massive intermediate-mass AGB star, or (3) the estimates of the temperatures in the convective shell that are given by extant models are too low by, say, 10 or 15 percent.

The absence of carbon stars more luminous than $M_{BOL} \sim -6$ is suggested to be due primarily to the fact that $\sim 10^6$ yr of AGB evolution is necessary to produce surface $C/O > 1$, rather than to be due to the burning of dredged-up carbon into nitrogen at the base of the convective envelope during the interpulse quiescent hydrogen-burning phase. Thus, the positive correlation between the nitrogen and helium abundances in planetary nebulae is perhaps primarily a consequence of the second dredge-up episode rather than a consequence of processes occurring during the thermally pulsing phase.

I. THE OBSERVATIONAL FACTS

A. Galactic Studies

One of the drawbacks to attempts at understanding carbon stars from observations of stars in the Galaxy is the fact that very few of them are in clusters for which distances can be estimated. Nevertheless, much can be learned by piecing together all of the available evidence.

Abundances of Carbon, Nitrogen, and Oxygen. From a study of the Sun and of stars, HII regions, and molecular clouds in the vicinity of the Sun, one infers that the CNO elements are produced in abundances such that $O > C > N$. This is not to say that every star which contributes to the synthesis of these elements produces them in the stated order of abundances, but that the net result of the synthesis of these elements by all stars is responsible for the observed order in the interstellar medium and in unevolved stars formed from this medium.

That the observed order of abundances must be the result of an averaging of the contributions from many different types of stars is obvious from the very existence of carbon stars which are, by definition, stars in which $C > O$ at the surface and which are known to be contributing matter (with $C > O$) to the interstellar medium by way of a strong wind. As we shall see, the evidence is overwhelming that the relative abundance of C relative to O in carbon stars is due to synthesis of C in the stars themselves.

Neutron-Rich Isotopes. Two other sets of elements (actually, sets of isotopes) which prove to be of interest in connection with carbon stars are found from studies of neutron-rich isotopes in meteorites. These are the so called s-process and r-process isotopes which are thought to be made by successive neutron captures and β decays beginning with the seed nucleus ^{56}Fe. In one group, adjacent stable isotopes are related to one another by $n_i \sigma_i = n_{i-1}\sigma_{i-1} (1 + \Lambda/\sigma_i)^{-1}$, where n_i is the number abundance of the ith isotope, σ_i is its neutron capture cross section (at an energy of ~ 30 keV), and Λ is a universal constant whose value is of the order of (3-5) millibarns. In theoretical nucleosynthesis experiments, such a distribution can most easily be formed if the flux (or density) of neutrons is sufficiently low that a beta-unstable nucleus will decay before capturing a neutron -- hence the name s(low neutron capture)-process. In contrast, the isotopes in the other group display abundance patterns that can be reproduced theoretically most easily if neutron capture takes precedence over beta decay until neutron number reaches the magic values of 82 and 126. These isotopes are therefore called r(apid neutron capture)-process isotopes.

Over the past three decades, an examination of abundances of neutron-rich isotopes in Galactic S-stars (in which $C \sim O$) has shown that neutron-rich isotopes are formed in such stars and brought to their surfaces. Of particular interest is the occurrence in many S-stars of Tc. The most stable isotope ^{99}Tc has a half life of only 2 x 10^5 yr, whereas the progenitor of an S-star must have a main sequence lifetime in excess of a few times 10^7 yr. In the spectrum of Galactic Ba and CH stars, which are not as bright as AGB stars, one finds evidence for neutron capture activity, but recent radial velocity studies have shown that many of these stars are in binaries with a degenerate dwarf companion (McClure, Fletcher and Nemec 1980, McClure 1983a,b). The thought is that the neutron-rich isotopes may have been formed in the AGB precursor of the degenerate dwarf and transmitted by Roche-lobe overflow or by a wind to the star which now shows Ba or CH features. Certainly, the orbital separations of many of the binaries are large enough to have accommodated a star of AGB dimensions. The high degree of variability in the relative abundances of the neutron-rich isotopes from one star to the next suggests that the neutron source is highly variable in strength and duration from one erstwhile AGB companion to the next.

Tc in Miras. Most Mira variables are not at the same time carbon stars. However, approximately half of the brightest fifteen percent of them are (Cahn 1980). Furthermore, except for those of the smallest periods (and hence presumably of the lowest luminosities), many of them show Tc lines and the probability of finding such lines increases with the period (and therefore presumably with brightness) of the Mira (Little-Marenin and Little 1979). We infer, first, that perhaps the production of neutron-rich isotopes and certainly the probability of dredging these isotopes to the surface increases with increasing AGB luminosity and, second, that the conversion of the surface C/O ratio from less than 1 to larger than 1 may require several dredge-up episodes.

<u>Frequency of C-Stars as a Function of Metallicity</u>. As shown by
Blanco, McCarthy, and Blanco (1978) there is a pronounced gradient in
the space frequency of Galactic carbon stars, with none occurring in
the Galactic bulge. This is despite the existence of metal rich AGB
stars in the bulge (Whitford and Rich 1983, Frogel and Whitford 1983).
The simplest interpretation of these facts is that the larger the
abundance of Fe/H, the more difficult it is for an AGB star to develop
a surface ratio C/O $>$ 1. Whether this means that, with increasing
Fe/H, it is more difficult for dredge-up to occur or whether, because
of a larger initial abundance of oxygen, that more carbon must be
dredged-up in order for C $>$ O at the surface, or both, is not settled
by these observations.

In summary, from an examination of stars in our own Galaxy, we
have learned that (1) carbon and neutron-rich isotopes are made in
cool, bright AGB giants and brought to the surfaces of these giants;
(2) the brighter the giant, the greater is the likelihood of producing
and bringing to the surface freshly manufactured elements and isotopes; and (3) the larger C and O are to begin with, the harder it is
to achieve C $>$ O at the surface.

B. Magellanic Cloud Studies

The development of infrared technology over the past decade has
made it possible to take advantage of the fact that all of the stars
in each of the Clouds are effectively at the same distance from the
earth. This has permitted us (since we can now compare absolute
luminosities of real carbon stars with those of model AGB stars) to
establish with a certainty that AGB stars bring freshly synthesized
carbon and neutron-rich isotopes to their surfaces and that the conversion from O-star, through S-star, to C-star occurs at a definite
point along the AGB branch.

<u>Field Stars</u>. Surveys of selected fields in both clouds show that
carbon stars are, in general, confined to a very narrow interval in
bolometric magnitude: $-6 \lesssim M_{BOL} \lesssim -4$ (survey by Blanco, McCarthy, and
Blanco 1980; bolometric magnitudes by Richer 1981, Cohen et al 1981,
Frogel et al 1981, and Frogel and Cohen 1982). Extrapolations from
the surveyed regions suggest that there are altogether $\sim 1.4 \times 10^4$
carbon stars in both clouds (Blanco and McCarthy 1983) and that there
are only a handful of S-stars (Blanco, Frogel, and McCarthy 1981,
Lloyd-Evans 1983). This permits us to say that the S-star phase, when
C $\sim$ O, is a very transitory one, with the transition from C $\sim$ O x (1-
.05) to C $\sim$ O x (1 + 0.05) occurring in perhaps only one dredge-up
episode.

The long period variables (LPV's) in the Magellanic Clouds may be
assigned to one of two distinct sequences (Wood, Bessel, and Fox 1983)
according to whether or not they show evidence for overabundances of
the neutron-rich element Zr. Those in which the tracer ZrO appears to
be "normal" are in general brighter than $M_{BOL} \sim -7.3$, which corre-

sponds to the magnitude of a theoretical AGB model whose core has reached the Chandrasekhar mass of 1.4 $M_\odot$. Those in which strong ZrO bands appear define a sequence in the M_{BOL}-pulsation period plane that is consistent with theoretical AGB models. The observed sequence that is characterized by strong ZrO bands extends, within the uncertainties, up to the maximum brightness reached by theoretical AGB models. The lowest period representatives of this sequence tend to be carbon stars, but the brightest ones are not carbon stars. The number ($\sim$ 100) of the non-carbon star LPV's with $M_{BOL} \lesssim -6$ is about a factor of twenty or so less than the number of Cepheids in the Clouds and, since the progenitors of the LPV's are thought to be Cepheids with lifetimes of $\sim 10^6$ yr, we infer that stars which reach the AGB with $M_{BOL} \lesssim -6$ remain on the AGB for only $\sim 10^5$ yr and this, as it turns out, is only ten percent or so of the AGB lifetime expected if mass loss via a stellar wind were to occur at roughly the Reimers rate (1975).

The frequency of carbon stars in the SMC is about 3 times larger than it is in the LMC (Blanco, McCarthy and Blanco 1978, 1980; Blanco and McCarthy 1983) and, since the metallicity of SMC stars is on average much less (by perhaps a factor of 3-6) than the metallicity of LMC stars, we recover the result of studies of Galactic C-stars that, for whatever reason, AGB stars of low metallicity find it easier to become carbon stars than do AGB stars of higher metallicity.

<u>Stars in Globular Clusters.</u>   Searle, Wilkinson, and Bagnuolo (1980) have suggested an ordering of Magellanic Cloud clusters in an approximate age sequence and Cohen (1982) has shown that this sequence is equivalent to an ordering according to metallicity with, naturally enough, the oldest clusters having the lowest metallicity. Frogel and Blanco (1984, this volume) find that, in the oldest ($\sim 10^{10}$ yr) clusters, there are essentially no carbon stars. From this one may infer that, if the initial mass of an AGB star is quite low ($\lesssim 0.8\ M_\odot$), either dredge-up cannot occur, or the envelope mass of the star is lost by a wind before dredge-up can lead to C > O, <u>or both.</u>

Among stars on the AGB of an intermediate age cluster (0.3×10^9 yr $\lesssim$ age $\lesssim 3 \times 10^9$ yr), Frogel and Blanco find that there is a clear separation in absolute luminosity between the dimmer M-stars (O > C) and the brighter C-stars (C > O), with the luminosity at the transition point between M-stars and C-stars increasing with decreasing cluster age. This result has also been obtained independently by Lloyd-Evans (1983) and it is consistent both with the idea that dredge-up occurs on the AGB and with the theoretical result that the luminosity with which a model star reaches the thermally pulsing stage on the AGB increases with increasing mass (decreasing nuclear burning lifetime) of its main sequence progenitor (e.g., Becker and Iben 1979, 1980).

A remarkable result of the Frogel-Blanco survey is that in young clusters which contain Cepheids (age $\lesssim 10^8$ yr) <u>there are no C-stars.</u> And yet, theoretically, AGB stars which have Cepheid progenitors, and hence have masses on the order of $(4 - 6)$ M_0, are expected to dredge-up carbon with great facility (Iben 1975a). Even more astonishing than the absence of C-stars is the total absence of even M-stars brighter than about $M_{BOL} \sim -6$. Since a model AGB star reaches a luminosity given by $M_{BOL} \sim -7.3$ if its core mass reaches ~ 1.4 M_0, and since the rate at which a theoretical model brightens (in magnitudes per year) is independent of core mass, the inescapable conclusion is that a real AGB star leaves the AGB considerably before its core mass reaches 1.4 M_0. The absence of carbon stars in the field brighter than $M_{BOL} \sim -6$ can now also be attributed to the paucity of AGB stars brighter than this rather than to two other possibilities that have been cited in the past: (1) AGB stars brighter than $M_{BOL} \sim -6$ do not dredge-up carbon (contrary to theoretical indications) or (2) such stars convert dredged-up carbon into nitrogen as a consequence of proton captures at the base of the convective envelope during the quiescent, interpulse hydrogen-burning phase.

THE THEORY – OVERVIEW

A. Surface Composition Changes Prior to the AGB Phase

It is important to distinguish the nature of those surface composition changes which may occur prior to the AGB phase from those changes which are due solely to processes occurring during the AGB phase proper. This is not a completely straightforward task, as we cannot deduce from first principles the rates at which rotationally induced forms of mixing actually operate in real stars. Even if we were able to calculate the effectiveness of such forms of mixing, the fact that there is a large spread in rotational velocities among real stars will introduce a large spread in surface composition changes if these forms of mixing are of importance.

<u>Meridional Circulation and Turbulent Diffusion in Main Sequence Stars.</u> In principle, these processes should both increase in effectiveness with increasing rotation rate and might be expected to bring to the surface products of nuclear processing in the interior. In particular, one might anticipate some outward diffusion of products of CN-cycle burning and a reduction in C and an enhancement of N at the stellar surface. However, no evidence has yet been presented by spectroscopists to show that there is a tendency for C/N to decrease in any systematic way from high T_e to low T_e across the main sequence band and there is therefore no compelling evidence that rotationally induced mixing is of importance.

<u>Convective Dredge-Up on the Giant Branch.</u> In contrast, there is good observational evidence to support the theoretical indications that CN-cycle processed material is first brought to the surface as a star begins its initial climb upwards along the (first) red giant branch and the base of the convective envelope extends inward in mass to where C has been converted almost entirely into N (see Iben and Renzini 1984 for a review). The theoretical calculations provide pre-

cise quantitative predictions [surface C down by ~ 1/3 x and N up by ~ 2 x] and these predictions are consistent with the observations.

<u>Meridional Mixing on the Giant Branch in Globular Cluster Stars.</u> Kraft et al (1982) and Carbon et al (1982) find that, among red giant stars in the Galactic globular cluster M92, the abundance of carbon decreases with stellar luminosity. This decrease continues to luminosities much larger than the luminosity at which the first convective dredge-up is expected to have run its course. The best explanation for the observed effect is the one given by Sweigart and Mengel (1979) who point out that, in giants, the distance between (a) the region above the main hydrogen-burning shell where C has been converted into N, and (b) the base of the convective envelope, is quite small and argue that meridional circulation may carry C-depleted, N-enhanced material into the base of the convective envelope. Convective mixing will then carry some of this processed material to the surface. The phenomenon appears to be confined to stars in only a subset of clusters and does not appear to occur in stars in the field (Kraft 1979). One infers that close stellar encounters among stars in clusters of the largest spatial concentration may lead to (probably temporary or reversible) spin-ups that initiate the requisite meridional currents.

<u>Convective Dredge-up After Central Helium Exhaustion.</u> In model stars which, as a consequence of hydrogen-burning either before or during the core helium-burning phase, develop a hydrogen-exhausted core larger than about 1 M_0 by the time that the central helium abundance goes to zero, the base of the convective envelope extends briefly inward in mass into the region through which the hydrogen-burning shell has passed during the preceding core helium-burning phase. Therefore, as the model star once again climbs along the (second) red giant branch and the base of the convective envelope moves inward in mass, fresh ^{4}He and ^{14}N (produced at the expense of both ^{12}C and ^{16}O) are brought to the surface. Thus, the surface abundances of both ^{4}He and ^{14}N increase, whereas the surface abundances of ^{12}C and ^{16}O decrease. This second dredge-up phenomenon occurs only in models with initial main sequence masses larger than ~ (4-5) M_0 and less than ~ (8-9) M_0. Within this narrow range of masses, the calculations show that ^{14}N and ^{4}He enhancements as large as a factor of two occur (for N, over and above the factor of ~ 2 enhancement that occurs during the first dredge up episode), with the degree of enhancement increasing with increasing mass within this range (Becker and Iben 1979).

B. Surface Composition Changes During the Thermally Pulsing (TP)-AGB
    Phase

We have seen that, prior to the AGB phase, enhancements in the surface abundances of N and He and depletions in the surface abundances of C and O (the depletion of C being much more substantial than that of O) occur as a consequence of the convective dredge-up of matter that has experienced hydrogen-burning in varying degrees. The fact that many AGB stars exhibit a surface abundance of oxygen larger than the surface abundance of carbon, in spite of the fact that surface C is depleted when hydrogen-burning products are dredged to the surface, provides compelling evidence that such stars must somehow bring to their surfaces matter that has experienced helium-burning in the interior.

Abundance Changes in Nuclear Burning Regions. During about ninety percent of the AGB lifetime of a model star, helium-burning proceeds at a very low level, with effectively all of the energy which flows to the surface coming from hydrogen-burning in a thin shell (whose mass varies from $\sim 10^{-4}$ M_0 to 10^{-7} M_0 as the mass M_{CO} of the electron-degenerate CO core is varied from ~ 0.5 M_0 to ~ 1.4 M_0). If the mass of the stellar envelope is large enough and if convective mixing is efficient enough, some burning may take place at the base of the convective envelope and, if $M > (4-5)$ M_0, it is even possible that the abundance of ^{12}C decreases and the abundance of ^{14}N increases in the envelope between pulses (Iben 1975a, Renzini and Voli 1981). When the mass of the He-N zone laid down by the hydrogen-burning shell reaches about 100 times the mass of the burning shell, densities and temperatures in this zone become large enough to excite a helium-burning thermonuclear runaway. In the course of this runaway, the rate of energy generation by helium-burning can reach as high as (10^7-10^8) L_0 and temperatures at the base of the runaway zone can reach as high as $(300-400) \times 10^6 K$. During the thermal flash, nuclear energy is injected so rapidly that it is converted locally into thermal energy and thence into local expansion energy before it has a chance to leak out from the burning zone by radiative or convective "diffusion." The runaway is quenched as a consequence of the expansion and subsequent cooling.

At the peak of a thermal pulse (or helium shell flash, as it is often called), the entire region between the base of the helium-burning zone and the hydrogen-helium discontinuity is unstable to convection. After the pulse subsides and helium begins to burn quiescently, the abundance of ^{12}C left behind just below the hydrogen-helium discontinuity is about 20 percent by mass. Because burning in the convective zone is far from complete, the amount of ^{16}O left behind is much less than the amount of ^{12}C. In the course of the quiescent helium-burning phase, which lasts for roughly ten percent of the interpulse lifetime, the total amount of helium converted into carbon and oxygen equals the amount of helium-produced during the quiescent hydrogen-burning phase.

Of particular interest is the neutron-capture nucleosynthesis that occurs in the convective helium-burning shell during pulse peak. The nature of this nucleosynthesis is a function of, among other things, the source of neutrons. If the mass of the CO core is larger than a critical value M_{CO}^{crit}, temperatures in the convective zone become large enough for a long enough time that ^{14}N is converted completely into ^{22}Ne and a substantial fraction of this ^{22}Ne is converted into ^{25}Mg and a neutron. Most of the neutrons released are captured by ^{22}Ne, ^{25}Mg and the neutron capture progeny of these isotopes, but the number of neutrons left to be captured by ^{56}Fe and its progeny is precisely what is needed to produce heavy s-process isotopes in the solar-system distribution (Iben 1975b, Truran and Iben 1977). An essential aspect of the environment that produces this result is the overlap in mass between successive convective shells; this overlap ensures that, in any given shell, the fraction of matter which has experienced N neutron exposures is an exponentially declining function of N. The universal parameter Λ characterising the final distribution of s-process isotopes is essentially the average neutron capture cross section of the light elements from ^{22}Ne to, say, ^{27}Al.

Analysis of early studies (Iben and Truran 1978) suggested that $M_{CO}^{crit} \sim 0.95\ M_0$, but more recent calculations (Becker 1983) give hope that M_{CO}^{crit} may be as small as (0.75-0.80) M_0. It is worthwhile remarking that a full understanding of the theoretical properties of thermally pulsing AGB models will require substantial amounts of computer time on what are, even today, considered to be "supercomputers."

For smaller CO core masses, temperatures in the convective shell do not become large enough for more than a percent or so of ^{22}Ne to be converted into ^{25}Mg and a neutron (Becker 1980, Iben 1982). However, if the metallicity is low enough, ^{13}C is made available as a neutron source as a consequence of a process (Iben and Renzini 1983) to be described in a later section. The neutron densities which are created by the ^{13}C(α, n)^{16}O source are too large to produce neutron-rich isotopes in a distribution resembling the solar system distribution of s-process isotopes. They are also too small to produce, in one pulse, the classical r-process distribution. It remains to be seen if the combination of an intermediate strength neutron flux and an exponential distribution of exposures may perhaps produce the classical r-process distribution.

Dredge-Up Following Thermal Pulses. The rapid expansion and cooling that is initiated in the helium-burning zone extends beyond this zone into the hydrogen-rich layers, with the result that hydrogen-burning is extinguished. There is a sharp but temporary drop in the surface luminosity as a consequence of this extinction. However, even as the total rate of energy generation by helium-burning drops dramatically from its peak values, more and more of the energy produced by helium-burning makes its way outward by radiative diffusion toward the base of the convective envelope and then through the convective envelope, until the major source of surface luminosity becomes helium-burning. Expansion and cooling in the helium-burning layers is eventually reversed and not only does the rate of energy generation by helium-burning increase for a time, but the flux of energy passing through the base of the convective envelope also increases. The net result is that the base of the convective envelope moves inward in mass and, in AGB models of sufficiently large core mass, this base extends into the region where ^{12}C and neutron-rich isotopes have been produced (Iben 1975a, 1976). The freshly made nucleosynthetic products are then convected to the surface. The minimum core mass at which this form of dredge-up occurs depends sensitively on the core mass, the total mass, and the metallicity of the model star as well as on the choice of mixing length/scale height used in the standard algorithm for modeling convective flow (Wood 1981).

For model stars of metallicity appropriate to the SMC, the minimum core mass is on the order of $0.6 \ M_0$ (Iben and Renzini 1982a,b, Iben 1983) and a surface ratio $C/O > 1$ is achieved after the first or second dredge-up episode. At the end of this dredge-up episode, the stellar luminosity corresponds to $M_{BOL} \sim -5$, but the model then settles down into an extended phase of quiescent helium-burning during which the stellar luminosity drops slowly (over about 10 percent of the interpulse lifetime) to about $M_{BOL} \sim -4$ and then rises slowly over the remainder of the interpulse lifetime to reach $M_{BOL} \sim -4.7$ just before the onset of another pulse. Thus, the models suggest that carbon star characteristics will first appear at $M_{BOL} \sim -4$ and that there should be an overlap in M_{BOL} between carbon stars (with $M_{CO} \gtrsim 0.6 \ M_0$) that are decreasing in luminosity following a helium shell flash and M-stars (with $M_{CO} \lesssim 0.6 \ M_0$) that are increasing in luminosity after reaching the luminosity minimum following a helium shell flash. These results are consistent with the distributions of field M-stars and field C-stars in both the SMC and the LMC.

For stars of core mass larger than, say, $0.7 \ M_0$, dredge-up occurs readily, regardless of metallicity, ℓ/H, and total stellar mass (Becker 1983). However, the larger M_{CO} is at the start of the thermally pulsing AGB phase, the larger is the main sequence mass of the progenitor model star. Thus, the larger M_{CO}^{start}, the more ^{12}C-rich material must be dredged-up to produce a surface $C/O > 1$. In models with $M_{CO}^{start} \gtrsim 0.8 \ M_0$ (corresponding to a maximum luminosity prior to a thermal pulse of $M_{BOL} \sim -6$), roughly 10^6 yr of AGB evolution is required to achieve $C/O > 1$ (Iben and Truran 1978, Iben 1981, Renzini and Voli 1981).

Since M_{CO}^{start} decreases with decreasing main sequence mass (and increasing main sequence lifetime) of a progenitor star, the theory suggests that the minimum luminosity at which C-star characteristics can occur in a population of a given age will decrease as the age of the population increases. This, of course, is what is known to be the case among globular clusters in the Magellanic Clouds (Frogel and Blanco 1984, Lloyd-Evans 1983). Not only is the agreement between theory and observation qualitatively satisfactory, it is also consistent quantitatively.

The absence of luminous single carbon stars in the oldest clusters in the Clouds can be understood in terms of the theoretical results that: (1) $M_{CO}^{start} \sim 0.53 \ M_0$ for stars of initial mass less than $\sim 2 \ M_0$; (2) the oldest stars (age $\sim 10^{10}$ yr) have initial main sequences masses $\lesssim 0.8 \ M_0$; (3) dredge-up does not occur until $M_{CO} \gtrsim$

0.6 M_O; and (4) wind mass loss causes the hydrogen-rich envelope of an AGB star of such a low mass to evaporate before its luminosity exceeds the luminosity at the top of the first red giant branch in Galactic globular clusters ($M_{BOL} \gtrsim -3.5$, $M_{CO} \lesssim 0.53$). Such stars have therefore at best just reached the thermally pulsing AGB phase just before ceasing to exist as AGB stars.

The observed fact that the frequency of carbon stars among AGB stars with intermediate age progenitors decreases with increasing Z can be understood as the consequence of two effects. The first effect is the obvious one that, the larger Z (and hence presumably the larger O is to begin with), the more C-rich material must be dredged-up before surface C $>$ O. The second effect is that, the larger Z is, the more extended is the envelope and the smaller is the gas pressure relative to the radiation pressure. Since the radiative gradient at any point in the envelope is proportional to κ (L/M) ($1 + P_{gas}/P_{rad}$), the smaller P_{gas}/P_{rad}, the less likely is the adiabatic gradient to be smaller than the radiative gradient (Iben 1983a).

<u>Implication of the Paucity of Bright AGB Stars for s-Process Nucleosynthesis.</u> The absence in the Magellanic Clouds of carbon stars brighter than $M_{BOL} \sim -6$, coupled with the paucity of M-stars brighter than this, may have dramatic implications not only for our understanding of the origin of the s-process isotopes in the solar system but also for our understanding of Galactic nucleosynthesis in general. If one adopts the standard choice for the $^{22}Ne(\alpha,n)^{25}Mg$ reaction (Fowler, Caughlan, and Zimmerman 1975), then early studies suggest that ^{22}Ne is not converted substantially into ^{25}Mg and a neutron unless $M_{CO} > M_{CO}^{crit} \gtrsim 0.95$ M_O (Iben 1976, Truran and Iben 1977, Iben and Truran 1978). Do enough stars achieve core masses this large and maintain themselves as thermally pulsing AGB stars long enough to produce s-process isotopes in sufficient quantity to account for the solar-system s-process distribution on the assumption that this distribution is equivalent to the average Galaxy-wide distribution of s-process isotopes?

From Becker and Iben (1979) one has that, when $Z \sim 0.01$ and $Y \sim 0.28$, M_{CO}^{start} and initial main sequence mass M_{MS} are related by $M_{CO}^{start} = 0.85 + 0.053$ ($M_{MS} - 4$) when $4 \lesssim M_{MS} \lesssim 8$. Masses and luminosities are here and hereinafter in solar units unless otherwise specified. For all $M_{MS} \lesssim 2$, $M_{CO}^{start} \sim 0.53$ and $M_{CO}^{start} \sim 0.53 + 0.16$ ($M_{MS} - 2$) for $2 \lesssim M_{MS} \lesssim 4$.

In the absence of convective dredge-up following thermal pulses, core mass grows according to $\dot{M}_{CO}$ (M_O yr^{-1}) $\sim 10^{-6}$ ($M_{CO} - 0.5$) so that, after t_{AGB} years of evolution, $\Delta M_{CO} \sim [exp (10^{-6} t_{AGB}) - 1]$ ($M_{CO}^{start} - 0.5$). Since the maximum core mass is 1.4 M_O and since, from the early studies, M_{CO} must exceed ~ 0.95 M_O if s-process isotopes are produced in the solar system distribution, one has that the maximum time which a star can spend producing s-process isotopes in this distribution is $\sim 7 \times 10^5$ yr. The abundance of s-process isotopes relative to solar is produced in a convective shell about 200 (Truran and Iben

1977). Assuming that dredge-up brings up a fraction λ of the mass that has been added to the He-N zone between pulses, the total amount of dredged-up material is approximately $\Delta M_{dredge} \sim \lambda \, \Delta M_{CO}$, and λ is on the order of $\sim 1/3$ when $M_{CO} \gtrsim 0.95$.

Let us next assume that the rate of star formation varies with mass according to $\frac{d}{dM} (\frac{dN}{dt}) = 1.3 \, \frac{1}{M^{2.3}}$, where we have normalized to a total birthrate of 1 star per year over the interval $M = 1 \rightarrow \infty$. A measure of the contribution of all stars (still assuming no mass loss) to the Galactic nucleosynthesis of solar system s-process isotopes may now be written as

$$\Delta_s \sim 200 \, \lambda \, 1.3 \, \{ \int_1^{1.4} \frac{dM}{M^{2.3}} \, (M-0.95) + \int_{1.4}^8 \frac{dM}{M^{2.3}} \, 0.45 \cong 68 \, \lambda \sim 23.$$

If mass were not lost, all stars initially more massive than 1.4 M_0 would become supernovae and thus the SN rate would be on the order of $\nu_{SNI} = 1.3 \int_{1.4}^8 \frac{dM}{M^{2.3}} = 0.58 \text{ yr}^{-1}$, which is approximately 60 times the observed SNI rate. It is clear that most real stars of initial mass less than 8 M_0 must lose their hydrogen-rich envelopes before their cores reach the Chandrasekhar mass.

Before pursuing quantitatively the consequences of mass loss let us explore just a bit further the demands of Galactic nucleosynthesis. Type I supernovae are thought to be the consequence of binary star evolution, they occur at the rate $\nu_{SNI} \sim 10^{-2} \text{ yr}^{-1}$ (see, e.g., Iben and Tutukov 1984), and they are thought to produce at least 0.8 M_0 of ^{56}Fe in the explosion (see, e.g., Woosley, Axelrod, and Weaver, 1984, Nomoto 1984). Since the overabundance (relative to solar) of ^{56}Fe in one gram of pure ^{56}Fe is ~ 1000x, a measure of the ^{56}Fe production rate in the Galaxy is

$$\Delta_{Fe} \sim 1000 \text{ x } 0.01 \text{ x } 0.8 \sim 8.$$

In the absence of mass loss, all intermediate mass stars of initial mass larger than 1.4 M_0 would also become SNeI and contribute 0.8 M_0 of ^{56}Fe each, leading to $\Delta_{Fe} \sim 58 \times 8 = 464$. Thus, we have an essential paradox: the actual ^{56}Fe production rate is a factor of about 60 smaller than it would be if all intermediate mass stars of initial mass larger than 1.4 M_0 were to lose no mass until they become supernovae, and yet, the estimated rate of production of solar system s-process isotopes in the absence of mass loss is only four times larger than the actual rate of ^{56}Fe production by SNeI with binary star progenitors. That is, abolishing single star progenitors of SNeI also abolishes the source of s-process isotopes in the solar system distribution.

We are now in a position to make the paradox even more dramatic. Assuming that mass is lost by real AGB stars at some fraction of the rate given by the semi-empirical Reimers expression (1975), it may be shown that the typical lifetime of an AGB star is on the order of $\sim$ 10^6 yr (Fusi-Pecci and Renzini 1976, Iben and Truran 1978, Renzini and Voli 1981). The paucity of real AGB stars brighter than $M_{BOL} \sim -6$ means that the actual lifetime of an AGB star with $M_{BOL} < -6$ must be much less than 10^6 yr and, from a comparison of the number of LPV's and Cepheids in the Magellanic Clouds one may estimate $t_{AGB} \lesssim 10^5$ yr for $M_{BOL} < -6$.

Note that, when $M_{CO} \sim 0.95$, the quiescent hydrogen-burning luminosity is $L \sim 6 \times 10^4$ (0.95-0.5), corresponding to $M_{BOL} \sim -6.3$. Thus, when M_{CO} exceeds ~ 0.95, the core can grow no more than $\Delta M_{CO} < 10^5 \times 10^{-6}$ (0.95-0.5) < 0.05 before the hydrogen-rich envelope evaporates. Being generous and assuming that all stars that develop an initial core mass as large as $M_{CO}^{start} \sim 0.9$ (corresponding to an initial main sequence mass of ~ 5 M_0) produce s-process isotopes in the solar system distribution we have that

$$\Delta_s' \gtrsim 200 \; \lambda \times 0.05 \times 1.3 \int_5^8 \frac{dM}{M^{2.3}} \sim 0.2.$$

Since $\Delta_s' \ll \Delta_{Fe}$, we might conclude that, if the cross section for the $^{22}Ne(\alpha,n)^{25}Mg$ reaction is correct and if the early estimates of the maximum temperatures achieved in the convective helium-burning shell are correct, then either there is another process than the one we have envisioned for the Galaxy-wide production of s-process isotopes in the solar system distribution, or the solar system distribution of these isotopes has been produced by an isolated massive AGB star which injected its nucleosynthesis products into the matter out of which the Sun was born. This latter interpretation is consistent with the recent arguments of Olive and Schramm (1982).

An alternative interpretation is that the cross section for the $^{22}Ne(\alpha,n)^{25}Mg$ reaction has been significantly underestimated. Suppose, for example, that the $^{22}Ne(\alpha,n)^{25}Mg$ reaction goes to completion in all AGB stars with progenitor main sequence masses in the range $(1.5-3)M_0$ and that they produce s-process isotopes in the solar system distribution as core mass increases, on average, by ~ 0.3 M_0. Then

$$\Delta_s'' \sim 200 \times 1/3 \times 0.3 \times 1.3 \int_{1.5}^4 \frac{dM}{M^{2.3}} \sim 8.5.$$

The proximity of this number to Δ_{Fe} allows one to argue that the solar system distribution of s-process isotopes is equivalent to the Galactic one. To the best of the author's knowledge there has been no effort on the part of the experimental physicist to address the possibility that the $^{22}Ne(\alpha,n)^{25}Mg$ cross section at 30 keV has been underestimated by one or two orders of magnitude or even by a factor of 2!

Still another possibility is that the maximum temperatures reached at the base of the convective shell have been underestimated by a considerable amount. Certainly, these maximum temperatures have been rising as the total number of thermal pulses has slowly increased over the past few years (see the discussions by Becker 1981 and by Iben and

Renzini 1983). For example, from a selection of the data available before 1976, Iben and Truran (1978) constructed the approximation $T_{CSB}^{max} \sim [3.1 + 2.85 (M_{CO}-0.96)] \times 10^8$ K. It has now become clear, however, that, even for core masses as small as ~ 0.65 M_0, T_{CSB}^{max} exceeds 3×10^8 K after only a dozen or so pulses (Iben 1983b, Becker 1983) and that the maximum temperatures achieved after 20 or so pulses at larger core masses have still not reached asymptotic values (Becker 1983, Chieffi and Iben 1983). Considering the fact that, at $T \sim 3 \times 10^8$ K, the $^{22}Ne(\alpha,n)^{25}Mg$ reaction rate is proportional to $\sim T^{23.4}$, it is conceivable that, even with the currently accepted cross section, M_{CO}^{crit} may drop from the value $M_{CO}^{crit} \sim 0.95$ M_0 suggested by data available five years ago. That is, only a ten percent increase in T_{CSB}^{max} will lead to a factor of 10 increase in the neutron producing reaction rate, and this might translate into $M_{CO}^{crit} \sim 0.70$ M_0. It is abundantly clear that progress in answering the crucial questions about nucleosynthesis in AGB model stars has been hampered by the lack of sufficient computer power in the hands of interested scientists.

<u>Implication of the Paucity of Bright AGB Stars for the Nitrogen Abundance in Planetary Nebulae.</u> The observed short lifetime of high luminosity AGB stars also has ramifications for the abundance of nitrogen in planetary nebulae. The first attempt to account for the observed relationship between N and He in planetary nebulae assumed that the effects of third dredge-up episodes could be neglected (Kaler, Iben, and Becker 1979) and found reasonable agreement between observation and theory. In this picture, most of the correlation between N and He is due to the effect of the second dredge-up episode which is experienced by massive intermediate mass stars. However, during the thermally pulsing AGB phase, third dredge-up episodes cause the abundance of nitrogen in the convective envelope to decrease, since there is no ^{14}N in the dredged-up material (Iben and Truran 1978, Becker and Iben 1980), and this decrease will persist unless enough of the simultaneously dredged-up ^{12}C is converted into ^{14}N by burning at the base of the convective envelope during the interpulse phase (Renzini and Voli 1981, Becker and Iben 1980).

The fact that there are no carbon stars brighter than $M_{BOL} \sim -6$ has on occasion been attributed to the effective burning of dredged-up carbon. However, in the absence of such burning, it still requires $\sim 5 \times 10^5$ yr for surface C to exceed surface O (Iben and Truran 1978, Becker and Iben 1980, Renzini and Voli 1981, Iben 1981) and, since the observations reduce the available lifetime to only $\sim 10^5$ yr, the absence of bright carbon stars may not be invoked as a demonstration of burning. It may very well be that the observed correlation between N and He in planetary nebulae is, after all, due almost entirely to the second dredge-up!

<u>Activation of the $^{13}C(\alpha,n)^{16}O$ Neutron Source in Low Mass, Low Z AGB Stars.</u> Although the maximum temperatures in the convective shells of low mass thermally pulsing AGB stars can reach 300×10^6 K when $M_{CO} \gtrsim 0.65$ M_0, the short duration of the high temperature phase pre-

vents more than about a percent of the ^{22}Ne from being burned (Becker 1981, Iben 1982, 1983b), provided, of course, that the currently used α-capture cross section is correct. However, even if this cross section were ten times larger than currently used, burning would be incomplete and the solar-system distribution of s-process isotopes would not result (Truran and Iben 1977).

Another potential source of neutrons is the $^{13}C(\alpha,n)^{16}O$ reaction (Cameron 1955) and, ever since the early work of Schwarzschild and Härm (1967) and of Sanders (1967), it has been assumed that somehow hydrogen may be injected into the ^{12}C-rich convective helium-burning shell and that the reaction $^{12}C(p,\gamma)^{13}N(\beta^+\nu)^{13}C$ will produce the ^{13}C that can then immediately act as a neutron source. Subsequent calculations have not provided support for this envisioned scenario (see, in particular, the entropy argument of Iben 1976, 1982), but another sequence of events has been found to occur in low-mass, low-metallicity AGB stars (Iben and Renzini 1983). In this sequence, semiconvective mixing during the dredge-up phase brings small amounts of hydrogen into the outer edge of the ^{12}C rich region <u>after</u> shell convection has died down. When hydrogen-burning is rekindled, the major result of burning in the region of initially small hydrogen abundance and large ^{12}C abundance is ^{13}C. Then, <u>after</u> the extended quiescent hydrogen-burning phase, when a new helium shell flash is triggered, the <u>already prepared</u> ^{13}C is engulfed by the convective shell. The effective rate of release of neutrons by the $^{13}C(\alpha,n)^{16}O$ reaction is governed more by the rate at which ^{13}C enters the convective shell than by the cross section for the α-capture reaction (Iben 1983a). The neutrons are released when the convective shell has attained only half of its maximum size and the temperature at the base of the shell is only about 150×10^6 K.

This process appears to work only in AGB models of low total mass, low core mass, and low Z (Iben 1983a,b), but considerably more theoretical exploration is required before its dependences on all parameters have been properly elucidated. As of this writing, it would appear that the ^{13}C neutron source will be highly variable from one AGB model to another and this should be reflected in a diversity in the distribution of neutron-rich isotopes from one real AGB star to the next.

REFERENCES

Becker, S.A. 1981, in <u>Physical Processes in Red Giants</u>, eds. I. Iben, Jr. and A. Renzini (Dordrecht: Reidel), p. 141.
______ 1983, in progress.
Becker, S.A. and Iben, I. Jr. 1979, Ap. J., <u>232</u>, 831.
______ 1980, Ap. J., <u>237</u>, 111.
Blanco, V.M., Frogel, J.A., and McCarthy, M.F. 1981, P.A.S.P., <u>93</u>, 532.
Blanco, B.M., McCarthy, M.F., and Blanco, V.M. 1978, Nature, <u>271</u>, 638.
Blanco, V.M. and McCarthy, M.F. 1983, preprint.

Blanco, V.M., McCarthy, M.F., and Blanco, B.M. 1980, Ap. J., 242, 938.
Cahn, J.H. 1980, Space Sci. Rev., 27, 457.
Cameron, A.G.W. 1955, Ap. J., 121, 144.
Carbon, D.F., Langer, G.E., Butler, D., Kraft, R.P., Suntzeff, N.B.,
    Kemper, E., Trezger, C.F., and Romanishin, W. 1983, Ap. J., 49,
    207.
Chieffi, A. and Iben, I. Jr. 1983, in progress.
Cohen, J.G. 1982, Ap. J., 258, 143.
Cohen, J.G., Frogel, J.A., Persson, S.E., and Elias, J.H. 1981, Ap.
    J., 249, 481.
Frogel, J. and Blanco, V.M. 1984, in Observational Tests of Stellar
    Evolution Theory, eds. A. Maeder and A. Renzini (Dordrecht:
    Reidel), p. 175.
Frogel, J. and Cohen, J.G. 1982, Ap. J., 253, 580.
Frogel, J. A., Cohen, J. G., Persson, S. E., and Elias, J. H. 1981, in
    Physical Processes in Red Giants, eds. I. Iben, Jr. and A.
    Renzini (Dordrecht: Reidel), p. 159.
Frogel, J. and Whitford, A.E. 1982, Ap. J. Lett., 259, L7.
Fusi-Pecci, F. and Renzini, A. 1976, A. and Ap., 46, 447.
Iben, I. Jr. 1975a, Ap. J., 196, 525.
_____ 1975b, Ap. J., 196, 549.
_____ 1976, Ap. J., 208, 165.
_____ 1977, Ap. J., 217, 788.
_____ 1981, Ap. J., 246, 278.
_____ 1982, Ap. J., 260, 821.
_____ 1983a, Ap. J. Lett., 275, _____.
_____ 1983b, in progress.
Iben, I. Jr. and Renzini, A. 1982, Ap. J. Lett., 263, L188.
_____ 1983, Ann. Rev. Ast. and Ap., 21, 271.
_____ 1984, Physics Reports, in press.
Iben, I. Jr. and Truran, J.W. 1978, Ap. J., 230, 980.
Kaler, J. B., Iben, I. Jr., and Becker, S. A. 1978, Ap. J. Lett., 224,
    L63.
Kraft, R. P. 1979, Ann. Rev. A. and Ap., 17, 309.
Kraft, R. P., Suntzeff, N. B., Langer, G. E., Carbon, D. F., Trefzger,
    Ch. F., Fried, E., and Stone, R. P. S. 1982, P.A.S.P., 94, 55.
Iben, I. Jr. and Tutukov, A. V. 1984, Ap. J. Suppl., Feb. 1 issue.
Little-Marenin, I.R. and Little, S.J. 1979, A. J., 84, 1374.
Lloyd-Evans, T. 1983, M.N.R.A.S., 205, _____.
McClure, R.D. 1983a, Ap. J., 268, 264.
_____ 1983b, preprint.
McClure, R.D., Fletcher, J.M., and Nemec, J.M. 1980, Ap. J. Lett.,
    238, L35.
Nomoto, K. 1984, in Stellar Nucleosynthesis, eds. C. Chiossi and A.
    Renzini (Dordrecht: Reidel).
Olive, K. and Schramm, D.N. 1982, Ap. J., 257, 276.
Reimers, D. 1975, Mem. Soc. R. Sci. Liege, 6e Ser., 8, 369.
Renzini, A. and Voli, M. 1981, A. and Ap., 94, 175.
Richer, H.B. 1981, Ap. J., 243, 744.
Sanders, R. H. 1967, Ap. J., 150, 971.
Schwarzschild, M. and Harm, R. 1967, Ap. J., 150, 961.

Searle, L., Wilkinson, A., and Bagnulo, W.G. 1980, Ap. J., 239, 803.
Sweigart, A. V. and Mengel, J. G. 1979, Ap. J., 229, 624.
Truran, J.W. and Iben, I. Jr. 1977, Ap. J., 216, 797.
Whitford, A.E. and Rich, M. 1983, Ap. J., in press.
Wood, P.R. 1981, in Physical Processes in Red Giants, eds. I. Iben,
    Jr. and A. Renzini (Dordrecht: Reidel), p. 135.
Wood, P.R., Bessel, M.S., and Fox, M.W. 1983, Ap. J., 272, 99.
Woosley, S. E., Axelrod, T. S., and Weaver, T. A. 1984, in Stellar
    Nucleosynthesis, eds. C. Chiosi and A. Renzini (Dordrecht:
    Reidel), p. 263.

DISCUSSION

Blanco: You referred to the difference in the ratio of carbon and M-type
stars found when one compares the LMC with the SMC. This ratio was defined by Blanco, McCarthy and Blanco as the proportion between the number
of carbon stars with sufficiently strong near-infrared CN bands to be
detected in unwidened 2350 $\overset{\circ}{A}$/mm spectra and the number of M6 or later
stars classified in similar spectra on the basis of TiO-band strengths.
The C/M ratio defined in this way was found by us to be about 2 in the
LMC and about 15 in the SMC. These ratios may be found to be somewhat
different by other observers depending on how they judge the strength
of the relevant molecular bands. The important thing, however, is that,
regardless of who determines the C/M ratios, an appreciable difference
is found between the two clouds, and if the galactic nuclear bulge is
included, a much smaller value for C/M, namely about 10^{-3}. If earlier
M-type stars than M6 are included in computing the C/M ratios of the
clouds, and we have done this down to type M2, C/M becomes smaller in
each cloud and the difference between the LMC and the SMC becomes
smaller. This is indeed what is expected from AGB theory.

Wing: As you pointed out, recent work on the clusters of the Magellanic
Clouds has shown clearly that the brightest carbon stars can be associated with the brightest and coolest members of an intermediate-age
population. But the conclusion that very old clusters do not contain
carbon stars may need to be qualified. In such clusters, M stars have
been found at the tip of the giant branch, but there may be carbon stars
at fainter absolute magnitudes and bluer colors. Carbon stars like that-
the R stars - do occur in the direction of the galactic center, and towards the galactic poles, and in ω Centauri; they would appear at V = 17
or fainter in the clouds.

Iben: I agree, but these are a different kind of carbon stars, not to be
confused with thermally pulsing AGB stars. They may rather result from
mass exchange in binary systems.

SELECTED TOPICS ON THE EVOLUTION OF LOW AND INTERMEDIATE MASS STARS

Alvio Renzini, Dipartimento di Astronomia, Bologna, Italy

ABSTRACT

Among the many aspects of the evolution of low and intermediate mass stars, two representative topics are selected for this review: the question why stars become red giants, and the problem of the age determination of galactic globular clusters. Concerning the first topic, it is shown that this happens because of a thermal instability taking place in the stellar envelopes, and the physical nature of this instability is identified. Several ramifications of these findings are then briefly mentioned. Concerning the second topic, the Oosterhoff-Sandage effect is briefly described, together with its implications for the age estimates and the problems rised by its interpretation in terms of evolutionary models. In this connection, it is suggested that an enhanced overall abundance of the elements CNO and Ne may solve these problems, although further evolutionary calculations are required before reaching firm conclusions. It is also amphasized for both topics the important role played by the metal contribution to the radiative opacity at middle temperatures.

1. INTRODUCTION

The evolution of low and intermediate mass stars offers a number of exciting aspects and problems. First comes the question of understanding *everything* happening to such stars, from their pre-main sequence stage to their final fate, of determining the diversity of their evolutionary behaviours depending on initial mass and composition, and then of comparing theoretical predictions with observations. This includes the problem of assessing the *accuracy* and *adequacy* of current canonical models in all evolutionary stages, and therefore of identifying the role, if any, played by those physical processes which are left out in the canonical approximation (like rotation, magnetic fields, mixings of non-convective origin, effects of the stellar environment, etc.). The accuracy of cano-

A. Maeder and A. Renzini (eds.), Observational Tests of the Stellar Evolution Theory, 21–40.

nical models clearly depends upon the accuracy of the input physics used
in constructing such models, which includes opacity, nuclear reaction rates, neutrino losses, equation of state, etc. It should be emphasized
that the question of the accuracy of the models must logically be asked
before that of the adequacy of the canonical approximations. Otherwise,
when a discrepancy between model predictions and observations is perceived,
one would not know whether to blame the insufficient accuracy of the input
physics, or the inadequacy of the canonical assumptions, thus producing
confusing answers to ill-posed questions.

The second major aspect of stellar evolution theory (in particular
for what concerns low and intermediate mass stars) refers to its applications in the broader astrophysical context, when theoretical evolutionary
models are used to infer several global properties of various stellar aggregates. For example, age determinations, star formation histories, chemical and spectral evolution of galaxies, and the like. Certainly, before
using the models it would be preferable to have first successfully checked
their reliability. But, how to resist the temptation of employing such a
powerful tool, when astronomers are challenged by so many fascinating
questions? After all, the use of the tool refines the tool itself.

As mentioned at the beginning, the number of problems posed by/to
the theory of stellar evolution is virtually illimitate. Therefore, for
this review I have singled out just a few topics, chosed as interesting
examples illustrating some of the situations mentioned above. These (and
other) aspects and applications of stellar evolution theory are also discussed in a recent series of reviews by Icko Iben and myself (Iben & Renzini 1983, 1984; Renzini & Iben 1984).

Section 2 of this paper deals with an old question: *why do stars become red giants?* In my opinion this question did not yet receive a satisfactory answer in the literature. I will then present my own explanation
for this phenomenon, and will also try to show that its better understanding may effectively help in solving several other open problems. Finally, Section 3 deals with the Oosterhoff-Sandage effect in galactic globular clusters, and with the question of determining the age of these oldest
stellar aggregates. In closing this section the problems posed by the
presence of unexpected gaps on the main branches of some cluster C-M diagrams will be briefly discussed. In both sections reference will be often
made to the "metal opacity at middle temperatures". With this expression
one means the contribution provided to the total radiative opacity by the
elements heavier than helium, at temperatures between roughly half a million and five million Kelvins. Indeed, in this temperature range, bound-
bound and bound-free transitions in the few last ionization stages of
these elements provide an important contribution to the total radiative
opacity (cf. Renzini 1977, Fig. 2.3).

2. ON WHY STARS BECOME RED GIANTS

Color-magnitude diagrams (for instance for stars in the solar neighborhood) show many stars near the main sequence, many red giants, but very few stars at intermediate temperatures. This is the well known "Hertzsprung Gap". Theoretical models for stars roughly more massive than 1.2 $M_\odot$, and near-solar metallicity, experience a rapid expansion shortly after the beginning of hydrogen-shell burning, and on a short (thermal) timescale reach red giant dimensions. Following their arrival on the Hayashi track, thermal balance is restored and the evolution proceeds on a nuclear timescale. So, there is excellent agreement between theory and observations: the existence of the Hertzsprung Gap is clearly reproduced. But, why models (stars) behave as they do?

2.1 Three illuminating cases

The answer to this question is rather complex, and requires a number of preliminary steps. I shall first discuss three cases which will help a clearer formulation of the problem.

Case 1: Models with (Y, Z) = (0.28, 0.01) and M = 3 to 7 $M_\odot$. Let us consider the evolutionary tracks shown in Fig. 1b in Becker et al (1977). These three models behave in a very similar way: following central hydrogen exhaustion (point 2), the surface luminosity increases until a relative maximum is reached, then the luminosity drops while the envelope is rapidly expanding (on a thermal timescale). The drop in surface luminosity continues until the model approaches the Hayashi track, envelope convection penetrates deep into the interior, and a relative minimum in luminosity is reached (point 4). While models evolve from the relative maximum to the relative minimum in luminosity, their envelopes depart dramatically from thermal equilibrium: a relevant fraction of the luminosity produced in the interior being absorbed in the envelope to sustain envelope expansion (cf. Iben 1965). Following point 4 thermal equilibrium is restored. One can conclude that these models become red giants because of a thermal runaway taking place in the envelope. Note that qualitatively similar runaways take also place following the central helium exhaustion. Models of smaller mass (but more massive than $\sim$1 $M_\odot$) also experience the thermal runaway shortly after the central hydrogen exhaustion (cf. Mengel et al 1979).

Case 2: Models with (Y, Z) = (0.28, 0.001) and M = 3 to 7 $M_\odot$. Fig. 1a in Becker et al (1977) shows what happens to the evolutionary tracks when the metal abundance is decreased by a factor of 10, compared to the previous case. The 3 $M_\odot$ model behaves qualitatively as in the previous case, i.e. it suffers the envelope thermal runaway, and ignites helium at the center as a red giant. In the 5 $M_\odot$ model, the thermal runaway has just started

when it is aborted by the central He ignition (point 5). The model experiences the runaway only following the central He exhaustion. This latter phase is qualitatively similar to the corresponding situation at
higher Z. Finally, in the 7 $M_\odot$ model the thermal runaway does not even
begin: one can say that the central He ignition anticipates (and then
prevents) the onset of the thermal runaway. As in the 5 $M_\odot$ model, a thermal runaway develops only after the central He exhaustion. From the comparison of these two cases one can conclude that the abundance of heavy
elements plays a crucial role in determining whether model stars become
red giants, and in which evolutionary stage they do so. More specifically, the abundance parameter Z affects the evolution of stars through the
rate of hydrogen burning (CNO cycle), and through the radiative opacity
in the middle temperature regime (cf. Section 1). Since the thermal runaway takes place in the envelope, one can tentatively identify in the opacity (rather than in the nuclear burning) the leading cause for the different behaviour of the models at different Z's. This conclusion will
be reinforced by the examination of the next case.
Case 3: Models of massive stars with unconventional middle temperature
opacities. Stothers & Chin (1977), Bertelli et al (1984) and Greggio
(1984) have computed evolutionary sequences for massive stars, using unconventional (higher) opacities at middle temperatures. In their models
more massive than ~ 20 $M_\odot$, the envelope thermal runaway starts while hydrogen is still being burned at the center, thus producing core hydrogen burning red giants. These numerical experiments are physically interesting,
independently of the question whether such unconventional opacities are
correct or not.

2.2 First inferences

From the three cases discussed above we can derive the following
preliminary conclusions:
1) At least in these cases, stars become red giants because of the onset
of a thermal instability in the envelope (runaway envelope expansion).
2) Reaching the so-called Schönberg-Chandrasekhar limit is neither a necessary nor a sufficient condition for stars to become red giants. In
fact, in Case 2 (5 and 7 $M_\odot$ models) the S-C limit is reached but models
do not become red giants (if not after core He exhaustion), and in Case 3
models become red giants well before reaching the S-C limit. The traditional misconception, associating the approach to the S-C limit with the
tendency of the models to expand to red giant dimensions, arises just
from Case 1 (near-solar Z), when it happens that these two events take
place rather close in time to each other. Indeed, the Schönberg and Chandrasekhar theorem, as originally stated (Schönberg & Chandrasekhar 1942),
shows only the propensity of the core to contract rapidly, not that of

the envelope to expand (Eggleton & Faulkner 1981).
3) The metal opacity at middle temperatures must play a crucial role in
determining the onset of the envelope thermal instability. Indeed, by
varying such opacity (either artificially or varying Z) the thermal instability can be prevented (cf. Case 2 with Case 1), or triggered at an earlier evolutionary stage (cf. Case 3 with Case 1). Therefore:
4) The first occurrence of the thermal instability cannot be unambiguously associated with any particular evolutionary stage, i.e. with any particular structure of the core. It may happen during the stage of core
hydrogen burning (Case 3), or during the early shell hydrogen burning
stage (Case 1), or delayed to the double-shell burning stage (Case 2, 7
$M_\odot$ model). The lesson to draw from these considerations is that, if we
want to understand why stars become red giants, our attention must primarily be focused on the behaviour and properties of star envelopes, rather than to what happens to stellar cores, i.e. we must identify the
physical origin of these thermal instabilities in stellar envelopes.

2.3 The physical nature of the thermal instability

     The main factor driving the evolution of stars is certainly repre-
sented by the continuous nuclear transmutations taking place in the deep
interior. Indeed, such transmutations affect the mean molecular weight
(and then the equation of state), the opacity, and the rate of the nuclear
reactions, thus causing the stellar structure to continuously readjust
in order to maintain a quasi-static configuration. In all canonical models (e.g. those discussed above) these transmutations are confined to
the deep interior: i.e. they do not extend to the envelope. Therefore,
the envelope *follows* the evolution of the core: it is just called to readjust itself to the changing structure of the core, and primarily to the
changing luminosity being provided by the core. There are not so many
possible readjustements: basically, the envelope can only either contract
or expand. The question then becomes: how does a stellar envelope react
to the changing luminosity being provided by the core?

     To answer this question let us start by considering one of the basic
equations of stellar structure:

$$L_r \;=\; 4\pi r^2 \times F_r \;=\; 4\pi r^2\,(4acT^3/3\kappa\rho)(dT/dr) \tag{1}$$

which simply says that the radiative luminosity (at any given mass coordinate M_r) is given by the radiative flux F_r times the area through which
this energy flows. All the symbols have here their traditional meaning.
Let us now focus our attention on a radiative layer located at a given
mass coordinate M_r, and suppose to expand this layer in such a way as to
maintain hydrostatic and radiative equilibrium. Then the area $4\pi r^2$ increa-

ses and the flux F_r decreases, but we don't a priori know whether their
product (the transmitted luminosity L_r) will increase or decrease. This
will depend on several details. However, one can differentiate Eq. (1)
to get (to first order in $\delta\log r$):

$$\frac{\delta\log L_r}{\delta\log r} \;=\; W(M_r) \;\simeq\; 4(1 - A) + (4 - A)\alpha - A\beta \qquad (2)$$

where $A = -(\delta\log T/\delta\log r)$, $\alpha = (\partial\log \kappa/\partial\log \rho)_T$, and $\beta = -(\partial\log \kappa/\partial\log T)_\rho$.
The function $W(M_r)$ describes how the transmitted luminosity L_r varies in
response to a local infinitesimal expansion or contraction $\delta\log r$. While
the utility of Eq. (2) is somewhat limited by the fact that the quantity
A cannot be precisely determined from elementary arguments, nevertheless
this relation clearly shows that the size and sign of $W(M_r)$ are crucially
dependent on the opacity derivatives α and β. This is further strength-
ened by the fact that (in actual evolutionary models) $A(M_r)$ is close to
unity (to within $\pm\sim10\%$) over most of the stellar envelope, which is equivalent to saying that evolutionary changes in the envelope are *locally*
virial to a fairly good approximation, as noted a long time ago by Iben
(1965). So, the three terms at the r.h.s. of Eq. (2) are of comparable
magnitude. Another, though equivalent way of looking at Eq. (2) is to
interpret the differentials $\delta\log L_r$ and $\delta\log r$ as the actual changes (at
any given M_r) between two consecutive evolutionary models, i.e. $W(M_r)$ can
also be defined as:

$$W(M_r) \;=\; (d\log L_r/dt)\cdot(d\log r/dt)^{-1} \qquad (3)$$

where the evolutionary time derivatives are evaluated at any given M_r.
In this way, the function $W(M_r)$ can easily be evaluated for any given model belonging to an evolutionary sequence.

After these premises, let us see how the function $W(M_r)$ can be used
to clarify the nature of the envelope thermal instability. A $W(M_r) > 0$
implies that, following an infinitesimal expansion, the radiative flux
drops by an amont which is less than the corresponding increase in the
surface area, and then the transmitted radiative luminosity increases.
The opposite holds when $W(M_r) < 0$, i.e. a local expansion implies a decrease in the transmitted luminosity. At this point one can easily realize that a thermal instability can appear when $W(M_r) < 0$. In fact, in
this case an envelope expansion implies a decrease in the transmitted
luminosity, energy is then trapped in the envelope causing further expansion and a further decrease of the transmitted luminosity: the result is
clearly a runaway expansion of the envelope. Such a runaway expansion
can be halted only if for some reason $W(M_r)$ can return to positive values.

The inverse function:

$$W(M_r)^{-1} = (\delta \log r / \delta \log L_r) \tag{4}$$

has also an interesting physical meaning. Indeed, it describes how an
envelope layer reacts to a variation of the luminosity impinging on the
inner side of that layer, i.e. an increase in the luminosity $\delta \log L_r$ produces an expansion $\delta \log r$ given by $W(M_r)^{-1} \delta \log L_r$.

Model calculations (Renzini & Chieffi 1984) show that in the envelope of core hydrogen burning models (with standard opacities) $W(M_r)$ is
a *positive*, slowly radially decreasing function of M_r. Since the luminosity being provided by the burning core is secularly increasing during
this phase, Eq.s (2) and (4) imply that the envelope must expand, and
with a rate which increases outwards (since $W(M_r)$ decreases from the core
to the surface). The fact that $W(M_r) > 0$ also ensures that thermal equilibrium can be maintained in the envelope: the increasing luminosity of
the core forces the envelope to expand, but by such an expansion the luminosity transmitted by the envelope also increases, and near equality
can be maintained between the core luminosity and the luminosity transmitted by the envelope and radiated away from the surface. One can conclude that envelopes with $W(M_r) > 0$ are thermally stable.

However, this favourable situation does not necessarily apply to all
evolutionary conditions. During the main sequence phase most of the envelope is at relatively high temperatures ($T > \sim 5\ 10^6$ K), where electron
scattering dominates and the opacity derivatives α and β are vanishingly
small. However, as the star expands, an increasing fraction of its envelope cools below $\sim 5\ 10^6$ K, the metal opacity increases, and α and β tend
towards the Kramers values 1 and 3.5, respectively. Therefore, for every
M_r in the envelope, $W(M_r)$ secularly decreases as the evolution proceeds.
This trend is maintained during the early hydrogen-shell burning phase,
until (for models like those of Case 1 in Section 2.1) $W(M_r)$ vanishes,
at first near the surface, and then becomes negative, with the point where
$W(M_r) = 0$ moving *rapidly inward* in mass (Renzini & Chieffi 1984). This
coincides with the onset of the thermal runaway in the envelope. Indeed,
what happens when $W(M_r) \to 0$? From Eq. (4) one has that $\delta \log r / \delta \log L_r$
$\to \infty$, i.e. a small increase in the luminosity emanating from the central
regions triggers a catastrophic expansion of the envelope. Moreover,
further expansion leads to negative values of $W(M_r)$, an event already
discussed above, together with its implications. The runaway will be
quenched later, by the inward penetration of convection, when then Eq.s
(1) and (2) lose their validity. One can really say that convection
saves the envelope from being dynamically ejected!

It is most important to emphasize again that $W(M_r)$ first vanishes
near the surface. This clarifies several interesting aspects of the problem: that the thermal instability arises in the envelope itself (not in

the deep interior), that the behaviour of the core works just as a trig-
ger for this instability, that the opacity derivatives α and β play a
crucial role in determining the onset of the instability, which explain
the different behaviour of the models in the Cases 1, 2 and 3 described
in Section 2.1.

2.4 Consequences

This identification of the physical reason "why stars become red
giants" has a variety of astrophysically interesting ramifications.
Some of these implications are here briefly mentioned, acknowledging that
many details need to be worked out before achieving a deeper quantitative
understanding of the mentiond phenomena, and even a more rigorous formulation of some of the following statements themselves. More on these topics will be presented in a future paper (Renzini & Chieffi 1984).

1) Stars (more massive than $\sim 1\ M_\odot$) become red giants because of a
thermal instability in the envelope, whose onset is primarily controlled
by the behaviour of the opacity at middle temperatures, i.e. by α and β.

2) The cool temperature boundary of the so-called main sequence band
coincides with the locus where $W(M_r)$ vanishes at the surface of evolutionary models. Therefore, the width of this band is expected to be (highly?) sensitive to the metal contribution to the opacity at middle temperatures, i.e. again to α and β. Ultimately, the thermal instability can
be described as a runaway partial recombination of the heavy ions in the
envelope.

3) In the models lying in the so-called Hertzsprung Gap the function
$W(M_r)$ is negative and/or close to zero. When $W(M_r)^{-1}$ is very large, any
small change in the luminosity emitted by the core can lead to large changes in the stellar radius (effective temperature). This implies that when
$W(M_r) \simeq 0$ small changes in the input physics may lead to large changes in
the morphology of the tracks. One should always bear this in mind when
comparing theoretical sequences with observations: in other words, the
accuracy of theoretical effective temperatures is rather low in this region of the HR diagram.

4) The thermal instabilities associated with the beginning and termination of the core helium burning loops (cf. Fig. 1 in Becker et al
1977) have essentially the same physical origin, and can be explained in
terms of the behaviour of the function $W(M_r)$. In particular, the runaway
envelope *contraction* experienced by some models during the core helium
burning stage is another manifestation of the same type of envelope thermal instability, here working in the *reverse* direction.

5) From points 3) and 4) it follows that the theoretically predicted distributions of the periods of Cepheids (cf. Becker et al 1977) must be very sensitive to the metal opacity (through α and β), and to small changes in the input physics as well.

6) From points 2), 3) and 4) it follows that theoretical blue to red ratios for the number of massive stars (a popular topic at this meeting) is expected to be very sensitive to α and β and to small changes in the input physics as well. This suggests extreme caution when comparing such ratios with the observations.

7) Finally, it is of great importance to realize that in models lying in the Hertzsprung Gap most of the material in the stellar envelope is at temperatures between 1 and 5 million K. In several cases, up to 70 % of the stellar mass has temperatures in this range. Still, practically all existing evolutionary models have been constructed using tables giving the opacity at only three temperatures in this range, just where α and β suffer the largest and most significant changes. Even worse, interpolation schemes which fit to these tables may give large discontinuities in α and β at tabular ρ and T values, thus adding further uncertainties to the model radii. The use of analytical approximations has at least the advantage of providing continuous derivatives of the opacity.

In conclusion, two are the main lessons to be drawn from these considerations: a) theoretical stellar radii (effective temperatures) cannot be very accurate (in several circumstances), and b) the full understanding of why models behave as they do is essential for their fruitful and correct use in the comparison with the observations. In other words, a simplistic use of the models may produce astrophysical nonsense.

3. THE OOSTERHOFF-SANDAGE EFFECT AND THE AGE OF GALACTIC GLOBULAR CLUSTERS

The determination of the age and helium abundance of galactic globular clusters is certainly the topic in which the connection between stellar evolution theory and cosmology is most straightforward. Deriving cluster ages from a comparison of observational C-M arrays with theoretical models (isochrones) is apparently an easy game. However, this is not the case if one wants to derive ages with the accuracy requested in order to be interesting for cosmology. There are many methods which have been used to determine cluster ages. Each of them requires many intermediate steps to get the age starting from the "brute" C-M diagram of a cluster. But all of them ultimately rely on theoretically established relationships between the characteristics of the main sequence turnoff and age.

The method which has been most widely adopted is the isochrone-fitting technique. However, I think that this method has so many disadvantages that its further use should be strongly discouraged. The main flaw is that, when fitting theoretical and observed C-M *shapes*, one gives at least equal weight to colors (temperatures) as does to luminosities. But colors (temperatures), in the relevant region from the turnoff to the red giant branch, are strongly affected by the uncertain treatment of the envelope convection (cf. Iben & Rood 1970a), and are much more sensitive to metallicity than they are to age (Iben & Renzini 1984). While at least part of the first difficulty can be removed by a proper choice of the mixing-length parameter ℓ/H (as done recently by Vandenberg 1983), there remains the metallicity problem: a variation in Fe/H by 0.3 dex (a rather optimistic estimate of current uncertainties) affects the turn-off temperature by the same amount produced by an age variation up to ~ 4 Gyr (1 Gyr = 10^9 years). Moreover, uncertainties in reddening corrections, color-temperature transformations, bolometric corrections, and quite possible systematic photometric errors at the faint end of the sequences, further complicate the game. For these reasons (and others as well) one should always prefer a cluster dating method in which the use of stellar effective temperatures is reduced to a minimum.

A method with these characteristics was indeed developed in the late 60's (e.g. Iben & Rood 1970b, see also Iben 1974, Renzini 1977, Iben & Renzini 1984). This method relies on a theoretically established relationship between the luminosity of horizontal branch (HB) stars (at a given temperature), and the composition (Y, Z) of the main sequence progenitors of these stars. It further relies on a theoretically established relationship between the luminosity at cluster turnoff and age (which also involves Y and Z). From photometric observations one can obtain the difference in luminosity between stars at the cluster turnoff and HB stars at the *same color* (same temperature). From spectroscopic observations one can estimate Z. Estimating Y by various methods then allows one to solve for age. Note that in this case stars with the same color are compared, and so uncertainties in convection treatment, reddening, color-temperature transformations, bolometric corrections and the like are largely avoided. The sensitivity to Z is also significantly reduced (Iben & Renzini 1984).

However, while avoiding the mentioned uncertainties, this method requires the luminosity of HB stars to be very accurately reproduced by the models. Indeed, an error of one tenth of a magnitude in the HB luminosity level translates into an error of 1-2 Gyr in the derived cluster age. Independent checks of the accuracy of HB models are thus necessary for assessing the accuracy of the derived ages. Fortunately enough, RR Lyrae variables are frequently present in large numbers in globular clusters thus allowing a consistency check between pulsation theory and evolutio-

nary models, and intimately connecting to each other the study of RR Lyrae variables and the determination of cluster ages (cf. Sandage 1970,
Iben 1971, 1974). This rather intricated connection is the subject of
the following discussion.

3.1 Three empirical facts

It is worth starting this discussion by recalling three crucial observational facts.

1) *The Oosterhoff effect*. It has been known for a long time that the a-
verage period of globular cluster RR Lyraes pulsating in the fundamental
mode ($<P_{ab}>$) shows a remarkable bimodal distribution (Oosterhoff 1939, Van
Agt & Oosterhoff 1959, cf. also Cacciari & Renzini 1976, for a more recent compilation of the data). Clusters with RR Lyraes are either of
Oo. type I ($<P_{ab}> \simeq 0^d.55$, prototype M3), or of Oo. type II ($<P_{ab}> \simeq 0^d.65$,
prototype M15). Obviously, an Oosterhoff type cannot be assigned to
those clusters with too few RR Lyraes (if any). However, these clusters
can be devided into two further groups: those lacking RR Lyraes either
because their HB is too blue (BHB clusters, prototype M13), or too red
(RHB clusters, prototype 47 Tuc). In this way, galactic globular clusters
are classified into four natural groups, according to their HB morphology,
and the properties of their RR Lyraes.

2) *The Sandage effect*. Sandage (1982) has recently discovered that the
period of RR Lyraes at a given effective temperature is a decreasing function of the metallicity of the parent cluster, i.e.:

$$\Delta \log P \simeq -0.116\, \Delta [Fe/H] . \tag{5}$$

Note the different sign convention with respect to Eq. (2) in Sandage
(1982). The reality of this effect is convincingly demonstrated by the
presence of a similar *period shift* when comparing variables with the same
amplitude or with the same light curve shape. Moreover, effectively the
same period shift is found by Lub (1977) for *field* RR Lyraes in the fundamental mode, and by Kemper (1982) for overtone variables.

By coupling the *empirical* relation (5) with the theoretical Period-
Luminosity-Mass-Temperature relation for RR Lyrae stars, Sandage then
derives:

$$\Delta M_{bol}^{RR} \simeq 0.35\, \Delta [Fe/H] \tag{6}$$

for the variation of the bolometric magnitude of the RR Lyraes with the
metallicity of the parent cluster. As we shall see later, the relations
(5) and (6) are of great importance for the determination of the age of
globular clusters.

3) *The HB morphology*. Having defined four natural groups of clusters, it is interesting to examine the average metallicity of the clusters in each group. This was accomplished by Renzini (1981, 1983) using the metallicities given by Zinn (1980), with the somewhat surprising result that the average metallicity $<$[Fe/H]$>$ in each group of clusters increases along the sequence OoII, BHB, OoI, RHB, assuming respectively the values -2.1, -1.7, -1.5, and $\sim$-0.8. The typical overlap in [Fe/H] between one group and the next is 0.2-0.3 dex, quite consistent with the overlap being (mostly) due to observational errors. We see, then, that with increasing metallicity, the HB first moves to the blue (from OoII to BHB clusters), then this trend is reversed with the HB entering again the RR Lyrae instability strip (from BHB to OoI clusters), and eventually exiting from the opposite side of this strip (from OoI to RHB clusters). Renzini (1981, 1983) and Iben & Renzini (1984) briefly discuss the possible origin of this non-monotonic behaviour, and emphasize the relevance of this empirical fact for understanding the old question of the so-called "Second Parameter", (apparently?) required to explain the poor correlation between HB morphology and cluster metallicities.

What matters here is that this non-monotonic behaviour can naturally account for the heretofore puzzling dichotomy in $<P_{ab}>$ between the two Oosterhoff groups. In fact, the existence of a gap in metallicity between OoII and OoI clusters (around [Fe/H] = -1.7, the range occupied by BHB clusters), coupled with Sandage's period shift effect, automatically ensures the existence of the discontinuity in $<P_{ab}>$ between the two groups. More precisely, inserting Δ[Fe/H] = 0.6 into Eq. (5) one derives Δlog P = -0.07, which is just the difference in log $<P_{ab}>$ between the two groups! This merely empirical explanation of the Oosterhoff effect is still valid even if the spectroscopic metallicities of Pilachowski et al (1983) are adopted, since the [Fe/H] gap between OoII and OoI clusters remains almost unchanged (Iben & Renzini 1984).

3.2 A theoretical embarrassment

What remains to be understood in terms of evolutionary models is the origin of the period shift effect, Eq. (5). Indeed, current HB models (e.g. Sweigart & Gross 1976) predict a negligible period shift, Δlog P $\simeq$ -0.02 Δ[Fe/H] , if all clusters have essentially the same helium abundance (Sandage 1982, Renzini 1983, Iben & Renzini 1984). In an attempt to reproduce Eq. (5) with current models, Sandage was then forced to suggest an anticorrelation of the helium abundance with metallicity, i.e. ΔY $\simeq$ -0.07 Δ[Fe/H] , with metal poor clusters resulting richer in helium. As acknowledged by Sandage, "the sense is against intuition", then prompting a search for alternative interpretations.

An obvious limitation of Sweigart & Gross HB sequences is that they are available for only one vale of [CNO/Fe] , namely [CNO/Fe] = 0, while there is clearly no a priori guarantee that elemental proportions in the halo should be the same as in the sun. However, adopting [CNO/Fe] = -a[Fe/H] , with a = const., does not solve the problem, as noted by Renzini (1983) on the basis of the models of Rood & Seitzer (1981): the period shift remains unconfortably small.

This discrepancy between the predicted and observed period shifts indicates that, for some reason, current HB models are not sensitive enough to [Fe/H] . Therefore, let us see in more detail how the period shift is related to [Fe/H] when pulsation and evolution theories are jointly used. Indeed, from pulsation theory one gets, at fixed temperature:

$$\Delta \log P = 0.84\ \Delta \log L - 0.68\ \Delta \log M \qquad (7)$$

where L and M are the luminosity and mass of the RR Lyraes. From evolution theory, we can now relate $\Delta \log L$ and $\Delta \log M$ to a variation of metal abundance Z:

$$\Delta \log L = \{(\partial \log L/\partial M_c)_{YZ}(\partial M_c/\partial \log Z)_Y + (\partial \log L/\partial \log Z)_{M_c Y}\}\ \Delta \log Z \qquad (8)$$

$$\Delta \log M = \{(\partial \log M/\partial M_c)_{YZ}(\partial M_c/\partial \log Z)_Y + (\partial \log M/\partial \log Z)_{M_c Y}\}\ \Delta \log Z \qquad (9)$$

where M_c is the core mass at the helium flash, and all the partial derivatives can be evaluated from evolutionary sequences. For example, when the partial derivatives are obtained from the models of Sweigart & Gross (1976, 1978), then (8) and (9) inserted into (7) give the too small period shift which has been reported above.

Therefore, the suspicion is that some of these derivatives may be in error, because of possible inaccuracies in the input physics, or for some other reason. Renzini, Sweigart & Tornambè (1983) have then undertaken a project aimed at checking these derivatives and their dependence on variations in the relevant input physics. Some preliminary results of this study are now reported here.

We first concentrated on the core mass at the helium flash. The value of M_c is known to depend on the neutrino losses from the core that models experience while ascending the red giant branch (e.g. Sweigart & Gross 1978). However, these sequences were computed using a rate of neutrino losses for the plasma process (the only relevant process in this case) which does not include the contribution of neutral-current interactions. In order to take this into account we have then computed several red giant branch sequences, for various values of Z, and increasing the neutrino losses by a factor F_ν over the rate of Beaudet et al (1967), which does not include the effect of neutral currents. The result is:

$$(\partial M_c/\partial \log Z)_Y \simeq -0.009 - 0.004\ (F_\nu - 1). \qquad (10)$$

Hence, by increasing the neutrino losses the metallicity dependence of M_c increases, and this goes in the desired direction to produce a period shift. However, according to Ramadurai (1976) a value $F_\nu \simeq 1.5$ is appropriate for taking neutral currents into account, and when inserting this value into (10) the resulting change is rather modest and largely insufficient to produce a period shift effect of the size demanded by (5).

Therefore, we next considered HB models, and in particular the third derivatives appearing in Eq.s (8) and (9). Following an early suggestion (Renzini 1983), we have computed zero age HB sequences with an artificially increased metal opacity at middle temperatures. (Note that only the metal opacity is increased, not the other contributions!) By increasing the metal opacity the luminosity of a HB model of given mass, core mass, and composition is not greatly affected, it decreases just a little. But the effective temperature is considerably decreased, and the net result is a shift of the zero age HB locus to lower temperatures (for a given luminosity) and to lower luminosities (for a given temperature). The mass of the models within the instability strip is also considerably decreased. The size of these effects increases with metallicity, and then, as expected, the size of the period shift is sensitive to the adopted opacity. However, the required increase over the Los Alamos metal opacities is at least by a factor of 5, if relation (5) is to be reproduced. This increase may appear as unrealistically large, i.e. it is unlikely that Los Alamos metal opacities are in error by such a large factor. Fortunately, there is another, more attractive way of achieving essentially the same result.

Indeed, the metal opacity at middle temperatures is mostly contributed by the elements CNO and Ne. Therefore, an increase in the abundance of these elements (relative to iron) will produce an increase in the metal opacity by nearly the same factor, and then in the size of the period shift. According to the previous numerical experiment, we estimate that a period shift of the correct size should be obtained by adopting [CNONe/Fe] $\simeq 0.7$. However, more work is needed to confirm this suggestion: first, new opacity tables with enhanced CNO and Ne abundances must be constructed, and, second, HB models using these tables have to be computed.

Concerning opacity tables, still a few considerations are in order. The total opacity at middle temperatures is highly non-linear with respect to abundances. Below $Z(\text{total}) \simeq 10^{-3}$ other opacity sources dominate, and variations in metallicity give small changes in the total opacity. The contrary happens above $Z(\text{total}) \simeq 10^{-3}$, when the metals (whichever their relative proportions) dominate the total opacity. Therefore, any trend established for $Z(\text{total}) < \sim 10^{-3}$ cannot be safely extrapolated at higher metallicities!

It is also worth emphasizing that, in order to get the proper period shift, [CNONe/Fe] should be both *positive* and nearly *constant* in passing from OoII to OoI clusters, i.e. for [Fe/H] < $\sim$-1.3. Indeed, as mentioned before, a decreasing trend in [CNO/Fe] with increasing [Fe/H] does not solve the problem. Is all this reasonable? There is actually some observational support for this being the case: for instance, Sneden et al (1979) find [O/Fe] $\simeq$ 0.5 for -2.3 $\lesssim$ [Fe/H] $\lesssim$ -0.5 in a sample of field metal poor dwarfs (see also Barbuy 1983, Pilachowski et al 1983, and references therein). Also from a theoretical point of view one may expect an overabundance of at least O and Ne in halo stars and clusters, since these elements are synthetized in short living massive stars, while an important fraction of Fe may come from type I supernovae having long-living progenitors. One can tentatively conclude that the [CNONe/Fe] ratio offers a viable possibility for explaining the origin of Sandage's period shift effect, although further theoretical and observational studies are required before reaching a firm settlement of the issue.

3.3 The age of globular clusters

Let us first consider the question of the age spread within the globular cluster family. From theoretical isochrones (e.g. Ciardullo & Demarque 1977, Vandenberg 1983) it follows that, at *fixed age*, [CNONe/Fe], and Y, the luminosity of the main sequence turnoff decreases with increasing metallicity, following the relation:

$$\Delta M_{bol}^{TO} \simeq 0.37 \, \Delta [Fe/H] . \tag{11}$$

From the near identity of the numerical coefficients in (6) and (11), it follows that the luminosity difference between the HB and the turnoff, $\Delta M_{TO}^{RR}(bol) = M_{bol}^{TO} - M_{bol}^{RR}$, should be almost independent of metallicity, if all globular clusters are coeval and have nearly the same Y and [CNONe/Fe] (Sandage 1982). Since observationally $\Delta M_{TO}^{RR}(bol)$ is actually constant to within the observational errors (= $3^m.4 \pm 0.2$), one can infer that clusters are indeed coeval (Sandage 1982). Obviously, this statement applies to the clusters for which a photometrically accurate C-M array is presently available, and will have certainly to be revised if, in the future, other clusters will be found to have significantly different values of $\Delta M_{TO}^{RR}(bol)$. However, the current uncertainty in this quantity (± 0.2 mag) implies an uncertainty of about 3.5 Gyr in each age determination, and therefore age differences of this size could not have been detected, although it is somewhat encouraging that no trend whatsoever is apparent in $\Delta M_{TO}^{RR}(bol)$ vs metallicity (cf. Sandage 1982).

It is certainly of great astrophysical interest to reduce this uncertainty. I think that this can only be achieved by using linear detec-

tors (e.G. CCD's) and observing as many clusters as possible. Further
photographic photometry of cluster turnoffs should rather be discouraged,
since with this technique the turnoff luminosity can hardly be determined
with an accuracy any better than the current value ($\pm\sim$0.2 mag), and since
linear detectors are already widely available.

Let us now turn to the question of the absolute age of the clusters.
To get the age one needs a *zero-point* for the luminosity of the HB.
Sandage proceeds on empirical grounds, adopting M_{bol}^{RR} = 0.8 for the RR
Lyraes in the cluster M3, and then derives an average age for the eight
studied clusters of 17$\pm$2 Gyr, the age of each individual cluster being
uncertain by $\pm$3.5 Gyr. Iben & Renzini (1984) proceed on theoretical
grounds, and argue that HB models of very low metallicity are expected to
be quantitatively more reliable, since in these models uncertainties in
opacity and [CNONe/Fe] have smaller effects. Therefore, they adopt low
metallicity HB models (Z = 0.0001) as providing the zero point for the HB
luminosity, and further adopt Y = 0.23, as recently obtained by Buzzoni
et al (1983) for a sample of 15 well studied clusters. This gives an age
of 16$\pm$3.5 Gyr for the cluster M92, and near equality for the age of the
other clusters follows from the argument given above, i.e. from Eq.s (6)
and (11). Additional technical details can be found in Iben & Renzini
(1984).

Having already mentioned the future observations which may help in
deriving more accurate ages, one has to do the same on the theoretical
side. The understanding of the period shift effect in terms of evolutionary models is the first challenge. Preliminarly, one would like to establish whether or not an increased [CNONe/Fe] ratio is the correct explanation for this phenomenon. If so, also isochrones with various values of [CNONe/Fe] will have to be computed. If not, it is unlikely that
other changes in the input composition parameters and/or in the input
physics may produce the desired effect, and the adequacy of the canonical
assumptions will inevitably be called into question.

3.4 Two puzzling gaps

The C-M diagram of a few globular clusters exhibits one or two gaps
in the distribution of stars on the main branches. Typical is the case
of the C-M diagram of NGC 6752 (Cannon 1981), which shows a well defined
gap at the base of the giant branch (hereafter SGB gap), and another wide
gap on the blue HB (hereafter HB gap). Other examples of clusters with
a SGB gap are ω Cen (Da Costa & Villumsen 1981) and NGC 288 (Buonanno et
al 1983b). Canonical theoretical models do not predict the existence of
a SGB gap, i.e. of an acceleration of the evolutionary rate when stars
meet the red giant branch. The luminosity of this gap corresponds to an

evolutionary stage at which the mass of the helium core (or, equivalently, the mass location of the hydrogen burning shell) has reached ~ 0.2 $M_\odot$. The presence of the gap may perhaps indicate that the composition profile around $M_r = 0.2$ $M_\odot$ is not the one predicted by the models, and then that some sort of mixing may have occurred in the vicinity of this region, during the previous evolutionary stages. I understand that Pierre Demarque has recently investigated this problem in some detail, so I hope that he will briefly mention his results during the discussion.

The most straightforward interpretation of HB gaps is in terms of a bimodal distribution for the mass of HB stars in a cluster. The physical origin of the bimodality, however, remains obscure (Norris 1981). Renzini (1983) has noticed that NGC 6752 and the other clusters with a HB gap (NGC 2808, Harris 1974; NGC 1851, Stetson 1981; and M15, Buonanno et al 1983a) share the common property of having unusually high central densities, and suggested that *tidal collisions* in such dense cluster cores could be responsible for the bimodal distribution of HB masses. However, Buonanno et al (1983b) have recently found that the very low density cluster NGC 288 also has a well defined HB gap. Therefore, either tidal collisions have nothing to do with HB gaps, or there are also other processes able to produce such gaps.

In any case, whatever the origin of the HB gaps, it is worth recalling that very hot HB models have such a small envelope mass ($\sim$ a few 10^{-2} $M_\odot$, cf Sweigart & Gross 1976), that even a modest mass loss rate during the core helium burning phase (of the order of a few 10^{-10} $M_\odot$/yr) is able to reverse the direction of their evolution (from redward to blueward). The possibility of such an effect should be taken into account when dealing with very hot HB stars. Indeed, even if mass loss can not by itself produce a blue HB gap, it can prevent such a gap from being filled up by evolutionary effects.

In conclusion, the origin of the two gaps still remains rather misterious, and this somewhat veils our confidence on current determinations of the age and helium abundance of globular cluster stars.

REFERENCES

Barbuy,B. 1983, Astron. Astrophys. 123, 1
Beaudet,G., Petrosian,V., Salpeter,E.E. 1967, Ap. J. 150, 979
Becker,S.A., Iben,I.Jr., Tuggle,R.S. 1977, Ap. J. 218, 633
Bertelli,G., Bressan,A.G., Chiosi,C. 1984, Astron. Astrophys. (in press)
Buonanno,R., Corsi,C.E., Fusi Pecci,F. 1983a, Astron. Astrophys. Suppl. 51, 83
Buonanno,R., Corsi,C.E., Fusi Pecci,F., Alcaino,G., Liller,W. 1983b, Ap. J. (in press)

Buzzoni,A., Fusi Pecci,F., Buonanno,R., Corsi,C.E. 1983, Astron. Astro-
     phys. (in press)
Ciardullo,R.B., Demarque,P. 1977, Trans. Yale Univ. Obs. Vol. 33
Cacciari,C., Renzini,A. 1976, Astron. Astrophys. Suppl. 25, 303
Cannon,R.D. 1981, Astrophysical Parameters for Globular Clusters, ed.
     A.G.D. philip, D.S. Hayes (Schenectady: L. Davis), p. 501
Da Costa,G.S., Villumsen,J.V. 1981, Astrophysical Parameters for Globular
     Clusters, ed. A.G.D. Philip, D.S. Hayes (Schenectady: L. Davis),p.527
Eggleton,P.P., Faulkner,J. 1981, Physical Processes in Red Giants, ed.
     I. Iben Jr., A. Renzini (Dordrecht: Reidel), p. 179
Greggio,L. 1984, Stellar Nucleosynthesis, ed. C. Chiosi, A. Renzini (Dor-
     drecht: Reidel), p. 137
Harris,J.G. 1974, Ap. J. Lett. 192, L161
Iben,I.Jr. 1965, Ap. J. 142, 1447
Iben,I.Jr. 1971, Publ. Astron. Soc. Pacific, 83, 697
Iben,I.Jr. 1974, Ann. Rev. Astron. Astrophys. 12, 215
Iben,I.Jr., Renzini,A. 1983, Ann. Rev. Astron. Astrophys. 21, 271
Iben,I.Jr., Renzini,A. 1984, Physics Reports (in press)
Iben,I.Jr., Rood,R.T. 1970a, Ap. J. 159, 605
Iben,I.Jr., Rood,R.T. 1970b, Ap. J. 161, 587
Kemper,E. 1982, Astron. J. 87, 1395
Lub,J. 1977, Thesis (Univ. of Leiden)
Mengel,J.G., Sweigart,A.V., Demarque,P., Gross,P.G. 1979, Ap. J. Suppl.
     40, 733
Norris,J. 1981, Ap. J. 248, 177
Oosterhoff,P.Th. 1939, Observatory, 62, 104
Pilachowski,C.A., Sneden,C., Wallerstein,G. 1983, Ap. J. Suppl. 52, 241
Ramadurai,S. 1976, M.N.R.A.S. 176, 9
Renzini,A. 1977, Advanced Stages in Stellar Evolution, ed. A. Maeder, P.
     Bouvier (Geneva), p. 149
Renzini,A. 1981, Astrophysical Parameters for Globular Clusters, ed.
     A.G.D. Philip, D.S. Hayes (Schenectady: L. Davis), p. 347
Renzini,A. 1983, Mem. Soc. Astron. It. 54, 335
Renzini,A., Chieffi,A. 1984 (in preparation)
Renzini,A., Iben,I.Jr. 1984, Physics Reports (in preparation)
Renzini,A., Sweigart,A.V., Tornambè,A. 1983 (in preparation)
Rood,R.T., Seitzer,P.O. 1981, Astrophysical Parameters for Globular Clus-
     ters, ed. A.G.D. Philip, D.S. Hayes (Schenectady: L. Davis), p. 369
Sandage,A. 1970, Ap. J. 162, 841
Sandage,A. 1982, Ap. J. 252, 553
Schönberg,M., Chandrasekhar,S. 1942, Ap. J. 96, 161
Sneden,C., Lambert,D.L., Whitaker,R.W. 1979, Ap. J. 234, 964
Stetson,P.B. 1981, Astron. J. 86, 687
Stothers,R., Chin,C.-W. 1977, Ap. J. 211, 189

Sweigart,A.V., Gross,P.G. 1976, Ap. J. Suppl. 32, 367
Sweigart,A.V., Gross,P.G. 1978, Ap. J. Suppl. 36, 405
Van Agt,S., Oosterhoff,P.Th. 1959, Ann. Sterrewacht Leiden, 21, 253
Vandenberg,D.A. 1983, Ap. J. Suppl. 51, 29
Zinn,R. 1980, Ap. J. Suppl. 42, 19

DISCUSSION

Cox: 1) Can you get even more effect in the core mass sensitivity with
Z by further changing neutrino losses?
2) Are the higher CNO abundances relative to iron consistent with recent
higher O abundances that we have heard about?

Renzini: Yes. But the question is which is the correct rate of neutrino
losses. A careful reexamination of the plasma process including the
neutral current interaction is certainly worthwhile.
2) Yes, it is.

Rood: Did you change the core mass when you change [CNO/Fe]?

Renzini: Yes, we tried to in an approximate way.

Rood: When I made ZAHB models with different [CNO/Fe] and M_{core} varying
as with the standard theory I find that d log(P')/d (CNO) is almost
zero.

Renzini: I should have a careful look to what you have done before
drawing any conclusion. My impression is that a period-shift may appear
only if [CNO/Fe] is positive and constant, but let me postpone my answer.

Itoh: We recently recalculated the neutrino-pair bremsstrahlung process
using the Weinberg-Salam theory of weak interaction. The paper will soon
appear in the Astrophysical Journal.

Renzini: This is very interesting, but the process dominating neutrino
losses in population II red giants is the plasma process.

Demarque: 1) Although I see the attractiveness of your treatment of the
ages of the globular clusters, I am concerned about the assumption that
the magnitude difference between the horizontal branch and turn-off stars
at the same colour is constant. It seems unclear at this point that the
observations support this statement. In addition, this assumption, together with your other arguments, forces all globular clusters to have
the same age. How then are we going to find out whether all globular
clusters have the same age?
2) In your presentation, you alluded to our unpublished work on the subgiant gap observed in globular cluster C-M diagrams. T. Armandroff and I

have in the last few months investigated the evolution of stars with a
mixed shell in the interior (an adhoc assumption). We find that <u>only</u>
if the mixing occurs in the subgiant phase, and <u>only</u> for certain thick-
nesses of the mixed shell, do we get a small gap approximately in the
right place of the luminosity function. It is unclear at this point what
relevance to the observed gap our calculations have.

<u>Renzini</u>: The constancy of the magnitude difference ΔM_{TO}^{RR} between the HB
and the turn-off is not an assumption, it is an observed fact for all
clusters with photometrically accurate C-M diagrams, including NGC 288
(Buonanno et al, Ap.J. in press) for which there were rumors for a
peculiarity in this respect. This fact, together with the Sandage period-
shift effect (another observed fact) ensures the constancy of the age
for the studied clusters, within the quoted uncertainty. Certainly, if
some day clusters will be found exhibiting substantially different values
of ΔM_{TO}^{RR} , then the influence will be that they have different ages.

<u>Hesser</u>: Concerning the gap in the blue HB in NGC 6752 and other clusters,
Russell Cannon mentioned the hypothesis put forth by Norris and his
collaborator that bimodal CN strengths might be correlated with the
presence of a BHB gap. Unfortunately they then tested the hypothesis by
observing NGC 2808 giants, which did not show the expected bimodal behavior! Thus, another observational clue about the gap apparently must
be discarded.

<u>Frogel</u>: Is your age spread for globulars still consistent with the
explanation for the "2nd parameter" effect in the distant halo objects
as being due to these distant clusters having formed later than in closer
clusters?

<u>Renzini</u>: As I said, the current uncertainty in the age determination of
individual clusters is of several billion years. It is also known that
the HB morphology is expected to change rather dramatically when the
age is changed by just 1 or 2 billion years. Therefore, if a second parameter is required for distant clusters, the age remains a reasonable
candidate.

SOME INVESTIGATIONS ABOUT THE CARSON OPACITIES

Arthur N. Cox and Russell B. Kidman
Theoretical Division, Los Alamos National Laboratory
University of California
Los Alamos, New Mexico 87545 USA

Accurate opacities for stellar composition mixtures are needed for all studies of stellar structure, evolution, stability, and pulsation. In several cases it appears that larger opacities in the range of temperature near one million kelvin would assist is resolving some current discrepancies between observations of stars and some theoretical predictions. Opacities published by Carson, Mayers, and Stibbs (1968) and more recently modified and available informally, have this larger opacity in this temperature region compared to the widely used Los Alamos opacities. See the tables of Cox and Stewart (1965, 1970ab) and Cox and Tabor (1976). It is therefore of great interest to see if the actual cause of the differences between these two sets of opacities can be found and discussed.

Three problem areas where increased opacities would be welcome are: the observed broadening of the upper main sequence that can be produced with larger opacities that tend to expand the stars; the existence of the double-mode Cepheids and their anomalously low period ratios which can be predicted to be lower, as observed, if opacities are larger; and the small sensitivity of the low mass population II horizontal branch luminosity to the metal content of their compositions that would be more effective if their opacity were increased. Several other problems that could be solved by larger opacities have been widely discussed, but we feel that they are not justifiably an opacity problem. The conclusion of our considerations are that the Thomas-Fermi method for getting opacities used by Carson and his collaborators does not produce values appreciably different from those obtained without this method at Los Alamos, and that these persistent astrophysical problems must be solved in other ways. We here propose a possible error in the Carson opacities, and, further, we mention another that seems to be the correct reconciliation between these two opacity sets.

Figure 1 displays the logarithm of the ratio of the Carson to the Los Alamos opacities versus temperature and density for his mixture C312. This mixture has a hydrogen mass fraction X of 0.73, a helium mass

A. Maeder and A. Renzini (eds.), Observational Tests of the Stellar Evolution Theory, 41–46.
© *1984 by the IAU.*

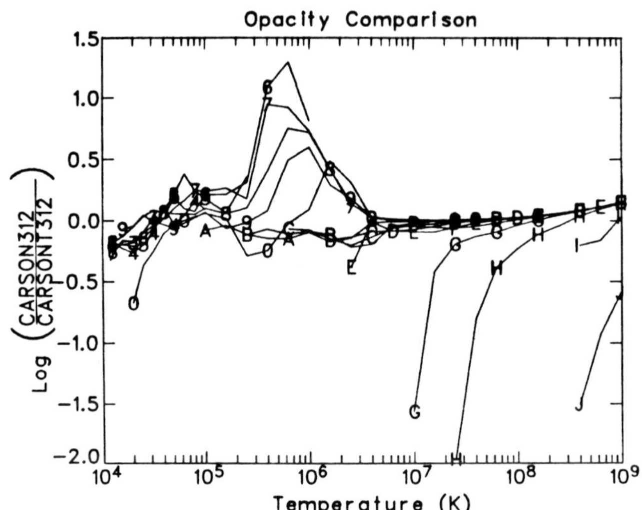

Figure 1. This comparison of the Carson and Los Alamos opacities gives
    the logarithm of the ratio of the Carson values to the Los Alamos
    Opacity Library values for integral values of the logarithm of
    the density from -12 to +7 versus temperature.

fraction Y of 0.25 and the other element mass fraction of 0.02. The
points for higher densities (the 0 and the letters) that are
significantly below the others are due to the fact that the Carson
values include the effects of electron conduction which is not present
in the Los Alamos Opacity Library (Huebner et al. 1977) generated
table. Differences between the two sets of opacities depend somewhat
on density, but mostly they occur in four temperature regions. The
Carson opacities are lower in the hydrogen ionization region up to
about 30,000 K. The helium second ionization region has higher
opacities by up to a factor of two for most densities in stars. The
well-known CNO bump where the Carson opacities are ten or more times
larger, depending on the density, occurs between 100,000 K and
5,000,000 K. Finally, above 100 million kelvin there is a slow
increase of the Carson opacities relative to the Los Alamos ones. We
investigate here the last point at one million kelvin on curve 6 which
is for a density of 10^{-7}.

A review of the absorption processes that can occur shows that, apart
from the dominant electron scattering, the only mechanism that is
operating to any noticeable extent here is the bound-free absorption of
the CNO elements, especially by oxygen because of its large abundance.
This immediately focuses the problem on the Thomas-Fermi method. In
this method there is a problem of getting the proper conservation of
electrons associated with each atom. The solution for the two
quantities, the electron density and the potential distributions, using
two equations, the Thomas-Fermi statistics for the electron density and

the Poisson equation for the potential, gives consistent values. However, when the Schrodinger equation is solved using this potential to discretize the energy levels of the bound electrons, the resulting electron numbers and the potential are no longer consistent. An iteration is required to get these values to be the same for both the potential and the energy level solutions. From a discussion by Cloutman (1973), and by inspection of subtle inconsistencies between quantities in Table 3 of Carson, Mayers, and Stibbs (1968), one can see that charge conservation is difficult to get and, further, energy levels can be greatly in error.

Merts in unpublished work has considered the application of the Thomas-Fermi method. He found that when he used the method to get occupation numbers and energy levels for stellar mixtures they differed only a little from those found by the usual Los Alamos methods. Actually, one finds that only K edges seem to contribute significantly to stellar mixture opacities, and the calculations to get their properties are just as accurately done with hydrogenic approximations for the atomic potential. A recent review of possible opacity increases near one million kelvin has been written by Magee, Merts and Huebner (1984).

Figure 2 shows the D versus u diagram used widely for opacity discussions. This comes directly from the current Los Alamos calculations for our one million kelvin and 10^{-7} g/cm^3 point. Here D is proportional to the absorption or scattering cross section multiplied by the cube of u, and u is the photon energy scaled by the temperature in energy units. At the top of the figure is plotted the weighting function that is used to get the Rosseland mean absorption integral. The line marked S is the dominant free electron scattering contribution to the opacity. The steps are the bound free edges for the K and L shells of carbon, nitrogen, and oxygen. Actually the K edge for carbon is at about u=6 and for nitrogen, at about u=8. The very abundant oxygen is not so ionized, and its hydrogen-like and helium-like K edges appear at just over u=10. The very weak K edge for hydrogen like helium with its ionization energy of 54 ev is at u=54/86=0.6. At even lower u one can see the minimum D which is the very small free-free absorption contribution. Some bound-bound (line) absorption in less completely ionized atoms such as neon can be seen at u greater than 10. Here our task is to somehow increase the mean absorption to produce the large Carson value.

This opacity increase can be accomplished either by increasing the numbers of L shell electrons in the CNO elements or by moving the K edges to much lower energies. Here we choose the latter direction, but the most recent work indicates that the L shell occupation numbers are incorrect in the Carson Thomas-Fermi formulation, and they are wrongly overabundant by a huge factor. All K shell energy levels are here divided by 5 to see what effect that will have on the opacities. As one can see, the K edge of oxygen then moves from u=10 to u=2, giving a great contribution to the opacity integral.

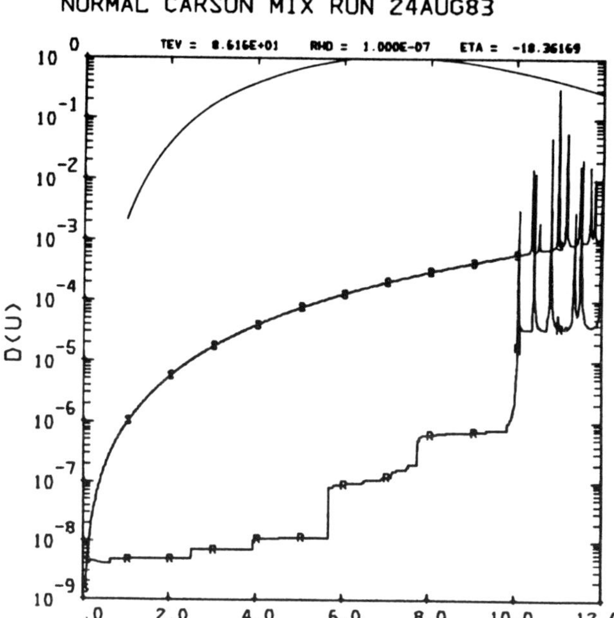

Figure 2. The quantity D, as defined in the text is plotted versus the
     scaled photon energy for the temperature of one million kelvin
     (86 ev) and for a density of 10^{-7} g/cm^3. The weighting function,
     the free electron scattering contribution, and the bound-free
     absorption edges for the CNO elements are apparent.

Figure 3 shows the Carson opacities for his C312 mixture with three
points plotted at 300,000, 500,000, and 1,000,000 K at the density line
6. They track the Carson values well. One further test of this
possible error shows, however, that this is not the correct solution
for fixing the Carson opacities. At a density of 10^{-4}, a thousand
times larger in density, moving the K edges by a factor of 5 gives an
opacity of 83 cm^2/g rather than the Carson value of 3.9. This x point
is also plotted on Figure 3, but to be a correct fix, it should have
fitted along the line labeled 9.

Currently Carson, Huebner, Magee, and Merts at Los Alamos are
collaborating to see how the implementing of the Thomas-Fermi method
for getting atomic models has been different between the Los Alamos,
where no appreciable effect was noticed, and the Carson opacity
programs.

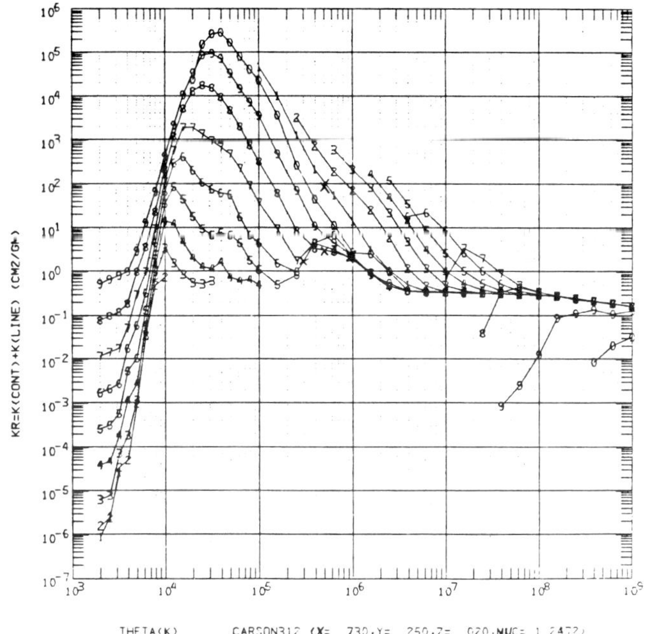

Figure 3. Carson opacities for the C312 mixture are plotted versus temperature for our 20 densities. Three points at 10^{-7} and one at 10^{-4} g/cm^3 are plotted as x points when our proposed fix for the K edge energies is used.

References

Carson, T.R., Mayers, D.F. and Stibbs, D.W.N. 1968, M.N.R.A.S. 140, 483.
Cloutman, L.D. 1973, Ap.J. 184, 675.
Cox, A.N. and Stewart, J.N. 1965, Ap.J. Suppl. 11, 22.
Cox, A.N. and Stewart, J.N. 1970a, Ap J. Suppl. 19, 243.
Cox, A.N. and Stewart, J.N. 1970b, Ap.J. Suppl, 19, 261.
Cox, A.N. and Tabor, J.E. 1976, Ap.J. Suppl. 31, 271.
Huebner, W.F., Merts, A.L., Magee, N.H., and Argo, M.F. 1977 Los Alamos Scientific Laboratory report LA-6760-M.
Magee, N.H., Merts, A.L., and Huebner, W.F. 1983, Ap.J. submitted.

DISCUSSION

Nissen: Recent abundance analyses of unevolved F-type main-sequence stars show that oxygen is overabundant in metal-poor stars by factors 5 to 10. Do you expect a major change of the opacities computed when such a non-solar oxygen abundance is taken into account?

<u>Cox</u>: Yes, I do. Such mixtures have been calculated. The effects on evolution tracks may not be very important, however, especially for population II stars. You need to ask people like Renzini, but I believe that higher oxygen is not all that important because Z is so small.

SOLAR CONSTRAINTS

Janine PROVOST
Observatoire de Nice, B.P. 139, 06003 Nice-Cedex, France

ABSTRACT.

Accurate tests of the theory of stellar structure and evolution are
available from the Sun's observations. The solar constraints are
reviewed , with a special attention to the recent progress in observing
global solar oscillations. Each constraint is sensitive to a given
region of the Sun. The present solar models (standard, low Z, mixed)
are discussed with respect to neutrino flux, low and high degree
five-minute oscillations and low degree internal gravity modes. It
appears that actually there do not exist   solar models able to fully
account for all the observed quantities.

1.   INTRODUCTION.

As the nearest star, the Sun provides the opportunity to test very
accurately the theory of stellar structure and evolution. Particularly
in the last years solar oscillation observations have considerably
progressed and their unusually high accuracy has even more constrained
the solar model, leading to still unresolved difficulties.

     Models of the Sun must in the first place satisfy the major usual
constraints, i.e. they must lead at the actual age of the Sun
($t_\odot$ = 4,6 10^9 years) to the present values of luminosity ($L_\odot$ = 3,86 10^{33}
erg/sec from Bahcall et al., 1982) and of radius ($R_\odot$ = 6.9599 10^{10} cm)
for the given mass of the Sun ($M_\odot$ = 1,989 10^{33} g). Models of the Sun are
more specifically constrained firstly by the present chemical composition
of the solar surface, which is assumed to reflect the initial composition
of the heavy elements, secondly by the observed abundances of nuclearly
processed elements, such a lithium and ^{3}He. This point will be considered subsequently in more detail by Schatzman. The most famous constraint
on solar models has been provided by the measure of the solar neutrino
flux by Davis and his collaborators. The measured value corresponds to
(2.1 + 0.3) SNU (Cleveland ,Davis and Rowley, 1981) (1 SNU corresponds
to 10^{-36} capture per target atom per second). It is only about 25 to 40

A. Maeder and A. Renzini (eds.), Observational Tests of the Stellar Evolution Theory, 47–65.

percents of the present theoretical prediction of the standard solar
model as recently reviewed by Bahcall et al. (1982).

The accurate measurements of the solar oscillation frequencies
made within the last ten years provide an additional independent
constraint on the solar model. The properties of these oscillations are
closely related to the internal structure of the Sun, the different
observed components of oscillations permitting to infer the properties
of different zones of the Sun. Since the discovery by Leighton, Noyes
and Simon (1962) of 5 mn solar oscillations, by measuring time variations
of the Dopplershift of a Fraunhofer line, quite interesting progress
has been made. The high degree five-minute acoustic modes with small
horizontal wavelengths, typically of 10 000 km, have been first
observed by Deubner (1975) and later by many observers (e.g. Deubner
et al., 1979; Harvey et al., 1983; Hill et al., 1983). The low degree
five-minute acoustic modes, i.e. with large horizontal wavelengths of
the order of the solar radius, have been recognised by Claverie et al.
(1979, 1981), Grec et al. (1980, 1983), Scherrer et al. (1982), Woodard
and Hudson (1983). This year, two important observational results have
been obtained : acoustic modes with intermediate degree have been
isolated by Duvall and Harvey (1983) and oscillations have been found
in the low frequency solar spectrum (160-360 mn) by different groups of
observers (Delache and Scherrer, 1983; Fröhlich, 1983; Van der Raay, 1983;
Severny et al., 1983) which have been interpreted as gravity modes.
Other oscillations with periods longer than 5 mn, detected by solar
diameter measurements or limb brightness fluctuations (e.g. Bos and
Hill, 1983; Rösch and Yerle, 1983) are not yet fully identified and
understood. Just as the well known 160 mn oscillation (e.g. Kotov et
al., 1983; Scherrer and Wilcox, 1983) which might be a gravity mode
if it is of solar origin.

An important fact is that all these considered solar constraints
concern different parts of the Sun, so that the whole solar model can
be tested. This is shown in figure 1 which represents the contribution
of the different layers of the Sun to different observed quantities.
The neutrino flux from bore decay, measured in the Davis experiment,
originates from the inner five percents of the solar mass. The mean
separation which characterizes the observed frequencies of low degree
acoustic modes originates from the outer ten percents of the solar mass.
Nevertheless, it will be shown in section 5 that other features of these
oscillations are closely related to the deep interior of the Sun. The
solar luminosity and the neutrino flux of the proton-proton reaction
(which could be detected by the ^{71}Ga experiment as discussed by Bahcall
et al. 1978) come mostly from an intermediate region. The"spacing"
period which characterizes the periods of low degree gravity modes
(see section 6) is mainly sensitive to the inner five percents of the
mass which determine also the neutrino flux.

The main purpose of this review is to point out the importance of
solar oscillations as a probe of the Sun's interior, without attempt
to consider this subject extensively. There exist many reviews on solar

oscillations : for their observations e.g. Deubner 1980, Stenflo 1980,
for their interpretations e.g. Christensen-Dalsgaard (1980, 1982),
Gough (1980, 1983b). More information can be found in the proceedings of
two recent meetings, the 66th IAU Colloquium, held in Crimea in 1981 and
the Europhysics Study Conference held at Catania in June 1983. In a
first part it is recalled how the neutrino problem has led to question
the solar standard model, and some ways of modifying this model are
briefly presented. Then the constraints on the solar structure, due to
observed global solar oscillations, are reviewed. The final discussion
shows that presently no model is able to account for all the solar
constraints.

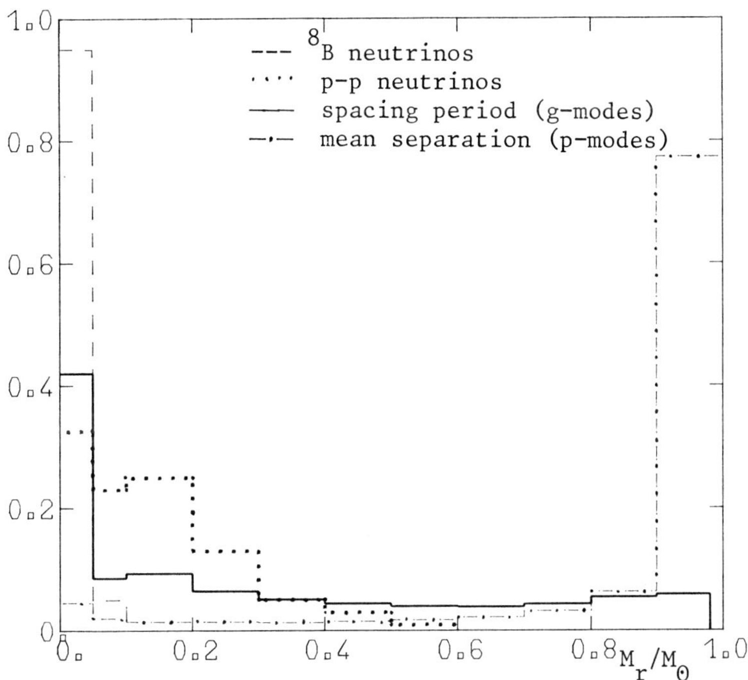

Figure 1 : Histogram of the fractional contributions to low degree
          p-mode mean separation ν_o, the low degree g-mode spacing
period P_o (computed for the standard solar model of Schatzman and Maeder,
1981) the ^{8}B neutrino flux and the neutrino flux from the p-p reaction
(taken from Bahcall, 1981), as a function of $M_r/M_\odot = \int_o^r 4\pi \rho\, r^2\, dr/M_\odot$

2. STANDARD SOLAR MODELS AND THE NEUTRINO PROBLEM.

Solar models are computed with usual simplifying assumptions, i.e.
spherical symmetry, thermal and hydrostatic equilibrium at each time of
the evolution, and a uniform chemical composition at the zero age. The

convection zone which extends on 30% of the solar radius is crudely
described with the mixing length theory. Other macroscopic motions
are ignored, as well as the possibility of chemical mixing in the solar
core. The initial solar model is specified by the chemical composition
parameters Y and Z, i.e. helium and heavy elements contents, and the
mixing length parameter λ, which characterizes the efficiency of
convective transport. No direct information on Y and λ is available.
A discussion of protosolar helium abundance has been given recently by
Gough (1983a). A usual calibration of a solar model consists in adopting
either the value of Z or the value of Z/X most directly determined from
spectroscopic measurements (Z $\cong$ 0.02 or Z $\cong$0.023 from Ross Aller 1976).
Y and λ are then fitted to adjust present luminosity and radius at the
actual age of the Sun. So calibrated standard models (e.g. Bahcall et
al., 1982; Christensen-Dalsgaard, 1982; Ulrich, 1982...) correspond to
Y $\cong$ 0.25. This helium abundance is not in conflict with the content of
the galaxy and with the prediction of cosmological theories. These
models have an extended convective zone (about 30% of the radius). But
they predict a too large neutrino flux compared to the measured one.
According to Bahcall et al. (1982) who have estimated all the uncertainties that could affect the neutrino flux, the predicted value is
(7.6 + 3.3) SNU, which has to be compared to the observed one
(2.1 + 0.3) SNU (Cleveland et al., 1981). The so-called neutrino problem
has led to a careful examination of the uncertainties of standard models.
Many improvements of the physical inputs have been made, which concern
the equation of state, the opacity and the nuclear cross sections,
without reducing noticeably the neutrino flux.

   The production rate of neutrinos detected by the Davis experiment,
which come essentially of ^{8}B decay, depends very strongly on the central
temperature of the Sun (F $\cong$ T$_c^{20}$ according to Iben, 1969).Thus,the main
idea is to bring down the central temperature in order to put the
predicted flux in agreement with the observed one. This can be obtained
in particular by increasing the proportion of hydrogen available for
nuclear burning. So,the solar luminosity can be generated in the core
at a lower temperature. As shown by Iben (1969),a possibility to lower
the neutrino flux is to lower the helium content. But the adjusting of
the neutrino flux requires extremely low Z (Abraham,Iben, 1971; Bahcall
and Ulrich, 1971), uncompatible with the observed surface chemical
abundances of the Sun. To avoid this difficulty "dirty" solar models
have been computed by Christensen-Dalsgaard et al. (1979) with low Z in
the interior, hence a low neutrino flux, but assuming that the outer
layers have accreted heavy elements during the evolution of the Sun and
thus have Z $\cong$ 0.02 at the surface. Low helium and low heavy elements
contents are associated with a shallow convection zone : this is a
consequence of a lower opacity which requires reducing the efficiency
of the convection to adjust the computed radius to the solar radius.
Another way to lower the predicted neutrino flux is provided by some
sort of mixing of the solar core, which brings more hydrogen into the
central region. Several attempts have been made (as for example by
Ezer and Cameron, 1968; Shaviv and Beaudet, 1968; Bahcall et al., 1968),
but without a discussion of the nature of the mixing. The physical

mechanisms of mixing in stellar evolution are discussed by Schatzman
(1984). The basic assumption is the existence of some sort of marginal
instability, driven for example by differential rotation, which generates turbulence . The turbulent transport can be described by a turbulent
diffusion coefficient, proportional to microscopic viscosity, and to a
pseudo-Reynolds number which measures the efficiency of the mixing, as
shown by Schatzman (1977) and Genova and Schatzman (1979). Diffusion
mixed solar models with a normal chemical composition (Y = 0.25,
Z = 0.02) have been computed by Schatzman and Maeder (1981). The neutrino
flux is highly reduced by turbulent diffusion mixing, by a factor
2.6 to 8, when the pseudo-Reynolds number R_e^* varies from 50 to 200. For
these models, the convection zone depth is almost unchanged.

3. DESCRIPTION OF SOLAR OSCILLATIONS

In this section general properties of solar oscillations are rapidly
described. An extensive presentation of stellar oscillations is given
in Ledoux and Walraven (1951), Cox (1980) and Unno et al. (1979). Oscillations of a star are usually studied assuming spherical symmetry of the
equilibrium model; rotation, magnetic field and all horizontal inhomogeneities are neglected. The observed small amplitude oscillations can
be studied in a linear approximation. The radiative dissipative time
being much larger than the characteristic time of the motion, the
adiabatic approximation can be used, at least for eigenfrequencies
studies. Global solar oscillations are standing waves, whose motion can
be decomposed into components varying sinusoïdally in time with a
frequency ν. A physical quantity related to the wave can be described
by : $X = f_n(r) \, Y_\ell^m (\theta,\phi) \cos 2 \pi \nu t$, where $Y_\ell^m (\theta,\phi)$ is the spherical
harmonic of azimutal order m and degree ℓ. Each oscillation can thus be
described by the radial order n, the degree ℓ and the azimutal order m,
which represent respectively the number of zeros along a solar radius,
on a sphere of radius r and along a circle parallel to the equator. At
each position r, it is possible to define a local radial wavenumber
$k_r \cong n/r$ and a local horizontal wavenumber

$k_h \cong \ell/r$ ($k_h = \dfrac{\sqrt{\ell \, (\ell+1)}}{r}$ due to the properties of the spherical harmonics).
As long as rotation is neglected, the frequencies are independent of
m.

    Two sorts of waves exist in a compressible self gravitating medium:
the acoustic waves (p-modes) are driven by the pressure fluctuation
produced by compression, and they depend on the structure of the sound
speed c ; the gravity waves (g-modes) are driven by the buoyancy force
and they depend on the structure of the Brunt-Väissälä frequency

$$N \left(N^2 = g\left(\frac{1}{\Gamma_1} \frac{1}{p} \frac{dp}{dr} - \frac{1}{\rho} \frac{d\rho}{dr} \right) \right).$$ These different modes are trapped in

different regions of the Sun and their frequencies depend evidently on
the structure of the trapping zones. Acoustic modes are reflected at the
surface where the density scale height becomes shorter than the

wavelength. Due to the increase of the sound speed from the surface to
the centre, they are reflected at the level where their frequencies
become equal to the acoustic local frequency (or Lamb frequency)
$S_\ell = k_h c = \sqrt{\ell(\ell+1)}\ c/r$. Thus five-minute acoustic modes are trapped in
an outer part of the solar envelope if they are of a high degree (the
outer 10% of radius if $\ell > 200$); they penetrate to the bottom of the
convective zone if the degree ℓ is about 40 ; low degree oscillations
are reflected near the centre of the Sun and their frequencies depend
on the structure of the sound speed across the whole Sun. Gravity modes
are trapped in convectively stable regions $(N^2 > 0)$ and they are eva-
nescent in convection zones. Thus,solar gravity modes can be trapped
either in the radiative interior or in the solar atmosphere. Their frequencies must be smaller than the maximum of the Väissälä frequency in
the trapping region, i.e. their period must be larger than 1/2 hour,
for the modes trapped in the interior, and larger than 4 mn for modes
trapped in the upper atmosphere. Dziembowski and Pamjatnyk (1978) have
shown that the observation of large degree gravity modes trapped in the
interior is unlikely, because of the rapid decay of their amplitude
through the convection zone. It is not the case for low degree gravity
modes which have been recently observed (see section 6). Despite not yet
identified atmospheric modes, some observed photospheric motions could
be interpreted in terms of gravity waves (Hill et al., 1982). In the
next sections the different observed oscillations, high and low degree
acoustic modes and low degree gravity modes will be analysed in more
details in relation with the solar structure.

4. FIVE MINUTE OSCILLATION OF HIGH DEGREE : A CONSTRAINT ON THE CONVECTION ZONE.

Five minute solar oscillation has been spatially resolved into different
components of horizontal wavelengths of 5 000 to 20 000 km by Deubner
(1975). The two dimensional power spectrum obtained in a horizontal
wavenumber-frequency plane shows that the power is maximum along well
defined ridges. Each ridge corresponds to oscillations with the same
radial order n ($n \lesssim 10$) and varying ℓ from 140 to 700. The interpretation of these oscillations as acoustic modes trapped below the solar
surface, first suggested by Ulrich (1970), has been confirmed by an
approximate agreement between observations (Deubner, 1975; Deubner,
Ulrich and Rhodes, 1979) and theoretical p modes calculations (Ando and
Osaki, 1975, 1977; Ulrich and Rhodes, 1977). As these oscillations are
trapped in the outer 10 percents of the radius, their frequencies
depend essentially on the structure of the sound speed in the convective
zone. The roughly parabolic shape of the ridge in the $(k - \omega)$ plane is
predicted with a simple model of the convection zone, i.e. an adiabatically stratified polytrope (Gough 1978). This model shows also that
the frequencies are proportional to the adiabatic temperature gradient.
Studies of the sensitivity of the theoretical five-minute frequencies
to the structure of the Sun have shown that the frequencies depend
essentially on the mixing length parameter λ, hence on the depth of the
convection zone (Berthomieu et al., 1980; Lubow et al., 1980). The

frequencies decrease when λ increases with a saturation at large λ.
The best fit with the observations is obtained for models with $\lambda \cong 2$
and a convection zone depth about 200 000 km (see for example Fig. 1
in Berthomieu et al., 1980). Such large convection zones are related
to a normal helium abundance for standard models as well as for diffusive mixed models. Thus high degree five-minute oscillations constrain
the solar model to have a large convective zone.

Improvements in the accuracy of the observations allow to infer
more precisely the structure of the convection zone. From Harvey, Duvall
and Rhodes observations, Duvall (1982) has found that most of the information about the solar interior contained in high degree five-minute
modes can be reduced to a dispersion relation of the form
$(n + 3/2)\pi/\omega = f(k/\omega)$, where n is the radial order, k the horizontal
wave number and $\omega = 2\pi \nu$. Such a relation is obtained for a polytrope
of index 3, only if it is adiabatically stratified (Gough, 1978;
Christensen-Dalsgaard, 1980). This led Gough (1983a) to the conclusion
that Duvall's result is an observational evidence of the nearly adiabatic stratification of most of the convection zone. This confirms some
features of the structure given by the mixing length theory, despite
its crudeness. Information on the usually ignored macroscopic motions
in the convection zone can be obtained from observations like those of
Hill, Toomre and November (1983), which have revealed significant
change in the position of the ridges in the $(k - \omega)$ diagram for different days. The position of the ridges are sensitive to large scale
subphotospheric velocity and temperature fields (Gough and Toomre,1983).
Variations of about 100 m/sec are obtained from one day to the other and
may be due to a giant cell convective pattern according to Gough et
al. (1983).

5. FIVE MINUTE OSCILLATIONS OF LOW DEGREE : AN INFORMATION ON THE WHOLE
    SUN.

Low degree five-minute oscillations will provide a more important constraint on the solar interior, because propagating nearly vertically
($\ell \ll n$) they penetrate deeper towards the centre of the Sun. Such
oscillations have been observed by the Birmingham group (Claverie et
al., 1979, 1981) measuring variations in the Dopplershift of a
Fraunhofer line in light integrated over the entire solar disk. The
temporal spectrum shows a sequence of almost uniformly spaced peaks with
a mean separation of 68 μ Hz and amplitudes of about 20 cm/sec. Subsequent observations have improved the frequency resolution, which is
about 2 μ Hz for the observations of Grec, Fossat and Pomerantz (1980,
1983) at the South Pole; the Birmingham group has reached a resolution
of about 0.1 μ Hz by combining 3 months of data from two different
stations of observations at Tenerife and Hawaï. Similar spectra have
been obtained by Woodard and Hudson (1983) for the total irradiance
fluctuations. Due to averaging over regions where the oscillations have
opposite signs, the sensitivity of such observations decreases rapidly
when the degree ℓ increases (Dziembowski,1977; Christensen-Dalsgaard

and Gough 1982), and only modes with $\ell \lesssim 3$ can be observed in light
integrated on the solar disk. With differential observations made at
Stanford (mean on a central region minus concentric annulus), modes
with ℓ = 3, 4, 5 have been observed (Scherrer et al., 1982).

Identification of observations requires a comparison with accurately numerically computed theoretical frequencies. Nevertheless some
main features of their properties are rather well reproduced by asymptotic results. The eigenfrequencies satisfy the following asymptotic
relation, derived by Tassoul (1980), neglecting the perturbation of the
gravitational potential :

$$\nu = \nu_o (n+\ell/2+\varepsilon) - A\, \nu_o^{2}(\ell(\ell+1)+\delta)/\nu \qquad (1)$$

$$\text{with } \nu_o = 1/(2\int_o^{R_\odot} (dx/c))$$

$$A = \{\, c(R_\odot) / R_\odot - \int_o^{R_\odot} \frac{dc}{dr}\, \frac{dr}{r} \,\} / 2\pi^2\, \nu_o$$

In this formula, n and ℓ are the radial order and the degree of the
mode. ν_o is the inverse of the sound travel time across a solar diameter. The main contribution to ν_o comes from the regions where the
sound speed is small, i.e. from the envelope and from the atmospheric
layers (see Figure 1). The atmosphere must be taken into account in
numerical calculations, its contribution to ν_o being about 5%. ε is a
constant related to the effective polytropic index of the solar surface.
A is a positive constant related to the variation of the sound speed
across the solar radius and is very sensitive to the gradient of sound
speed in the central region. δ is a more complex integral on the model
(δ is negative in the particular isothermal problem as shown by Gough
1983b). As n is much greater than ℓ, the second term in the asymptotic
formula is a correction to the quantity $\nu_o(n+\ell/2+\varepsilon)$. If ℓ increases by
2 and n decreases by 1, the frequency is almost unchanged. This proper-
ty explains that the observed power spectrum has been resolved into
almost equidistant peaks, corresponding to even and odd ℓ values, the
mean separation $\nu_o/2$ corresponding to 68 μ Hz. In fact, due to the
corrective term, there is a small splitting between frequencies of modes
with a given value of n + $\ell/2$, which permits to identify the degree ℓ
of the mode. Grec et al. (1980) have been able to detect this small
splitting, with a continuous 120 hours record of observations obtained
at the South Pole. They found that $\overline{\nu_{n,o} - \nu_{n-1,2}}$ = 9,4 μ Hz   and
$\overline{\nu_{n,1} - \nu_{n-1,3}}$ = 15.3 μ Hz. More than 100 modes ($0 \leqslant \ell \leqslant 5$, $12 \leqslant n \leqslant 35$,
$1.88 \leqslant \nu_{m\,Hz} \leqslant 5.08$) have been identified by different groups of
observers with an average accuracy of 2 μ Hz. Now the solar model is
constrained not only to reproduce the mean separation ν_o, but its
theoretical frequencies must be equal to the observed frequencies within
2 μ Hz.

The observed frequencies sequence is in rather good agreement with that predicted by standard solar models with a normal chemical composition (Z = ≅ 0.02, 0.21 $\lesssim$ Y $\lesssim$ 0.28) as computed by many groups (Christensen-Dalsgaard and Gough, 1980, 1981; Shibahashi and Osaki, 1981; Scuflaire et al., 1981, 1982; Gabriel et al., 1982; Noels et al., 1983; Shibahashi et al., 1983; Ulrich and Rhodes, 1983). Attempts to improve this agreement by variation of chemical composition (Christensen-Dalsgaard and Gough, 1981; Shibahashi et al., 1983) or by making various adjustments of the solar model (different sort of mixing, high strength internal magnetic fields as done by Ulrich and Rhodes, 1983), have not succeeded. Between observed and computed frequencies there remain significant discrepancies, up to 15 μ Hz, larger than the observational errors (2 μ Hz). This can be seen for instance in Figure 2 of Shibahashi et al. (1983) where their theoretical frequencies and the observational results are compared. The systematic differences reach up to 10 μ Hz, most of the theoretical frequencies being too small, which is a general result obtained in other works. Thus, present standard solar models have too small frequencies compared to the observed ones, and a little too large mean separations. This result has led to a careful examination of some possible source of uncertainties : numerical errors by computing the frequencies are discussed by Christensen-Dalsgaard (1982) and Noels et al. (1983). Uncertainties in the input physics, as equation of state, nuclear cross sections, opacities,can change the frequencies of about 1 μ Hz, as estimated by Ulrich and Rhodes (1983). The most important modification of frequencies, about 10 μ Hz, is obtained varying the position of the outer boundary condition, but it does not improve the fit between observed and computed frequency sequences. So, despite the remaining uncertainties in the computation of equilibrium models and of the oscillation frequencies (Christensen-Dalsgaard, 1982), the conclusion is that something is missing, or wrong, in the standard theory of solar models (Noels et al., 1983; Scuflaire et al., 1981; Ulrich and Rhodes, 1983).

According to the asymptotic formula, a way to increase the frequencies could be to decrease the value of A, hence the splitting value between the frequencies of modes with given n + ℓ/2. Table I indicates the splitting values for the observations, for standard models, for low Z models and for some mixed models. It appears that the fit with the observations is not improved for low Z models and mixed models. A better agreement is obtained with standard models which give generally the smallest theoretical splitting. The splitting is increased in low Z models or in mixed models. This is explained by the fact that for these models the sound speed in the central region varies more rapidly than in the standard model, due to the more gradual variation of the mean molecular weight relatively to the temperature variation.

The continuity between low and high degree 5 mn modes has been obtained recently by observations of Duvall and Harvey (1983). They have succeeded in connecting the low degree range to the high degree one, by projecting longitudinally averaged Doppler measurements onto zonal harmonics. The power spectrum they obtained looks like Deubner's spectrum, but corresponds to degree ℓ < 140. An important consequence is

that the order n of the low degree modes becomes unambiguously identified (Gough, 1983c). A comparison between observed frequencies and
those computed by Ulrich and Rhodes (1983) for degree ℓ up to 40 shows
the greatest discrepancy of order 15 μ Hz for degree ℓ about 20, the
theoretical frequencies being too small. The higher degree modes that
sample the convection zone agree relatively well. That leads Duvall
and Harvey to the conclusion that there exists a substantial difference
between the model and the Sun below the convection zone.

To summarize, five-minute oscillations trapped in the convection
zone are relatively well described by current standard models, whereas
oscillations which penetrate deeper into the radiative inner zone cannot
be accounted for by these models : it seems to be a problem of the
present description of the radiative interior of the Sun. An independent
constraint on this radiative interior has been provided recently by
gravity modes observations.

6.   LOW DEGREE GRAVITY MODES : A CONSTRAINT ON THE RADIATIVE INTERIOR

The recent analysis of the low frequency solar spectrum obtained from
velocity measurements (Delache and Scherrer, 1983; Severny et al., 1983;
Van der Raay, 1983) or from the data of the ACRIM radiometer on board
of the SMM (Fröhlich 1983) has shown the existence of low frequency
global solar oscillations ranging in periods of 160-360 mn, which have
been interpreted as low degree solar gravity modes. The low frequency
spectrum is very complex; it presents a number of sets of lines with
separations about 11.57 μ Hz corresponding to day sidebands, which
come from the nightly gaps in the data. To eliminate these sidebands,
Delache and Scherrer have used an iterative peak removal technique to
find and remove the peaks one at a time; they obtained 14 peaks with
frequencies in the range 45-105 μ Hz.

The identification of these oscillations has been made using the
theoretical properties of low degree gravity modes. In fact, the period
P of high order low degree gravity modes can be approximated by an
asymptotic relation. For a solar model with a radiative inner zone, P
is given by the formula, derived by Vandakurov (1967) and Tassoul and
Tassoul (1968), without perturbing the gravitational potential :

$$P = \frac{P_o}{\sqrt{\ell(\ell+1)}} \ (n + \ell/2 - 1/4) \tag{2}$$

$$\text{with } P_o = 2\pi^2 / \int_o^{r_c} dr \ \frac{N}{r}$$

n and ℓ are the order and degree of the mode, P_o is a characteristic
frequency which depends only on the Brunt-Väissälä frequency N, i.e.
on the density stratification, in the radiative interior (r_c is the
position of the base of the convection zone). Numerical calculations
in complete solar models show that neglecting the perturbations of the

	Y	Z	$\overline{\nu_{n,0} - \nu_{n-1,2}}$	$\overline{\nu_{n,1} - \nu_{n-1,3}}$	
CLAVERIE et al. 1981			8.3		Observations
GREC et al. 1983			9.4	15.3	
CHRISTENSEN-DALSGAARD	0.25	0.02	10	16	
and GOUGH 1980	0.19	0.04	13	17	
NOELS, SCUFLAIRE and	0.28	0.02	9.3		
GABRIEL 1983	0.27	0.018	9.4		
SHIBAHASHI, NOELS	0.23	0.02	10.5		
and GABRIEL 1983	0.229	0.018	6.6		
SCHATZMAN, MAEDER 1981	0.25	0.02	8.8	15.6	$R_e^* = 0$ $n = 22$
ULRICH, RHODES 1983	–	0.021	8.3	16.3	
	0.27	0.018	9.2	16.0	$n = 22$
		0.005	9.7	17.5	
CHRISTENSEN-DALSGAARD 1982	0.27	0.02	15	22	Homogeneous
SCHATZMAN,MAEDER 1981	0.25	0.02	12.4	20.5	$R_e^* = 100$ $n = 22$
ULRICH, RHODES 1983a	–	–	15.4	20.3	0.05MO mixed,n=22
	–	–	17.3	19.7	0.05MO mixed,n=17
ULRICH, RHODES 1983b	–	–	13.4	–	diffusive mixed n = 17

Table I : Splitting values of low degree five-minute oscillations for
the observations and for some standard, low Z, and mixed
models. The observation values from Grec et al. are power weighted
mean values; other values are arithmetic mean values, corresponding to
the observed frequency range, except when the radial order is quoted.

gravitational potential has little influence on the periods of low
degree gravity modes, at this order of approximation, and that the
asymptotic formula (2) is a good approximation for orders larger than
about 10. It can be seen on Figure 2, where are plotted,as a function
of $n + \ell/2 - 1/4$,the periods of gravity modes multiplied by the square
root of $\ell(\ell+1)$, numerically computed by Berthomieu et al.(1983) for
2 different models of Schatzman and Maeder (1981). The points corresponding to a given model and to different ℓ values are on a straight
line with a slope P_0. Asymptotic expressions for the periods show that
for a given degree ℓ, the periods are equidistant, the difference
between the periods of modes with two consecutive radial orders being
$P_0/\sqrt{\ell(\ell+1)}$. In what follows, P_0 will be referred to as "spacing" period.

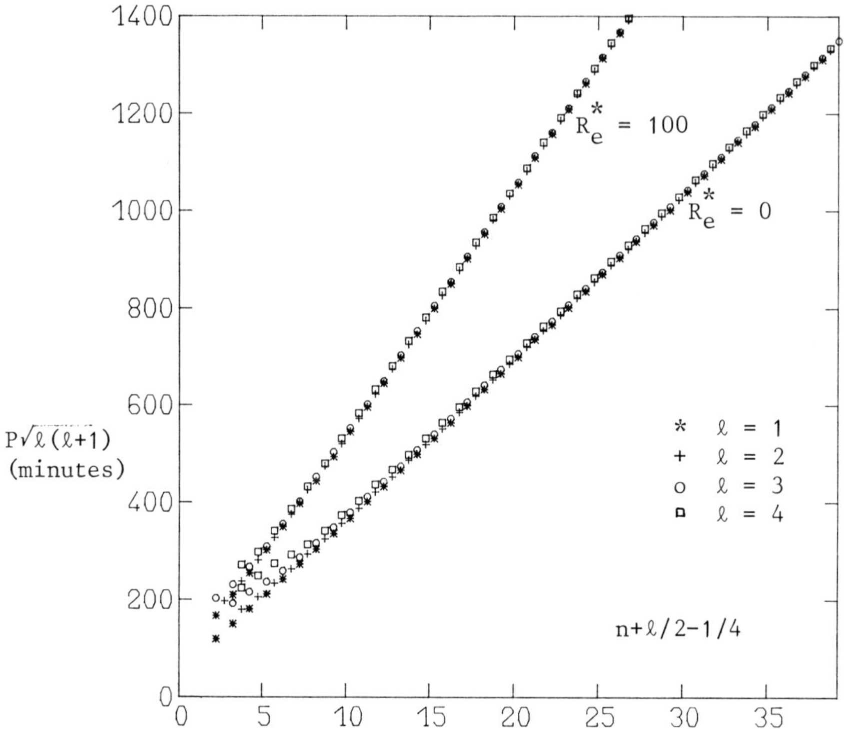

Figure 2 : Periods (multiplied by $\sqrt{\ell(\ell+1)}$) of the solar gravity
modes for the models of Schatzman and Maeder (1981)
versus $n+\ell/2 - 1/4$. (From Berthomieu et al. (1983)).

It is this property that has led to the identification of gravity
modes. The value of the spacing period P_0 derived from observations is
of the order of 38.6 mn (Delache and Scherrer 1983; Fröhlich 1983). It
is suggested that it could be higher, of the order of 41 mn (Van der
Raay, 1983). The amplitude of modes identified by Delache and Scherrer
are compatible with an equipartition of energy between the different
modes (Delache, 1983).

How this result constrains the solar interior has been discussed
by Berthomieu, Provost and Schatzman (1983). In table II the chemical
composition, convection zone depth, neutrino flux and estimated spacing
period of gravity modes are given for different models. It appears that
standard models which produce a too large neutrino flux give a too low
value of P_0, of the order of 35 to 36 mn. Low Z models, which produce a
lower neutrino flux (but possess a shallower convection zone) correspond
to larger values of P_0 of the order of 37 to 38 mn, as can be estimated
from published gravity modes periods. Berthomieu et al. have shown that
turbulent diffusion mixing, a physical mechanism very efficient to lower
the neutrino flux, increases significantly the periods of gravity modes
as can be seen in table II. It is also seen in Figure 2, where the slope

of the straight line corresponding to the Schatzman-Maeder model with pseudo Reynolds number $R_e^* = 100$, is much larger than for the model without mixing. Thus, turbulent diffusion enhances greatly the spacing period of gravity modes. This effect is much greater than the effect of chemical composition change. The variation of the spacing period as a function of R_e^* can be obtained by an interpolation formula. It is almost linear for $R_e^* \lesssim 100$, with $\partial P_o/\partial R_e^* \cong 0.2$ mn. The main result is, that a rather moderate turbulent diffusion mixing ($R_e^* \cong 25$) is required to account for the observed spacing period of low degree gravity modes. Thus, the observation of solar gravity modes severely constrains the degree of mixing in the core of the Sun.

Y	Z	H_c 10^5km	N SNU	P_o min	
					SOME STANDARD MODELS
0.236	0.02	1.6	10.3	35	Iben and Mahaffy 1979
0.25	0.02	1.7	5.2	36	Christensen-Dalsgaard et al. 1979
0.25	0.02	1.6	11.96	34.3	Schatzman and Maeder $R_e = 0$ 1981
					HELIUM DEFICIENT MODELS
0.173	0.01	1.	4.4	37	Iben and Mahaffy 1979
0.186	0.004	1.1	2.3	38	Christensen-Dalsgaard et al. 1979
0.159	0.001	0.62	1.7	37.7	" " " " "
					TURBULENT DIFFUSION MIXED MODELS
0.25	0.02	1.77	2.38	52	Schatzman,Maeder $R_e^* = 100$ 1981
0.25	0.02	1.7	1.43	58.7	Schatzman,Maeder $R_e^* = 200$ 1981

Table II :   Initial abundances, depth of the convective zone, neutrino flux and estimated spacing period P_o for some solar models (from Berthomieu et al., 1983).

7. DISCUSSION

Now the compatibility of all the considered solar constraints and the limits they impose on different solar models will be discussed. Concerning the chemical surface abundances, observations of lithium depletion yields a constraint on the convection zone depth. This depth must not be too large, since the temperature at the bottom of the convection zone must be smaller than the lithium burning temperature by an amount which depends on the physical mechanisms carrying lithium from the bottom of the convection zone to the lithium burning zone, i.e. convective overshooting or turbulent diffusion,or both. Lithium observation puts limits on the efficiency of mixing as it is shown by Schatzman (1984) and by Baglin and Morel (1984). Their results imply for the Sun

a convection zone depth of the order of 200 000 km and a pseudo-Reynolds
number about 60, just below the convection zone. Observation of the ^{3}He
at the solar surface cannot be explained without some transport of this
isotope from ^{3}He peak, located deep in the interior, towards the
bottom of the convection zone. The observed ^{3}He abundance limits also
the efficiency of the mixing to a value of the pseudo-Reynolds number of
about 40 (Schatzman, 1984). The measured neutrino flux requires a
sufficiently low central temperature, which can be obtained by providing
more nuclear fuel to the central region compared to the standard model,
i.e. for models with extremely low heavy elements content, or for mixed
models. To account for the neutrino flux, the pseudo-Reynolds number
must be about 100, or smaller (Schatzman, 1984).

To satisfy the constraint imposed by high degree five-minute
oscillations, a solar model must have a large convection zone, about
200 000 km, which excludes low Z models. There is a problem with low
degree 5 mn oscillations, since their frequency spectrum is not reproduced in all details by any existing models. The observed fine structure,
i.e. the splitting of the modes, constrains the mixing to be sufficiently small. The recent observations of intermediate degree 5 mn oscillations seems to localize the origin of the problem below the convection
zone, according to Duvall and Harvey (1983), but this remains to be
confirmed by further theoretical work. Finally, solar gravity modes
provide a probe of the radiative interior of the Sun. It is difficult
to account for their spacing period within the context of standard
models or even low Z models, except for rather unrealistic models. Their
observation limits the mixing to a rather moderate value ($R_e^* \cong 25$), too
small to fit the observed neutrino flux. It is clear that the simple
model with a constant pseudo-Reynolds number in the whole Sun seems to
present a contradiction between the values of R_e^* derived from neutrino
flux and from gravity modes. More elaborate models, taking into account
recent results on rotational instabilities, as given by Spruit et al.
(1983) and Zahn (1983), and with different chemical compositions, are
necessary to discuss both the neutrino flux and the oscillations,
particularly the spacing period of gravity modes. This is partly
discussed by Law, Knobloch and Spruit (1984).

8.   FUTURE PROSPECTS.

The above discussion shows that no present solar models are able to
account for all the observed solar constraints. In fact, solar oscillations contain a great amount of information on the solar interior. In
principle it should be possible to use the inverse method developed in
terrestrial seismology by Backus and Gilbert (1968), to infer from the
observation of solar oscillations the physical conditions inside the
Sun. Helioseismology requires to know a sufficiently large spectrum of
oscillations, representing all parts of the Sun. At the present time, well
identified modes with periods ranging from 5 mn to 160 mn are missing.
They probably have a very low surface amplitude, as indicated by present
5 mn observations. Thus their detection will require very long sequences

of observation to pick up the oscillation frequencies out of the noise.

Another important question concerns the dynamics of the stellar
interior, specially the distribution of angular velocity, which is of
fundamental importance for the understanding of stellar structure and
evolution (Tassoul, 1984). Moreover, the knowledge of the internal
solar rotation will permit to compute the resulting solar oblateness and
consequently the modification of the solar gravity field, which is
needed to test the general relativity. Solar oscillations will provide,
in a near future, the possibility to measure the solar rotation. In fact,
their frequencies are splitted by the rotation by a very small amount,
which can be detected only with a long time sequence of observations.
The rotational splitting provides an average value of the rotation on
the trapping region of the oscillation. Two groups have reported
detection of 5 mn splitting (Claverie et al., 1981 and Hill et al.,
1982) and they have found that the Sun rotates 2-6 times faster in the
centre than at the surface. However these observations and their
interpretation are still not clearly understood and the existence of
an intense rotating magnetic field in the solar core has been proposed
and discussed by Dicke (1982), Gough (1982) and Isaak (1982). Very
recently, splitting of low degree gravity modes has been reported by
Delache and Scherrer from Stanford observations and by Fröhlich from
SMM data. It is about $1.22\,\mu$ Hz and corresponds also to an indication
of fast rotation in the centre of the Sun, about 3 times the surface
rotation.

In conclusion,more observations on a long time continuous sequence
are needed, both to complete the observed solar spectrum of oscillations and
to resolve rotational splitting. This will be possible with projects of
combining data from many stations at the Earth surface,as proposed independently by Fossat and Isaak, a project of observations at the South
Pole within a balloon as ,proposed by Fossat ,and maybe,in a long range,
by spatial observations. Finally, the Sun is a normal main sequence
star and the promising preliminary results of helioseismology make way
for stellar seismology, which would yield new tests for the theory of
stellar evolution and greatly improve our knowledge on stellar structure
and evolution.

I am grateful to Gabrielle Berthomieu for constructive comments and
to Philippe Delache and Evry Shatzman for useful discussions. I would
like to thank all colleagues who sent information prior to publication.

REFERENCES.

Abraham Z. and Iben I.,Jr : 1971, Astrophys. J. 170, 157.
Ando H. and Osaki Y. : 1975, Publ. Astron. Soc. Japan 27, 581.
Ando H. and Osaki Y. : 1977, Publ. Astron. Soc. Japan 29, 221.
Backus G.E. and Gilbert J.F. : 1968, Geophys. J.R. Astron. Soc. 16, 169.
Baglin A. and Morel P. : 1984, this volume, p. 529.
Bahcall J.N., Bahcall N.A. and Ulrich R.K. : 1968, Astrophys. Letters
    2, 91.
Bahcall J.N. and Ulrich R.K. : 1971, Astrophys. J. 170, 593.
Bahcall J.N., Cleveland B.T., Davis R. Jr, Dostrovsky I., Evans J.C. Jr,
    Frati W., Friedlander G., Lande K., Rowley J.K., Stoenner R.W. and
    Weneser J. : 1978, Phys. Rev. Letters 40, 1351.
Bahcall J.N. : 1981, Proceedings of Neutrino-81 (Maui,Hawaï), R.J.Cence
    editor.
Bahcall J.N., Huebner W.F., Lubow S.H., Parker P.D. and Ulrich R.K. :
    1982, Rev. of Modern Physics, 54, 767.
Berthomieu G., Cooper A.J., Gough D.O., Osaki Y., Provost J. and
    Rocca A. : 1980, Non Radial and Non Linear Stellar Pulsation, eds.
    H.A. Hill and W.A. Dziembowski, Springer Verlag Berlin, p. 307.
Berthomieu G., Provost J. and Schatzman E. : 1983, submitted to Nature.
Bos R.J. and Hill H.A. : 1983, Solar Phys. 82, 89.
Christensen-Dalsgaard J. : 1980a, Proc. Vth European Regional Meeting
    in Astronomy, Institut d'Astrophysique, Liège.
Christensen-Dalsgaard J. : 1980b, Mon. Not. R. Astron. Soc. 190, 765.
Christensen-Dalsgaard J. : 1982a, COSPAR Meeting, Ottawa, to be published.
Christensen-Dalsgaard J. : 1982b, Mon. Not. R. Astron. Soc. 199, 735.
Christensen-Dalsgaard J., Gough D.O. and Morgan J.G. : 1979, Astron.
    Astrophys. 73, 121.
Christensen-Dalsgaard J. and Gough D.O. : 1980, Nature 228, 544.
Christensen-Dalsgaard J. and Gough D.O. : 1981, Astron. Astrophys.
    104, 173.
Christensen-Dalsgaard J. and Gough D.O. : 1982, Mon. Not. R. Astron.
    Soc. 198, 141.
Claverie A., Isaak G.R., McLeod C.P. and Van der Raay H.B. : 1979,
    Nature 282, 591.
Claverie A.,Isaak G.R., McLeod C.P., Van der Raay H.B. and Roca
    Cortes T. : 1981, Nature 293, 443.
Cleveland B.T., Davis R., Jr, and Rowley T.K. : 1981, in "Weak Inte-
    ractions as Probes of Unification", edited by G.B. Collins,
    L.N. Chang and J.R. Fience (AIP Conference Proceedings n° 72),p. 322.
Cox J.P. : 1980, Theory of Stellar Pulsation, Princeton University
    Press.
Delache Ph. : 1983, Proceedings of the Study Conference "Oscillations
    as a Probe of the Sun's Interior". Catania. Ed. G. Belvedere and
    L. Paterno. To be published.
Delache Ph. and Scherrer Ph. : 1983, Nature, in press.
Deubner F.L. : 1975, Astron. Astrophys. 44, 371.
Deubner F.L., Ulrich R.K. and Rhodes E.J. Jr : 1979, Astron. Astrophys.
    72, 177.
Dicke R.: 1982, Nature 300, 693.

Duvall T.L. : 1982, Nature 300, 242.
Duvall T.L. and Harvey J.W. : 1983, Nature 302, 24.
Dziembowski W.A. : 1977, Acta Astr. 27, 203.
Dziembowski W.A. and Pamjatnykh A.A. : 1978, "Deuxième Assemblée
     Européenne de Physique Solaire. Pleins feux sur la Physique
     solaire", J. Rösch ed. (Paris CNRS) p. 135.
Ezer D. and Cameron A.G.W. : 1968, Astrophys. Lett. 1, 177.
Fröhlich C. : 1983, Proceedings of the Study Conference "Oscillations
     as a probe of the Sun's Interior". Catania. Ed. C. Belvedere and
     L. Paterno. To be published.
Gabriel M., Scuflaire R. and Noels A. : 1982, Astron. Astrophys. 110, 50
Genova F. and Schatzman E. : 1979, Astron. Astrophys. 78, 323.
Gough D.O. : 1978, Proceedings of the "Workshop on Solar Rotation".
     Catania. Ed. G. Belvedere and L. Paterno.
Gough D.O. : 1980, Lectures Notes in Physics, Springer Verlag Heidelberg,
     125, 273.
Gough D.O. : 1982, Nature 298, 350.
Gough D.O. : 1983a, To appear in Proceedings Primordial Helium Workshop,
     ed. P. Shaver, European Southern Observatory, Garching 1983.
Gough D.O. : 1983b, To appear in "Physics Bulletin"
Gough D.O. : 1983c, Nature 302, 18.
Gough D.O., Hill F. and Toomre J. : 1983, Proceedings of the Study
     Conference "Oscillations as a Probe of the Sun's interior". Catena.
     Ed. G. Belvedere and L. Paterno. To be published.
Gough D.O. and Toomre J. : 1983, Solar Physics 82, 41.
Grec G., Fossat E. and Pomerantz M. : 1980, Nature 288, 541
Grec G., Fossat E. and Pomerantz M. : 1983, Solar Phys. 82, 55.
Harvey J.W., Rhodes E.J. and Duvall T.L. : 1983, in preparation.
Hill F., Toomre J. and November L.J. : 1983, Solar Phys. 82, 411.
Hill H.A., Goode P.R. and Stebbins R.T. : 1982, Astrophys. J. 256, L17.
Hill H.A., Bos R.J. and Goode P.R. : 1982, Phys. Rev. Letters 49, 1794.
Iben I. Jr : 1969, Ann. Physics 54, 164.
Iben I. Jr and Mahaffy J. : 1976, Astrophys. J. 209, L39.
Isaak G.R. : 1982, Nature 296, 130.
Kotov V.A., Severny A.B., Tsap T.T., Moiseev I.G., Efanov V.A. and
     Nesterov N.S. : 1983, Solar Phys. 82, 9.
Law, W.Y., Knobloch E. and Spruit H.C. : 1984, this volume, p. 523.
Ledoux P. and Walraven T. : 1958, Lecture Notes in Physics, Springer
     Verlag Berlin 51, Chap. 4.
Leighton R.B., Noyes R.W. and Simon G.W. : 1962, Astrophys. J. 155, 474.
Lubow S.H., Rhodes E.J. and Ulrich R.K. : 1980, Lecture Notes in Physics,
Springer Verlag Heidelberg, 125 , 300.
Noels A., Scuflaire R. and Gabriel M. : 1983, preprint.
Rösch J. and Yerle R. : 1983, Solar Phys. 82, 139.
Ross J.E. and Aller L.H. : 1976, Science 191, 1223.
Schatzman E. : 1977, Astron. Astrophys. 56, 211.
Schatzman E. : 1984, this volume, p. 491.
Schatzman E. and Maeder A. : 1981, Astron. Astrophys. 96, 1.
Scherrer Ph. and Wilcox J.M. : 1983, Solar Phys. 82, 37.
Scherrer Ph., Wilcox J.M., Christensen-Dalsgaard J. and Gough D.O. : 1982
     Nature 297, 312.

Scuflaire R., Gabriel M. and Noels A. : 1981, Astron. Astrophys. 99, 39.
Scuflaire R., Gabriel M. and Noels A. : 1982, Astron. Astrophys.
    113, 219.
Severny A.B., Kotov V.A. and Tsap T.T. : 1983, Proceedings of the Study
    Conference "Oscillations as a Probe of the Sun's Interior" -
    Catania. Ed. G. Belvedere and L. Paterno. To be published.
Shaviv G. and Beaudet G. : 1968, Astrophys. Lett. 2, 17.
Shibahashi H., Noels A. and Gabriel M. : 1983, Astron. Astrophys. 123,
    283.
Shibahashi H. and Osaki Y. : 1981, Publ. Astron. Soc. Japan 33, 713.
Spruit H.C., Knobloch E. and Roxburgh I.W. : 1983, to appear in Nature.
Stenflo J.O. : 1980, Proc. Vth European Regional Meeting in Astronomy,
    Institut d'Astrophysique , Liège.
Tassoul J.L. and Tassoul M. : 1968, Ann. Astrophys. 31, 251.
Tassoul M. : 1980, Astrophys. J. Suppl. 43, 469.
Tassoul J.L. : 1984, this volume, p. 475.
Ulrich R.K. : 1970, Astrophys. J. 162, 993.
Ulrich R.K. : 1982, Astrophys. J. 258, 404.
Ulrich R.K. and Rhodes E.J. : 1977, Astrophys. J. 218, 521.
Ulrich R.K. and Rhodes E.J. : 1983, Astrophys. J. 265, 551
Ulrich R.K. Rhodes E.J., Tomczyk S., Dumont Ph. J. and Brunish W.M. :
    1983b, AIP Conference Proceedings, n° 96, Ed. H.C. Wolfe.
Unno W., Osaki Y., Ando H. and Shibahashi H. : 1979, Non Radial
    Oscillations of Stars, University of Tokyo Press.
Vandakurov Yu.V. : 1967, Astr. Zh. 44, 786.
Van der Raay H.B. : 1983, Proceedings of the Study Conference "Oscilla-
    tions as a Probe of the Sun's Interior" - Catania. Ed. G. Belvedere
    and L. Paterno. To be published.
Woodard M. and Hudson H. : 1983, Solar Phys. 82, 67.
Zahn J.P. : 1983, Saas Fee Lectures, 13th Advanced Course of the
    Swiss Society of Astronomy and Astrophysics. Ed. B. Hauck and
    A. Maeder.

DISCUSSION

Cox: What is the effect of the non-adiabaticity on the periods? They
may be important for the longer periods.

Provost: The effect of the non-adiabaticity is to decrease the frequencies of low degree 5 mn oscillations of about 5μHz, according to
Christensen-Dalsgaard and Frandsen (1983). For gravity modes, the modification of periods could be larger since they have longer periods and
non-adiabatic exchanges have time to operate. However, the maximum amplitude of g-modes occurs in the radiative inner region where the non-
adiabaticity due to radiation is quite negligible. An exact non-adiabatic
computation is necessary to conclude. As far as I know, there do not
exist published results, except some stability studies; these studies do
not indicate both adiabatic and non-adiabatic periods.

<u>Hesser</u>: Could you please repeat how much faster the interior is thought to rotate, as a result of the study of the rotational splitting of the g-modes?

<u>Provost</u>: The rotational splitting of g-modes reported by Delache and Scherrer (1983) and Fröhlich (1983) is of the order of 1.22 μHz. This indicates that the interior of the Sun rotates about three times faster than the solar surface.

<u>Zahn</u>: In your discussion you consider the parameter R_e^* characterizing the turbulent diffusion as a constant throughout the Sun. But it could well depend on depth. Is it possible to refine the determination of R_e^* to make it depth-dependent? For instance, to which part of the Sun are the g-modes most sensitive (since they yield the smallest value of R_e^*)?

<u>Provost</u>: Models with constant R_e^* are a first approach in the construction of mixed models and more elaborated models are needed. Depth-dependent values of R_e^* can be theoretically obtained assuming marginal instability (Spruit et al, 1983, Zahn, 1983). In fact it is possible to estimate different values of R_e^* at different depths of the Sun from the observations. For instance, a value of R_e^* for the region below the convection zone and above the lithium-burning zone can be determined from the observation of lithium depletion, as discussed by Baglin and Morel (1983). In the same way, an estimate of R_e^* for the inner five percents of the mass could be obtained from the observed low degree gravity modes periods.

RESONANT THREE-WAVE INTERACTIONS AND AN APPLICATION
TO SOLAR G MODE OSCILLATIONS

D. B. Guenther and P. Demarque
Yale University Observatory, New Haven, CT 06511, USA

We present here some initial results of a long-term
investigation of resonant three-wave interactions in the solar interior,
which indicate that these nonlinear interactions take place in the sun
and are, in fact, responsible for the observed g mode spectrum.
Resonant three-wave interactions, as the name implies, involve the
coupled interaction of three waves which satisfy the resonant condition

$$\omega_o = \omega_1 + \omega_2,$$

where ω_o, ω_1, and ω_2 are the frequencies of the three waves. Two of the
waves, mode 0 and 1 for example, couple together to produce a third beat
wave which has a frequency equal to the frequency difference of the first
two waves. This wave is normally very weak. If its frequency happens to
equal the frequency of an oscillation mode of the sun, the wave can be
pumped by resonance to very large amplitudes. In the sun, resonant
interactions would take place if three natural oscillation modes of the
sun satisfy the resonance condition.

The observed p mode spectrum is complex and, therefore, it
would be difficult to test for the existence of resonant interactions of
these modes. The observed g mode spectrum, on the other hand, consists
of only a few modes. We are assuming that the 160-minute oscillation is
a g mode and that the recently identified modes by Delache and Scherrer
(1983) are also g modes. We believe that the amplitudes of these
selected few modes have been enhanced by resonant interaction so that
they may be observed at the sun's surface. Several authors (see papers
in Dziembowski and Gough 1983) have already made similar suggestions to
explain the existence of the 160-minute oscillation, but almost no
detailed work has been done in this area.

We have solved the pulsation equations (Guenther 1983)
describing resonant interactions for many low degree triads of g modes.
The particular interactions investigated were calculated from adiabatic
frequency data for low degree g modes of the calculations of
Christensen-Dalsgaard, Gough, and Morgan (1979). Modes were selected
from this data which satisfy the resonance condition within a frequency
tolerance of 2.5 microHz (see Guenther and Demarque 1983 for a

A. Maeder and A. Renzini (eds.), Observational Tests of the Stellar Evolution Theory, 67–70.
© *1984 by the IAU.*

discussion of this tolerance). After calculating the coupling
coefficients for the selected modes the equations were solved. We did
not find anything specifically unusual about interactions involving g
modes with frequencies near 104 microHz (periods = 160 minutes), and are
forced to conclude that no single resonant interaction of three g modes
is likely to account for the outstanding amplitude of the 160-minute
oscillation. However, we have found some evidence which indicated that
the 160-minute oscillation may be due to the collective interactions of
many resonant three-wave interactions. This can be seen in the resonant
count diagram (figure 1), which we will now describe.

The diagram is a plot of the number of possible three-wave
interactions involving two waves and their beat wave versus the beat
wave's period. The numbers along the ordinate correspond to the number
of possible resonant interactions of the lowest degree g modes which
satisfy the resonance condition within a frequency tolerance of 2.5
microHz (similar curves are obtained using finer tolerances). The modes
included in the calculation are marked with short vertical lines. All
the modes calculated by Christensen-Dalsgaard et al., based upon a
standard solar model, are included. Higher degree and order modes are
probably not important because their complex pulsation shapes prevent
strong coupling (this was found in our detailed calculations, Guenther
1983).

The curve contains several pronounced peaks, one of which is
located at 160 minutes. It is apparent that the separation of g modes
for the sun favor resonant interactions near 160 minutes.
Interestingly, the tentatively identified g modes of Delache and
Scherrer, which fall within the range of the diagram, correspond to the
other peaks. We interpret this as evidence that many resonant
three-wave interactions work together to stimulate the few modes which
fall on the peaks of the diagram.

We note that the correspondence is not precise. The
frequencies used to generate the curve have not been fine tuned to match
the observed g mode spectrum, therefore some shift may be expected when
more accurate frequencies are calculated. In light of our
interpretation of the observed g modes, a complete spectrum of g modes
will not be observable and it will therefore be difficult to compare the
complete theoretical spectrum with the sparse observed spectrum. The
initial data of Delache and Scherrer indicates that the actual spectrum
of solar g modes have lower frequencies than those used to generate the
curve and, hence, the curve will be shifted to the left. We will be
examining this in more detail in the near future.

The resonant count diagram, along with Delache and Scherrer's
data is suggestive that resonant interactions are responsible for the
selective enhancement of g modes. The enhanced amplitudes allow these
few modes to be observed at the surface of the sun. The diagram and our
detailed calculations indicate that the enhancements are the result of
several resonant three-wave interactions, i.e. no single resonant
three-wave interaction can account for the 160-minute oscillation, for

example. The use of this diagram for testing the mass distribution in
the inner parts of models of the sun is being investigated.

        We wish to thank Bernard Durney and Richard B. Larson for
their useful comments. This research was supported in part by grant
AST80-23743 from the National Science Foundation.

References

Christensen-Dalsgaard, J., Gough, D. O., and Morgan, J. G. 1979.
        Astron. Astrophys. 73, 121.
Delache, P. and Scherrer, P. H. 1983. Nature, submitted.
Dziembowski, W. and Gough, D. O. eds., 1983. Solar Phys., Vol. 82,
        Proceedings of Crimean Symposium on Solar Oscillations.
Guenther, D. B. 1983. Ph.D. dissertation, Yale University.
Guenther, D. B. and Demarque, P. 1983. Astrophys. J. Letters,
        submitted.

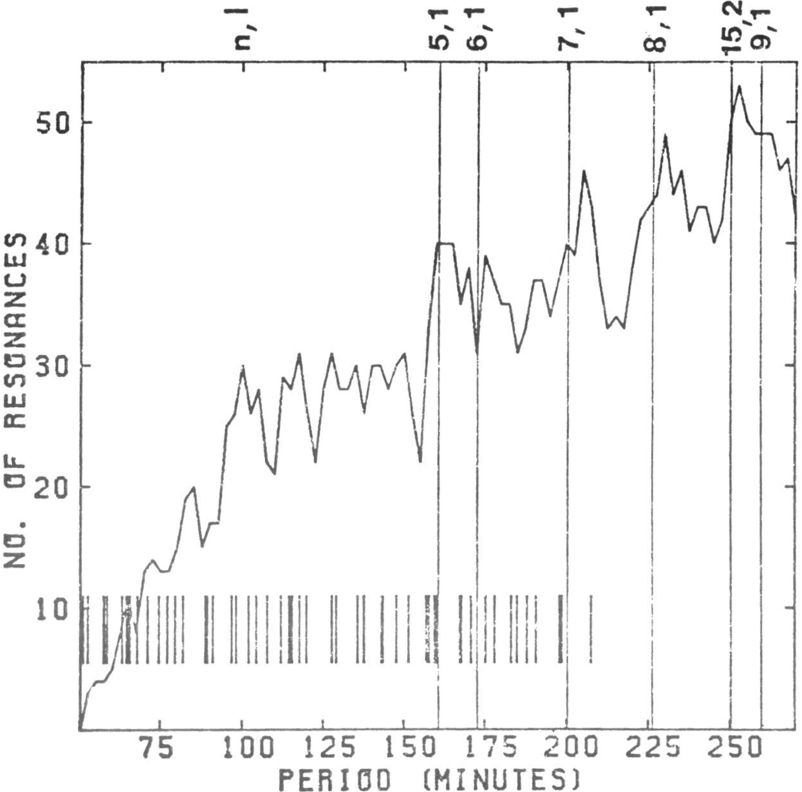

Figure 1. The resonant count diagram. The short vertical lines mark
the positions of the modes calculated for a standard solar model, and
the long vertical lines mark the position of the g modes identified by
Delache and Scherrer. These tentative identifications are shown along
the top of the diagram.

DISCUSSION

<u>Hesser</u>: Is it intuitively obvious how the interior mass distribution would have to be modified to bring about agreement between theory and observation?

<u>Demarque</u>: No. We have, of course, tried to determine, at least in a rough way, how to modify the mass distribution of the standard model to achieve agreement with the observed modes, but the available information is insufficient.

<u>Zahn</u>: Have you taken the selection rules between modes into account, in combining the frequencies?

<u>Demarque</u>: We are aware of these selection rules, although they have not been taken into account in the diagram I have shown. On the average, the selection rules will reduce the total number of resonances by a factor of two.

<u>Cox</u>: What is the range of ℓ in the g-mode list of Gough and Christensen-Daalsgaard? Would the coupling peak be even better fitted if ℓ were limited to 1, 2 or 3?

<u>Demarque</u>: The Christensen-Dalsgaard, Gough and Morgan list contains modes with $\ell < 5$. I don't know the answer to your second question, but it would be easy to check.

CONVECTIVE MOTIONS AS AN INDICATOR OF SOLAR STRUCTURE

R. Van der Borght and P. Fox
Monash University, Clayton (Vic.) Australia

Most stars contain regions which are convectively unstable and one
of the more daunting tasks facing astrophysics today is to find a
satisfactory theoretical formulation of turbulent energy transport in
stars. Various theories have been proposed, such as the mixing-length
formalism and its extensions, and it would be most useful if one could
test the accuracy of such models in view of their importance in the
theory of stellar structure and evolution.

Fortunately, the outer layers of the sun are convectively unstable
and the sun is near enough for some of its surface characteristics, such
as granulation and supergranulation, to be observed. Granules have an
average cell diameter of 2,000 kms, horizontal and vertical velocities
of the order of 1 km/sec, an intensity modulation of 15%, a temperature
difference of about 900° K between ascending and descending currents and
an average lifetime of approximately 20 minutes. On the other hand,
supergranules have an average horizontal extent of 30,000 kms,
horizontal and vertical velocities of .3 to .5 km/sec, have no
observable intensity modulation and a lifetime of approximately 20
hours.

Such a wealth of information should allow us to probe the structure
of the sun's outer layers and ultimately to test the validity of
existing models. Unfortunately, in order to do this one must be able to
model fairly accurately large-scale convective motions in a highly
stratified compressible layer.

Numerous attempts have been made to study thermal convection within
the Boussinesq and even the anelastic approximation but, since the depth
of the convective layer is very much larger than the pressure scale
height, a satisfactory model of granulation and supergranulation should
be based on the fully compressible equations. Such equations, for
layers with polytropic structure, were derived some time ago
(Van der Borght 1977) within the framework of the one-mode approximation
and have lately (Van der Borght and Fox 1983) been integrated in an
attempt to model granulation. The results are very encouraging and

A. Maeder and A. Renzini (eds.), Observational Tests of the Stellar Evolution Theory, 71–74.
© *1984 by the IAU.*

velocities of the order of 0.97 km/sec, an intensity modulation of 16.37% and an e-folding time of 6 minutes are obtained in a medium for which the Prandtl number is 0.2.

Unfortunately, such models make it necessary to fit a polytropic structure to the model and requires the introduction of average values for the buoyancy, eddy diffusivity and eddy viscosity with a resultant loss in accuracy. The depth dependence of these quantities is ignored and the models are not accurate enough to enable us to compare the accuracy of competing models.

The basic hydrodynamic equations can be written (Van der Borght 1980).

Continuity equation:

$$\frac{\partial \rho}{\partial t} + \frac{\partial}{\partial x_i} (\rho x_i) = 0 \tag{1}$$

Equation of motion:

$$\frac{\partial}{\partial t} (\rho u_i) + \frac{\partial}{\partial x_j} (\rho u_i u_j + \delta_{ij} p - P_{ij}) + \delta_{i3} g\rho = 0 \tag{2}$$

where the viscous tensor P_{ij} is defined as follows

$$P_{ij} = \mu \left(\frac{\partial u_i}{\partial x_j} + \frac{\partial u_j}{\partial x_i} - \frac{2}{3} \delta_{ij} \frac{\partial u_\ell}{\partial x_\ell} \right) \tag{3}$$

and μ is the viscosity.

Energy equation:

$$\frac{\partial}{\partial t}[\rho (E + \tfrac{1}{2} u_\ell u_\ell)] + \frac{\partial}{\partial x_j} \left\{ \rho u_j \left(E + \frac{u_\ell u_\ell}{2} + gz \right) \right.$$

$$\left. + p u_j - K \frac{\partial T}{\partial x_j} - u_i P_{ij} \right\} = 0 \tag{4}$$

where E is the internal energy per unit mass and K is the conductivity.

State equation:

$$p = R \eta \rho T \tag{5}$$

where η is the inverse of the molecular weight.

For a given model (e.g. Böhm 1963, Kohl 1966) the values and variation with depth of the conductivity K, internal energy E and inverse molecular weight η are known. The only quantity in the above equations which is left unspecified is the eddy viscosity μ. One could, for instance, assume that the viscosity is constant or that the kinematic viscosity is constant (models A and B of Graham and Moore, 1978). It would be even better if the exact values were given by the theory of turbulence, if one were available. In any case, as mentioned above, enough observed characteristics are available to distinguish not only between the accuracy of various models but also to establish the best law of variation of viscosity with depth.

Solving the full three-dimensional equations is out of the question for the moment, due mainly to the numerical complexities. But, since granulation and supergranulation exhibit a periodic structure, it is to be expected that the one-mode approximation would yield fairly accurate results. The fully compressible single mode equations which take into account the variation with depth of the degree of ionization, thermal diffusivity, eddy kinematic viscosity and buoyancy have been derived but space prevents us from giving them in this paper. With such equations the characteristics of a particular model can be fully taken into account and the resultant thermal convection can be studied in detail without the need of ad-hoc assumptions, except for the value and depth dependence of the turbulent kinematic viscosity.

Trial integrations of this complicated system of differential equations have been carried out and the results are very encouraging. They confirm earlier results based on the polytropic approximation (Van der Borght and Fox 1983) and show how sensitive the results are to the value adopted for the Prandtl number. The adoption of a more accurate variation with depth of the buoyancy leads to more realistic distributions of the velocity and the temperature perturbation. It seems likely that work of this kind will not only be useful in comparing models of convective regions but will help in our understanding of the turbulent processes in such regions.

References

Böhm, K.H., Astrophys. J., 137, 881 (1963).
Graham, E. and Moore, D.R., Mon. Not. R. Astron. Soc., 183, 617 (1978).
Kohl, K., Z. f. Ap., 64, 472 (1966).
Van der Borght, R., Proc. Astr. Soc. Aust., 3, 91 (1977).
Van der Borght, R., J. Comp. Appl. Math., 6, 283 (1980).
Van der Borght, R. and Fox, P., Proc. Astr. Soc. Aust., 5 (1983) (to
    appear).

DISCUSSION

R. Cayrel: Do your computations allow you to estimate the amount of overshooting above the granulation layer?

Van der Borght: Yes, the upper boundary for the numerical integrations can be set well above the unstable region.

TWO NONSTANDARD SOLAR MODELS: WITH MIXED INTERIORS AND WITH ENHANCED HEAVY ELEMENT OPACITY

A. A. Pamyatnykh and I. K. Sapsaj
Astronomical Council of the USSR Academy of Sciences

Abstract. We present and briefly discuss basic characteristics of a standard solar model as well as those for two series of nonstandard models computed with Paczyński's stellar evolution code.

The solar models with mixed interiors seem to be more preferable, comparingly with the standard ones, as concerns solar neutrino experiments and solar pulsation problem - particularly, that of 160^m-pulsations (see, for example, Ezer and Cameron, 1968; Hoyle, 1975; Schatzman and Maeder, 1981; Gavryuseva et al., 1983). The mixing may be caused by existence of convective core which in it's turn may be related to opacity enchanced for some reason.

We have computed the standard solar model and two series of the nonstandard ones. In the first series we have postulated the solar interiors to be continuously mixed during the evolution from zero age main sequence. It was assumed implicitly that the timescale of the mixing is of order of 10^8 years or less. At the same time, the mixing was considered as not convective, i.e. it has no explicit influence on heat transfer which was computed in diffusion approximation, with Schwarzschild's criterion for convection. In the second series of models we have studied the effect of enchanced heavy element opacity, as it was suggested by Simon (1982) for models of double-mode and beat Cepheids. The enchanced heavy element opacity was imitated by use of opacity tables for increased heavy element abundances Z = 0.04 and 0.10, while the standard value Z = 0.02 was left in the rest of computations. Such a procedure is somewhat analogous to the one used by Simon (1982). The corresponding opacities were interpolated/extrapolated in Cox-Stewart tables (1969).

All the computations were performed with use of stellar evolution code by Paczyński (1970). In all cases we were computing evolution of 1 Mo star with heavy element abundan-

A. Maeder and A. Renzini (eds.), Observational Tests of the Stellar Evolution Theory, 75–77.

ce Z = 0.02 from zero age main sequence to the present age of the Sun. The initial hydrogen abundance X and the mixing length parameter α = l/Hp were adjusted to give solar luminosity and radius. Basic characteristics of the models computed are listed in Table 1. The standard model (no postulated mixing, standard opacity for Z = 0.02) is almost identical to the one computed by Shibahashi and Osaki (1981) also by using Paczyński's program.

Table 1. Basic characteristics of the models computed. (In all models the heavy element abundance is Z = 0.02. Mmix and Lmix - mass and luminosity of mixed core. $Z\!\!\ae$ - value of Z in opacity calculations. r_β and T_β - radius and temperature at the base of the convective envelope whose mass is Mconv. Other symbols have their usual meaning.)

Model	stand.	with mixed interiors				"Z-opacity"	
Mmix/M	0	0.1	0.2	0.3	0.5	no mixing	
Lmix/L	0	0.554	0.804	0.910	0.987	no mixing	
$Z\ae$	0.02	standard value				0.04	0.10
X	0.77	standard value				0.704	0.623
α=1/Hp	1.30	standard value				1.48	1.67
L/L$\odot$	0.985	0.966	0.978	0.979	0.977	0.989	1.008
R/R$\odot$	0.994	0.987	0.991	0.986	0.975	0.995	1.005
Age,10^9y	4.50	4.42	4.50	4.53	4.60	4.41	4.42
T_c, 10^6K	15.75	15.57	15.32	14.99	14.66	16.44	17.79
ρ_c,c.g.s.	193.0	144.1	128.6	117.4	108.3	208.6	246.7
X_c	0.369	0.546	0.605	0.648	0.689	0.298	0.154
Mconv/M	0.017	standard value				0.030	0.048
r_β/R	0.727	standard value				0.696	0.669
T_β, 10^6K	1.987	standard value				2.432	2.988

The models with mixed interiors are similar to corresponding models by Shaviv and Beaudet (1968). In agreement with Ezer and Cameron (1968) and Shaviv and Beaudet (1968), the evolutionary tracks of such models in HR diagram practically coincide with the track for the standard model having the same initial chemical composition. Also the parameters of the convective zone in the standard model and in the mixed ones are the same. With increasing mass of mixed core the central temperature and density are decreasing. Such changes may be favourable for the solution of the neutrino problem and for the interpretation of 160^m-pulsations (see references above). The interesting feature of the mixed models is the existence of small convective core in the models with Mmix = 0.1 and 0.2 M$\odot$ (Mconv.core = 0.0017 and 0.0013 M$\odot$, respectively). The convective core appeared at the age of ~$4\cdot10^9$ years, shortly before the present stage. In the models with a more extended mixing the convective core appeared at later stages. It should be noted that even in the models without convective core the stability against convection is rather weak: the radiative temperature gradi-

ent in the center of the ZAMS-model is only 4.5% smaller
than the adiabatic temperature gradient; for the models of
the present Sun with Mmix = 0.3 and 0.5 Mo this difference
is 1.5% and 2.5%, respectively. (For the standard model of
the present Sun the difference of gradients is about 25%.)
The appearance of convective core may enchance the efficiency of postulated mixing mechanisms.

The increase of the heavy element opacity causes the increase of central temperature and density in the models and
the decrease of hydrogen abundance in the center. At temperatures of order of 10^5-10^7K the opacity in the model with
$Z\!\!\mathit{æ}$ = 0.10 is 1.5-3.5 times higher than in the standard model
(the maximal difference lies at temperature ~10^6K). Even at
such an opacity increase the convection in the core is absent.
The difference between the radiative and adiabatic gradients
in the center of the initial model and the present model
with $Z\!\!\mathit{æ}$ = 0.10 is 13% and 19%, respectively. But then the
enchanced opacity causes the increase of convective zone
whose depth increases from 0.273 Ro for the standard model
to 0.331 Ro for the model with $Z\!\!\mathit{æ}$ = 0.10. The temperature at
the base of convective envelope increases in this case from
$2.0\cdot10^6$ to $3.0\cdot10^6$K. This model may be suitable for the solution of Li/Be problem and for the interpretation of the spectrum of 5^m-oscillations (see, for example, Scuflaire et al.,
1982, and references there). However, due to high central
temperature this model has difficulties (when there is no
convective core) in respect to the neutrino problem. Moreover, the hydrogen abundance in outer layers (X = 0.623)
seems to be rather low when compared with observations.

All the computations were performed on the ES-1033 computer in the Astronomical Council of the USSR Acad.Sci. We
are indebted to P.A.Kuzurman for technical assistance.

<u>References.</u>

Cox A.N., Stewart J.N.: 1969, Nauchnye Informatsii 15,
          pp.4-103.
Ezer D., Cameron A.G.W.: 1968, Astrophys.Lett. 1, pp.177-179.
Gavryuseva E.A., Kopysov Yu.S., Zatsepin G.T.: 1983,
          Solar Phys. 82, pp.209-213.
Hoyle F.: 1975, Astrophys.J.(Letters) 197, pp.L127-L131.
Paczyński B.: 1970, Acta Astronomica 20, pp.47-58.
Schatzman E., Maeder A.: 1981, Astron.Astrophys. 96, pp.1-16.
Scuflaire R., Gabriel M., Noels A.: 1982, Astron.Astrophys.
          113, pp.219-222.
Shaviv G., Beaudet G.: 1968, Astrophys.Lett. 2, pp.17-19.
Shibahashi H., Osaki Y.: 1981, Publ.Astron.Soc.Japan 33,
          pp.713-719.
Simon N.R.: 1982, Astrophys.J.(Letters) 260, pp.L87-L90.

INTERMEDIATE MASS STELLAR MODELS: AGES OF MAGELLANIC CLOUD STAR CLUSTERS

Phillip J. Flower
Clemson University

Ages of six intermediate-age Large Magellanic Cloud star clusters have been estimated using the time dependent behavior of the luminosity of stellar interior models of red giants. All clusters studied, NGC 1783, NGC 1868, NGC 1978, NGC 2121, NGC 2209, and NGC 2231, were found to have ages $< 10^9$ yr. It is concluded that there is currently no substantial evidence for a major cluster population of large, populous clusters $> 10^9$ yr in the Large Magellanic Cloud.

The distributions of red giants on the six cluster color-magnitude diagrams were compared to a grid of 33 stellar evolutionary tracks, evolved from the main sequence through core-helium-exhaustion and up the asymptotic giant branch, spanning the expected mass range (2-3 solar masses) and metallicity range ($-0.2 \leq$ [Fe/H] ≤ -1.2) for intermediate-age Large Magellanic Cloud clusters. The faintest model core-helium-burners decreased in luminosity with decreasing mass; thus, model red giant luminosities decreased with age.

Although Cannon (1970) indicates that the mean M_V of the red "clump" giants reaches a limiting magnitude of +1, many galactic intermediate-age clusters exhibit red clump giants much fainter; e.g., the faintest red giants in NGC 559 reach M_V = +2.5, in NGC 1245, NGC 2477, and NGC 3496 they reach $M_V \simeq$ +2.

Since the current main sequence photometry is generally still too inaccurate to obtain reliable Magellanic Cloud cluster ages with main sequence turnoffs, the red giant models have been used to estimate cluster ages (the red giant photometry is more accurate than the necessarily fainter main sequence photometry). These red giant ages are compared in Table 1 to the main sequence termination ages of Hodge (1982), Olszewski (1983), and Flower et al. (1983), the AGB ages of Mould and Aaronson (1982), and the integrated spectra ages of Rabin (1982). The red giant ages agreed with the main sequence termination ages; both techniques are based on cluster color-magnitude diagrams (CMD).

A. Maeder and A. Renzini (eds.), Observational Tests of the Stellar Evolution Theory, 79–81.

TABLE 1   LMC CLUSTER AGE ESTIMATES (10^9 yr)

Cluster	Red Giants	MS Termination	Mould and Aaronson	Rabin	[Fe/H] Estimate	SWB Type
NGC 2209	0.8	0.7	3	1.5-2.5	-1.1	III-IV
NGC 1868	0.75	0.3	-	-	-1.2	-
NGC 2231	0.55	1.2	-	-	-1.3	V
NGC 1783	0.46	>0.2	3	4	-0.5	V
NGC 1978	0.45	0.7	2	>6	-0.5	VI
NGC 2121	0.39	0.4	4	5	-1.0	VI

In every instance of an age estimate for a cluster by both CMD and AGB dating techniques, the CMD ages are always significantly (factor of 3 or more) lower. The AGB ages are, however, very sensitive to the choice of the mass loss parameter used in the AGB evolutionary models (Renzini 1982). Increasing it from 0.45 (Mould and Aaronson 1982) to 1.0 will reduce all AGB age estimates to less than 10^9 yr. The AGB ages are also susceptible to statistical fluctuations in the luminosity function (in many clusters only one star defines the age).

Because the integrated spectra ages (Rabin 1982) are based on integrated light models that ignore the contribution of red clump giants, the extremely large ages (based on the strengths of the Balmer lines) are probably entirely unreliable. The working assumption of Rabin (1982) is that the blue part of the spectrum is dominated by the contribution of the brightest main sequence stars, those at the cluster turnoff. However, color-magnitude diagrams of populous LMC intermediate-age star clusters show a large number of bright, core-helium-burning red giants. In clusters like NGC 1978 (Olszewski 1983) and NGC 2121 (Flower et al. 1983), the red clump giants are at least a magnitude brighter than the brightest cluster main sequence stars. At colors near (B-V) $\sim$ 1, these giants are as bright in the blue as the main sequence stars.

The integrated light models used by Rabin (1982) were constructed from grids of stellar evolutionary models that did not include core-helium-burning giants; thus the important contribution of these giants to the cluster integrated light was ignored. Although such models may be applicable for sparce galactic clusters, they are inappropriate for populous galactic clusters like NGC 2158 and NGC 7789 and for the populous LMC clusters.

The strength of the red giant contribution to the integrated light from a cluster is a function of the number of cluster red giants, a stochastic effect, the magnitude of which is unknown to the observer and of the cluster metallicity (Flower and Jones 1983); the lower the metallicity, the brighter the core-helium-burning giants relative to the brightest main sequence stars.

Although it is clear that deep (CCD) cluster color-magnitude
diagrams are needed to confirm the relatively low ages obtained for
the LMC clusters listed in Table 1, it is equally clear that integrated
photometric/spectroscopic dating techniques have serious flaws.

REFERENCES

Cannon, R.:   1970, Monthly Notices Roy. Astron. Soc. 150, pp. 111.
Flower, P.J., Geisler, D., Hodge, P., Olszewski, E., and Schommer, R.:
    1983, Ap. J. December 15.
Flower, P.J. and Jones, J.H. 1983, in preparation.
Hodge, P.:   1982, in *Astrophysical Parameters for Globular Clusters*,
    IAU Colloq. No. 68, A.G. Davis and D.S. Hayes, eds. (L. Davis Press:
    Schenectady) pp. 207.
Mould, J. and Aaronson, M.:   1982, Ap. J. 263, pp. 629.
Olszewski, E.:   1983, Ph.D. Thesis, University of Washington.
Rabin, D.:   1982, Ap. J. 261, pp. 85.
Renzini, A.:   1982, in *Astrophysical Parameters for Globular Clusters*,
    IAU Colloq. No. 68, A.G. Davis and D.S. Hayes, eds. (L. Davis Press:
    Schenectady) pp. 251.

DISCUSSION

McCarthy: Can you tell us more about how you determine the faintest
giant star in a globular cluster?

Flower: For all the LMC clusters that I have dated there is a clear gap
between the core-helium-burners or clump giants and the cluster main
sequence. The distinctive red giant clump is the result of the clusters
being too young ($\lesssim 3 \times 10^9$ yr) to exhibit subgiant branches and being
very populated with red giants. Furthermore, most published data of LMC
clusters also provide data of nearby star fields; thus the field contribution to the cluster CMD's can be accurately evaluated. This greatly
improves the distinctiveness of the red giant clump.

COMPARISONS BETWEEN OBSERVATIONAL COLOR-MAGNITUDE DIAGRAMS AND SYNTHETIC CLUSTER DIAGRAMS FOR YOUNG STAR CLUSTERS IN THE MAGELLANIC CLOUDS

Stephen A. Becker[1], Grant J. Mathews[2], and Wendee M. Brunish[1]
[1]Los Alamos National Laboratory
[2]Lawrence Livermore National Laboratory

Young star clusters ($<3 \times 10^8$ yr) in the Magellanic Clouds (MC) can be used to test the current status of the theory of stellar evolution as applied to intermediate and massive stars. The color-magnitude diagram of many young clusters in the MC shows, unlike the case of clusters in our Galaxy, large numbers of stars in both the main sequence and post main sequence evolutionary phases. Using a grid of stellar evolution models, synthetic cluster H-R diagrams are constructed and compared to observed color-magnitude diagrams to determine the age, age spread, and composition for any given cluster. In addition, for those cases where the data is of high quality, detailed comparisons between theory and observation can provide a diagnostic of the accuracy of the stellar evolution models. Initial indications of these comparisons suggest that the theoretical models should be altered to include: a larger value for the mixing length parameter (α), a larger rate of mass loss during the asymptotic giant branch (AGB) phase, and possibly convective overshoot during the core burning phases.

1.   INTRODUCTION

Our grid of stellar evolution models spans the range of $3 < M_*/M_\odot < 40$ in mass, and $0.20 < Y < 0.36$ and $0.001 < Z < 0.03$ in composition(see Becker and Iben 1979, 1980; Becker 1981, Brunish and Truran 1982 a,b; and Becker and Brunish 1983). A detailed description of our method of constructing synthetic cluster diagrams is given in Becker and Mathews (1983). The conversions of Kurucz (1979) and Flower (1977) are used to convert luminosity into absolute visual magnitude and temperature into B-V. For composition determinations, we have elected to link Y and Z together via the Lequeux et al. (1979) relation:

$$Y = (0.228 \pm 0.004) + (2.83 \pm 0.60)Z$$

although we do have the ability to vary Y and Z independent of each other.

A. Maeder and A. Renzini (eds.), Observational Tests of the Stellar Evolution Theory, 83–87.

2. COMPARISONS BETWEEN THEORY AND OBSERVATION

In Figures 1 a-b synthetic cluster diagrams are plotted for two different compositions for six different ages. In both cases, one can see
that the location of the bluest core He burning giants (i.e., those located at the tip of the blue loop) provides a very sensitive age indicator. This behavior can be used in many cases to obtain an initial
estimate of the age and composition of a cluster based on the location
of its blue loop tip.   In Figure 2 this procedure has been applied to
six clusters in the LMC and one cluster in the SMC.

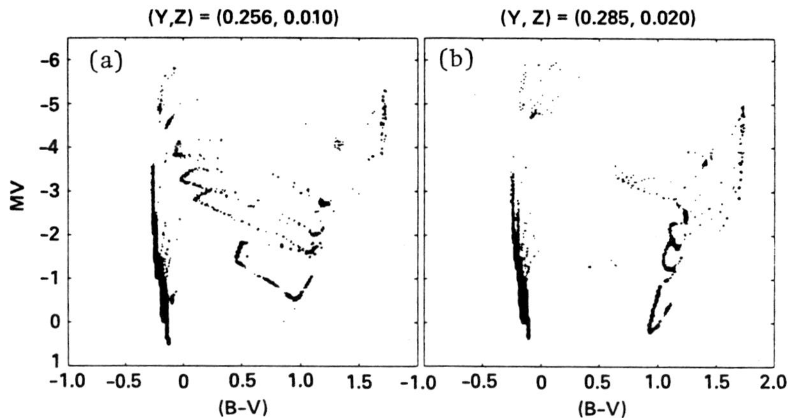

Figure 1.   Synthetic H-R diagrams as a function of age (15,
30, 60, 90, 120, 240 x 10^6 yr) for two different
chemical compositions.

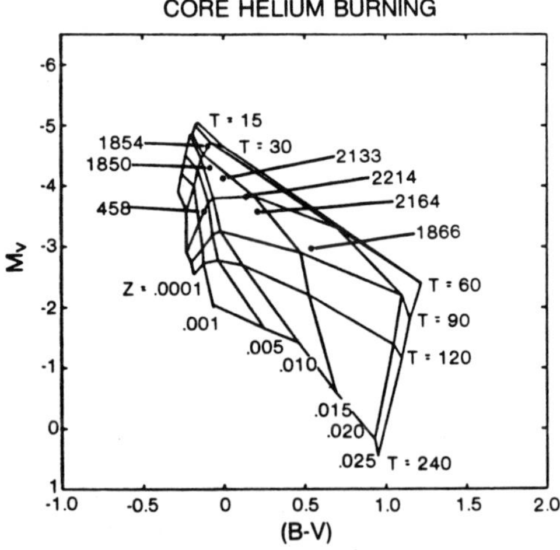

Figure 2.   The location of the blue loop tip as a function of
age (in 10^6 yr) and composition (Y is obtained
from Lequeux et al. 1979). NGC 458 is in the SMC,
the other clusters are in the LMC.

Using these estimates as a guide, whole cluster fits between synthetic
H-R diagrams and color-magnitude diagrams can then be constructed. This
procedure is necessary in order to obtain the most information possible,
i.e., the age, age spread, and composition of a given cluster as well as
feedback on the accuracy of the stellar evolution models. For example,
in Figures 3a-b and 4a-b, we compare the observations with our best
theoretical fits for the cases of NGC 1866 and NGC 2214.

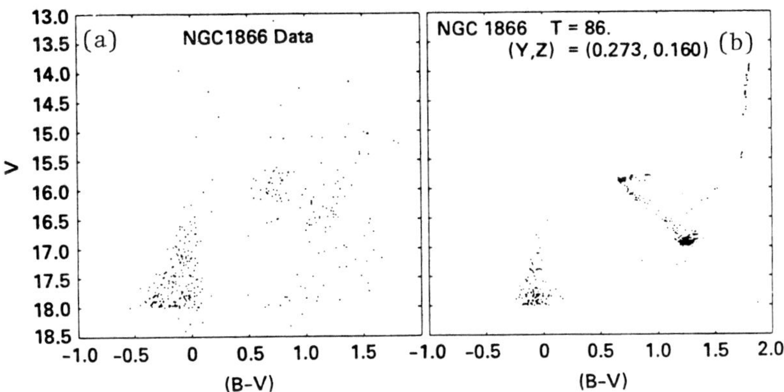

Figure 3.   Comparison of the observed and theoretical H-R
             diagrams for the LMC cluster NGC 1866. The data is
             from Arp and Thackery (1967), Robertson (1974),
             Walker (1974) and Flower (1981).

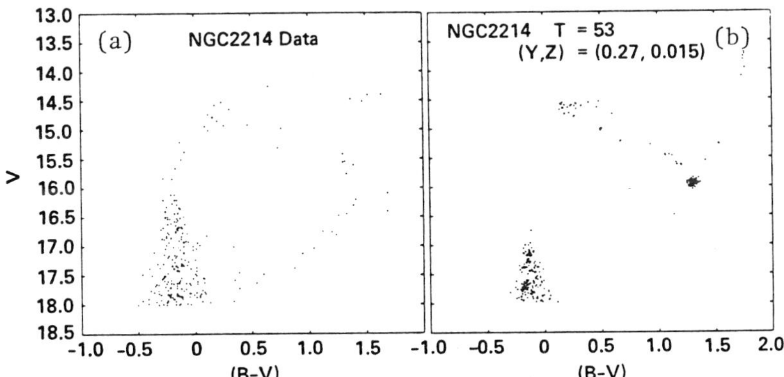

Figure 4.   Comparison of the observed and theoretical H-R
             diagram for the LMC cluster NGC 2214. The data is
             from Robertson (1974).

In Table 1 we summarize the results to date of our whole cluster fits
with regard to their age and composition.

TABLE 1

Cluster	Age	Composition
NGC 1866	86×10^6 yr	$(Y,Z) = (0.273, 0.016)$
NGC 2214	53×10^6 yr	$(Y,Z) = (0.27, 0.015)$
NGC 1854	30×10^6 yr	$(Y,Z) = (0.27, 0.015)$
NGC 2100	10×10^6 yr	$(Y,Z) = (0.28, 0.02)$

Finally, observational feedback in the above cases suggests the following changes in the theoretical models: 1) based on the location of the
red giant branch and the AGB, the value of α should be increased to
$\approx$1.5; 2) based on the observed truncation at high luminosity of the AGB,
the rate of mass loss for models in the AGB phase should be increased;
and 3) based on the observed densities of stars located on the main sequence and the red giant region, some additional effect like convective
overshoot (or perhaps rotation or binary evolution) needs to be considered. Work on other clusters in both the LMC and the SMC as well as on
theoretical models is currently in progress.

REFERENCES

Arp, H.C., and Thackery,A 1967, Astrophys J. 149, pp. 73-89.
Becker, S.A. 1981, Astophys. J. Suppl. 45, pp. 475-505.
Becker, S.A., and Brunish, W. M. 1983, Astrophys. J. to be submitted.
Becker, S.A., and Iben, I. Jr. 1979, Astrophys. J. 232, pp. 831-853.
__________________. 1980, Astrophys. J. 237, pp. 111-129.
Becker, S.A., and Mathews, G.J. 1983, Astrophys. J. 270, pp. 155-168.
Brunish, W.M., and Truran, J. W. 1982a, Astrophys. J. 256, pp. 247-258.
__________________. 1982b, Astrophys. J. Suppl. 49, pp. 447-468.
Flower, P.J. 1977, Astron. Astrophys. 54, pp. 31-39.
__________________. 1981, Astrophys. J. (Letters) 249, pp. L11-L14.
Kurucz, R.L. 1979, Astrophys.. J. Suppl. 40, pp. 1-340
Lequeux, S., Pembert, M., Rayo, J. Serrano, A. and Torres-Pembert, S.,
     1979, Astron. Astrophys. 80, pp. 155-166.
Robertson, J. W. 1974, Astron. Astrophys. Suppl. 15, pp. 261-309.
Walker, M.F. 1974, Mon. Not. Roy. Astr. Soc. 169, pp. 199-209.

*This work performed under the auspices of the U.S. Department of Energy
     Contract #W-7405-ENG. 36.*

DISCUSSION

<u>Richer</u>: Do any of the clusters that you have looked at have carbon stars in them? If not, can you say anything about the ages of the clusters that do have carbon stars in them?

<u>Becker</u>: None of the clusters we are studying or are about to study have been observed to have carbon stars in them. It appears that clusters younger than about 3×10^8 years are unable to form them. This is consistent with my thermal pulse models in which models $> 3\,M_\odot$ don't become carbon stars, but rather they evolve into N rich M supergiants on the AGB.

THE ROLE OF THE RADIATION PRESSURE GRADIENT IN GIANT AND SUPERGIANT STAR EVOLUTION

Wendee M. Brunish, Arthur N. Cox, Stephen A. Becker,
and Keith H. Despain
Los Alamos National Laboratory,
Los Alamos, New Mexico 87545

Since some of the earliest evolutionary calculations it has been found that post main sequence stars become red giants (e.g. Sandage and Schwarzschild, 1952). However the exact physical processes that lead to and determine the rate of redward evolution are not completely understood.

We hypothesized that the redward evolution might be due to an increase in radiation pressure somewhere in the star that causes the layers above it to be pushed outward, resulting in an expanded envelope and a cooler surface temperature. If the radiative luminosity somewhere in the star approached the Eddington limit, the outer layers would obviously expand. However, due to the presence of gas pressure, the critical value for expansion would be somewhat less than the Eddington limit.

We define the ratio of the radiative luminosity to the critical Eddington luminosity δ, where

$$\delta(r) = \frac{L_{rad}(r)}{L_{crit}(r)} = \frac{L_{rad}(r)\ K(r)}{4\pi cG\ M(r)} \ .$$

Presumably, then, when δ somewhere in the star reaches some critical value, δ_{max}, the layers above it will expand.

We found that the maximum value of δ increased with the initial mass of the star, which is understandable since $\beta = (P_{gas}/P)$ decreases and radiation pressure plays a larger role (see Table 1). In massive stars, $(M_i > 15\ M_\theta)$ a sharp increase in δ appeared shortly after hydrogen exhaustion in the core. The position of the increase coincided with the base of the convective shell which is associated with hydrogen burning shell. Also, this behaviour coincided with a change in sign of the slope of dr/r (the change in radius) as the contraction slowed and began to turn around due to the increased radiation pressure (see Figure 1).

A. Maeder and A. Renzini (eds.), Observational Tests of the Stellar Evolution Theory, 89–91.
© *1984 by the IAU.*

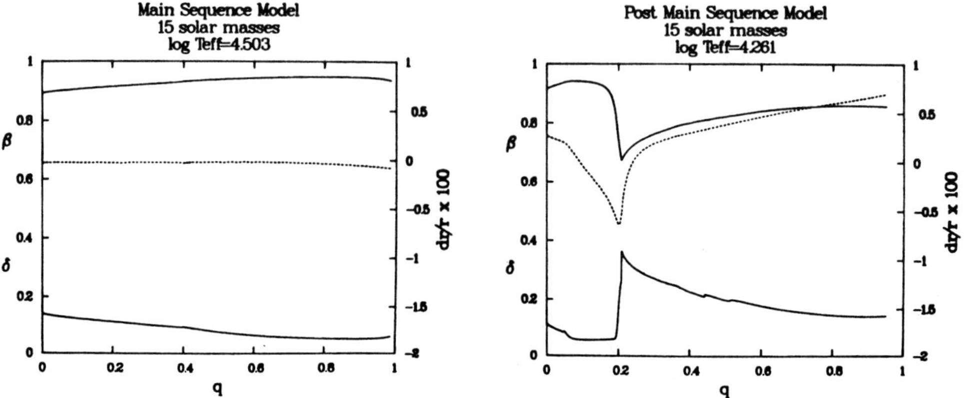

Figures 1a and b. Plotted are β, δ and dr/r × 100 versus mass fraction q for a 15 $M_\odot$ main sequence and post main sequence models. The upper solid curve is β and the lower solid curve is δ. The dotted (center) curve represents dr/r × 100.

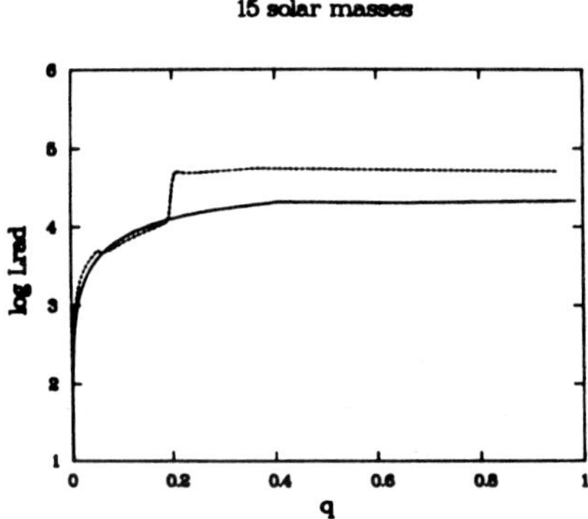

Figure 2. The log of the radiative luminosity plotted versus mass fraction q for 15 $M_\odot$ models. The solid line shows the luminosity profile for the main sequence model (log T_{eff} = 4.503) while the dashed line represents the post main sequence profile (log T_{eff} = 4.261).

Table 1

M_i	(X_i, Z_i)	$\dot{M}_{ave}$ $(10^{-6}$ $M_\odot/yr)$	δ_{max}
40 $M_\odot$	(0.70, 0.02)	0.0	0.652
40 $M_\odot$	(0.70, 0.02)	2.3	0.536
20 $M_\odot$	(0.70, 0.02)	0.11	0.440
15 $M_\odot$	(0.70, 0.02)	0.0	0.369
15 $M_\odot$	(0.70, 0.02)	0.032	0.363
15 $M_\odot$	(0.69, 0.03)	0.011	0.346
15 $M_\odot$	(0.78, 0.02)	0.020	0.314
15 $M_\odot$	(0.799, 0.001)	0.060	0.309
9 $M_\odot$	(0.78, 0.02)	0.0	0.110
6 $M_\odot$	(0.70, 0.02)	0.0	0.05
0.6 $M_\odot$	(0.749, 0.001)	0.0	0.001

In stars with a high mass loss rate, when the mass of the hydrogen exhausted core exceeds some critical value, the peak in δ disappears and the star contracts and returns to the blue.

In low and intermediate mass stars the peak (though much smaller) in δ still appears and also coincides with the base of the hydrogen burning shell. However in intermediate mass stars undergoing blue loops, the peak in δ does not disappear (or even become smaller). We must conclude that in these stars radiation pressure does not play a critical role and that gas pressure may be much more important.

The increased energy generation due to the combination of the contraction of the hydrogen exhausted core and ignition of the hydrogen burning shell leads to a sharp rise in the radiative luminosity (see Figure 2). We are investigating further how the physical conditions (i.e. pressure, temperature, density) in the interior cause this and how they affect the rate of redward evolution and subsequent blueward evolution in massive stars with high rates of mass loss.

Reference

A. Sandage and M. Schwartzschild, 1952, Ap. J. 116, 463.

STELLAR EVOLUTION AS EVIDENCED BY MEMBERS OF OPEN CLUSTERS NEAR THE
TURN-OFF FROM THE MAIN SEQUENCE

D. L. Crawford
Kitt Peak National Observatory
Tucson, AZ 85726 USA

In many ways, this paper can be thought of as an introduction to
that of Nissen's, and others, which contain high quality, new data for
open clusters, particularly for work that includes observations of stars
near the turn-off from the main sequence.

The comparison of observed data for stars in open galactic clusters
with the predictions of theoretical models has always been one of the
most fruitful methods and constraints in developing our understanding of
stellar evolution. In most cases, the stars in the clusters have the
same age, same chemical composition, and other properties, while
differing in stellar mass, and hence in surface temperature and in the
relative state of evolution on and away from the main sequence.

In most cases in the past, the comparisons have been made on the
basis of (mostly) smoothed sequences drawn onto the diagrams of observed
parameters such as magnitude and color. The general fit, main sequence
turn-off location, gaps, etc. have all been useful in fits. Recently,
improved observational accuracy and the quantity of photometric and
other data allow a detailed comparison for individual stars. While
problems remain (adequate corrections for or elimination of the effects
of variable reddening, nonmembership, etc.), potentially we can learn
much from the effort to fit all member stars, whatever their location on
such diagrams, into the existing theories.

Analysis of data for stars on the main sequence, and those not so
far above the turn-off, offers considerable advantages relative to
brighter, more evolved stars. There are less problems due to
evolutionary changes on stellar structure. Problems still do exist, of
course: the question of membership, binaries, and the fact that the
stars are fainter and observational error generally larger.

Due largely to new techniques, the potential for new and better
data is great, and the future looks very exciting - it's the age that
Schatzmann called "High Precision Astrophysics" earlier in this
meeting. New telescopes and new instruments, wider wavelength coverage,

A. Maeder and A. Renzini (eds.), Observational Tests of the Stellar Evolution Theory, 93–94.
© *1984 by the IAU.*

from the X-ray through the UV to the near and far infrared. We are
getting more photons, better detectors, multiplexing, and improved data
handling and reductions particularly. We will be working fainter, on
more objects, and with better accuracy.

Now we even hear of 4 and 5-m telescopes being called "small", or
back-up, instruments. Meanwhile, due to improved detectors and the
technology, our existing small (1-m class or so) telescopes are becoming
"large" telescopes. It's a new world, and should greatly improve our
understanding of Stellar Evolution.

STRÖMGREN PHOTOMETRY OF F-TYPE STARS IN M67 AND NGC3680

P.E. Nissen
Institute of Astronomy
University of Aarhus

Photoelectric uvby photometry has been obtained for about 30 F-type stars in each of the open clusters M67 and NGC3680. For some of the stars the H-beta index was also observed. The photometry was carried out with the Danish 1.5 m telescope at La Silla by the aid of a four-channel uvby photometer and a two channel H-beta photometer.

The stars were selected to cover the upper main sequence, the turn-off region, and the lower sub-giant branch. The corresponding range in magnitude is V = 14.5 to 12.0. The majority of the stars were observed on 3 different nights. The errors of the Strömgren photometric indices, b-y, m_1, c_1, and β are of the order of 0.01 magnitudes. The error of the V magnitude is 0.03.

By comparing the observed β, (b-y) values with the standard β, $(b-y)_0$ relation for nearby F-type stars (Crawford, 1975) an interstellar reddening of E(b-y) = 0.02 is found for M67 and E(b-y) = 0.03 for NGC3680. The corresponding reddening in B-V is E(B-V) = 0.03 for M67 and E(B-V) = 0.04 for NGC3680. The value for M67 is significantly lower than the value of E(B-V) = 0.06 found by Eggen and Sandage (1964) from the two-colour (U-B), (B-V) relation.

The metal abundance of the clusters are derived from the dereddened $(b-y)_0$ - m_0 diagram, using the calibration of this diagram in terms of [Me/H] by Nissen (1981). Both clusters are found to have a metal abundance very close to that of the Sun. The rms dispersion in [Me/H] is 0.10 only.

Having determined accurate values of the interstellar reddening and metal abundance for the clusters it is possible to make a detailed comparison of the colour-magnitude diagrams with theoretical isochrones. For M67 a very good fit is obtained to the shape of the isochrones of Hejlesen (1980) for Z = 0.02, including the position of the gab between the upper turn-off and the lower sub-giant branch. The age of M67 is found to be between 3 and $4 \cdot 10^9$ y, depending on the choice of the values of the mixing-length parameter, the helium abundance, and the zero-point

A. Maeder and A. Renzini (eds.), Observational Tests of the Stellar Evolution Theory, 95–96.
© *1984 by the IAU.*

for the calibration of (b-y) in terms of T_{eff}.

The stars in NGC3680 show a much larger dispersion in the $(b-y)_0 - V_0$ diagram than the stars in M67. The distribution tends to be bimodal with one sequence of stars around the isochrone corresponding to an age of $1.5 \cdot 10^9$ y and the other sequence around an isochrone corresponding to $2.5 \cdot 10^9$ y. The $c_0 - (b-y)_0$ diagram of NGC3680 shows that this bimodal distribution cannot be explained in terms of binaries alone. Apparently other parameters - most likely rotation - have rather dramatic effects on the position of early F-type stars in the colour-magnitude diagram. A similar conclusion has recently been reached by Twarog (1983) from uvby-β photometry of the open cluster NGC752.

References

Crawford, D.L.: 1975, Astron. J. 80, 955
Eggen, O.J., Sandage, A.R.: 1964, Astrophys. J. 140, 130
Hejlesen, P.M.: 1980, Astron. Astrophys. Suppl. 39, 347
Nissen, P.E.: 1981, Astron. Astrophys. 97, 145
Twarog, B.A.: 1983, Astron. J. 267, 207

DISCUSSION

R. Cayrel: Do you consider that the results you have reported settle for ever the question of the alledged supermetallicity of M67?

Nissen: The Strömgren $m_0 - (b-y)_0$ diagram for M67 clearly indicates a solar metal abundance. Thus, at least in the case of the elements affecting m_1 (e.g. the iron-peak elements) M67 is not supermetal-rich.

Mermilliod: The peculiar aspect of the main sequence of NGC 3680 is probably related to the temperature of the cluster turn-up, which corresponds to the one of the F0-F5 gap described by Böhm-Vitense and Canterna, appearing at the onset of the convection in the atmosphere.

Janes: I will be reporting tomorrow on more M67 observations which are in excellent agreement with most of the results of this paper. In particular, I find a metallicity very close to that of the Sun.

STATISTICAL ANALYSIS OF THE RED GIANT DISTRIBUTIONS OF OLD OPEN CLUSTERS

G. Barbaro and L. Pigatto
Istituto di Astronomia, Università di Padova
Osservatorio Astronomico, Padova

In order to verify the agreement with the theory of the observed HR diagram as a whole, 38 open clusters with $(B-V)_{o,t} \geq 0.10$ at the turnoff, have been analyzed with particular regard to the RGB luminosity function.

Theoretical isochrones derived from Ciardullo and Demarque (1977) and evolutionary tracks for central He-burning calculated by Sweigart and Gross (1976) have been used, adopting chemical composition $X=0.700$ and $Z=0.001$, 0.004, 0.01.
The RGB of isochrones have been corrected for a mixing length parameter $\alpha=1.5$ (Ciardullo and Demarque (1979)). The conversion tables in the M_V, $(B-V)$ plane by Morton (1969), Morton and Adams (1968) for the V class and by Johnson (1966) and Lee (1970) for the III and I classes of luminosity have been adopted.
The analysis has been performed in two steps:
- comparison of the HR diagram with isochrones
- statistical check of the fit between the observed luminosity distri<u>bu</u>tions of red giant stars and the theoretical ones by means of the Kolmogoroff test.
Two different behaviours come out when the theoretical luminosity functions of the RGB are compared with observed ones:
- clusters with $(B-V)_{o,t} > 0.35$ (Group I) fit well both the MS parameters M_V, $(B-V)_o$ at the turnoff point and the RGB luminosity functions,
- clusters with $(B-V)_{o,t} \leq 0.35$ (Group II) never agree with the RGB lumi<u>no</u>sity functions here considered. (Table 1 reports $(B-V)_{o,t}$ and metal abundance from literature of the clusters). This behaviour is brought <u>in</u>to evidence in Fig. 1, where the magnitude of the weakest red stars of the base of the RGB is plotted against $(B-V)_{o,t}$ both for theoretical iso<u>o</u>chrones of metal abundance $z=0.01$, 0.001 and for the clusters of the two groups. In Fig. 2 the observed hystogram of the red giant stars distribu<u>u</u>tion of NGC 7789, one of the most populous cluster of Group II, (the other clusters of the group have very similar distributions), compared with the theoretical distribution (RGB and He-burning phases) for the age and metal abundance compatible with the cluster ones, clearly shows the disagreement consisting in a evident lack of stars at the base of RGB.
It is obvious that Group II does not agree with the classical evolution

A. Maeder and A. Renzini (eds.), Observational Tests of the Stellar Evolution Theory, 97–100.
© *1984 by the IAU.*

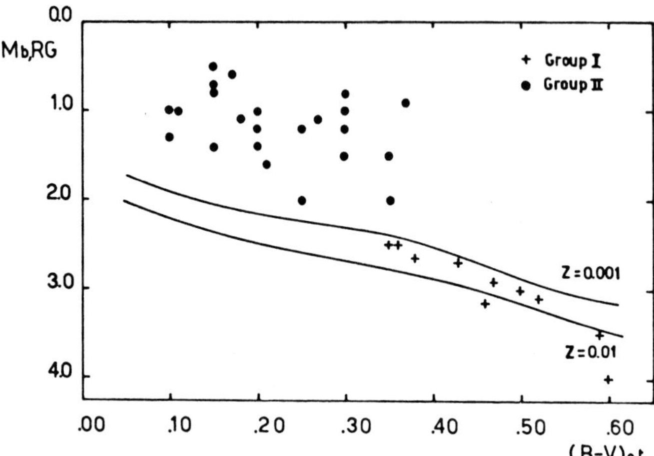

Figure 1 – Magnitude of the faintest red giant stars $(M_{b,RG})$ versus (B-V)o at turnoff of the clusters. The solid lines represent values for the isochrones of metal abundance z=0.01 and z=0.001.

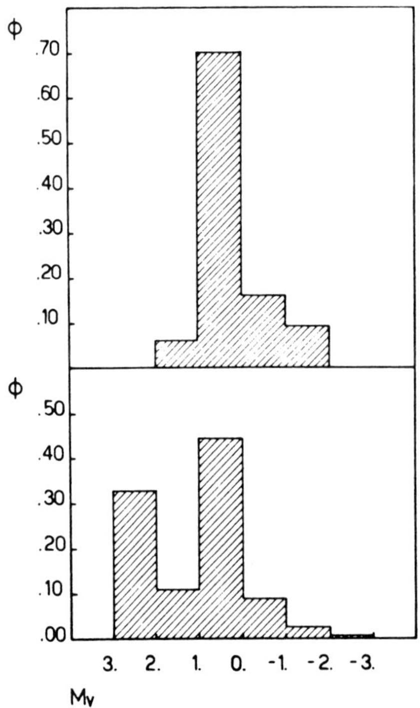

Figure 2 – Luminosity function of RGB of NGC 7789 (top) and for the theoretical isochrone (t=2 Gyrs, z=0.01) (bottom).

TABLE 1

| GROUP I | | | GROUP II | | | | | |
NGC	$(B-V)_{o,t}$	z_{obs}	NGC	$(B-V)_{o,t}$	z_{obs}	NGC	$(B-V)_{o,t}$	z_{obs}
188	0.60	0.013	IC166	0.35	--	3960	0.21	0.008
2141	0.50	0.006	752	0.35	0.010	5822	0.11	0.020
2158	0.35	0.005	1245	0.20	0.023	IC465	0.37	0.013
2204	0.36	0.008	1342	0.17	0.008	6633	0.10	0.010
2243	0.47	0.008	IC361	0.25	--	IC4756	0.20	0.015
2420	0.50	0.005	Hyades	0.10	0.025	6811	0.15	--
2506	0.38	0.007	1817	0.15	0.021	6819	0.30	0.019
2682	0.46	0.010	2236	0.18	--	6939	0.30	--
3680	0.38	0.012	2360	0.30	0.012	6940	0.25	0.022
6791	0.59	--	2477	0.20	0.016	7039	0.10	--
7142	0.43	0.005	Praesepe	0.15	0.024	7062	0.20	--
MEL66	0.52	0.004	2660	0.27	0.019	7762	0.30	--
			2818	0.15	--	7789	0.30	0.010

scheme, according to which the clusters should exhibit extended red giant branches as a consequence of the electron degeneracy developed during the shell H-burning phase, but their behaviour is rather reminiscent of that of younger clusters, whose evolved stars have a mass $M > 2.2\ M_\odot$ and igni te helium quietly in a non-degenerate core.

The hypothesis we make is that models in which the overshoot of the con vective core is taken into account, could explain this discrepancy. In fact, as shown by Maeder (1975, 1976), the effect of overshoot for low mass stars, is to increase the mass of the convective core before and du ring the shell H-burning phase. It could happen that for a mass $1.5 < M \leq 2.2\ M_\odot$, i.e. in the mass range we are concerned, the mass frac tion q_{core} becomes larger than the Schönberg-Chandrasekhar limit q_{S-C} before the core becomes degenerate: in this case the core begins to con-tract and evolves in the region of non degeneracy, giving rise to a quiet onset of He-burning.

Preliminary results of a model of $1.8\ M_\odot$, X=0.700, Z=0.02, by Bertelli and Bressan (paper in progress) show that, for a conservative overshoo-ting parameter $\lambda=1/H_p=1$, the core behaves as described above and the evo-lutionary path of the model is qualitatively similar to that of a $3\ M_\odot$.

REFERENCES

Bertelli, G., Bressan, A.: private communication.
Ciardullo, R.B., Demarque, P.: 1977, Yale Trans., Vol. 33.
Ciardullo, R.B., Demarque, P.: 1979, Dud.Obs.Rpt. 14, 317.
Johnson, H.L.: 1966, Ann.Rev.Astron.Astrophys. 4, 193.
Lee, T.A.: 1970, Astrophys. J. 162, 217.
Maeder, A.: 1975, Astron. Astrophys. 43, 61.
Maeder, A.: 1976, Astron. Astrophys. 47, 389.
Morton, D.C.: 1969, Astrophys. J. 158, 629.
Morton, D.C., Adams, T.F.: 1968, Astrophys. J. 151, 611.
Sweigart, A.V., Gross, P.G.: 1976, Astrophys. J. Suppl. 32, 367.

DISCUSSION

<u>Mermilliod</u>: The different trends in the two age groups may result from
the different morphology of the giant region. In the younger age group,
most giants are in the clump phase, while in the older one, there is a
full giant branch. Thus, the faintest giant in both groups does not
correspond to the same evolutionary state.

X-RAY SURVEY OF THE PLEIADES: DEPENDENCE OF X-RAY LUMINOSITY ON STELLAR AGE

G. Micela[*], S. Sciortino[*], S. Serio[*†], G.S. Vaiana[*†], L. Golub[†], F.R. Harnden[†], R. Rosner[†]

[*] Osservatorio Astronomico di Palermo
[†] Harvard-Smithsonian Center for Astrophysics

The study of X-ray emission of stellar clusters, allows to decouple the influence of some individual stellar parameters, as initial conditions, composition and age, on the stellar X-ray luminosity function.

In order to be studied in the soft X-ray band, a cluster must be sufficiently near for its stars to be detected in "normal" observations times (10^3 - 10^4 sec); this means that the cluster must have a maximum distance $\leq$ 150 pcs. The clusters which meet this requirement are only a few, namely: the Hyades, Ursa Major, Coma and the Pleiades.

A detailed study on the central region of the Hyades has been done by Stern et al. (1981). They have detected X-ray emission above a threshold of $10^{28.5}$ ergs/sec from $\sim$ 50% of the cluster stars. The median X-ray luminosity for dwarfs G Hyades stars resulted to be $\sim$ 30 times the luminosity of the Sun which is $\sim$ 1 order of magnitude older. Since the Pleiades are even younger than Hyades, a survey of this cluster can improve our knowledge of the dependence between X-ray luminosity and stellar age.

We report here preliminary results from an Einstein X-ray survey of the Pleiades. We have analysed, using the standard Einstein Observatory software a $1° \times 1°$ exposure centered over one of the more luminous stars of the cluster (20 TAU, [B7-III]), taken with Imaging Proportional Counter (IPC) (Giacconi et al., 1979) which is sensitive to X-rays in the energy band .15 - 4.0 KeV with a energetic resolution ($\Delta E/E$) of $\sim$ 1 at 1.0 KeV and a spatial resolution of $\sim$ 1'.

This field contains $\sim$ 62 cluster members out of a total of $\sim$ 270 stars with magnitude lower than 14^m. (Hertzsprung, 1947).

The exposure time of the observation sets a detection treshold of $\sim$ $10^{29.5}$ ergs/sec. With this threshold we have detected 17 distinct X-ray sources; 16 sources are identified with a cluster stars within a distance less than 1'. The probability of a chance identification is $\leq$ 2 10^{-3}. X-ray emission from 2 (out of 8) B stars, 1 (out of 9) A star, 3 (out of 6) F stars, 8 (out of 19) G stars, 2 (out of 20) K stars has been detected. The brightest X-ray sources is Hz 303[‡] (spectral type G1), which has Log L_x $\sim$ 30.3.

We give in Table 1 the X-ray luminosities, together with the optical properties, of the detected sources.

The estimated error on the values of the X-ray luminosity is $\sim$ 40% compounded by a statistical error ranging from 10 % to 30%, sistematic errors in instrument calibration < 20% (Harnden et al., 1979), error in the individual cluster member

---

[‡] In the following will use the numeration of Hertzsprung, 1947.

101

A. Maeder and A. Renzini (eds.), Observational Tests of the Stellar Evolution Theory, 101–104.
© 1984 by the IAU.

TABLE 1

X-ray Source #	L_x [ergs/sec] 10^{29}	Counterpart Hz II #	Sp	m_V	B - V	Note●
1E 0340.9+2404	7.2	193	G7 †*	11.29	+0.81	
1E 0341.1+2406	6.3	263	G8 †*	11.54	+0.88	
1E 0341.3+2356	19.0	303	G9 †*	10.48	+0.89	
1E 0341.4+2437	18.0	320	G5 *	11.04	+0.88	
1E 0341.5+2425	9.3	345	G8 *	11.65	+0.85	
1E 0342.2+2419	5.5	563	B6 *	4.31	-0.11	19 TAU
1E 0342.6+2355	6.6	708	G0 *	10.13	+0.62	
1E 0342.6+2408	6.0	686	K2 †*	13.62	+1.04	f v
1E 0342.7+2403	4.2	761	G1 *	10.55	+0.67	
1E 0342.7+2428	8.1	727	F9 *	9.70	+0.55	v
1E 0343.3+2402	4.4	956	F0 †	7.96	+0.32	d
1E 0343.3+2347	8.1	980	B6IV *	4.18	-0.06	23 TAU
1E 0343.5+2416	8.7	1032	G8 *	11.34	+0.86	v
1E 0343.6+2411	5.1	1100	K3 *	12.16	+1.15	f
1E 0343.7+2426	4.3	1122	F4 *	9.29	+0.46	
1E 0344.4+2426	14.0	1384	A2 *	7.66	+0.21	
1E 0344.6+2413	7.5					

● v indicates variable star, f flare star, d binary system.

† Spectral type determined from B-V values (Johnson & Mitchell, 1958; Jones, 1973; Landolt, 1979; Stauffer, 1980) corrected for reddening, using as mean E(B-V)=0.04 (Crawford & Perry, 1976).

* Spectral types based on spectroscopic data (Mendoza, 1956; Wilson, 1963; Herbig, 1962; Kraft & Greenstein, 1969).

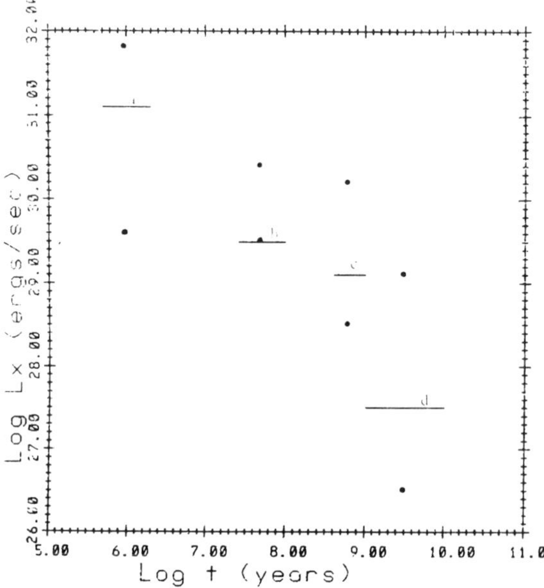

Fig. 1 - Dependence of median X-ray luminosity from age for different samples of G stars : a) pre-main sequence stars (Ku & Chanan, 1979; Feigelson & De Campli, 1981); b) main sequence G stars in the Pleiades (present work); c) main sequence G stars in the Hyades (Stern et al., 1981); d) local disk population G dwarfs (Vaiana et al., 1981; Topka et al., 1981; Rosner et al., 1981). Solid line indicates the median value and the error bar represents the uncertainty in age determination. The range of observed luminosities is indicated by ●: the lower limit is always fixed by the best detection treshold for each group.

distance $<$ 3%, and a sistematic error in converting counts to flux $<$ 20% due to the assumed hydrogen column density and source temperature (N_H = $10^{20.3}$ atoms/cm^2, T = $10^{6.5}$ °K).

Only 5 stars ($\sim$ 3% of the stars with comparable limiting magnitude) in the Hyades survey have been detected as X-ray sources with a luminosity above the threshold for the present Pleiades survey. Since the Pleiades are $\sim$ 1 order of magnitude younger than the Hyades, this different behaviour can be attributed to the age difference of the two clusters.

Since have been detected in X-rays $\sim$ 42% of dwarfs G the value of the median of the X-ray luminosity function is not far from $10^{29.5}$ ergs/sec. We have plotted in figure 1 this value together with the median of the X-ray luminosity of T Tauri stars, of main sequence G stars in the Hyades, of local disk population G dwarfs. This plot provide evidence of a dependence of the level of the X-ray emission for G stars from stellar age. Fitting a relationship of the type $L_x \propto \tau^{-\beta}$, β is of the order of 1. The absence of sources identified with M stars, except perhaps the one source without optical counterpart[‡], may indicate a dependence of X-ray luminosity from age more complex than a simple law of monotonic decrease for all spectral types. In fact, in the nearby sample, the median X-ray luminosity of M stars is higher than that of G stars, while in the Pleiades the upper limit to the X-ray luminosity of M stars is lower than the median luminosity of G stars.

We acknowledge the support of Ministero Pubblica Istruzione, Piano Spaziale Nazionale and CRRNSM.

REFERENCES

Allen, C.W., 1976, *Astrophysical Quantities* (London: Athlone)
Binnedijk, K.L., 1946, *Ann. Leiden Obs.*, **19**, Part 2.
Crawford, D.L., Perry, C.L., 1976, *Astron.J.*, **81**, 419.
Feigelson, H.C., DeCampli, W.M., 1981, *Astrophys.J. Lett.*, **243**, L89.
Giacconi, R et al., 1979, *Ap.J.*, **230**, 540.
Harnden, F.R., Jr., Branduardi, G., Elvis, H., Gorenstein, P., Grindlay, J., Pye, J.P., Rosner, R., Topka, K., Vaiana, G.S., 1979, *Astrophys.J. Lett.*, **234**, L51
Herbig, G.H., 1962, *Astrophys.J.*, **135**, 736.
Hertzprung, E., 1947, *Ann. Leiden Obs.*, **19**, Part 1A.
Johnson, H.L., Mitchell, R.I., 1958, *Astrophys.J.*, **128**, 3.
Jones, B.F, 1973, *Astrophys.J. Suppl.*, **9**, 313.
Kraft, R.P., Greenstein, J.L., 1969, S.S. Kumar (ed.), *Low Luminosity Stars*, Gordon and Breach, New York, p.65.
Ku, W.H., Chanan, G.A., 1979, *Astrophys.J. Lett.*, **234**, L59.
Landolt, A.U., 1979, *Astrophys.J.*, **231**, 468.
Mendoza,V.E.E., 1956, *Astrophys. J.* **123**, 54.
Rosner, R., Avni, Y., Bookbinder, J., Giacconi, R., Golub, L., Harnden, F.R., Jr., Maxson, C.W., Topka, K., Vaiana, G.S., 1981, *Astrophys.J.*, **249**, L5.
Stauffer, J.R., 1980, *Astron.J.*, **85**, 1341.
Stern, R.A., Zolcinski, M.C., Antiochos, S.C., Underwood, J.M., 1981, *Astrophys.J.*, **249**, 647.
Topka, K.P., 1980, *Thesis.*
Topka, K.P., et al., 1982, *Astrophys.J.*, **259**, 677.
Vaiana, G.S., et al., 1981, *Astrophys.J.*, **245**, 163.
Wilson, O.C., 1963, *Astrophys.J.*, **138**, 832.

[‡] The optical catalog is complet to $m_V < 14^m$, i.e. to late K stars.

DISCUSSION

<u>Richer</u>: Did you detect any X-ray sources that were not visible as stars
on the plates? Did you detect the supposed white dwarf member of the
Pleiades in the X-ray region?

<u>Micela et al</u>: One of the X-ray source detected in our X-ray observation
is not identified with a Pleiades member. However, the published optical
catalogue is complete until 14th magnitude (i.e. late-K main sequence
stars). The nature of the unidentified X-ray source should be object of
more detailed investigation to clear if we are looking at an M main-
sequence star belonging to the cluster, or a field star or an object of
different nature.

THE $\log L/\log T_e$ DIAGRAM FOR INTERMEDIATE MASS POP I RED GIANTS

M. Grenon[1], J.C. Mermilliod[2]
[1]Observatoire de Genève
[2]Institut d'Astronomie de l'Université de Lausanne

1. INTRODUCTION

The best empirical evolutionary tracks are provided by open cluster
HR diagrams. The scarcity of giants in young clusters implies that diagrams of clusters having similar ages and metallicities have to be summed.
Using UBV data, Mermilliod (1981a) has divided the nearby open clusters
into 14 age groups, and has described the properties of the composite
diagrams $M_v/(B-V)$ and $M_{bol}/\log T_e$. Several features complicate the
structure of composite HRD, namely the presence of double stars which
often mimic extended blue loops and the contamination by non-member stars
difficult to identify as such in distant clusters. In order to refine
the red giant observed evolutionary properties, cluster giants have been
remeasured in the Geneva photometry and with the spectrovelocimeter
CORAVEL.

2. THE PHOTOMETRY

For the present analysis 179 giants belonging to 29 clusters have
been retained. The colour excesses are determined from Geneva colours of
early-type star cluster members or from UBV data. [M/H] ratios are obtained for non-binary stars using Grenon's (1978, 1982) calibrations. This
ratio is very sensitive to reddening errors and is used for checking the
consistency of the adopted distance moduli and colour excesses. These
latter parameters have been homogenized and Mermilliod's age grouping
slightly modified. The cluster metal abundances appear, in the mean,
solar to slightly metal-rich. The empirical isochrones may be compared
with models built with $Z = .02$. The exceptions are the Hyades generation
(without NGC 2539) with [M/H] +.15 or $Z = .03$, and NGC 752 and 7789 with
$Z = .01$. The stellar T_{eff} are derived from Geneva colours using the 1982
calibration by Grenon. The luminosities are deduced from m_v, (m–M), E_{B-V}
and a B.C. adapted from Bell et al.(1978) and Cohen et al.(1978).

A. Maeder and A. Renzini (eds.), Observational Tests of the Stellar Evolution Theory, 105–107.

3. THE RADIAL VELOCITIES

All program stars have been measured with CORAVEL, in both hemisphere
Most stars have been monitored during several years in order to check
the stability of RV or for the obtention of orbits. As σ_{VR} is less than
1 km/sec, the RV is now the best membership criterion. Only 5 program
stars were rejected as non-members whereas 41 have been discovered or
confirmed as binaries. The present rate of multiple systems among giants
is at least 23 %.

4. THE STELLAR AGES AND MASSES

The cluster ages are obtained by fitting the observed upper main
sequence HRD with Maeder and Mermilliod's (1981) isochrones. A mean age
is deduced for each generation. The minimum mass of a red giant is derived
from that of subgiants using the same models. The evaluation of the ratio
of the number of red giants on that of MS stars in a two mag. interval
below the evolutionary gap leads to an estimate of the maximum initial
mass for the most evolved giants, assuming a Salpeter mass function.

5. THE $\log L / \log T_e$ DIAGRAM

Empirical isochrones for single
giants with age from 10^8 to $3.2 \cdot 10^9$ y.
are now defined.Four generations are
shown in Fig. 1. That of NGC 752, a =
$1.4 \cdot 10^9$, shows the classical features
of old clusters, i.e. an extended GB
and a clump of core-He burning stars
($\log L$ 1.62, $\log T_e$ 3.673). The Hyades
and NGC 3532 generations show an elon-
gated concentration centered at $\log T_e$
3.69 and $\log L$ 1.85 and 2.25 respectively with few stars on the AB. Their
ages are $6.6 \cdot 10^8$ and $2.7 \cdot 10^8$y. The
first blueward loop appears for the
NGC 2516 generation, a = $1.1 \cdot 10^8$y. and
seems more extended than predicted by
Becker's (1981) models. The AB is very
thin and well defined up to $\log L$ 3.9.
Mermilliod's (1981b) relations between
the mean M_{bol} and $\log T_e$ of giant concentration versus the age remain valid.

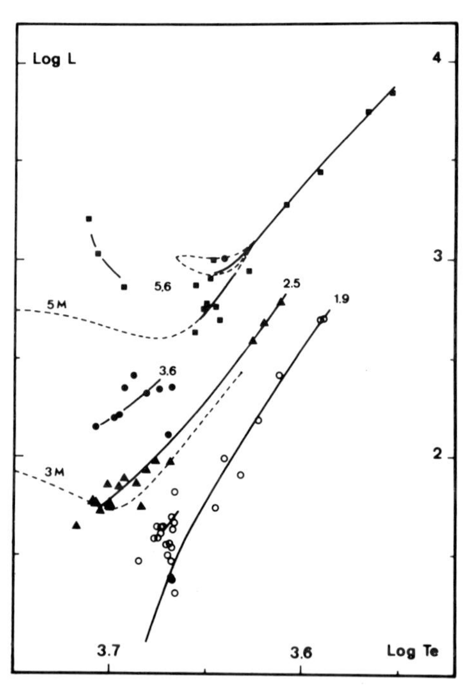

Fig. 1 HR diagram for single red
giants of four age groups:

o : NGC 752   ■ : NGC 2516+Pleiades
▲ : Hyades   —— : Becker's models
● : NGC 3532

REFERENCES

Becker, S.A. 1981, Ap.J.Suppl. 45, 475.
Bell, R.A. et al. 1978, A.A.Suppl. 34, 229.
Cohen, J.G. et al. 1978, Ap.J. 222, 165.
Grenon, M. 1978, Publ. Obs. Genève 5, sér. B.
Grenon, M. 1982, IAU Coll. 68, 393.
Maeder, A., Mermilliod, J.C. 1981, A.A. 93, 136.
Mermilliod, J.C. 1981a, A.A.Suppl. 44, 467.
Mermilliod, J.C. 1981b, A.A. 97, 235.

DISCUSSION

Weidemann: In your HR diagram the brightest giant of the 5 M$_\odot$ turn-off mass group is located at about log L/L$_\odot$ = 3.9. Similarly NGC 1866 with the same turn-off mass (5 M$_\odot$) in the LMC shows the giant branch populated only up to log L/L$_\odot$ = 4.0, corresponding to core masses of only 0.7 M$_\odot$. Do you think that your ensemble is well enough populated to exclude red giant evolution to much higher luminosities (and thereby final masses)?

Mermilliod: Although we have plotted data from several clusters, the sample is too small to draw definitive conclusions concerning the evolution to higher luminosities. In addition, it would be worth to consider the difference in chemical composition between our Galaxy and the LMC.

GRAIN SEDIMENTATION AND MAIN SEQUENCE EVOLUTION

John Lattanzio
Department of Mathematics
Monash University
Clayton, Victoria, Australia

ABSTRACT. Under the assumption that grain sedimentation can lead to
an enhancement of heavy elements in the central regions of proto-stars,
we present the evolution of stellar models with metal-rich cores. The
resulting isochrones can explain the "double-gap" clusters NGC2420 and
NGC2506, without destroying the agreement for other clusters.

Previous evolutionary calculations have generally assumed a
homogeneous abundance distribution in initial main sequence models.
This assumption may not be valid. If theories of grain sedimentation
are correct then we may expect an enhancement of grain material in
the core of pre-stellar clouds (see Horedt 1973, Prentice 1976,
Krautschneider 1977, Flannery and Krook 1978, etc.). Recent collapse
calculations have cast doubt upon the ability of the Hayashi phase to
homogenize the star (Winkler and Newman 1980, Stahler et.al. 1980a,b).
Thus any abundance inhomogeneity caused by grain sedimentation may
survive to the ZAMS.

In this study we have followed the evolution of models of
various masses. The models have an envelope of $(X,Z) = (0.70, 0.02)$
and an inner core of 2% by mass with $Z = 0.10$ and the same X/Y as the
envelope. Convective dilution of the core with the lower metallicity
surrounds has been followed in detail (see Lattanzio 1983), with
opacities taken from the Los Alamos code (Huebner et.al., 1977).

During contraction to the main sequence, CN cycling leads to a
convective core covering the inner 20-30% by mass, which reduces
the core $Z \simeq 0.025-0.03$, a value maintained throughout subsequent
evolution. Details of the evolutionary calculations may be found
in Lattanzio (1983), but the main effect is an increase in the time
of core H burning. For $M \gtrsim 1.4$ this is due to a hydrostatic readjust-
ment associated with the increased CNO abundances in the core. For
lower masses the cause is the larger convective core, due to the
higher opacity of the metal-rich material. Figure 1 shows that the
resulting H burning lifetimes are no longer monotonic. There are two

A. Maeder and A. Renzini (eds.), Observational Tests of the Stellar Evolution Theory, 109–112.

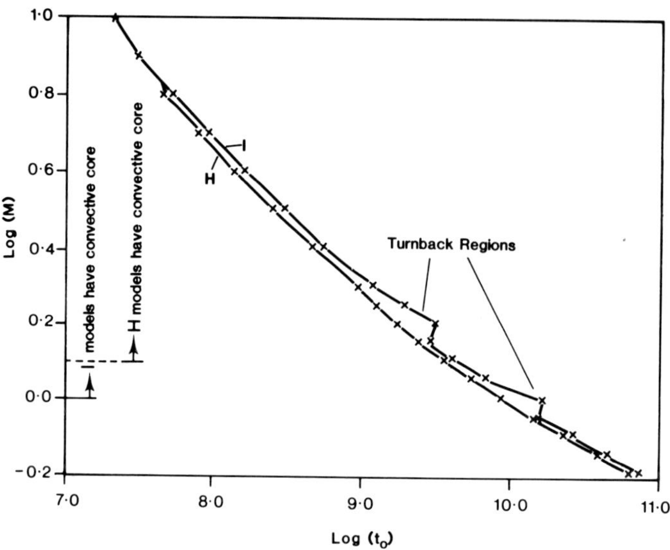

Figure 1. Plot of time of core H exhaustion (t_0) against mass. Crosses show results for both homogeneous (H) and inhomogeneous (I) models.

"turn-back" regions, and the important question is the nature of the isochrones for ages corresponding to these regions. A line of constant time will, in the turn-back region, cross the H exhaustion curve at three places. There will be two regions of the H exhaustion curve that lie to the right of the time line. These regions represent stars burning H in their core, and will thus heavily populate the isochrone. Regions to the left of the time line represent stars having just exhausted core H, and thus will be sparse in stars when plotted in the HR diagram. As there are two such regions we would expect two gaps to appear in the isochrone.

An isochrone for log(age) = 9.45, in the turn-back region, is shown in Figure 2. We do indeed find two gaps in the resultant distribution. Observations of NGC2420 (McClure et. al. 1978) and NGC2506 (McClure et. al. 1981) also show two gaps. These two clusters are metal deficient relative to the models presented. Using the expressions of Patenaude (1978) we can estimate the age at which we would expect two gaps to occur in clusters with the metallicity of NGC2420 and NGC2506. We find log(age) = 9.67, which agrees will with previous age estimates.

Figure 1 also explains the scarcity of "double gap" clusters, as only those with ages in the turn-back region will show two gaps. Finally, a comparison between the gaps in observed colour-magnitude diagrams and in theoretical isochrones has been performed (see Lattanzio 1983) in the manner outlined by Maeder (1974). The discrepancies found by Maeder are found to decrease when recent opacity tables are used, and may further decrease for initially inhomogeneous models.

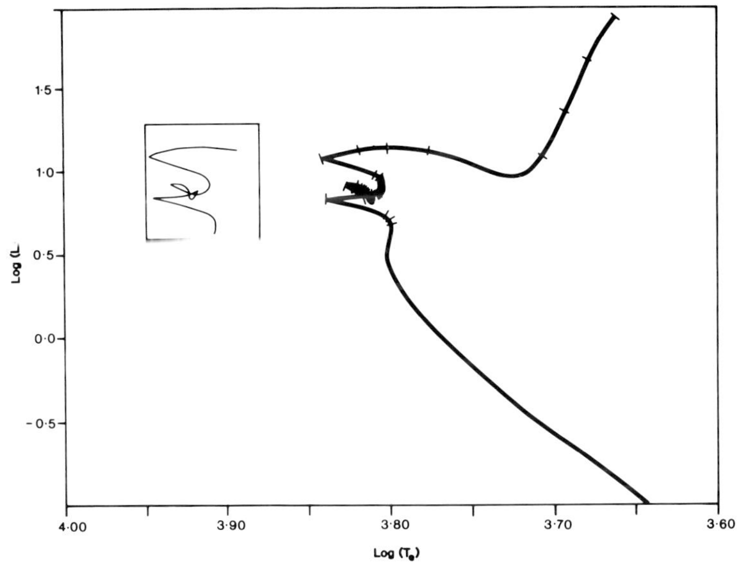

Figure 2.   Isochrone for inhomogeneous models at log(age) = 9.45.   The thick portion of the line represents a region of high star density. Ticks are placed every 0.005 $M_\odot$ in the region of the gaps.   This region is repeated in the inset.

Thanks to the following people for their assistance: A.J.R. Prentice, R.A. Gingold, P.R. Wood, W.F. Huebner and M.F. Argo.   Thanks also go to the Ian Potter Foundation and the IAU for support.

References

Flannery, B.P. and Krook, M., 1978, *Astrophys. J.*, *223* p.447.
Horedt, Gp., 1973, *Mon. Not. R. astr. Soc.*, *163*, p.285.
Huebner, W.F., Merts, A.L., Magee, N.H., Jr. and Argo, M.F., 1977. *Astrophysical Opacity Library*, Los Alamos Scientific Laboratory report LA-6760-M.
Krautschneider, M.J., 1977, *Astr, Astrophys. 57*, p.291.
Lattanzio, J.C., 1983, *Mon. Not. R. Astr. Soc.*, submitted.
Maeder, A., 1974, *Astr. Astrophys.*, *32*, p.177.
McClure, R.D., Newell, E.B. and Barnes, J.V., 1978, *Publs. Astr. Soc. Pacif.*, *90*, p.170.
McClure, R.D., Twarog, B.A. and Forrester, W.T., 1981, *Astrophys. J.*, *243*, p.841.
Patenaude, M., 1978, *Astr. Astrophys.*, *66*, p.225.
Prentice, A.J.R., 1976, *Astr. Astrophys.*, *50*, p.59.
Stahler, S.W., Shu, F.H. and Taam, R.E., 1980a, *Astrophys. J.*, *241*, p.637.
Stahler, S.W., Shu, F.H. and Taam, R.E., 1980b, *Astrophys. J.*, *242*, p.226.
Winkler, K.-H. A. and Newman, M.J., 1980, *Astrophys. J.*, *238*, p.311.

DISCUSSION

R. Cayrel: Have you computed the actual sedimentation of grains during the pre-main sequence stage?

Lattanzio: Those calculations have been performed by Krautschneider (1977), Flannery and Krook (1978) and others. My calculations begin where theirs finish.

Cox: Did you include the pre-main sequence burning of deuterium and He^3? They could lead to convection and mixing.

Lattanzio: No, but the mixing produced by burning deuterium and He^3 will be much less than that resulting from CN equilibration, which was included.

ROTATION OF B-TYPE STARS AS AN INDICATOR OF STAR FORMATION MECHANISMS

B.N.G. Guthrie
Royal Observatory, Edinburgh

ABSTRACT

Differences in the distributions of v sin i for various samples of B-type stars may reflect different mechanisms of star formation. The overall distribution of v sin i for late B-type stars in clusters at low galactic latitudes is bimodal, but normal class V field stars may have a Maxwellian distribution of v. The spin axes of stars in tightly bound clusters may be preferentially aligned perpendicular to the galactic plane. Early B-type stars in the youngest subgroups of associations may have a Maxwellian distribution of v, but there is a prominent excess of slow rotators among stars in older subgroups and field stars. This excess cannot be entirely accounted for by tidal or magnetic braking during main-sequence evolution.

1.   INTRODUCTION

In a previous study of late B-type stars (Guthrie 1982), it was found that cluster and field stars differ markedly with regard to their distribution of projected rotational velocities v sin i. The overall distribution of v sin i for stars in clusters at low galactic latitudes is bimodal with a scarcity of values between 80 and 160 km s^{-1}, and the distribution for the rich cluster NGC 2516 is sharply peaked. Thus there is probably a tendency for the spin axes of stars in tightly bound clusters to be aligned perpendicular to the galactic plane; such stars may have been formed with initial values around 230 km s^{-1} through collapse and fragmentation of gas clouds. The distribution of v sin i for field stars is not bimodal and does not depend on galactic latitude; the data are consistent with their spin axes being randomly orientated and their initial values of v being Maxwellian. Field stars may have been formed in loose clusters and associations through turbulence in gas clouds. For both cluster and field stars the proportion of slow rotators increases with age; this is partly due to tidal braking in binaries and rapid braking of magnetic stars through interaction with interstellar clouds.

A. Maeder and A. Renzini (eds.), Observational Tests of the Stellar Evolution Theory, 113–115.

Early B-type stars are now discussed in a similar way. B0 - B5
stars may be unique among 0-, B- and A-type stars in having a high
proportion of very slow rotators (Wolff, Fdwards, and Preston 1982,
henceforth WEP), but their short main-sequence lifetimes afford
little opportunity for braking. Early B-type stars occur in young
associations as well as in the field.

2.   B0 - B5 STARS IN ASSOCIATIONS

OB associations often contain a young, compact cluster or subgroup
embedded in nebulosity and older, more dispersed subgroups (Blaauw
1964). Such subgroups may be recognized in Ori OB1, Cep OB3, Lac OB1,
and Sco OB2, and the NGC 2264 and h and χ Persei clusters constitute
young subgroups of Mon OB1 and Per OB1 respectively. Values of v sin i
for B0 - B5 stars in these subgroups were found in the references
compiled by Uesugi and Fukuda (1982) and in the paper on NGC 2264 by
Vogel and Kuhi (1981). Table 1 gives, for each subgroup, the total
number N of B0 - B5 stars with known v sin i, the number N_{45} of these
stars which have v sin i $\leqslant$ 45 km s^{-1}, and the mean value of v sin i.
The overall ratio N_{45}/N is significantly higher for the older subgroups
(31/120 as compared with 7/90 for the young subgroups). The overall
distribution of v sin i for the 90 stars in the young subgroups is
consistent with a Maxwellian distribution of v and random orientations
of the spin axes. The distribution of v sin i for the stars in the
older subgroups is similar except for the prominent excess of values
$\leqslant$ 45 km s^{-1}.

Table 1. Rotation of B0 - B5 stars in associations

Subgroup	Earliest MK type	N	N_{45}	Mean v sin i (km s^{-1})
(a) Young subgroups				
Ori OB1 Nebula	06	10	1	157
Cep OB3 young	07	9	0	206
NGC 2264	07	7	0	146
Per OB2	08	11	1	133
Lac OB1b	09	10	2	141
Upper Scorpius	09.5	27	1	185
h and χ Persei	B0.5	16	2	162
(b) Older subgroups				
Ori OB1c Outer Sword	09	22	11	70
Ori OB1b Belt	09.5	16	4	180
Ori OB1a Northwest	B0.5	31	6	171
Cep OB3 old	B0	14	0	145
Lac OB1a	B1.5	13	3	149
Upper Centaurus	B1.5	24	7	127

3. B0 - B5 FIELD STARS

WEP measured v sin i for 70 per cent of the B0 - B5 stars in the Bright Star Catalogue having declinations $> -45^{\circ}$ and luminosity classes III - V. They published their values of v sin i for the 308 stars included in the uvbyβ photometric catalogue compiled by Philip et al. (1976). Excluding members and possible members of clusters and associations, there are 103 class V stars and 80 stars of classes III and IV. There is a significant bias against Be stars in WEP's published sample, probably due to the exclusion of stars with $\beta \leqslant 0.195(u - b) + 2.5$ from the photometric catalogue. This bias was allowed for by adding 13 class V Be stars and 5 Be stars of classes III and IV with suitable distributions of v sin i. The overall ratio N_{45}/N for the 116 class V field stars is 25/116 and the distribution of v sin i does not differ significantly from that for the 120 stars in the older subgroups of associations ($\chi^2 = 6.6$ for seven degrees of freedom). However, the 85 field stars of classes III and IV have a lower mean value of v sin i (91 km s^{-1}) and a higher N_{45}/N ratio (37/85).

4. DISCUSSION

The B0 - B5 sample studied by WEP, which included both field stars and stars in associations, was magnitude-limited. Thus low-luminosity class V stars were under-represented. By distinguishing stars according to their luminosity classes and association membership, it has been shown that the proportion of slow rotators among class V field stars is high, although not quite as high as for stars of classes III and IV. A high proportion of small values of v sin i is also found for members of the older subgroups of associations, but not for stars in the youngest subgroups. This poses a severe problem, since the older subgroups have ages of only $\sim 10^7$ yr. The stars in the older subgroups may have shed angular momentum rapidly when they left the nebulosity in which they were embedded; alternatively the stars in the older subgroups may have been formed in a different way from those in the youngest subgroups. Unlike late B-type stars in clusters and associations, early B-type stars in associations do not have bimodal distributions of v sin i.

REFERENCES

Blaauw, A.: 1964, Ann. Rev. Astron. Astrophys. 2, pp. 213 - 246.
Guthrie, B.N.G.: 1982, Mon. Not. R. astr. Soc. 198, pp. 795 - 810.
Philip, A.G.D., Miller, T.M., and Relyea, L.J.: 1976, Dudley Obs. Rep. 12.
Uesugi, A. and Fukuda, I.: 1982, "Revised Catalogue of Stellar Rotational Velocities", Kyoto University.
Vogel, S.N. and Kuhi, L.V.: 1981, Astrophys. J. 245, pp. 960 - 976.
Wolff, S.C., Edwards, S., and Preston, G.W.: 1982, Astrophys. J. 252, pp. 322 - 336.

A COMPOSITE COLOR-MAGNITUDE DIAGRAM FOR A NUMBER OF VERY POOR CLUSTER-
INGS IN THE MILKY WAY

L.O. Lodén
Astronomical Observatory
UPPSALA, Sweden

There are indications of the presence of numerous stellar clusterings in
the Milky Way which are so poor and unconspicuous that they become detected merely by accident. The true frequency of such objects is probably considerable as there is reason to believe that only a minor fraction has been detected.

The candidate objects in the present investigation have been revealed by
couples of apparently identical spectra appearing with so small angular
separation on objective-prism plates that they form a conspicuous configuration at visual inspection. The observed frequency of such coinci-
dences significantly exceeds the expected random frequency. In rather
few cases it has been impossible to detect additional clustering members but the few solitary couples found may be interpreted as the ultimate remnant of a dynamically disintegrating cluster, formed by its most
massive stars.

Tentative V vs. B-V diagrams have now been drawn for a number of candidate objects in the Carina-Crux-Centaurus region. Each object consists
of the coincidence couple and a few stars (10-30) in its nearest surrounding, for good reasons suspected to be fellow clustering members.
Subsequently, all diagrams have been superimposed in order to form some
kind of a blurred "average loose clustering diagram". In an idealized
case, the relevant points should form a distinct pattern in the diagram
whilst the irrelevant ones should tend to be more homogeneously distributed. A recent detail investigation of a selection of candidate objects, including studies of scanned slit spectra, gives an indication of
an extremely high frequency of multiplicity. As a practical consequence
of this fact one has to consider the occurrence of multiple stars as responsible for a considerable part of the scattering of the points in the
diagram.

In order to standardize the diagrams with respect to the position of the
sequences, an A0-age-zero point has been indicated in each diagram, basically by means of the objective-prism spectra of relevant stars. At the
superposition every such point has been placed at the same spot in the

117

A. Maeder and A. Renzini (eds.), Observational Tests of the Stellar Evolution Theory, 117–118.
© *1984 by the IAU.*

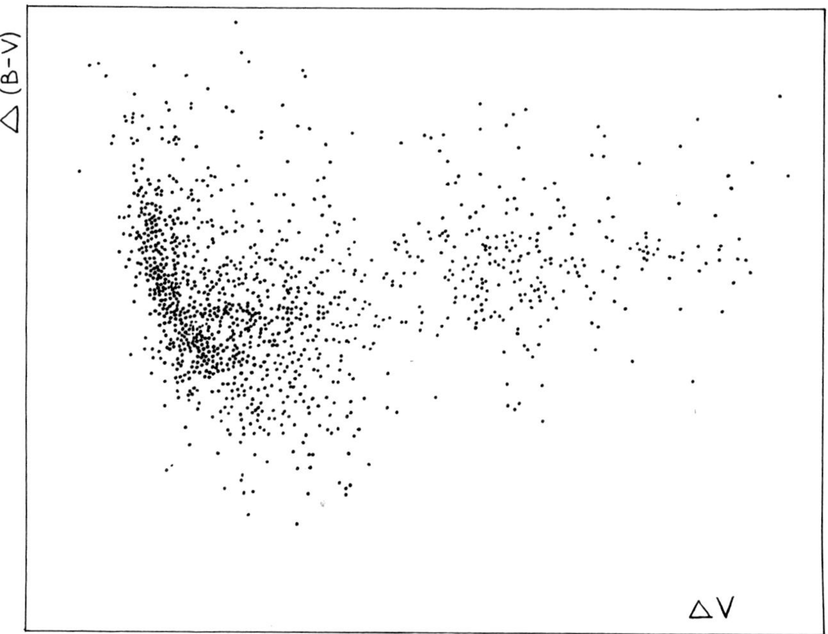

diagram. The final result is shown in the figure above which represents about 50 candidate objects and 1400 stars.

By means of the figure one may then draw some conclusions which are particularly interesting with respect to problems concerning stellar evolution, for instance:

1. A majority of the candidate objects are closely related to ordinary open clusters (a few of them might be considered as an apparent intermediate configuration between multiple system and open cluster – typical "micro-cluster").

2. As the true frequency of such objects must be very high, it is reasonable to believe that most stars (if not all) are born within clusters or associations of various types.

3. The "typical" turn-off point indicates a "typical" age of about 150 million years and a range from 50 to 200 million years.

4. There is a pronounced deficiency of stars on the main sequence below A2 in these clusterings (as in ordinary open clusters).

An interesting question is: To what extent is the difference in cluster appearance conditioned by a difference in the state of dynamical disintegration?

UBV PHOTOMETRY OF NGC 581

A. Osman and A. Khalil
Helwan Observatory, Helwan, Egypt

A. Aiad and M. Marie
Astronomy Department, Cairo University, Cairo, Egypt

Abstract

NGC 581 was studied using new plates taken by the Kottamia 74"
telescope and measured by the Helwan Iris photometer. A mean value of
3.13×10^7 years and 2334 pcs were found for the age and distance of the
cluster respectively. The cluster has an evolved giant branch.

INTRODUCTION

NGC 581 is a young, bright and round open cluster of Trumpler type
II3m in Cassiopia region. Previous photometric investigations (see references below) have contributed discrepant true distances ranging from
1200 to 3110 pcs. In an attempt to overcome such spread in distances
and to investigate the variable reddening across this young cluster, UBV
photometry has been carried out using plates collected during October
1979 with the 74" Kottamia reflector of plate scale 22".5/mm.

OBSERVATIONS AND MEASUREMENTS

Table 1 summarizes the observational material. The plates were
measured in Helwan using the Askania Iris photometer of Becker's type
and calibrated with the photoelectric sequence of Hoag et al (1961).

No of plates	Band	Emulsion+Filter	Exposure time (min)
4	U	103 O+UG2	30
6	B	103$_a$O+GG13	10
4	V	103 D+GG14	14

Table 1 : Observational material

A. Maeder and A. Renzini (eds.), Observational Tests of the Stellar Evolution Theory, 119–120.

The mean probable error in a single observation is found to be ±0.03 mag.
in both B and V and ±0.05 mag. in U. All magnitudes fainter than U =
14.60, B = 14.36 and V = 13.76 are considered of less weight since they
are obtained by extrapolating the standard sequence.

RESULTS

The two colour magnitude diagrams V-(B-V) and V-(U-B) are plotted
in Figs 1 and 2. Large dots, circles and small dots represent, respectively, probable physical members, additional members supported by their

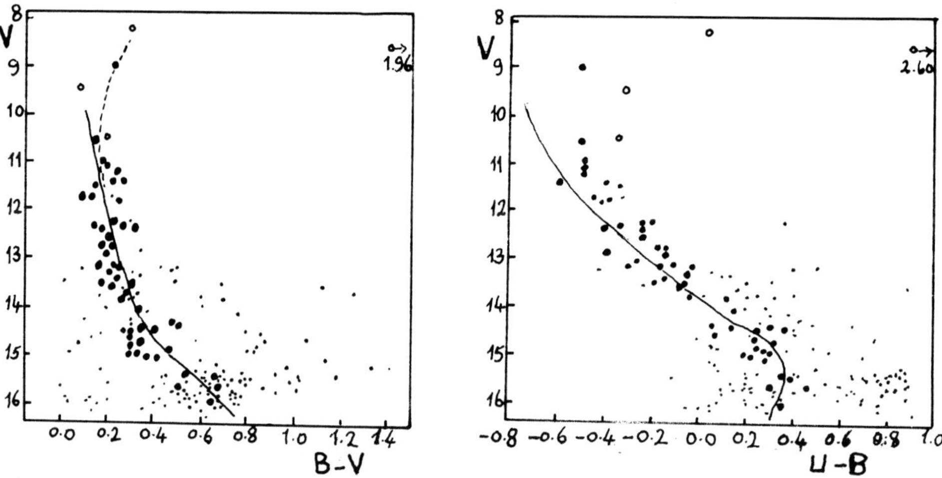

Fig. 1 : Colour-magnitude diagram   Fig. 2 : Colour-magnitude diagram
         V-(B-V)   V-(U-B)

proper motions (Oja, 1966) and field stars. Some stars at the upper end
of the main sequence have evolved slightly. By the fitting method of the
standard ZAMS of Schmidt-Kaler (1965) to the apparent C-M diagrams we
obtained: m-M = 13.12+0.12 (m.e) (m-M) = 11.84+0.12 E_{B-V} = 0.40
True distance = 2334+132 pcs   Age = 3.13x10^7 years   Apparent radius =
4!25 Linear diameter = 5.8 pcs.
Comparing the positions of the stars in both C-M diagrams 53 stars could
be separated as probable physical members. The membership of only 45 of
these stars has been confirmed by their proper motions, Oja (1966). For
full details of this work see for the same authors J. Astron. Soc. of
Egypt (1984).

REFERENCES

Barkatowa, K.A. (1950) Z. Astrophys. 27, 181
Johnson, H.L. (1961) LOB 5 No 8
Kruspan, E. (1959) Z. Astrophys. 48, 1
McCuskey, S.W. and Houk, N. (1964) AJ 69, 412
Moffat, A.F.J. (1972) Astron. Astrophys. Suppl. 7, 355
Oja, T. (1966) Ark. Astr. 4, 14
Purgathofer, A. (1961) Z. Astrophys. 52, 51
Sagar, R. and Joshi, U.C. (1978) Bull. of Astr. Soc. of India 6, 12
Steppe, H.L. (1974) Astron. Astrophys. Suppl. 15, 91.

THE LUMINOSITY FUNCTION OF THE SMC CLUSTERS NGC152 AND K3

M. Kontizas, E. Kontizas and N. Stewart
University of Athens, Observatory of Athens,
University of Edinburgh.

INTRODUCTION

In a previous investigation of the Luminosity Functions
(LF) it was found that some SMC clusters show significant
characteristics like the very large number of stars at
$M_V = 0.0$mag and a gap just below the Horizontal Branch (HB)
like feature of their colour-magnitude (c-m) diagram
(Kontizas and Kontizas, 1982).

NGC152 is a very unusual cluster and its c-m diagram
has been studied by several authors (Kontizas, 1976, 1980;
Stewart, 1980; Hodge, 1981). From a dynamical study of the
SMC clusters (Kontizas, 1983) NGC152 was difficult to be
classified as a disc or halo cluster being like the old open
clusters but having dynamical parameters of a true globular.
The next cluster K3 is one of the most populous, halo globular of the SMC often studied by Gascoigne (1966, 1980)
Walker (1972) and Stewart (1980).

OBSERVATIONS

The LF of NGC152 and K3 have been produced from V and
B plates taken with the ATT 3.8 m telescope in Australia.
Photographic photometry of all good stellar images were
carried out with an irisphotometer in circular areas around
each cluster and in two corresponding adjoining fields
(Stewart, 1980). Star counts in the same areas have provided the completion factors and the derived LFs of each
cluster field were normalised to equal areas. The bolometric correction curve used was that adopted for our galactic
globulars extrapolated to the bright red end by the measurements of the red stars studied by Mould and Aaronson (1980).

A. Maeder and A. Renzini (eds.), Observational Tests of the Stellar Evolution Theory, 121–122.
© *1984 by the IAU.*

DISCUSSION

From the LFs of the clusters NGC152 and K3 and their adjoining fields the following points can be outlined.

a)   The LFs of both fields are very similar and seem to be typical of the halo SMC stellar content (Brück and Marsoglu, 1978).

b)   The very red bright stars in both clusters are cluster members and not field stars.

c)   The LF of the cluster NGC152 shows two characteristic features (1) a clear gap below the giant branch clump and (2) the number of stars at the top of the main sequence ($M_V = 1.00\pm0.5$ mag) is smaller than the number of stars in the clump. This is an indication that this cluster is younger than K3 and seems to fit the isochrone $t = 5\times10^8$ years given by Faulkner and Cannon (1973) for the old open clusters.

d)   The cluster K3 exhibits a LF where the red giants and the HB clump are much more populated than the field giving evidence of being younger than the old halo SMC stars, but older than NGC152.

REFERENCES

Brück, M. T., and Marsoglu, A. : 1978, Astron. and Astroph.
        68, 193.
Faulkner, D. J., and Cannon, R.D. : 1973, Ap. J., 180, 435.
Gascoigne, S. C. B. : 1966, MNRAS, 134, 59.
Gascoigne, S. C. B. : 1980, IAU Symp. No 85 (ed. J. Hesser)
        pp. 305.
Hodge, P. W. : 1981, Ap. J., 247, 894.
Kontizas, M. : 1976, PhD Thesis, University of Edinburg.
Kontizas, M. : 1980, Astron. and Astrophys. Ser. 40, 151.
Kontizas, M. : 1983, Astron. and Astrophys. (submitted).
Kontizas, M. , Kontizas E. : 1982, Astron. and Astrophys.
        108, 314.
Mould, J., and Aaronson, M. : 1982, Ap. J., 263, 629.
Stewart, N. : 1980, PhD Thesis, University of Edinburgh.
Walker, M. :1972, MNRAS, 159, 379.

OBSERVATIONS OF LOW MASS STARS IN CLUSTERS: SOME CONSTRAINTS AND PUZZLES FOR STELLAR EVOLUTION THEORY

R.D. Cannon
Royal Observatory, Blackford Hill, Edinburgh EH9 3HJ.

1. INTRODUCTION

This review will attempt to do two things: (i) discuss some of the data which are available for testing the theory of evolution of low mass stars, and (ii) point out some problem areas where observations and theory do not seem to agree very well. This is of course too vast a field of research to be covered in one brief review, so I shall concentrate on one particular aspect, namely the study of star clusters and especially their colour-magnitude (CM) diagrams. Star clusters provide large samples of stars at the same distance and with the same age, and the CM diagram gives the easiest way of comparing theoretical predictions with observations, although crucial evidence is also provided by spectroscopic abundance analyses and studies of variable stars. Since this is primarily a review of observational data it is natural to divide it into two parts: (i) galactic globular clusters, and (ii) old and intermediate-age open clusters. Some additional evidence comes from Local Group galaxies, especially now that CM diagrams which reach the old main sequence are becoming available. For each class of cluster I shall consider successive stages of evolution from the main sequence, up the hydrogen-burning red giant branch, and through the helium-burning giant phase.

2. GLOBULAR CLUSTERS

2.1 The overall picture

The conventional wisdom is that globular clusters are massive, independent, gravitationally bound systems containing up to a million stars, that they are of great antiquity with ages of around 15 billion years, and that they were born from chemically homogeneous material over a timescale very short compared with their ages. It is usually assumed that the original (generally low metal abundance) chemical composition can be determined spectroscopically from the surface layers of luminous, highly evolved, stars.

123

A. Maeder and A. Renzini (eds.), Observational Tests of the Stellar Evolution Theory, 123–138.
© *1984 by the IAU.*

A cluster CM diagram thus provides an instantaneous picture of the points of evolution reached in a given time by stars of different masses, and this can be compared directly with theoretical isochrones in the equivalent diagram of luminosity versus temperature, provided that those parameters can be transformed to or from observed magnitudes and colours.

Recent work has raised uncertainties concerning the validity of nearly all of the fundamental assumptions about globular clusters, but it seems that those assumptions are probably still approximately true for most of the clusters most of the time.

2.2   The main sequence

The position of the zero age main sequence (ZAMS) is expected to be a function of chemical composition. Unfortunately, unless some other feature is used to fix a cluster's distance, the absolute magnitude of a globular cluster ZAMS cannot be determined, while one of the most important parameters is helium abundance which is not directly observable. An independent measure of distance can be found for some clusters from the properties of RR Lyrae variables (e.g. Sandage, 1982), but it can only be applied to clusters with enough variables and its reliability depends on the adequacy of pulsation theory as well as on having very precise observational data. The usual procedure is to adopt a chemical composition and to use unevolved main sequence stars to determine cluster distance, either via an empirical calibration of the ZAMS-abundance relation (Sandage, 1970;   Carney, 1980) using nearby subdwarfs with trigonometrical parallaxes, or by appealing directly to theoretical predictions (e.g. Simoda & Iben, 1968; VandenBerg, 1983).

The main sequence turn-off is used to determine the second major parameter, a cluster's age, and hence also the masses of stars evolving up the giant branch. Since there is no other way of determining these two parameters, there is again little scope for checking the predictions of stellar evolution theory. Furthermore, even the mass is not well-determined since for a given turn-off position the mass is once more a strong function of the helium abundance. Luckily the age determination itself is not much affected by the uncertainty in helium abundance.

Some consistency checks on the theory are nevertheless possible. The overall shape of the theoretical isochrone for the presumed chemical composition of the cluster should match the observed shape. Only recently have the many theoretical and observational parameters begun to be well enough known to make this a useful test, and even so VandenBerg (1983) effectively used the turn-off shape to fix the best value of yet another uncertain parameter, the convective mixing length in his models. Unfortunately very few clusters have observationally well-determined turn-offs, since these require accurate photometry for large samples of faint (usually $V > 18$) stars (cf Cannon, 1981).

There are also two external 'consistency checks' or constraints on the ages found for globular clusters, both depending on theories unconnected with stellar evolution. Some dynamical models for the formation of our Galaxy (Eggen, Lynden-Bell & Sandage, 1962) predict that the very extended spherical galactic halo must have formed over a relatively short period of time, so that the globular clusters should all be virtually the same age. On the other hand, Rood & Iben (1968) showed that the spread in formation times of the globular clusters could well exceed 10% of their present age, while Searle & Zinn (1978) argued for a possibly much larger spread. The most recent determinations (Sandage, 1982; VandenBerg, 1983) do make all clusters coeval, with an age of around 17.10^9y, although an alternative metallicity-dependent age spread has also been proposed (Carney, 1980; Demarque 1980).

The second external constraint follows from the very great age found for globular clusters: an age of 17.10^9y is close to the maximum permissable within the framework of Big Bang cosmology if Hubble's constant is 55 km s^{-1} Mpc^{-1} (Sandage, 1982), and is inconsistent with larger values of H_o. If H_o is as high as 100 km s^{-1} Mpc^{-1}, it follows that the age of the Universe is no more than 10^{10}y and that either stellar evolution theory or Big Bang cosmology is badly wrong. Of course no-one would claim that stellar evolution theory combined with globular cluster observations will yield ages more accurate than say 10% (see Cannon, 1982 for a review), but an error of 40% in the ages implies either an unfortunate combination of several errors all acting in the same direction, or a fundamental flaw.

2.3 Giant branch evolution

Giant branch evolution has been regarded as basically understood since the work of Hoyle & Schwarzschild (1955), although there is still an entertaining debate about the underlying physical mechanisms (e.g. Eggleton & Faulkner, 1981; Weiss, 1983). Unfortunately, once again a direct comparison between theory and observations is rather difficult. In this case the problem is that the temperature of a red giant model is determined by the extent of the convective envelope, which is a strong function of the rather arbitrary 'mixing length' parameter. Thus although there is a clear dependence of giant branch colour (as measured by the (B-V)o,g parameter of Sandage & Smith, 1966) on overall metallicity, any theoretical interpretation depends on some assumption relating metallicity, opacity and convection.

Fortunately the luminosities, and hence the rates of evolution, of red giants do not change much with mixing length. Thus a direct comparison is possible between theoretical rates of evolution and observed numbers of stars; this has been used for example as one method for determining the helium abundance of globular clusters (Simoda & Kimura, 1968; Faulkner, 1972).

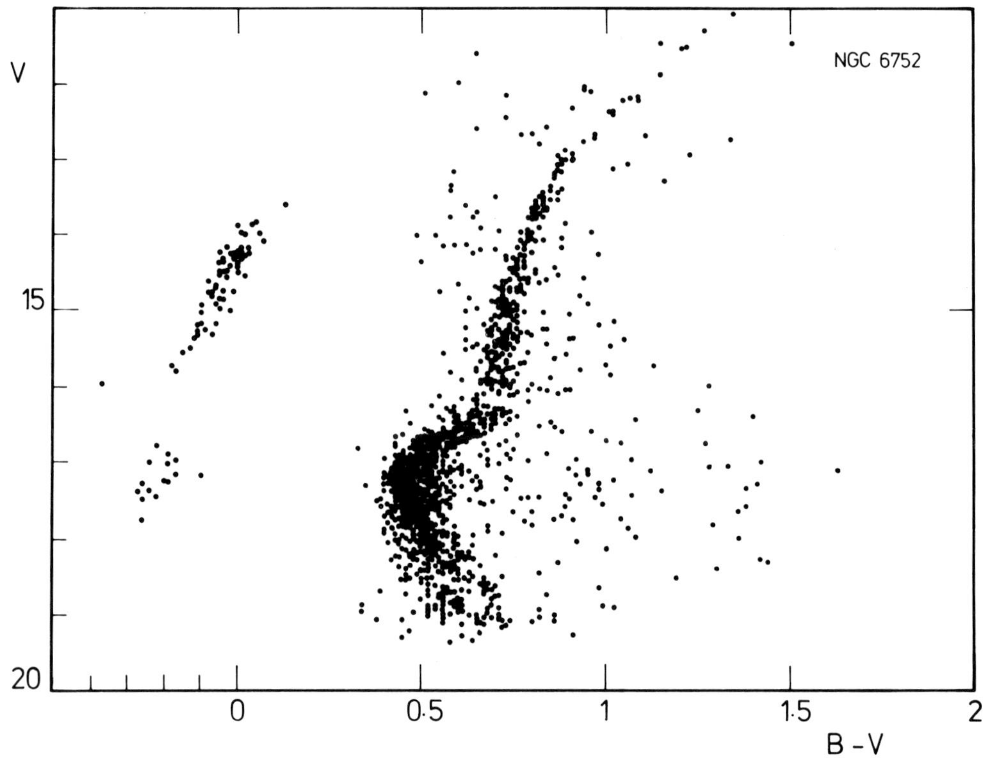

Fig. 1   The CM diagram of NGC 6752

There is one clear conflict between theory and observation, right at the base of the giant branch.   In 1972 Cannon & Lee found a small gap, about 0.1 mag wide, at V = 16.2 in NGC 6752.   Since the CM diagram of this cluster illustrates some other points discussed subsequently, and since the data are regrettably still unpublished, it is shown here as Fig. 1.   Careful checking, involving remeasurement and re-calibration of all stars within one magnitude of the gap, showed that it is certainly a real feature.   A similar gap may be present in some other clusters, but in no case are the data yet completely convincing;   equally there are no clusters for which one can definitely say no such gap exists, since accurate photometry is needed for large samples of faint stars.   There is no theoretical prediction of a corresponding fast phase of evolution which would give rise to such a gap, although possible clumps and gaps are predicted at higher luminosities when the hydrogen-burning shell encounters composition discontinuities left behind by previous convective zones (Sweigart & Gross, 1978).   Unfortunately the predicted perturbations in the number density of stars seen are generally of order only ten percent, and such fluctuations would be swamped by stochastic variations in the relatively small numbers of red giants observed in most clusters.

2.4 The horizontal branch

It is generally accepted that globular cluster stars move rapidly on to the horizontal branch (HB) after the 'helium flash' at the tip of the giant branch, following the classic work of Faulkner (1966) and the subsequent detailed core-plus-shell energy source evolutionary models of Sweigart & Gross (1976). There is however a major hiatus in the theory, in that the details of the very rapid onset of helium burning and lifting of degeneracy have not been fully worked out, and there may be considerable amounts of core-envelope mixing and mass loss during the helium flash. Recent hydrodynamical calculations by Deupree & Cole (1983), give considerably different results from the earlier 'static' predictions. Thus the two principal parameters determining the initial location of a star on the HB, the core mass and the envelope mass, are both uncertain, while the HB lifetimes may be wrong by as much as 50 percent.

The original Faulkner (1966) and Faulkner & Iben (1966) theory explained the existence of the HB and gave the correct evolutionary lifetimes (as required to explain the numbers of HB stars seen) of order 10^8y, and was also consistent with the general correlation between metallicity and morphology (i.e. most metal poor clusters have blue HBs whereas relatively metal rich clusters usually have red HBs). However, this latter consistency may have been fortuitous since there was no explanation for the now well-established anomalous 'second-parameter' clusters such as NGC 7006 (Sandage & Wildey, 1967) and NGC 288 (Cannon, 1974), while Iben & Rood (1970) showed that there must be a spread in mass along the HB. Rood (1973) and Renzini (1977) showed that quite small variations in the essentially arbitrary mass loss parameter could lead to great variations in HB structure. The debate on the 'second parameter' is continuing with still no single simple explanation, or rather, an embarrassing abundance of plausible explanations, able to accommodate all of the data (e.g. Freeman & Norris, 1981), and perhaps one should say that it is all too easy to explain a wide range of HB morphology by rather minor variations in any of several parameters (Rood & Seitzer, 1981).

One particular aspect of this problem is the great variation in the morphology of the HB in different clusters. NGC 1851 (Stetson, 1981) has separate bunches of blue and red HB stars, with relatively few stars in between, and the same is true of NGC 2808 (Harris, 1974, 1978), and M4 (Lee, 1977a). The NGC 6752 HB (Fig. 1) is even more clearly split into two groups, but this time both are at the high temperature blue end of the branch, while M15 (Buonanno, Corsi & Fusi Pecci, 1981a) has a break at intermediate colours. This type of structure can be contrasted with that in M3 (Sandage, 1953) or M5

(Buonanno, Corsi & Fusi Pecci 1981b) which have long almost uniformly populated HBs or with IC 4499 (Fourcade, Laborde & Arias, 1974); Cannon & Lloyd, in preparation) where the HB is so concentrated that almost all of the stars lie in the RR Lyrae instability strip. The variations in HB structure are so great that each cluster is unique and 'afficionados' can readily identify their favourite clusters from the appearance of the HB alone. This great variation is perhaps not surprising in view of the sensitivity of HB models to small changes in several parameters (Rood & Seitzer, 1981), and it may well be that the striking differences have no great physical significance. What is more surprising is that the horizontal branch should be such a prominent feature of all globular clusters, since the theoretical models show that in order to reach the higher temperature end of the HB it is necessary for a star to have a very small hydrogen-rich envelope. If there is no mass loss on the giant branch, then blue horizontal branches can only appear in clusters having ages in a narrow interval, which would mean that the present is a favoured epoch. It is perhaps philosophically more acceptable, as well as being consistent with the predicted difference in masses between subgiants and HB stars, if there is a mass loss mechanism such that nearly but not quite all of the hydrogen rich envelope of globular cluster red giant stars is lost prior to or during the helium flash.

Direct confirmation of the relatively low masses for HB stars in globular clusters has come recently from the discovery of RR Lyrae variables with beat periods in M15 (Sandage, Katem & Sandage, 1981). Cox, Hodson & Clancy (1983) have used these to deduce masses of around 0.6 $M_\odot$, whereas VandenBerg (1983) and others find that their progenitor red giant stars leave the main sequence with masses around 0.8 $M_\odot$ if the helium abundance is near Y = 0.23. However more direct evidence for mass loss from pre-HB stars comes from the older open clusters, discussed below.

Returning to the unusual extremely blue HB of NGC 6752, recent optical and IUE ultraviolet observations by Caloi et al. (1983) show that these stars are indeed the same as the field subdwarf OB stars of Greenstein & Sargent (1974), which the latter had postulated must be extreme HB stars of a type not up till then known in any globular cluster. Such very hot BHB stars do exist in some other clusters, including M13 (Simoda & Tanikawa, 1972), but are certainly not present in comparable numbers in some others including NGC 288 (Buonanno, Corsi, Fusi Pecci & Alcaino, 1983)) and ω Cen (Cannon & Stewart, 1981). In most clusters the samples of faint stars which have been measured are too small to say whether or not extreme BHB stars are present. The very clear gap between the two groups of BHB stars in NGC 6752, which occurs at the same temperature as a gap in the field star HB population noted by Newell & Graham (1976), can perhaps be understood in terms of two different types of evolution. The hotter group being almost on the helium main sequence may evolve towards higher luminosities whereas the cooler group will follow the usual HB evolution back towards the asymptotic red giant branch (AGB), as postulated by Heber et al. (1983) for a sample of field sdO and sdB stars. On the other hand, Norris et al. (1981) have speculated that the bimodal HB of NGC 6752 may be related to the bimodal distribution of CN band strengths seen on the giant branch; the same may be true for M4 which also has both a bimodal (but cooler) HB and a bimodal CN band strength distribution (Norris, 1981), although NGC 2808 does not seem to fit into this pattern (Norris & Smith, 1983).

2.5 The asymptotic giant branch

A comprehensive review of the AGB and later stages of evolution has been given very recently from a theoretical point of view by Iben & Renzini (1983), so only a few observational topics are discussed here. In the CM diagram, the AGB is more clearly separated from the first (hydrogen-burning) giant branch in some clusters than in others, but it always becomes difficult to separate the two as the luminosity increases. Many clusters show apparent clumps and gaps on both giant branches, but it is very difficult to be sure of the reality of these features since the samples of stars are usually small (even when every star in a cluster is measured: most globulars simply do not contain a large enough number of red giants) (Lee, 1977a,b,c,d). Detailed modelling has been carried out by Gingold (1974,1976) but only for one metallicity. Theory indicates that the luminosity at the tip of the AGB, without mass loss, would be much higher than that of the first giant branch. This result has been used e.g. by Ciardullo & Demarque (1978) as another argument that very significant mass loss must occur on the RGB and/or AGB. It may well be that many of the brightest red giants in clusters (which are often those which have spectroscopically determined abundances) are highly evolved AGB stars which have under-gone both mass loss and mixing. Such stars may include the very rare CH stars found in ω Cen (Dickens, 1972); it certainly seems that the brighter carbon stars which are relatively common in dwarf spheroidal galaxies are AGB stars. If the AGB and normal giant branch could be reliably and completely separated, their luminosity functions would put constraints on the helium abundance, but once again the statistical fluctuations mean that this method is not very successful in practice (Green, 1980); the best results seem to come from comparing the total number of giants above the HB with the number of HB stars (the 'R-method', reviewed recently by Buzzoni et al., 1983).

2.6 Late stages of evolution

Planetary nebulae are too rare in globular clusters to contribute significantly to an understanding of that phase of evolution (M15 contains the only example known (Pease, 1928; Peimbert, 1973)), but there are considerable numbers of 'UV bright' (Zinn, Newell & Gibson, 1972) or 'supra-horizontal branch' stars which are generally taken to be in rapid advanced stages of evolution. There are also the occasional sdO stars such as one each in NGC 6752 (the star with V = 16.00, B-V = -0.36 in Fig 1; Caloi et al., 1983) and NGC 6397 (Searle & Rodgers, 1966); these may either be evolving from the AGB

to the white dwarf region, or directly from the BHB. The difficulty
is that no cluster contains more than one or two such stars, so that
it is difficult to plot out evolutionary tracks or calculate lifetimes. However their presence in globular clusters does give a good
determination of their luminosities, ages and original masses, so that
it should be possible to fit them into a stellar evolution framework.

The final white dwarf stage should be very highly populated, and
any determination of the numbers of white dwarfs would be immensely
valuable for checking evolution theory, for giving estimates of the
original luminosity function for higher mass stars in globulars, and
for providing an important missing parameter in dynamical models of
clusters since the total mass involved may be very considerable.
Direct searches for white dwarfs have been carried out in NGC 6752
(Richer, 1978) and in M4 (Richer et al., 1981), two of the nearest
globular clusters. These did find a population of faint blue stars
but they are considerably brighter ($M_v \sim 10$) than the expected bright
end of the white dwarf cooling curve and so it is not yet certain
whether these really are white dwarf cluster members. An intriguing
new possibility is that the compact remnants of old massive cluster
stars may now be detectable as X-ray sources. In addition to the
well-known bright X-ray sources in globular clusters, a new population
of weak X-ray sources has recently been discovered in several clusters
by Hertz & Grindlay (1983). These are interpreted as due to accretion
on to white dwarfs which have formed binary systems with late-type
dwarfs in the clusters. However, a proper study of white dwarfs will
have to await the advent of the Space Telescope; the exciting
possibilities have been outlined by Castellani (1979). Since in many
respects the theory of cooling white dwarfs is simpler and probably
better understood than even that of ordinary main sequence stars, the
white dwarfs might eventually give the best determination of cluster
distances and reddenings.

2.7 Other problems

Two major areas (at least!) have been omitted from this review.
One is the problem of abundance variations <u>within</u> globular clusters.
The first indications that there might be significant variations in
the abundances of stars within a cluster came from the demonstration
of an exceptionally broad giant branch in ω Cen (Cannon & Stobie,
1973). Following this up has become a major industry, well-reviewed
recently by Kraft (1979) and by Freeman & Norris (1981). Part of the
debate centres on whether such surface abundance variations are
primordial, due to self-enrichment by mixing, or due to some sort of
accretion process; current evidence is conflicting and it may well be
that several mechanisms are operating. It is worth noting that ω Cen
is certainly unusual in this respect, perhaps because it is the
largest globular cluster in our Galaxy. Most other clusters have much
better-defined giant branches, although in many there are significant
variations in some spectral features such as CN band strengths.

The second problem concerns 'blue straggler' stars. These were first found in M3 by Sandage (1953), and although some have probably been found in other clusters, M3 remains much the clearest example. A major puzzle here is that similar stars certainly do not occur in comparable numbers in several other clusters (including NGC 6752, see Fig. 1). As with the extended BHB, it is necessary to remember that the total number of clusters where enough faint stars have been measured to prove the existence or otherwise of blue stragglers remains very small, certainly fewer than ten clusters. However, blue straggler stars do occur in significant numbers in most intermediate-age and old open clusters and so will be further discussed in the next section.

3.   OLD AND INTERMEDIATE-AGE OPEN CLUSTERS

3.1 General considerations

It is convenient to divide the older open clusters into two main classes: 'old' clusters with ages $\geq 3.10^9$y, and 'intermediate-age' clusters with ages of around $0.5 - 2.10^9$y. The first group, typified by clusters such as M67 (Racine, 1971) and NGC 188 (Eggen & Sandage, 1969), have CM diagrams somewhat similar to those of globular clusters, with a continuous subgiant and giant branch consisting of stars with masses $1.0 - 1.25$ M$_\odot$. The 'intermediate-age' clusters, first considered as a group by Arp (1962), have rather different CM diagrams with a Hertzsprung gap and a strong clump of red giants with presumed masses in the range $1.5 - 2.25$ M$_\odot$; typical examples are NGC 7789 (Burbidge & Sandage, 1958) and NGC 2477 (Hartwick, Hesser & McClure, 1972). Of course this is a rather artificial division, with clusters now known which cover a continuous range of ages and CM diagram types, and Patenaude (1978) has fitted theoretical isochrones to a number of clusters with well-determined CM diagrams.

There are several ways in which the study of the CM diagrams of old open clusters is harder than that of globular clusters: open clusters contain far fewer stars (often by several orders of magnitude) and are generally at lower galactic latitudes, so that it is difficult to get well defined sequences in the CM diagram, while contamination by field stars and differential reddening often cause problems. Thus there are areas where there may well be problems in reconciling observations with evolution theory, such as the shape of the subgiant branch, the location of the base of the giant branch and the width of the upper giant branch, but the observational data are not yet adequate for a definitive conflict with or confirmation of the theory. In this review I shall concentrate on just three problems.

3.2 The main sequence gap

A clear main sequence gap was first seen in M67 (Eggen & Sandage, 1964), and this remains the best-observed example. Qualitatively, the feature is readily explained as being due to the rapid gravitational contraction of stars following the exhaustion of hydrogen fuel in a convective core, prior to the ignition of a hydrogen-burning shell. However, early standard evolutionary models did not give a good quantitative fit: the observed gap occurs too near the top of the main sequence (at least in M67), while the theory predicts a significant bluewards shift as stars cross the gap whereas no colour shift is observed. The problem has been well reviewed by Maeder (1974). Several possible resolutions have been proposed: Prather & Demarque (1974) invoked a simple form of convective overshooting to give larger convective cores and hence brighter predicted gaps, while Maeder (1975, 1976) gave more extensive and elaborate theoretical models. However, Morgan & Eggleton (1978) showed that a reasonable fit to the observed M67 CM diagram could be obtained using conventional models when the effects of unresolved binary stars and observational errors were taken into account.

This uncertainty can probably never be resolved with reference to M67 alone; the greatest need is for better quality data (i.e. of higher accuracy as well as with less field star contamination) for a larger sample of clusters. There are unfortunately few clusters which are sufficiently populous ever to give a well-defined observational gap, but it is very difficult to avoid the conclusion that there should already be <u>some</u> evidence for a marked jump towards higher temperatures (i.e. with shifts of ~ 0.1 mag. in B-V) on the upper main sequences of some clusters, if conventional evolutionary theory is correct.

3.3 The red giant clump

The strongest feature in all intermediate-age cluster CM diagrams is a tight clump of red giants near M_V = +1, B-V = 1.0; a similar but less prominent feature appears in old open clusters as well. These 'clump giants' were identified by Cannon (1970) as the Population I analogues of horizontal branch stars in globular clusters, i.e. stars which had passed through the helium flash and were powered by both a helium-burning core and a hydrogen-burning shell. This interpretation was confirmed with theoretical models by Faulkner & Cannon (1973), who showed that both the location and the lifetime of the clump phase were correctly predicted. Furthermore, the constancy of the clump luminosity for a wide range of cluster ages (and hence of stellar masses) is direct evidence that stars with masses less than 2.25 $M_\odot$ do indeed have electron degenerate cores and hence undergo a helium flash when their cores reach some critical mass which is almost independent of the total mass.

Although the observed clump in most clusters shows no structure, the theoretical models show that it should in fact be almost a vertical sequence in the CM diagram, running parallel to the Hayashi fully-convective boundary. This has not yet been clearly verified observationally; there are very few clusters which contain enough red giants, and those which are rich enough either have inaccurate photometry or suffer from differential reddening (e.g. Cannon, 1971), or both. There is one related puzzle which has not yet been resolved either observationally or theoretically: most relatively metal rich globular clusters such as 47 Tuc have a red horizontal branch which is short but nevertheless covers a range in temperature, i.e. is truly horizontal, while intermediate-age open clusters seem to have a red giant clump which is more vertical than horizontal. At some stage there must be a transition between these two types, and there are indications that the oldest open clusters such as NGC 188 may have a range in B-V colour among clump-type giants.

In old clusters with populous red giant clumps, the clump stars are generally spatially less concentrated towards the cluster centre than stars on the upper main sequence. This provides more evidence that significant mass loss must occur during the red giant phase of evolution, after which dynamical relaxation leads to the now lower-mass clump giants becoming more spread out (Hawarden, 1975).

3.4 Blue stragglers

Most old and intermediate-age clusters contain 'blue straggler' stars lying well above the main sequence turn-off. In many cases the cluster membership of these stars is unambiguously established by both their proper motions and their spatial distribution. Although most clusters contain only a few blue stragglers, they can be quite numerous. Most striking is the case of NGC 188, which was originally thought to be a relatively young cluster (Barkhatova, 1956) on the basis of a CM diagram for only the brighter stars, until Sandage (1962) showed that there was a well-defined turn-off at much fainter magnitudes. Indeed, the blue stragglers are as numerous as the red giants in NGC 188, an observation which may have far-reaching implications for population synthesis studies of galaxies.

In a still regrettably unpublished thesis study (Cannon, 1968), I used proper motions to obtain 'pure' samples of blue stragglers in several clusters, and showed that almost all of these stars do lie within the hydrogen-burning main sequence band, and are less than two magnitudes above the turn-off. Furthermore, at least in the oldest clusters the blue stragglers are very strongly concentrated towards the cluster centres, and hence are presumably relatively massive. All of these data indicate that blue stragglers are hydrogen-burning stars up to twice as massive as current turn-off stars and giants. The two leading theories to explain their continued existence on the main sequence are that either they are binary systems which have undergone mass exchange (McCrea, 1964), or they have for some reason managed to mix themselves (Wheeler 1979; Saio & Wheeler, 1980).

There are difficulties with both of these explanations. If mixing is the right answer, there has to be some rather arbitrary reason why a minority of stars become mixed while the majority do not.

On the other hand, a number of attempts to detect binary systems via radial velocity variations have produced conflicting results, mostly negative (e.g. Stryker & Hrivnak, 1983).

4. SUMMARY

Globular clusters provide a number of very significant constraints on stellar evolution theory: for example, their ages presumably have to be less than that of the Universe as a whole; mass loss must occur to explain the horizontal branch; some specific abundance variations are predicted as a consequence of internal mixing. There are also a number of intriguing puzzles: what causes the gap observed at the base of the giant branch in NGC 6752? Which abundance variations are due to mixing and which are primordial or due to accretion? What (and how many?) parameters control horizontal branch type, why are some HBs bimodal, and indeed why do we see blue HBs at all? How common are binary systems in globular clusters, and are these needed to explain either blue stragglers or X-ray sources?

Open clusters place fewer constraints and set fewer puzzles, mainly because the data are less convincing. However, a strong constraint on internal convection is set by the existence or absence of gaps on the upper main sequence, while the red giant clump gives evidence on the size of the core at the time of the helium flash. Some outstanding puzzles are: why is no colour shift observed across main sequence gaps? Why is the red giant branch not better defined, both below and above the clump? What _are_ blue stragglers?

Space did not permit a discussion of the Magellanic Clouds and the dwarf spheroidal satellite galaxies, but exciting new prospects are being opened up there, especially through the use of two-dimensional TV and CCD detectors to obtain superb very deep CM diagrams. Earlier expectations that the stellar populations of the dwarf spheroidals must be similar to galactic globular clusters have been shown to be quite wrong, first through the discovery of ubiquitous carbon stars (Aaronson, Olszewski & Hodge, 1983), and most recently through the direct demonstration that the Carina dwarf has a dominant intermediate-age population (Mould & Aaronson, 1983). Nevertheless, dwarf spheroidal galaxies do contain RR Lyrae variable stars and one (Ursa Minor: van Agt, 1967) has a globular-like blue horizontal branch. Either such features can occur in systems with higher-mass stars, or the dwarfs contain a mixture of stellar populations. Thus these systems open up new domains in the age (or stellar mass) versus metallicity domain, where stellar evolution theory can be checked. No doubt these very nearby galaxies will soon raise as many constraints and puzzles as have star clusters in the past.

REFERENCES

Aaronson, M., Olszewski, E.W. & Hodge, P.W.: 1983, Astrophys.J. 267, p.271.
Arp, H.C.: 1962, ApJ 136, p.66.
Barkhatova, K.A.: 1956, Ast.Zhurnal 33, p.850.
Buonanno, R., Corsi, C.E. & Fusi Pecci, F.: 1981a, In "Astrophysical Parameters for Globular Clusters", IAU Coll. 68, eds. A.G. Davis Philip & D.S. Hayes, p.551.
Buonanno, R., Corsi, C.E. & Fusi Pecci, F.: 1981b, Mon.Not.R.astr.Soc. 196, p.435.
Buonanno, R., Corsi, C.E., Fusi Pecci, F. & Alcaino, G.: 1983, Preprint.
Burbidge, E.M. & Sandage, A.: 1958, Astrophys.J. 128, p.174.
Buzzoni, A., Fusi Pecci, F., Buonanno, R. & Corsi, C.E.: 1983, Astron.Astrophys. 128, p.94.
Caloi, V., Cannon, R.D., Castellani, V., Danziger, J., Gilmozzi, R. & Hill, P.W.: 1983, submitted to Mon.Not.R.astr.Soc.
Cannon, R.D.: 1968, PhD Thesis, University of Cambridge.
Cannon, R.D.: 1970, Mon.Not.R.astr.Soc. 150, p.111.
Cannon, R.D.: 1971, Proc.Astr.Soc.Australia 2, p.25.
Cannon, R.D.: 1974, Mon.Not.R.astr.Soc. 167, p.551.
Cannon, R.D.: 1981, In "Astrophysical Parameters for Globular Clusters", IAU Coll. 68, eds. A.G. Davis Philip & D.S. Hayes, p.501.
Cannon, R.D.: 1982, IAU Highlights of Astronomy, ed. R.M. West, 6, p.109.
Cannon, R.D. & Stewart, N.J.: 1981, Mon.Not.R.astr.Soc. 195, p.15.
Cannon, R.D. & Stobie, R.S.: 1973, Mon.Not.R.astr.Soc. 162, p.207.
Carney, B.W.: 1980, Astrophys.J.Supp. 42, p.481.
Castellani, V.: 1979, In "ESA/ESO Workshop on Astronomical Uses of Space Telescope", Geneva, ed F. Macchetto, F. Pacini & M. Tarenghi, p.157.
Ciardullo, R.B. & Demarque, P.: 1978, In "The HR Diagram", IAU Symp. 80, eds. A.G. Davis Philip & D.S. Hayes, p.345.
Cox, A.N., Hodson, S.W. & Clancy, S.P.: 1983, Astrophys.J. 266, 94.
Demarque, P.: 1980, In "Star Clusters", IAU Symp 85, ed. J.E. Hesser, p.281.
Deupree, R.G. & Cole, P.W.: 1983, Astrophys.J. 269, p.676.
Dickens, R.J.: 1972, Mon.Not.R.astr.Soc. 159, p.7P.
Eggen, O.J., Lynden-Bell, D. & Sandage, A.: 1962, Astrophys.J. 136, p.748.
Eggen, O.J. & Sandage, A.: 1964, Astrophys.J. 140, p.130.
Eggen, O.J. & Sandage, A.: 1969, Astrophys.J. 158, 669.
Eggleton, P.P. & Faulkner, J.: 1981, In "Physical Processes in Red Giants", eds. I. Iben & A. Renzini, pub. D. Reidel, Dordrecht, p.179.
Faulkner, J.: 1966, Astrophys.J. 144, p.978.
Faulkner, J.: 1972, Nature Phys.Sci. 235, p.27.
Faulkner, J. & Iben, I.: 1966, Astrophys.J. 144, p.995.

Faulkner, D.J. & Cannon, R.D.: 1973, Astrophys.J. 180, p.435.
Fourcade,   C.R.,   Laborde,   J.R.   &   Arias,   J.C.:   1974,
     Astron.Astrophys.Supp. 18, p.3.
Freeman, K.C. & Norris, J.: 1981, Ann.Rev. Astron.Astrophys. 19,
     p.319.
Gingold, R.A.: 1974, Astrophys.J. 193, p.177.
Gingold, R.A.: 1976, Astrophys.J. 204, p.116.
Green, E.M.: 1980, In "Star Clusters", IAU Symp 85, ed. J.E. Hesser,
     p.441.
Greenstein, J.L. & Sargent, A.I.: 1974, Astrophys.J. Supp 28, p.157.
Harris, W.E.: 1974, Astrophys.J.Letts. 192, L161.
Harris, W.E.: 1978, Publ. Astron.Soc.Pacific 90, p.45.
Hartwick, F.D.A., Hesser, J.E. & McClure, R.D.: 1972, Astrophys.J.
     174, p.557.
Hawarden, T.G.: 1975, Mon.Not.R.astr.Soc., 173, p.223.
Heber,   J.,   Hunger,   K.,   Jonas,   G. & Kudritzki, R.P.:   1983,
     Astron.Astrophys. in press.
Hertz, P. & Grindlay, J.E.: 1983, Astrophys.J. 267, L.83.
Hoyle, F. & Schwarzschild, M.: 1955, Astrophys.J. Supp. 2, 1.
Iben, I. & Renzini, A.: 1983, Ann.Rev. Astron.Astrophys. 21, p.271.
Iben, I. & Rood, R.T.: 1970, Astrophys.J. 161, p.587.
Kraft, R.P.: 1979, Ann.Rev. Astron.Astrophys. 17, p.309.
Lee, S-W.: 1977a, Astron.Astrophys.Supp. 27, p.367.
Lee, S-W.: 1977b, Astron.Astrophys.Supp. 27, p.381.
Lee, S-W.: 1977c, Astron.Astrophys.Supp. 28, p.409.
Lee, S-W.: 1977d, Astron.Astrophys.Supp. 29, p.1.
McCrea, W.H.: 1964, Mon.Not.R.astr.Soc. 128, p.147.
Maeder, A.: 1974, Astron.Astrophys. 32, p.177.
Maeder, A.: 1975, Astron.Astrophys. 43, p.61.
Maeder, A.: 1976, Astron.Astrophys. 47, p.389.
Morgan, J.G. & Eggleton, P.P.: 1978, Mon.Not.R.astr.Soc. 182, p.219.
Mould, J. & Aaronson, M.: 1983, Astrophys.J. 273, p.530.
Newell, E.B. & Graham, J.A.: 1976, Astrophys.J. 204, p.804.
Norris, J.: 1981, Astrophys.J. 248, p.177.
Norris, J., Cottrell, P.L., Freeman, K.C. & Da Costa, G.S.:   1981,
     Astrophys.J. 244, p.205.
Norris, J. & Smith, G.H.: 1983, Astrophys.J. 275, p.120.
Patenaude, M.: 1978, Astron.Astrophys. 66, p.225.
Pease, F.G.: 1928, Publ. Astron.Soc.Pacific, 40, p.342.
Peimbert, M.: 1973, Mem.Soc.Roy.Sci.Liege, 6 Serie, 5, p.307, (18th
     Liege Colloquium).
Prather, M.J. & Demarque, P.: 1974, Astrophys.J. 193, p.109.
Racine, R.: 1971, Astrophys.J. 168, p.393.
Renzini, A.: 1977, In "Advanced Stages in Stellar Evolution", ed P.
     Bouvier, A. Maeder, Sauverny:Geneva Obs., p.151.
Richer, H.B.: 1978, Astrophys.J. 224, L.9.
Richer, H.B., Chan, E., Fahlman, G.G. & Hickson, P.:   1981, In
     "Astrophysical Parameters for Globular Clusters", IAU Coll. 68,
     eds. A.G. Davis Philip & D.S. Hayes, p.519.
Rood, R.T.: 1973, Astrophys.J. 184, 815.

Rood, R.T. & Iben, I.: 1968, Astrophys.J. 154, p.215.
Rood, R.T. & Seitzer, P.O.: 1981, In "Astrophysical Parameters for
     Globular Clusters", IAU Coll 68, eds. A.G. Davis Philip & D.S.
     Hayes, p.369.
Saio, H. & Wheeler, J.C.: 1980, Astrophys.J. 242, p.1176.
Sandage, A.: 1953, Astron.J. 58, p.61.
Sandage, A.: 1962, Astrophys.J. 135, p.333.
Sandage, A.: 1970, Astrophys.J. 162, p.841.
Sandage, A.: 1982, Astrophys.J. 252, p.553.
Sandage, A., Katem, B. & Sandage, M.: 1981, Astrophys.J.Supp. 46,
     p.41.
Sandage, A.R. & Smith, L.L.: 1966, Astrophys.J. 144, p.886.
Sandage, A. & Wildey, R.: 1967, Astrophys.J. 150, p.469.
Searle, L. & Zinn, R.J.: 1978, Astrophys.J. 225, p.357.
Searle, L. & Rodgers, A.W.: 1966, Astrophys.J. 143, p.809.
Simoda, M. & Iben, I.: 1968, Astrophys.J. 152, p.509.
Simoda, M. & Kimura H.: 1968, Astrophys.J. 151, p.133.
Simoda, M. & Tanikawa, K.: 1972, Pub.Astr.Soc.Japan 29, p.1.
Stetson, P.B.: 1981, Astron.J. 86, p.687.
Stryker, L.L. & Hrivnak, B.J.: 1983, DAO Preprint.
Sweigart, A.V. & Gross, P.G.: 1976, Astrophys.J..Supp. 32, p.367.
Sweigart, A.V. & Gross, P.G.: 1978, Astrophys.J..Supp. 36, p.405.
van Agt, S.: 1967, Bull.astr.Soc.Netherlands, 19, p.275.
VandenBerg, D.A.: 1983, Astrophys.J..Supp. 51, p.29.
Weiss, A.: 1983, Astron.Astrophys. 127, p.411.
Wheeler, J.C.: 1979, Astrophys.J. 234, p.569.
Zinn, R.J., Newell, E.B. & Gibson, J.B.: 1972, Astron.Astrophys 18,
     p.390.

DISCUSSION

Alcaino: Regarding the theoretical yet to be explained gaps in globular
clusters at the fainter end of the subgiant branches, I would like to
place the comment that in a recent C-M diagram of M4 now shown as a
poster paper, William Liller and myself have found a conspicuous gap
at about half a magnitude above the turnoff point. For checking purposes
the photometry has been redone at 1 mag. above and below the gap's limit
which has been confirmed. Furthermore, the gap originally discovered in
our joint work with Liller on NGC 288 published in 1980, has been independently confirmed by the work now in press of Buonanno and Corsi and
Fusi-Pecci (1984). It is therefore concluded that these gaps are real
and await a theoretical interpretation.

Richer: You made the point that it is usually very difficult to
distinguish between the RGB and the AGB. Is it not possible that we are
looking in the wrong colours (i.e. V, B)? We have some U photometry in a
number of globulars which shows a very nice separation between the RGB
and AGB in a (B, U-B) diagram.

Cannon: That is very interesting. I have not tried plotting other
varieties of C–M diagram but it would be extremely useful to have an
unambiguous way of separating the two branches.

Barlai: On the diagram type of HB versus $\Delta(B-V)$ the clusters M92, ω Cen,
M5, M15 and M3 form a group in the middle (and in the lower) part of the
figure. All these clusters contain RR Lyrae variables. Do the other
clusters of the diagram also contain variables or has this grouping been
formed by chance?

Cannon: The globular clusters with many RR Lyrae variables must occur
in the middle of the diagram because the ordinate of the diagram is the
Dickens horizontal branch type. In that empirical classification, purely
blue horizontal branches are called type 1 and purely red branches are
type 7. Intermediate types have horizontal branches spanning the RR
Lyrae instability strip.

SPECTROSCOPY OF MAIN-SEQUENCE AND SUBGIANT STARS IN GLOBULAR STAR
CLUSTERS

James E. Hesser
Dominion Astrophysical Observatory

Gretchen L.H. Harris
University of Waterloo

R.A. Bell
National Science Foundation

R.D. Cannon
Royal Observatory, Edinburgh

Since the mid-1970's it has been apparent that giant stars of
similar V and B-V within a "normal" globular cluster [i.e., one with a
narrow giant branch in its color-magnitude diagram (CMD)] exhibit a
perplexing range of strengths for such spectral features as CH, CN and
NH. This complex subject has been reviewed by Kraft (1979), McClure
(1979) and Freeman and Norris (1981). DDO photometry first revealed
star-to-star differences of CN strengths at $M_V > +1$, where observational
confusion between asymptotic and first-ascent giant stars is removed
(Hesser, Hartwick and McClure 1976, 1977; Hesser 1978). Subsequently,
we have sought to place observational constraints on possible mechanisms
by studying such questions as: At what M_V's do spectral differences
first become observable? Do spectral features other than those from
CNO-based molecules vary from star-to-star? Can small temperature or
gravity differences produce the observed ranges?

Spectra covering ~ 3750–4500 Å with 3–4 Å resolution have been taken
with the CTIO 4-m telescope, R-C spectrograph and SIT vidicon detector
(Atwood et al. 1979). To achieve a perspective on the range of
spectral characteristics of very faint (B ~ 18) globular cluster stars,
our philosophy has generally been to survey many stars to modest S/N
levels, rather than concentrating on achieving high S/N for only a few.
Our analysis relies heavily upon comparisons with synthetic spectra
computed using model atmospheres (e.g., Gustafsson et al. 1975; Bell
and Gustafsson 1978; Gustafsson and Bell 1979). Some of our results
follow.

NGC 6752: 26 stars, ranging from the main-sequence to the tip of
the giant branch, have been observed in this cluster, for which we
estimate [M/H]~ -1.5 (Bell, Hesser and Cannon 1984). To our knowledge
this is the first time that spectral observations made with the same
equipment for globular cluster stars exhibiting such a range of M_V have
been compared with model calculations; the overall agreement is very
gratifying. We find no evidence for star-to-star abundance differences

139

A. Maeder and A. Renzini (eds.), Observational Tests of the Stellar Evolution Theory, 139–142.

at any M_V of any elements except C and N, but our modest-resolution and modest-S/N data are not particularly suitable for studying the correlation between CN, Ca and Al abundances found by Cottrell and Da Costa (1981) among pairs of bright giants. Varying amounts of carbon depletion are inferred for all giants. Nitrogen enhancements >3 are inferred for the brightest giants. We observe a range of CN strengths among the faintest giants, but the sample is too small to ascertain if the bimodality seen in bright giants by Norris et al. (1981) persists to M_V~+2-3. At $(B-V)_0$~0.45 the main-sequence stars are too warm for CN or CH formation. Our spectra suggest that the observed color width of the CMD at faint levels is due largely, if not entirely, to observational scatter in the photometry, rather than to real temperature variations induced by metal-abundance.

NGC 1851: A dramatic range of CN band strengths has been found among bright giants in this [M/H]~-1.0 cluster (Hesser et al. 1982); and a marked range in CN strength persists to the M_V ~+2 limit of our survey.

NGC 104 (47 Tuc): DDO photometry first indicated the existence of CN strength differences at M_V ~+2 in this metal-rich globular. That finding was extended (with ~16 Å resolution spectra) to the base of the giant branch (Hesser 1978) and then to the turnoff region (Hesser and Bell 1980). New 3-4 Å resolution spectra include 11 dwarfs and show a range of CN strengths among both turnoff and subgiant stars (Bell, Hesser and Cannon 1983). Observational scatter in the CMD (Harris, Hesser and Atwood 1983, 1984), or in our spectra of dwarfs, seems too small to account for the observed range by temperature differences. Thus, we conclude that abundance differences are responsible. To interpret our spectra, we adopted, following Dickens, Bell and Gustafsson (1979), [M/H]~-0.8, a value intermediate between the lower values obtained from high-dispersion spectroscopy (Pilachowski, Sneden and Wallerstein 1983) and the higher-values favored by, e.g., DDO (Hesser, Hartwick and McClure 1977) or IR (Frogel, Cohen and Persson 1983) photometry. From comparison with synthetic spectra we deduce that the observed range of CN strengths could be produced by a star-to-star range of nitrogen abundances of ~5; this range is similar to that required to explain observations of highly evolved stars in the cluster (see, e.g., Dickens, Bell and Gustafsson 1979).

NGC5139 (ω Centauri) and NGC 6656 (M22): Photometric and spectroscopic observations of bright giant and HB stars in ω Cen, the most massive and luminous globular in the Galaxy, have firmly established that [M/H] ranges from -2.0 to ~-0.5, with very few stars having [M/H]>-1.0. Such a metallicity range is often inferred for giants in dwarf spheroidal galaxies. From our initial spectra for 11 stars having +2<M_V<+3.5, we found five probable members to have -1.0<[M/H]<-1.5, while a sixth has [M/H]~-0.5, i.e., as high as those deduced for any of the brighter stars in the cluster (Bell et al. 1981). Its nitrogen appears to be enhanced by as much as a factor of 3. Subsequent spectra bring to 40 the number of faint (B~18) subgiants observed in the Cannon and Stewart (1981) CMD. We are working as close

(~0.3 r_t) to the cluster center as crowding will permit, but half of the observed stars appear to be radial-velocity (V_r) non-members! Furthermore, 90% of the stars with B-V>0.8 are V_r non-members. Spectra of V_r members reflect real temperature and abundance differences, with a total range very similar to that observed for more highly evolved stars. The spectral differences, combined with the B-V range of 0.2 for confirmed members, show conclusively that the wide range of colors, etc. observed among giants in ω Cen originates near, if not on, its main sequence. We also suspect that there may be a range of [M/H] among subgiants at a given T_{eff}. Higher S/N spectra are required to substantiate our inference; if correct, it may indicate a range of ages among ω Cen stars. Finally, the suggestion that M22 may share some of the distinguishing anomalies of ω Cen (Hesser, Hartwick and McClure 1977, Hesser and Harris 1979) has been confirmed by Pilachowski et al. (1982) and Norris and Freeman (1983); thus, ω Cen is no longer unique in the Galaxy.

In summary, among the salient findings of our observational program are that: (1) All "normal" globular clusters studied have shown a range of CN, and often CH, strengths for stars within a magnitude or so of the turnoff. (2) Enhancements of nitrogen seem commonplace. (3) Some,and perhaps many, spectral differences among highly evolved stars of similar V, B-V originate on or near the main-sequence. Four possible ways in which a high nitrogen abundance can arise in class IV-V stars are differences in abundance at the time of star formation, mixing of stars during their evolution, accretion of material lost by other cluster stars, and accretion of material lost by a binary companion (Bell et al. 1981). No single explanation seems adequate to explain the observations. More and better data are clearly needed. (4) There is evidence that the behavior of spectral features due to CNO elements differs in Pop. I and II stars (Hesser, Hartwick, and McClure 1976; Kraft et al. 1982; Kraft 1983).

REFERENCES:

Atwood,B.A., Ingerson,T., Lasker,B.M. & Osmer,P.S.: 1979,
     Pub.A.S.P. 91, p.120.
Osmer,P.S.: 1979, Pub.A.S.P. 91, p.120.
Bell,R.A. & Gustafsson,B.: 1978, Astron.Ap.Suppl.34, p.229.
Bell,R.A., Harris,G.L.H., Hesser,J.E. & Cannon,R.D.:1981,Ap.J.249,p.637.
Bell,R.A., Hesser,J.E. & Cannon,R.D.: 1983, Ap.J. 269, p.580.
_____ :1984, Ap.J., submitted.
Cannon,R.D. & Stewart,N.J.: 1981, M.N.R.A.S. 195, p.15.
Cottrell,P. & DaCosta,G.S.: 1981, Ap.J. 245, p.L79.
Dickens,R.J., Bell,R.A. & Gustafsson,B.: 1979, Ap.J. 232, 428.
Freeman,K.C. & Norris,J.: 1981, Ann.Rev.Astr.Ast.19, p.319.
Frogel,J.A., Cohen,J.G. & Persson,E.: 1983, Ap.J., in press.
Gustafsson,B. & Bell,R.A.: 1979, Astr.Ap.74, p.313.
Gustafsson,B., Bell,R.A., Eriksson,K. & Nordlund,A.:1975,ibid.,42,p.407.
Harris,W.E., Hesser,J.E. & Atwood,B.A.: 1983, Ap.J.268, p.L111.
_____ :1984, Pub.A.S.P., submitted.

Hesser,J.E.: 1978, Ap.J. 223, p.L117.
Hesser,J.E. & Bell,R.A.: 1980, Ap.J. 238, p.L149.
Hesser,J.E., Bell,R.A., Cannon,R.D. & Harris,G.L.H.:1982,A.J.87,p.1470.
Hesser,J.E. & Harris,G.L.H.: 1979, Ap.J.234, p.513.
Hesser,J.E., Hartwick,F.D.A. & McClure,R.D.: 1976, Ap.J.207, p.L113.
    ____:1977, Ap.J.Suppl.33, p.471.
Kraft,R.P.: 1979, Ann.Rev.Astr.Ast. 17, p.309.
    ____:1983, Highlights of Astronomy 6, p.129.
Kraft, R.P., et al.: 1982, Pub.A.S.P. 94, p.55.
McClure,R.D.: 1979, Mem.Soc.Astr.Italiana 50, p.15.
Norris,J., Cottrell,P.I., Freeman,K.C. & DaCosta,G.S.:1981,Ap.J.244,p.205.
Norris,J. and Freeman,K.C.:1983,Ap.J.266,p.130.
Pilachowski,C.A., Sneden,C. & Wallerstein,G.: 1983, Ap.J.Suppl.52, p.241.
Pilachowski,C.A., Wallerstein,G., Leep,E.M. & Peterson,R.C.: 1982,
    Ap.J.263,p.187.

DISCUSSION

Demarque: You mentioned variations in C abundances from star to star.
Are these variations correlated to the observed variations in N
abundances?

Hesser: Our observed samples in 47 Tuc and NGC 6752 are small and the
S/N modest. Bearing those caveats in mind, it is my impression that in
NGC 6752 we see generally weaker G bands in stars with stronger CN bands;
for 47 Tuc this does not seem to be the case (see Table 4 of Bell,
Hesser and Cannon, 1983).

Wing: Have you tried computing synthetic B-V colors from the Bell-
Gustafsson model spectra to see what range in B-V can be produced by the
observed range in molecular band strength?

Hesser: The general question of colors of the models is discussed briefly
by Bell, Hesser and Cannon (1983). For the 47 Tuc dwarfs the UV CN-bands
are enhanced in some stars, but the 4216 Å bands are too weak to affect
the (B-V) color at observable levels. For extreme CN stars, such as those
observed on the NGC 1851 giant branch (Hesser et al, 1982), color
differences would be expected, but Roger Bell would have to answer your
question.

THEORETICAL RED-GIANT BRANCHES FOR GLOBULAR CLUSTERS

Don A. VandenBerg
University of Victoria

VandenBerg (1983a; hereinafter V) has recently computed a large grid of stellar evolutionary sequences from the zero-age main sequence to the base of the red-giant branch (RGB) and carried out extensive comparisons of the associated isochrones with published photometry of globular clusters (GCs). These calculations were based on the latest Los Alamos opacities and, in addition, made use of available model atmospheres for improved surface boundary conditions as well as for the transformation of the stellar models from the (M_{bol}, $\log T_{eff}$)- to the (M_V, B-V)-plane. In general, very encouraging agreement of the predicted and observed morphologies of cluster C-M diagrams was obtained.

This paper reports the continuation of these published sequences from the base of the RGB to the helium flash. Representative models with masses in the range of 0.8 to 0.9 $M_\odot$ were selected in order that the stars on the giant branches had ages of approximately 16 billion yr, so as to have consistency with the GC age estimates made by V. Initial numerical experiments indicated that a value of α = 1.6 for the ratio of the mixing length to the pressure scale height was needed to provide the best match of the Z = 0.0001 model sequence with the observations of M92. Sequences for the other assumed metallicities, Z = 0.0003, 0.001, 0.003, and 0.006, were then computed for the same value of the mixing-length parameter and overlayed directly on the observations, which were transposed to the $[M_V, (B-V)_o]$-plane using canonical values of distance moduli and reddening (see V). The resultant comparisons are illustrated in Figure 1. The overall good agreement indicates that current model atmosphere and interior calculations are able to provide a remarkably faithful reproduction of observed giant branches. The slight discrepancies which do occur at the base of the metal-poor sequences and at the tips of the metal-rich sequences are within the existing observational and theoretical uncertainties. (For instance, the mean lines of M92 and M13 fainter than $M_V \sim 1.5$ are based on relatively few observations and the predicted colors and bolometric corrections of stars cooler than 4000 K are poorly understood.) Note that the observations of 47 Tuc have come out approximately centered between $[Fe/H]$ = -0.5 and $[Fe/H]$ = -0.8.

A. Maeder and A. Renzini (eds.), Observational Tests of the Stellar Evolution Theory, 143–146.
© *1984 by the IAU.*

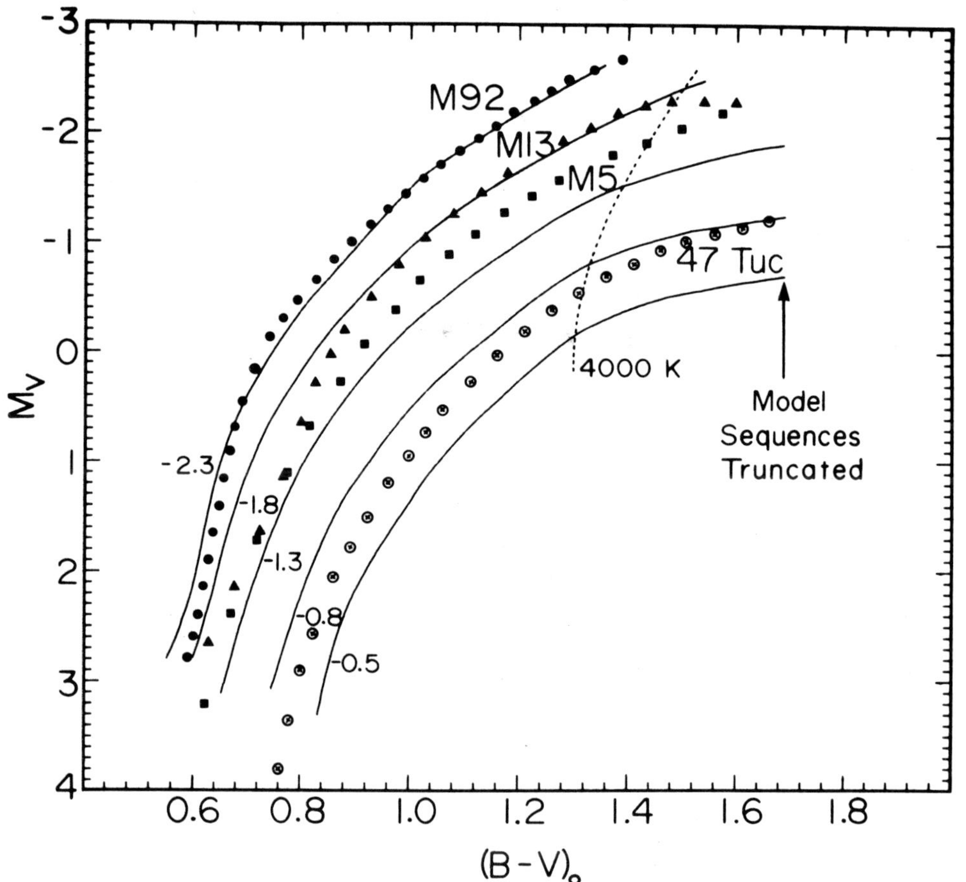

Fig. 1. – Comparison of computed model sequences (solid curves) with the
mean giant branches (symbols) of four galactic globular clusters. Given
at the lower left are the [Fe/H] values appropriate to the theoretical
RGB's, which are based on model atmospheres for surface boundary conditions, colors, and bolometric corrections to the left of the 4000 K locus.
Extrapolations of the atmospheric quantities were required for models
predicted to have T_{eff} < 4000 K. Due to uncertainties arising from TiO
absorption, model sequences were arbitrarily truncated at $(B-V)_o$ = 1.68.

        Figure 2 presents the same model sequences on the theoretical
plane along with the fiducial loci for three globular clusters as derived
by Frogel, Persson, and Cohen (1981; hereinafter FPC) using infrared
photometry. Although the calculations assume Y = 0.2, similar good
agreement of the models with the observations can be obtained for Y = 0.3
provided that α is reduced slightly. In any case, this work suggests
that the mixing-length parameter must be nearly constant, having a value
in the range of 1.45 < α < 1.65, in agreement with conclusions reached
by V from main-sequence turnoff comparisons. We find good consistency
between Figs. 1 and 2, particularly for M92 and M13, but less so for
47 Tuc. The latter should have been centered between Z = 0.003 and Z =

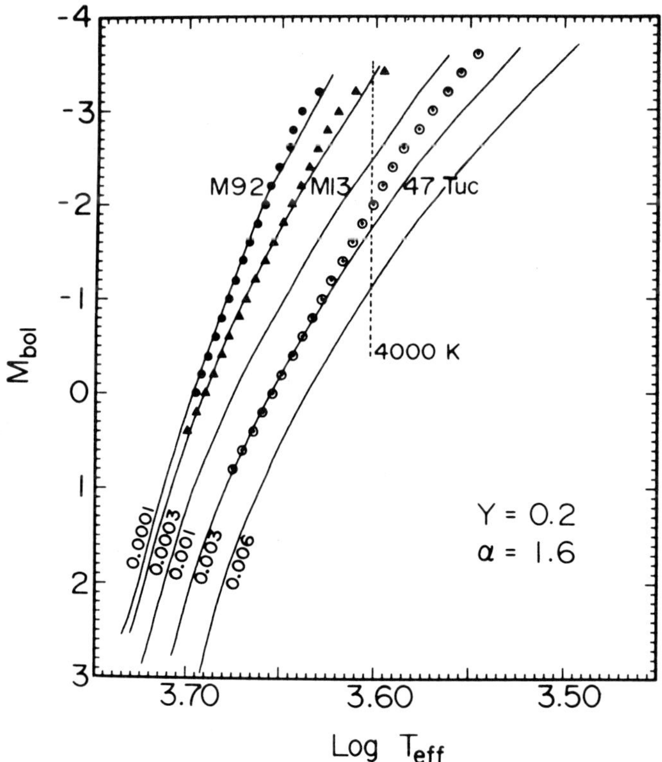

Fig. 2. - Comparison on the theoretical plane of the same model sequences shown in Fig. 1 (solid curves) with the mean lines (symbols) derived by FPC for M92, M13, and 47 Tuc.

0.006 (in Fig. 2) to be fully consistent with its location in Fig. 1, which may be suggesting that the B-V value at a given temperature (in the metal-rich regime) is underestimated by current model atmospheres. If this is the case, the models would tend to favor the possibility that the metallicity of 47 Tuc is closer to $[Fe/H] = -0.8$. The apparent discrepancy of about 150 K between the tip of the Z = 0.003 theoretical RGB and that of 47 Tuc reflects the present uncertainty in our understanding of the color-temperature relation for cool stars. A careful examination of that used by FPC suggests that their adopted relation is systematically too hot below $\sim$ 4000 K, which in turn would imply that the present models provide a more realistic representation of the temperatures of metal-rich giants (see VandenBerg 1983b).

REFERENCES

Frogel, J.A., Persson, S.E., and Cohen, J.G. 1981, Ap.J., 246, 842. (FPC)
VandenBerg, D.A. 1983a, Ap.J.Suppl., 51, 29. (V)
______________. 1983b, in preparation.

DISCUSSION

<u>Mestel</u>: You find that taking α = mixing length/scale height as a constant through a convection zone yields satisfactory results. Can you say how sensitive are the results to the actual value of α? Also, I recall that 25 years ago there was a question about whether one could simultaneously describe both type I and type II cluster giant branches just by metallicity variations, or whether different values of α are required.

<u>VandenBerg</u>: Changing α from 1.5 to 1.7, say, changes the predicted T_{eff} by $\gtrsim$ 150 K. Due to the steepness of the color-temperature relation for low gravity stars, the predicted B-V color is altered by nearly 0.1 mag; consequently large variations in α are almost certainly ruled out. Although I have not made detailed comparisons of isochrones with young Pop. I clusters, preliminary indications are that similar values of α ($\approx$1.5) also apply, so that cluster giant branches are indicative primarily of composition or age.

FAINT CCD PHOTOMETRY IN GLOBULAR CLUSTERS. II. COMPARISON OF THEORETICAL ISOCHRONES WITH THE GLOBULAR CLUSTERS M4 AND M15

Harvey B. Richer* and G.G. Fahlman*
Department of Geophysics and Astronomy
University of British Columbia
Vancouver, B.C. V6T 1W5
Canada

INTRODUCTION

Globular cluster work touches on virtually all branches of astronomical research. The age of the globular system (or any variations in age among individual clusters) has important cosmological implications as well as relating to the formation time of the halo of our galaxy. Star to star chemical inhomogeneity within a cluster may set important constraints on either mixing within the stars themselves or on the chemical inhomogeneity of the early universe. Metallicity variations among clusters may provide the clue to galaxy-wide enrichment processes, while the cluster color-magnitude diagrams themselves are a testing ground for virtually every facet of stellar evolution.

One important aspect of globular cluster research is photometry of individual stars within the cluster, but before it can provide the critical input needed to constrain some of the problems mentioned above, extremely high accuracy is required. This required accuracy is actually composed of 3 separate units, high precision (small internal scatter), no systematic errors, and deep enough so that a cluster turnoff and lower main sequence is well defined.

At the time this paper was written, a rather casual search of the literature indicated that 17 globular clusters had photometry available for stars below the main sequence turnoff. In most cases the scatter near the turnoffs and on the lower main sequences of these clusters exceeds $\delta(B-V) = 0.2$ magnitudes. This means that many of the most important cluster parameters are simply lost in the noise. Table 1 provides a graphic illustration of this; the entries coming from interpolation of VandenBerg's (1983) theoretical isochrones for a cluster of intermediate metallicity.

---

*Visiting Astronomer Cerro Tololo Interamerican Observatory which is operated by AURA, Inc., under NSF contract AST 74-04129.

A. Maeder and A. Renzini (eds.), Observational Tests of the Stellar Evolution Theory, 147–151.
© *1984 by the IAU.*

Table 1
Effect on Derived Cluster Parameters for
δ(B-V) = 0.2 Among Main Sequence Stars

Cluster Parameter	Minimum Detectable Star to Star Variation with δ(B-V) = 0.2
δ[Fe/H] among stars (δ Y = 0)	200%
δY among stars (δ[Fe/H] = 0)	300%
binary frequency in cluster	all knowledge lost
δage (uncertainty in derived age of cluster itself, assuming all other parameters perfectly known)	$\pm$ 6 x 10^9 years

Table 1 points out the need for the highest possible precision in the observations. With this in mind, we began a program of CCD observations of a number of globular clusters. The data discussed here was acquired at CTIO using the 4.0 meter telescope. We present results for two clusters of rather different metallicity: M4 and M15.

FUNDAMENTAL CLUSTER PARAMETERS

From our UBV CCD data of M4 and M15 we were able to derive: (1) the reddening to each cluster (from field stars in the cluster direction) and (2) an estimate of [Fe/H] for each cluster (from the ultraviolet excess of main sequence stars). (3) The distance moduli to the clusters were determined using the Sandage (1981, 1982) technique of assigning an absolute magnitude to the cluster RR Lyraes. We have taken account of the suggestion of Cox et al. (1983) that the zeropoint in the Sandage recipe may be too faint. (4) The age of each cluster was derived from a comparison with the recent theoretical isochrones of VandenBerg (1983 a,b) which were superimposed (not fitted) to our cluster color-magnitude diagrams. (5) In all cases the helium abundance (Y) was assumed to be 0.2. Table 2 provides a summary of our derived parameters for M4 and M15.

Table 2
Fundamental Parameters for M4 and M15

Parameter	M4	M15
E(B-V)	0.37 $\pm$ .06	0.12 $\pm$ .04
[Fe/H]	-.93 $\pm$ .31	-2.07 $\pm$ .31
(m-M)$_V$	12.50 $\pm$ .15	15.25 $\pm$ .15
age	15($\pm$1) x 10^9 years	16.5($\pm$1) x 10^9 years

The error in the age listed in Table 2 is based solely on the
scatter of the individual stars around the appropriate isochrone (see
Figures 1(a) and (b)) and does not include possible errors in the
reddening, [Fe/H], distance modulus, or Y.

COLOR-MAGNITUDE DIAGRAM AND ISOCHRONES

Figures 1(a) and (b) present our color-magnitude diagrams for each
cluster and the appropriate VandenBerg isochrones (1983 a,b). No
fitting of the isochrones was done to the data; we simply selected the
VandenBerg isochrones which most closely matched our derived metallicity
and shifted them so they had the same reddening and distance modulus
as the cluster. The agreement is quite excellent, particularly for M4
where the data is of much higher quality than that for M15. Field stars
were not removed from these diagrams. The straight line shown in the
lower left hand corner of the M4 diagram is a theoretical cooling curve
for 0.51 M$_\odot$ white dwarfs (Sweeney 1976). A number of promising cluster
white dwarf candidates are clearly present.

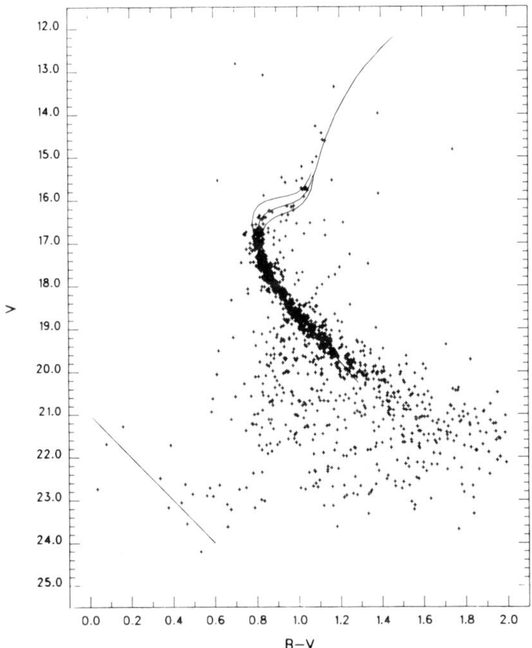

Figure 1 (a). Observed color-magnitude diagram of M4. The isochrones
shown are from VandenBerg (1983 a,b) for [Fe/H] = -1.0, Y = .20,
α = 1.5 (1.6 for giant branch) and ages 12, 15 and 18 billion years.
The isochrones were shifted in order to represent a cluster with
(m-M)$_V$ = 12.50 and E(B-V) = 0.37. The straight line in the lower left
hand corner of the diagram is a cooling curve for 0.51 M$_\odot$ white dwarfs
(Sweeney 1976).

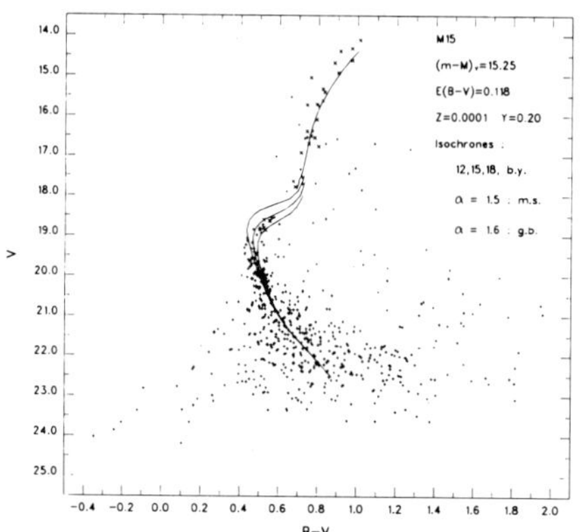

Figure 1 (b). Observed color-magnitude diagram of M15. The x's are
data taken from Sandage (1970) to illustrate the giant branch. The
isochrones shown are from VandenBerg (1983 a,b) for [Fe/H] = -2.30,
Y = .20, α = 1.5 (α = 1.6 for the giant branch) and ages 12, 15 and
18 billion years. The isochrones were shifted in order to represent
a cluster with $(m-M)_V$ = 15.25 and E(B-V) = 0.118.

SUMMARY

     Our CCD observations of M4 and M15 have provided precise and deep
photometry for individual stars in these clusters. A comparison between
our derived color-magnitude diagram and the theoretical isochrones of
VandenBerg (1983 a,b) have shown that these new isochrones provide a
very faithful representation of the observational data. To within an
error of $\pm$ 1 billion years, the ages of M4 and M15 are respectively
15×10^9 and 16.5×10^9 years.

     Further details concerning these observations can be found in
Richer and Fahlman (1984) and Fahlman and Richer (1984).

REFERENCES

Cox, A.N., Hodson, W., and Clancey, S.P.: 1983, Ap. J. 266, 94.
Fahlman, G.G., and Richer, H.B.: 1984, Ap. J., in press.
Richer, H.B., and Fahlman, G.G.: 1984, Ap. J., in press (February 1/84).
Sandage, A.: 1970, Ap. J. 162, 841.
Sandage, A.: 1981, Ap. J. 248, 161.
Sandage, A.: 1982, Ap. J. 252, 553.
Sweeney, M.A.: 1976, Astron. Astrophys. 49, 375.
VandenBerg, D.A.: 1983a, Ap. J. Suppl. 51, 29.
VandenBerg, D.A.: 1983b, private communication.

DISCUSSION

Demarque: Your beautiful results emphasize once more the importance of determining reliable metallicities for the globular clusters, if one wants to derive reliable ages for them.

McClure: If all the stars, which scatter widely in the fainter part of the CM diagram of M4, are counted in the luminosity function, does the luminosity function then turn over or not?

Richer: The direct answer to your question is no, for the following reasons. (1) Firstly, the errors in the photometry at V = 20 are about 0.03 magnitudes while they are about 0.20 at V = 24.0. These error estimates come from two tests: (a) We measured about 100 stars on each of the 6 V frames and 4 B frames and looked at the scatter in the observed values. (b) We also introduced stars of known magnitude into the frames and measured them. Both of them gave consistent results. The errors are thus too small to be able to move the stars that upper to scatter away from the main sequence back to the main sequence.
(2) There are, of course, field stars on the frames. In particular the stars seen to the blue of the M4 main sequence are contributed mainly from the nuclear bulge of the galaxy as M4 is seen projected against the bulge. The numbers and magnitudes of the stars seen are moderately consistent with current models of the bulge.

Penny: The accuracy of the age estimates is improved if one looks at the shape of the CMD locus from the turn-off to the SGB. I would estimate from the CMDs in my poster paper that they only permit a range of ±20 % in age, even when errors in the metallicity are allowed for.

SYSTEMATIC MAIN SEQUENCE PHOTOMETRY OF GLOBULAR CLUSTER STARS FOR AGE
DETERMINATION

Gonzalo Alcaino and William Liller
Instituto Isaac Newton, Ministerio de Educación de Chile,
Santiago, Chile.

The individual photometric study of the coeval stars in globular clusters
presents one of the best observational tests of the stellar evolution
theory. Our own globular cluster system provides fundamental clues to
the dynamical and chemical evolutionary history of the galaxy, and the
study of their ages give us a lower limit to the age of the galaxy as
well as to that of the universe. We have been undertaking a systematic
research program, and discuss herewith the ages deduced by fitting main
sequence photometry to theoretical isochrones of six galactic globular
clusters : M4, M22, M30, NGC 288, NGC 3201 and NGC 6397.

From fits of the isochrones of Ciardullo and Demarque (1977) trans-
formed to the observational plane $|M_V, (B-V)_0|$ of the observed Color
Magnitude Diagram (CMD) in the BV system, we deduced the following ages
each of which have an uncertainty of 3 or 4 Gys: NGC 6397, $|Fe/H|$ = -2.0,
age 17 Gys; NGC 288, $|Fe/H|$ = -1.6, age 15 Gys; NGC 7099 (M30), $|Fe/H|$ =
-2.16, age 15 Gys; NGC 3201, $|Fe/H|$ = -1.4, age 12 Gys; NGC 6656 (M22),
$|Fe/H|$ = -1.9, age 15 Gys and NGC 6121 (M4), $|Fe/H|$ = -1.3, age 12 Gys,
(mean (of 6) = 14.3 Gys. σ_s = 1.97, σ_{mean} = 0.88).

The CMD for M4 shown in Figure 1 has been obtained most recently.
At a distance of 2 kpc, it is the closest globular cluster to the sun.
Adopting $(m-M)_V$ = 12.52 (V_{HB} = 0.9), E(B-V) = 0.42, Z = 0.001 ($|Fe/H|$ =
-1.3), Y = 0.3, mixing length parameter α = 1.6, and using the improved
isochrones of VandenBerg (1983) an excellent fit is produced at 12 Gys.
This investigation presents as well the first results of RI main sequence photometry for a globular cluster. The greater baselines provided
by the additional pass-bands (i.e., B-I, B-R, V-I) would make possible
improved isochrone fittings and points up the urgent need for calculations of synthetic isochrones to be compared with observations. In the
figures described below we proceed to show some of the more relevant
examples.

Figure 2: the CMD of V vs B-R. The various branches lying above the
main sequence are well-defined owing to the greater color baseline provided by B-R (as compared with B-V). The gap between the subgiant (SGB)

A. Maeder and A. Renzini (eds.), Observational Tests of the Stellar Evolution Theory, 153–155.

and the Main Sequence Turn-Off (MSTO) is perhaps more conspicuous.
Figure 3: The CMD of V vs B-I. The separation between the SGB and
MSTO is dramatic: The bluest SGB stars are redder than the reddest MS
stars measured, and an extrapolation of the SGB to fainter magnitudes
misses the MSTO by a large margin. Figure 4: The CMD of I vs B-I.
The full range of I magnitudes (9.2) and B-I (3.5) are greater than for
any other magnitudes and colors. Again, the separation of the SGB and
MSTO is dramatic. The sequences are very well defined.

References:

Ciardullo, R. B., and Demarque, P.: 1977, Trans. Yale Univ. Obs., pp. 33.
VandenBerg, D.: 1983, Astrophys. J. Suppl. 51, pp. 29.

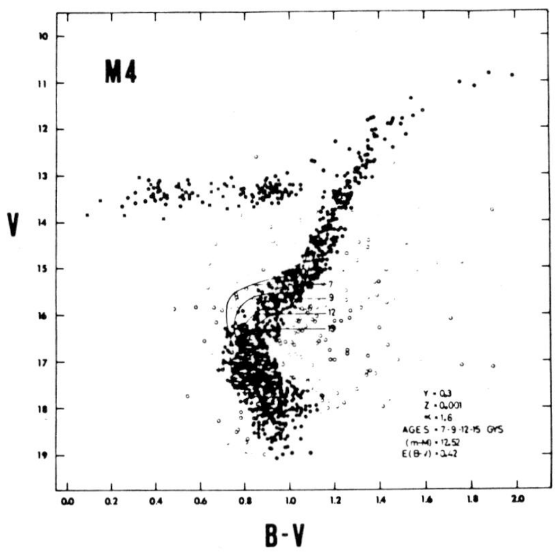

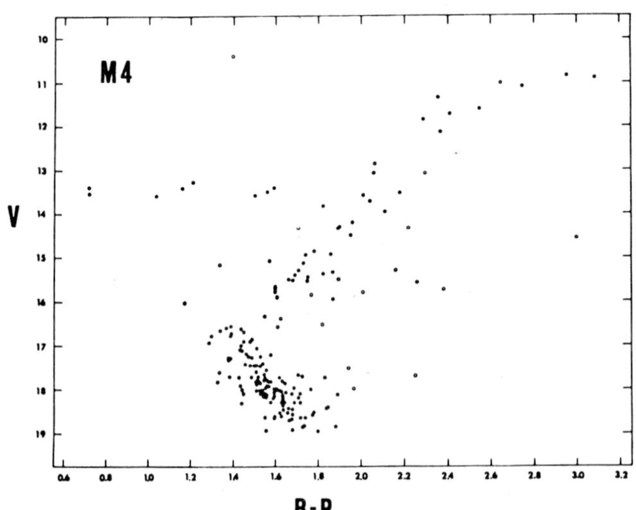

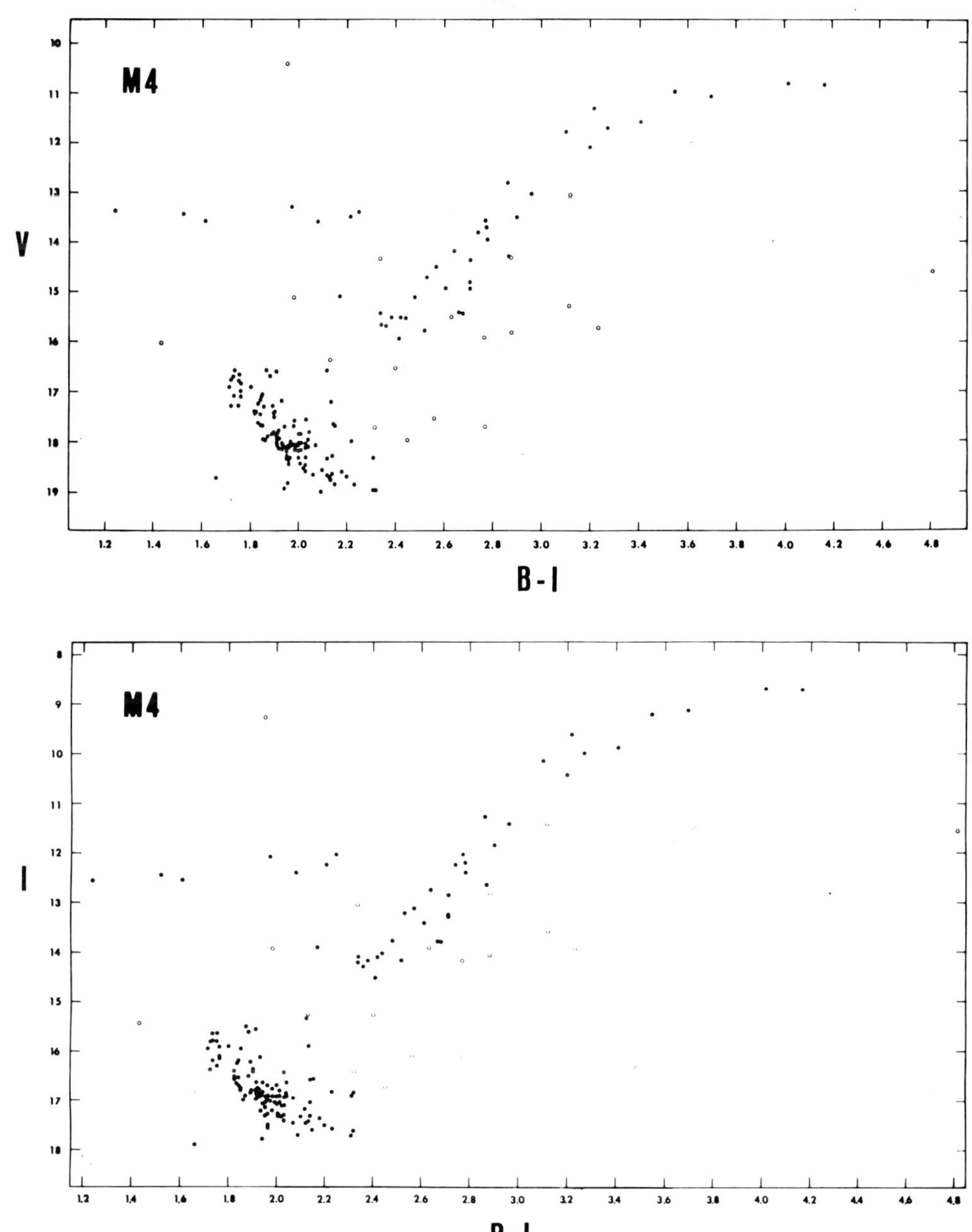
M4
V
B-I
M4
I
B-I

THE MAIN-SEQUENCES OF NGCs 288, 3201, 4590, AND 6809

A.J.Penny,   Royal Greenwich Observatory, U.K. and
              South African Astronomical Observatory

A globular cluster is an admirable object for the study of stellar
evolution and theoretical investigations of the isochrones in the
colour-magnitude diagram (CMD) have been numerous, with the most recent
being that of Vandenberg (1983,Ap.J.Suppl.51,29), but there are still
uncertainties such as rotation, abundance variations and gravitational
settling. The matching observations are in fair agreement. However only
a small number of clusters have been observed and the detailed matching
is poor. Most of the CMDs are from photoelectrically calibrated
photographic photometry which are affected by the variable background
found in clusters. Modern detectors suffer less from this problem and
are very sensitive. A programme has been underway on a number of
clusters using electronographs and CCDs with the SAAO 1.0m telescope and
the crowded-field Starlink Aspic software. This paper reports on
preliminary results from four of the clusters. The CCD observations were
done in collaboration with W.K.Griffiths of Leeds University.

Zero points for the CMDs came from existing photometry for NGCs 288
and 6809, and from CCD transfers for 3201 and 4590. The linearity and
colour equation of the electronographs and CCD were found from exposures
to standard stars. The CCD errors in particular were very small, with
errors of 0.004mag being possible on bright stars.

The resultant CMDs are shown in the figures. The data are from fields
of about 5' by 5', some 5' from the cluster centres, except for 3201
where additional bright stars were measured at the cluster centre, and
for 6809, where there are additional bright stars taken from Lee (197z).
Thus only for 288 and 4590 are the luminosity functions correct. For
each cluster a theoretical Vandenberg line has been fitted. An age of
17Gyr with Y=0.24 has been used with reddening and metallicities taken
from the literature. The adopted values were
(0.00,-1.0;0.21,-1.2;0.07,-1.9;0.08,-1.4). (I am indebted to
S.P.Caldwell for the metallicities of 288 and 6809.) The only free
parameter was the apparent distance modulus. As can be seen the fit is
good. 288 is not very old as has been suggested, cluster ages differing
by more than 3Gyr will not fit, and there is no sign of the discrepency

A. Maeder and A. Renzini (eds.), Observational Tests of the Stellar Evolution Theory, 157–158.
© *1984 by the IAU.*

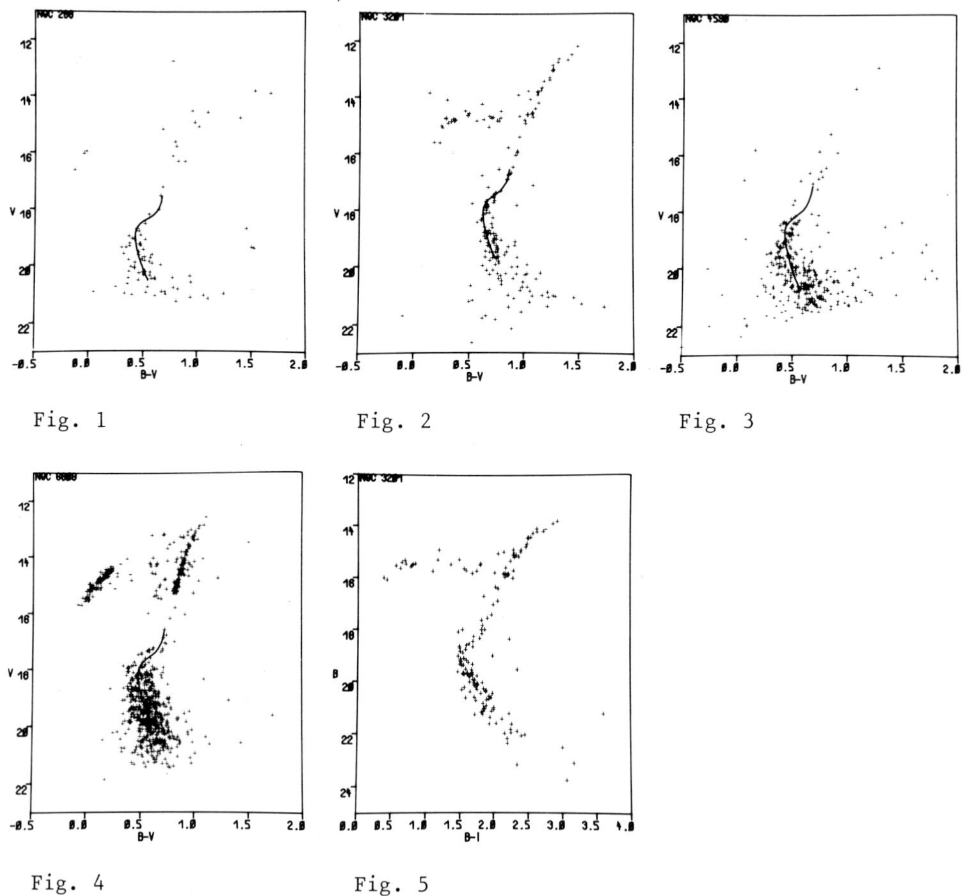

Fig. 1   Fig. 2   Fig. 3

Fig. 4   Fig. 5

at the faintest magnitudes with theory seen in some photographic CMDs.

Together with the horizontal-branch levels from previous studies, the vertical shift gives the respective H-B luminosities of (1.0;1.0;0.7:0.7) which are consistent with the relationship of Sandage (1982,Ap.J.252,553).

The B/B-I diagram of NGC 3201 shows how an extended colour baseline produces a more accurate CMD, and also illustrates the possibility of using other passbands, including the use of narrow-bands to investigate the main-sequence metallicities.

MONTE CARLO SIMULATIONS OF GLOBULAR CLUSTER RED GIANT BRANCHES

Deborah A. Crocker
Robert T. Rood
Department of Astronomy
University of Virginia

ABSTRACT

Simulations of globular cluster luminosity functions have shown that many of the features observed in actual clusters are probably statistical fluctuations. We suggest one new way of looking at the data which may ultimately lead to some information about the physical parameters of clusters.

I. METHOD

Rood (1984) has found that the time dependence of the evolution of a star along the first giant branch can be well approximated by $\log (L) = a + b \sinh^{-1} t$, where t is the time before flash in units of 5×10^7 years. The parameters a and b vary with composition and mass, and for a given star have different values above and below the luminosity function bumps predicted to occur when the H burning shell passes through the discontinuity left by the deepest penetration of the convective envelope. Thus uniformly distributed random times before flash yield $\log L$ and then $\log T_e$ and $\log g$ (which can be expressed as functions of $\log L$). Magnitudes and colors with gaussianly distributed random errors are then calculated. Details will be found in Crocker and Rood (1984).

II. RESULTS

(1) The dispersion of only 0.09 m difference between the bolometric magnitude of the brightest giant and the theoretical flash luminosity found by Frogel, et al. (1980) does not require any modification to standard evolution. Simulations with 45 stars in the upper 2.5 m of the RGB show a dispersion of 0.11 m, similar to the observed value. We do not find an increasing dispersion for the second and third brightest star as found by Frogel, et al.

A. Maeder and A. Renzini (eds.), Observational Tests of the Stellar Evolution Theory, 159–161.

(2) Comparison between theoretical models and observation should always be made in the observed quantities. Transforming quantities which contain random errors of measurement causes loss of information.

(3) Determination of Y at a level of $\Delta Y = 0.1$, and Z at a level of $\Delta \log Z = 1$ from luminosity functions requires samples of more than 600 stars in the upper 5 m of the RGB.

(4) Finding a statistically significant bump of the size theoretically expected in the differential luminosity function requires samples of more than 1000 stars.

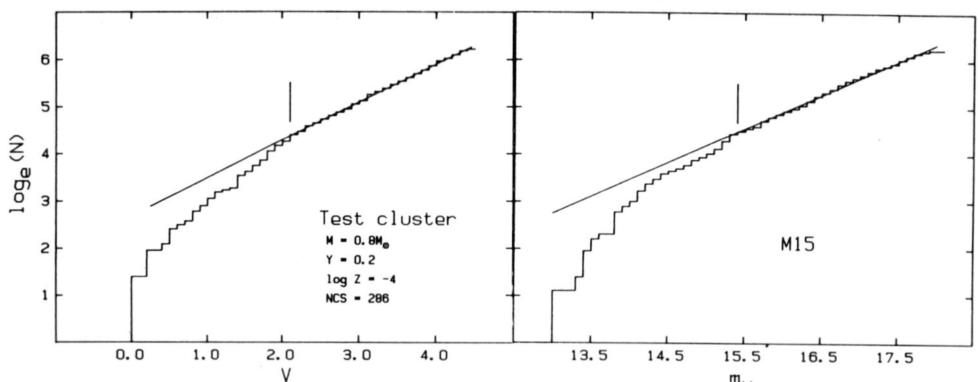

Figure 1. Log of the integrated V luminosity function. The left panel is for a simulated cluster; the right is taken from the data of Buonanno, et al. (1983). The vertical mark shows the theoretical bump location on the left and the hypothesized bump in M15 on the right. On the left V is magnitudes below the brightest star.

(5) It may be possible to locate the bump by using the break in slope of the evolution rate which occurs there. Using the log of the integrated luminosity function and some prejudice about the location of the break, it can be located in samples comparable to the largest cluster surveys (450 stars in upper 5 m). Figure 1 shows that the break in slope is visible in both simulated and real clusters.

REFERENCES

Buonanno, R., Buscema, C., Corsi, C. E. Iannicola, G., and Fusi-Pecci, F. 1983, Astr. Ap. Suppl., 51, 83.
Crocker, D. A., and Rood, R. T. (1984), in preparation.
Frogel, J. A., Persson, S. E., Cohen, J. G. 1981, in Astrophysical Processes in Red Giants, eds. I. Iben and A. Renzini (Dordrecht: Reidel), p. 55.
Rood, R. T. (1984), in preparation.

DISCUSSION

Rood: One of our results shows that the result of Frogel et al who find a dispersion of only 0.1 mag in the difference between M_{bol} (observed) and M_{bol} (theoretical flash) is pretty much what is expected on theoretical grounds. This gives an independent way of getting the distance to globular clusters using the brightest giant which should not be in error by more than 0.1 mag. (1 σ).

THE NGC 6752 HR-DIAGRAM

M. Grenon, A. Blecha
Observatoire de Genève

Deep electronographic photometry has been performed down to m_v 21.7 with the 90 mm McMullan camera on the Danish 1.54 m telescope at La Silla. Three exposures per colour were obtained. The plates have been scanned with the Geneva microdensitometer. The selected stars are isolated and are located in the outer corona. The magnitudes are obtained by the profile fitting method, cf. Blecha (1982), which enhances the dynamical range of the plate and reduces the noise due to crowding. The random error appears magnitude independent down to m_v = 21 with σ = .05 per colour. Below $m_v = 20$ a number of galaxies were identified and rejected according to their image profiles.

The structure of the resulting HR-diagram is complicated by the presence of non-member stars, namely the foreground dM with (B-V)$\geqslant$1.3 and V from 18 to 21, and the background stars belonging to the galactic bulge. The latter cause a broadening of the MS between V 18.0 and 19.5. The turn-off is located at V = 17.35 and B-V = 0.44 approximately. Between V 19.2 and 21.2 the MS slope becomes identical to that of Pop. I stars. For (B-V) larger than 0.6, the Carney (1979) sequence is brighter by 0.44 mag. One star is a possible blue straggler with V 16.76 and (B-V) +.32. The blue object, at V 21.4 and (B-V) 0.00 is probably a foreground white dwarf.

The distance modulus is obtained by comparison of the cluster unevolved main sequence with that of nearby subdwarfs having significant parallaxes. Small blanketing corrections have been brought on the subdwarf's M_v for reducing all metallicities to [M/H] = -1.30, the value observed for the red giants in NGC 6752 (Grenon, 1982). With E_{B-V} = 0.04, the distance modulus is now equal to 13.94 $\pm$ .08. This value is significantly larger than those commonly accepted for this cluster ranging from 13.0 to 13.5.

The HB apparent magnitude is somewhat uncertain as the blue branch does not extend up to the flat portion of the HB. With our modulus,

A. Maeder and A. Renzini (eds.), Observational Tests of the Stellar Evolution Theory, 163–165.
© *1984 by the IAU.*

M_V(HB) is estimated to 0.07 $\pm$ 0.18. If such a bright value were to be accepted for all metal-poor globular clusters, then the galactocentric distance based on their spatial distribution would be 1.28 times greater than that obtained using the classical 0.6 value for M_V(HB). The 1980 Harris' value of 8.5 kpc for R_0 would become 10.8 kpc.

For comparing the 6752 ridge sequence with the models the T_{eff} are transformed into (B-V)$_0$ using an ad hoc relation valid for [M/H] = -1.30. VandenBerg's BC for metal poor stars are adopted. The Ciardullo and Demarque isochrones with Y = .20, Z = 0.001 and α = 1.0 fit exactly the subdwarf M_V/(B-V) sequence as well as the slope of the 6752 MS below the turn off. A value α = 1.0 is not adapted for the SGB and GB stars. VandenBerg's (1983) isochrones with α = 1.6 underestimate the luminosity of the MS but describe better the shape of the GB. The observed location of GB stars suggests a value of 1.9 for the mixing length parameter. The luminosity of the flat portion of the SGB appears brighter than predicted by the models.

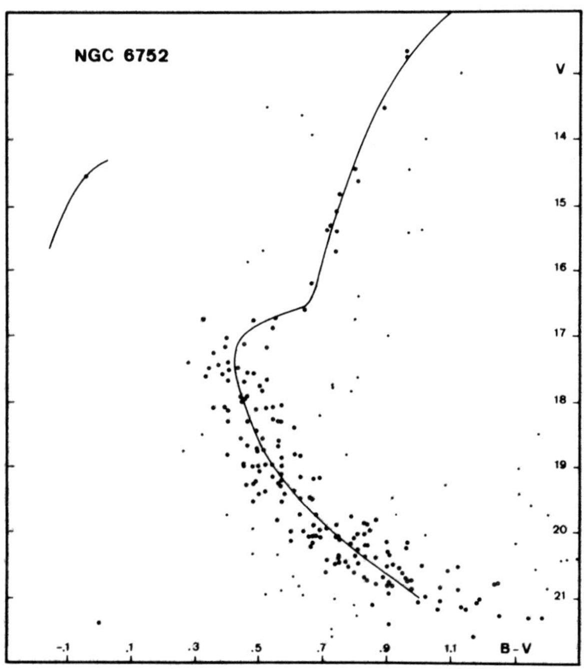

Fig. 1 : Solid lines: electronographic and p.e. ridge
sequences. Small dots: probable non-members

REFERENCES

Blecha, A. 1982, Bulletin Inf. CDS 22, 25.
Carney, B.W. 1979, Astron. J. 84, 515.
Ciardullo, R.B., Demarque, P. 1977, Yale Trans. 33
Grenon, M. 1982, IAU Coll. 68, 393.
Harris, W.E. 1980, IAU Coll. 85, 91.
VandenBerg, D.A. 1983, Astrophys. J. Suppl. 51, 29.

DISCUSSION

<u>Richer</u>: The distance modulus you derive for NGC 6752 puts the horizontal
branch at M_V = .07. This seems somewhat bright, and is caused, I think,
by the fact that you determined the distance by fitting the main sequence to the Carney subdwarf sequence. We show in Richer and Fahlman
(Ap.J., Feb 1, 1984) that the Carney sequence is incompatible with the
VandenBerg isochrones in the sense that the Carney subdwarfs are about
0.4 magnitudes <u>brighter</u> than the appropriate isochrone. A recent Phd
thesis (Dawson, Univ. of Victoria) shows that with a much larger sample
of stars good agreement with the VandenBerg isochrones can be obtained.

<u>Grenon</u>: Our cluster sequence has been fitted to that formed by a set of
nearby stars from Wooley's catalogue. Their parallaxes are rather
accurate and some of them have been recently improved. The discrepancy
with the VandenBerg main sequence luminosity is much larger than the uncertainty of the subdwarf MS location in the HRD.

A NON-EXPLANATION OF THE SANDAGE EFFECT: $[CNO/Fe] \neq 0$

Robert T. Rood
Department of Astronomy
University of Virginia

ABSTRACT

In the $\log T_e$, $\log P'$ plot for RR Lyrae variables Sandage (1981) has noted that at a given color the period of the variables decreases as cluster metalicity increases. In particular the variables in M3 were shifted in $\log P'$ by 0.065 days compared to M15. The only explanation he could find for this shift was that Y in M3 was 0.05 less than in M15 and thus anticorrelated with Z. I have investigated the possibility that the shift is due to variations in $[CNO/Fe]$. At least for the range of $[CNO/Fe]$ considered, it does not seem possible to explain the shift at constant Y.

Synthetic horizontal branches have been constructed following the techniques of Rood (1973) and Rood and Seitzer (1980). New ZAHB's with core masses as indicated by red giant calculations have been constructed as necessary. Evolution is only crudely approximated. For M15 the mean mass loss was adjusted so that 70% of the HB was blueward of the variables. For M3 it was adjusted so the number of blue HB was approximately the same as the number of RR Lyrae. Different assumptions have been made concerning the composition and are designated (assuming $Z_\odot = 0.02$):

Name	Y	[Fe/H]	[CNO/Fe]	$\langle \Delta M \rangle$	$\log (P')_{3.84}$
M15F	0.75	−2.0	0.0	0.100	−0.304
M15G	0.75	−2.0	0.48	0.147	−0.312
M3F	0.75	−1.6	0.0	0.135	−0.328
M3G	0.75	−1.6	0.25	0.160	−0.322
M3G80	0.80	−1.6	0.0	0.210	−0.378

Figure 1 shows the sort of color-magnitude diagram these simulations produce. ("Observational error" of 0.01 m in V and 0.02 m in B-V is included). Figure 2 shows the $\log (T_e)$,

A. Maeder and A. Renzini (eds.), Observational Tests of the Stellar Evolution Theory, 167–170.

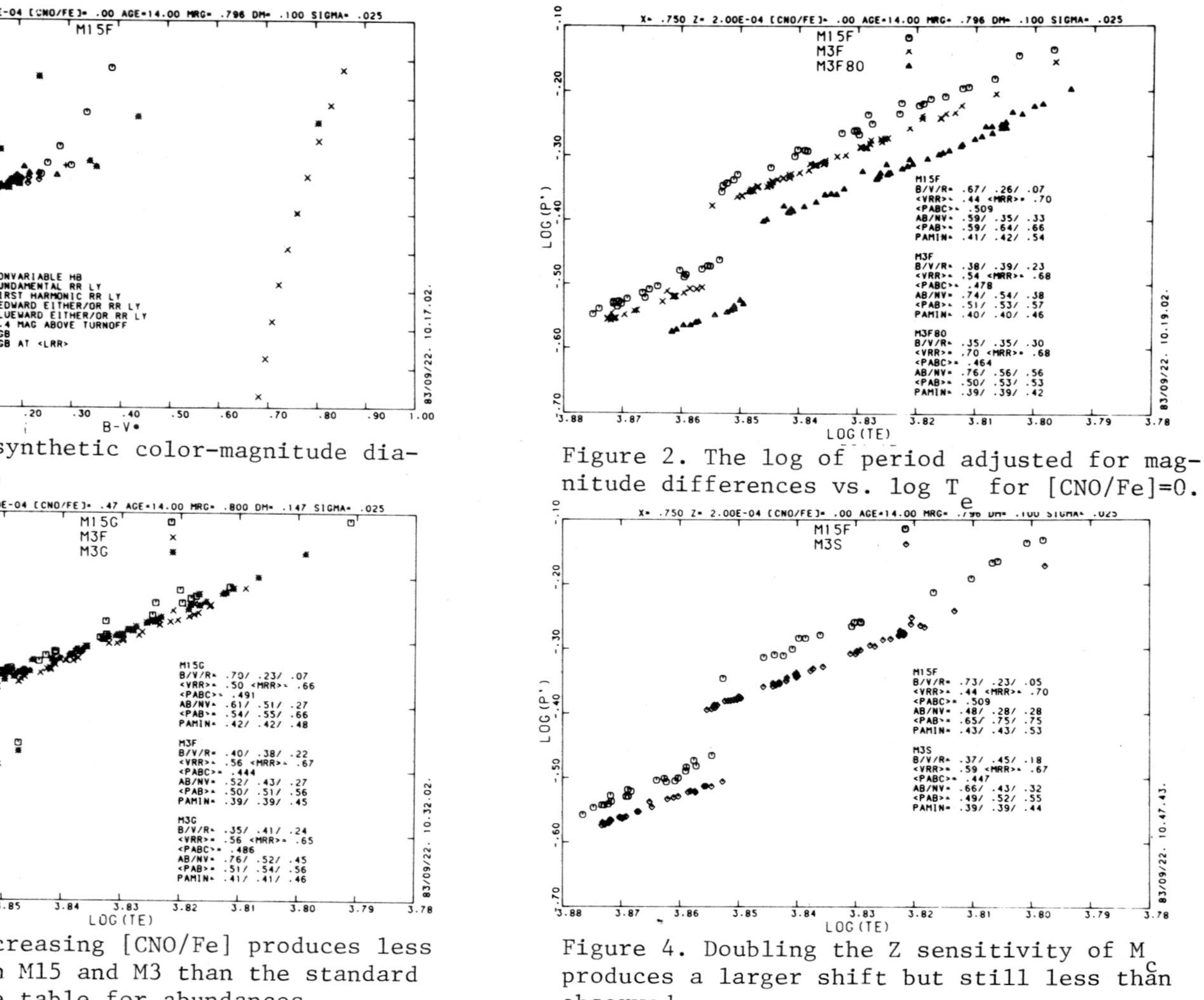

Figure 1. A synthetic color-magnitude diagram for M15.

Figure 2. The log of period adjusted for magnitude differences vs. log T_e for [CNO/Fe]=0.

Figure 3. Increasing [CNO/Fe] produces less shift between M15 and M3 than the standard case. See the table for abundances.

Figure 4. Doubling the Z sensitivity of M_c produces a larger shift but still less than observed.

log P' [= log (P) + 0.328 (V-V$_{AV}$)] diagram for the standard assumption
[CNO/Fe] = 0. The shift at constant Y (log P' at log T$_e$ = 3.84 is
given in the table) is about a factor of 3 too small, and Y(M15) =
Y (M3) + 0.05 produces the observed shift. Figure 3 shows experiments
with increased CNO. Both a CNO enhanced M15 vs. an unenhanced M3,
or vs. a slightly enhanced M3 (equivalent to CNO = CNO$_o$ + cons. x Fe)
produce even less shift than the standard case. These are small
increases in CNO compared to those suggested by Renzini at this meeting,
and there may be some threshold effect yet to be exposed. Finally
Figure 4 shows the effect of doubling the predicted Z dependence of
M core; the observed shift will only be achieved if d M core/d log Z
is 5 times the predicted value. At this point it does not appear that
a minor shift in the input physics will work.

The same basic conclusions hold for a wider range of [Fe/H]
for M3 and M15 than those reported here. The only explanation which
I have found is the helium shift. Yet, the helium shift explanation
has some difficulties beyond its implausibility. It predicts that the
blue edge of the instability strips should shift by Δlog T$_e$ ~ -0.01
in M3 compared to M15. This corresponds to a shift of roughly 0.03 m
in B-V which does not seem out of the question given the errors in
reddening. The difficulty arises when the theoretical diagrams are
adjusted assuming reddening errors lend to coincident blue edges,
the shift in log P' between M3F80 and M15F is then only slightly
larger than the standard case, and we are left with no explanation.
Further the lower Y for M3 would imply that the gap between turnoff
and HB would be 3.25 m rather than the observed 3.4 if M3 is the same
age as M15. Neither shifts in Y or CNO produce the mass difference
of -0.1M$_\odot$ between M15 and M3 suggested by Cox at this meeting. There
is really no satisfactory explanation of the period shift.

This work is partially supported by NSF 81-08418.

REFERENCES

Rood, R. T. 1973, Ap. J., 184, 815.
Rood, R. T., Seitzer, P. O. 1981, in Ap. Parms. of Globular Clusters,
      eds. A. G. D. Philip, D. S. Hayes, (Schenectady: L. Davis),
      p. 369.
Sandage, A. 1981, Ap. J., 248, 161.

DISCUSSION

Nissen: I would like to ask Renzini about the effect of varying [CNO/Fe];
he seems to reach quite different conclusions.

Renzini: Bob and I have discussed this point at length over the past
couple of days. I think we agree that more model calculations are required before concluding about [CNO/Fe] as the cause of the period-shift
effect. The reason is that Bob has not computed HB models for [Fe/H]
$\simeq$ -1.3 and [CNO/Fe] $\simeq$ 0.5-1.0. Opacities are rather non-linear with
abundance of heavy elements, in the interesting temperature range. This
may explain Bob's negative result. Also, the opacity tables used by Bob
assume [Ne/Fe] = 0 while I would rather expect Ne to follow oxygen rather
than iron (on nucleosynthetic arguments). Remember that Ne is at least
as important as oxygen, as far as the opacity is concerned.

REMARKS CONCERNING THE MASSES OF HORIZONTAL-BRANCH A STARS

A. G. Davis Philip

Van Vleck Observatory and Union College

D. S. Hayes

Kitt Peak National Observatory

Kristina D. Philip

Van Vleck Observatory

Studies are underway concerning the masses of horizontal-branch A stars. Hayes and Philip (1979) discussed masses determined for seven field HB stars using gravities and temperatures obtained from spectrophotometric scans. They found a mean mass of $0.6 M_\odot \pm 0.4$ for these stars. However, there was a strong trend of mass with temperature. In the present study four-color measures of over 150 FHB and BHB stars have been dereddened and plotted in the grid relating $(b-y)$, c_1 and $\log g$, T_{eff} (Philip and Relyea 1979). If one calculates a mean mass, again the figure of $0.6 M_\odot$ is obtained. A more detailed look at the data, however, reveals structure in the distribution of points in the various diagrams. A discussion of this result follows.

In the four figures solid triangles represent BHB stars in the globular cluster M 4 and hollow triangles represent the best observed FHB stars (an average of 44 observations each). The four-color indices c_1 and m_1 are plotted vs. the index $(b-y)$ in Figs. 1 and 2. Most of the points fall along a line in each diagram, but there are four dotted symbols in each figure that fall away from the main relation. In M 4 this main relation represents a line of constant mass in the c_1 vs. $(b-y)$ diagram. The dotted symbols were selected from the distribution of points in Fig. 3 in which the mass (calculated from the four-color indices transformed to $\log g$ and T_{eff}) is plotted versus T_{eff}. In this diagram the points scattered about two relations, marked A and B. The four most extreme points on relation B were dotted. In Fig. 4 (mass vs. $\log g$) it can be seen that these four points are the most extreme points in the graph.

A. Maeder and A. Renzini (eds.), Observational Tests of the Stellar Evolution Theory, 171–173.
© *1984 by the IAU.*

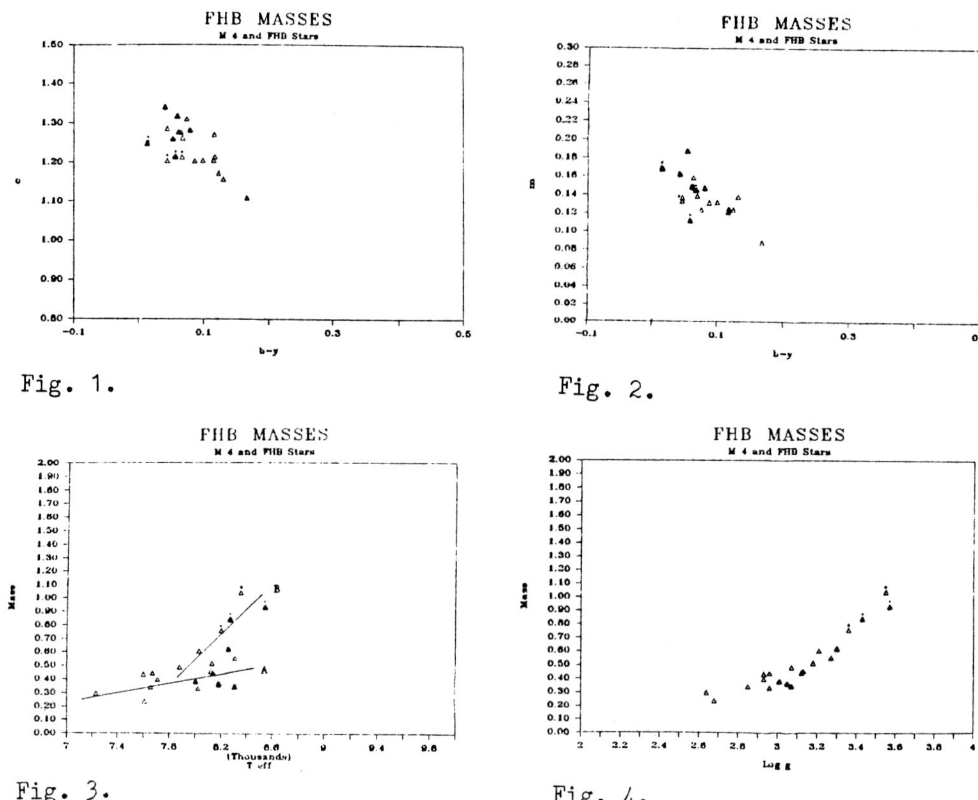

Fig. 1.   Fig. 2.

Fig. 3.   Fig. 4.

Our interpretation of the distribution of points in these four graphs is that some of the BHB stars in globular clusters and some of the FHB stars have higher masses (and lower c_1 and m_1 indices) than the majority of horizontal-branch A-stars. We plan to use the line of constant mass in the c_1 vs (b-y) diagram to calibrate the Population II grid relating c_1 and (b-y) to log g and T_{eff}. At present the relation A in Fig 1. has a slope and the masses predicted by the relation are too small; most probably the surface gravities are somewhat in error. There should be no trend of mass with temperature.

Additional four-color observations are planned of BHB stars in globular clusters and H β observations of the FHB stars in order to obtain additional data to confirm our results. HD 117880 is one of the FHB stars that has a high mass; it also has peculiar UV and IR colors.

REFERENCES

Hayes, D. S. and Philip, A. G. D. 1979 PASP 91, 71.
Philip, A. G. D. and Relyea, L. J. 1979 Astron. J. 84, 1743.

DISCUSSION

Richer: Do you place the magnitude of the horizontal branch of M4 at V = 13.6?

Philip: In the four colour system measures that I make, I measured the quantities b-y, c_1 and m_1. I do not reduce the y magnitude. For informational purposes the V magnitudes for the M4 BHB stars were taken from Greenstein's 1939 paper in the Ap.J.

Alcaino: Our determination of the horizontal branch position of M4 is at a visual observed magnitude of V_{HB} = 13.4.

Lub: Have you tried the later and improved version of the Kurucz models?

Philip: The Kurucz models used were the revised models reported in 1976 (Problems of calibration of multicolor photometry, Dudley Obs. Report no 14). The grid relating the four colour indices $(b-y)_o$ and $(c_1)_o$ to log g and T_{eff} for [Fe/H] = -1 was given in Philip and Releya (1979, Astron. J. 84, 1743).

LUMINOSITY FUNCTIONS FOR ASYMPTOTIC GIANT BRANCH STARS

Jay A. Frogel and V.M. Blanco
Cerro Tololo Inter-American Observatory

1. Introduction and Summary

To understand the evolution of carbon and oxygen rich asymptotic giant branch (AGB) stars it is important to <u>observationally</u> establish the relations between parameters such as luminosity, age, and metallicity. The existence of luminous (brighter than the top of the first giant branch) AGB stars in clusters in the Magellanic Clouds (Mould and Aaronson 1979; Frogel, et al. 1980) provides a nearly ideal situation in which these relations can be determined. Although metallicities and main sequence turn-off ages are known for only a fraction of the clusters, the ranking scheme devised by Searle, et al. (1980; SWB) indicates that age and metallicity are closely related to one another.

A search has been completed and infrared photometry obtained for luminous AGB stars in 35 clusters classified by SWB. This paper discusses the bolometric luminosities of the C and M stars found.

II. Observational Results

Magellanic Cloud clusters were surveyed for C and M stars with a transmission grating and thin prism at the prime focus of the CTIO 4-m reflector as described by Blanco, et al. (1980, BMB). A few additional luminous red stars were identified in the cores of the most crowded clusters with the aid of multicolor images obtained with a SIT vidicon system. Infrared photometry is available, either from the literature or from new data, for all stars found. These data allow bolometric luminosities to be obtained (Frogel, et al. 1980).

Luminous carbon stars were found in 18 of the 20 clusters of SWB types IV-VI searched. With few exceptions the faintest C star is more luminous than the brightest M star in the same cluster. This extends considerably the result of Frogel and Cohen (1982) based on fewer clusters. We define a "transition luminosity" for a cluster as a flux average of the brightest M and the faintest C star. If there are no M stars in a cluster, 0.7 is arbitrarily added to the bolometric

A. Maeder and A. Renzini (eds.), Observational Tests of the Stellar Evolution Theory, 175–178.
© *1984 by the IAU.*

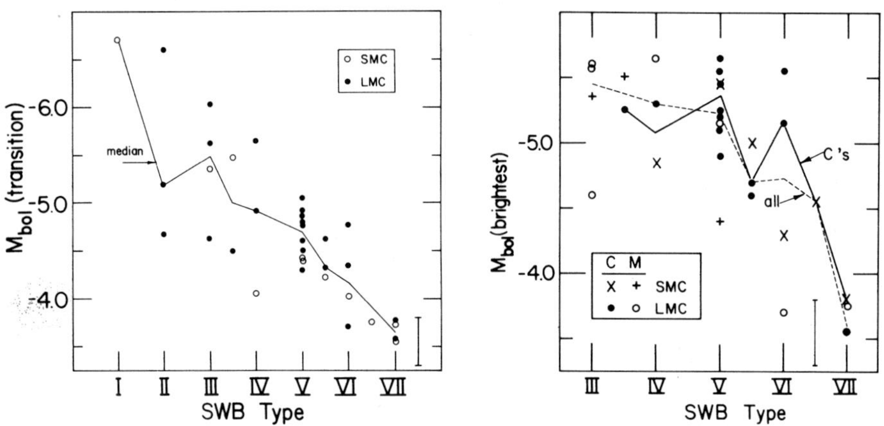

Figure 1.   The C-M star transition luminosity for each of the SWB
clusters surveyed.   The bar at the lower right is the range in
maximum luminosity observed for giants in Galactic globulars.

Figure 2.   The bolometric magnitude of the brightest C star (M star
if a cluster has no C's) in each of the SWB clusters in our survey.

magnitude of the faintest C star.   If there are no C stars, the
transition luminosity is taken as equal to that of the brightest M star.
It is apparant from Figure 1 that the transition luminosity increases
steadily as one goes to early (younger) SWB cluster types.   A similar
plot for clusters with main sequence ages (Hodge 1983) is less conclu-
sive because such ages are available for only a small number of clusters
in the sample and these tend to be systematically less than ages
determined by other means for the same clusters (Hodge 1981).

The mean bolometric magnitude of the 10 C stars in the group IV
clusters is only 0.04 brighter than that of the 38 C stars in the group
V clusters but is 0.37 mag brighter than that of the 17 C stars in the
group VI clusters.   The luminosity spread for each of the three groups
of C stars from these three cluster groups is 0.08-0.17 mags less than
for field C stars in the Clouds (Cohen, et al. 1981).   However, the
combined distribution of the cluster C stars is virtually identical
(mean magnitude and dispersion) to that of the field stars.

Hodge (1983) noted that cluster ages derived from the luminosity
of the brightest AGB star (e.g. Mould and Aaronson 1983) correlate
poorly with ages derived from other techniques.   Figure 2 displays
the same lack of correlation.   In fact, this figure suggests that for
SWB groups III-VI the maximum luminosity of an AGB star is independent
of age.   The vertical bar in the lower right of the figure indicates
the range in luminosity exhibited by the tips of the giant branches of
galactic globular clusters.   Tip luminosity increases with metallicity.
Only long period variables in the metal rich galactic globulars exceeds

the top of the bar (Frogel, et al. 1983). Hence the type VII clusters appear not to have luminous AGB stars, consistent with SWB's identifying them with galactic globulars.

III. Discussion

The dependence of the transition luminosity of SWB cluster type is qualitatively similar to Iben and Renzini's (1983, fig. 7) prediction of the dependence on age of the luminosity of the start of a star's thermally pulsing AGB phase. If ages of 0.1 and 10 Gyr are assigned to type III and VII clusters, respectively (Rabin 1982), the observed and predicted luminosities become quantitatively similar. The S stars discovered by Bessell, et al. (1983) in a number of the clusters of the present sample lie close to the transition luminosities of the clusters as would be expected if they represent the first observable evidence of mixing of processed material to the surface of a thermally pulsing AGB star (as was also noted by Bessell, et al.).

From the work reported on here and from inspection of similar data presented by Bessell, et al. (1983) and Mould and Aaronson (1983) it is obvious that in a population of AGB stars which can be characterized by a single age and metallicity, carbon stars are significantly more luminous than oxygen rich stars. We have shown that a summing of cluster C star luminosity functions reproduces that for Magellanic Cloud field stars quite well. It is likely that a significant range in age and metallicity amongst the field star contributes to the apparent similarity of the field C and M star luminosity functions. (BMB and Cohen, et al. 1981). Furthermore, the BMB survey specifically excluded stars earlier than M5. With the inclusion of M2-4 stars (Frogel and Blanco 1983) the mean of the field M star luminosity function becomes nearly 0.5 magnitudes fainter than that for field C stars. Thus, there seems to be no need to invoke nondeterministic mixing processes to explain C stars (Miller and Scalo 1982). Instead, it seems that the processes which are responsible for bringing carbon rich material to the surface of a star do so at a well defined luminosity. This luminosity is a function of stellar age and metallicity.

If the SWB cluster sequence provides an accurate age/metallicity ranking (Searle and Smith 1981), than the lack of dependence of the luminosity of the brightest AGB stars on cluster type (Figure 2) may be related to the problem of the "missing luminous C stars" (Iben 1981). The situation in Figure 2 could arise, for example, if true mass loss rates for the most luminous stars are considerably higher than normally assumed. In any case, Figure 2 is not consistent with the use of these luminous stars as age indicators.

Finally, on the basis of these data we consider it unlikely that luminous C stars can be converted back into luminous M stars (Iben 1981). If such a fate befell a significant fraction of the most luminous C stars, then the almost complete lack of overlap in luminos-

ity for the M's and C's in a cluster would require that the born-
again M star be of much lower luminosity than its predecessor. This
would violate the core-mass luminosity relationship.

REFERENCES

Bessell, M.S., Wood, P.R., and Lloyd Evans, T.: 1983, M.N.R.A.S. 202,
     p. 59.
Blanco, V.M., and Frogel, J.A.: 1983, Ap. J., to be submitted.
Blanco, V.M., McCarthy, M.F., and Blanco, B.: 1980. Ap. J. 242, p. 938.
Cohen, J.G., Frogel, J.A., Persson, S.E., and Elias, J.H.: 1981,
     Ap. J. 249, p. 481.
Frogel, J.A. and Cohen, J.G.: 1982, Ap. J. 253, p. 580.
Frogel, J.A., Cohen, J.G., and Persson, S.E.: 1983, Ap. J., in press.
Frogel, J.A., Persson, S.E., and Cohen, J.G.: 1980, Ap. J. 239, p. 495.
Hodge, P.: 1981, in IAU Coll. 68, Philip, A.G.D. and Hayes D.S., eds.
     L. Davis Press, p. 205.
Hodge, P.: 1983, Ap. J. 264, p. 470.
Iben, I., Jr.: 1981, Ap. J. 246, p. 278.
Iben, I., Jr., and Renzini, A.: 1983, Ann. Rev. Astr. Ap., in press.
Miller, G.E., and Scalo, J.M.: 1982, Ap. J. 263, p. 259.
Mould, J., and Aaronson, M.: 1979, Ap. J. 232, p. 421.
Mould, J., and Aaronson, M.: 1983, Ap. J., in press.
Rabin, D.: 1982, Ap. J. 261, p. 85.
Searle, L., and Smith, H.A.: 1981, in IAU Cool. 68, Philip, A.G.D.
     and Hayes, D.S., eds., L. Davis Press, p. 201.
Searle, L., Wilkinson, A., and Bagnuolo, W.: 1980, Ap. J. 239, p. 803.

DISCUSSION

<u>Mould</u>: One of the sources of scatter in Figure 2 must be the stochastic
population of the AGB. Less massive clusters have a lower probability
of approximating a star truly marking the AGB tip. Secondly, one should
note that the luminosity function of Long Period Variables in the
Magellanic Clouds and the luminosity function of photometrically selected red giants show that <u>some</u> stars with bolometric magnitude −6 are
present. Their numbers are reduced but not to zero.

<u>Frogel</u>: I agree that the Long Period Variables being found in the
Clouds are quite luminous. The problem is their scarcity.

ON THE TERMINATION OF ASYMPTOTIC GIANT BRANCH EVOLUTION

P. R. Wood and D. J. Faulkner
Mount Stromlo and Siding Spring Observatories,
Canberra, Australia

ABSTRACT

The evolution of planetary nebula (PN) nuclei has been studied at masses of 0.60, 0.70 and 0.76 $M_\odot$, and for the ejection of the PN at various phases of a helium shell flash cycle. The evolution at high luminosities takes longer for nuclei resulting from PN ejection at shell flash peak than it does for those resulting from ejection in the inter-flash phase. Comparison of our calculations with various observational results does not allow us to reach any definite conclusions regarding the phase of the shell flash cycle at which PN ejection occurs.

1. INTRODUCTION

The means by which asymptotic giant branch (AGB) evolution is terminated for low mass ($M \lesssim 5 M_\odot$) stars is generally assumed to be PN ejection, but the nature and duration of the ejection process are at present very speculative. There is a widely held belief that stars such as the OH/IR stars, which have high mass loss rates, are in the phase of PN ejection, but the cause of the high $\dot{M}$ is not known. One suggested cause is a switch in the mode of pulsation of a Mira variable from first overtone to fundamental, the large amplitude of the latter mode driving the high mass loss rate (Wood 1974; Tuchman, Sack and Barkat 1979). Such a mode switch would be expected to occur when the luminosity of the star first exceeds some critical value. Since any specific AGB luminosity is first attained during the surface luminosity rise associated with a helium shell flash (e.g., Wood and Zarro 1981), PN ejection by the above process should be initiated at a helium shell flash peak rather than in the quiescent interflash phase of lower surface luminosity. The purpose of the present study is to examine the evolution of PN nuclei in two cases, (i) when the ejection occurs at a helium shell flash, and (ii) when it occurs midway between flashes. A comparison with observational data for PN nuclei might then distinguish between the two possibilities.

A. Maeder and A. Renzini (eds.), Observational Tests of the Stellar Evolution Theory, 179–182.

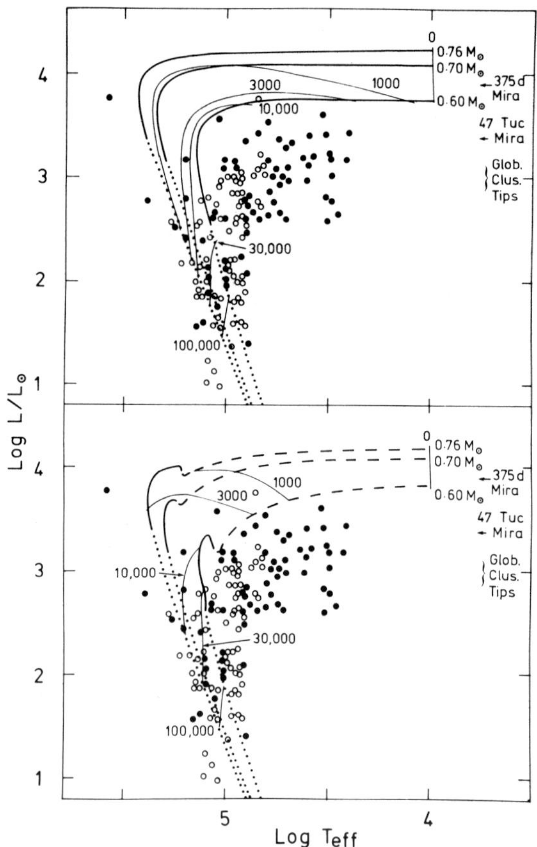

Fig. 1 Evolutionary tracks of PN nuclei of the masses shown when PN
ejection occurs midway between helium shell flashes (top), or at the
peak of a shell flash (bottom). Hydrogen burning phases are shown as
thick lines, helium burning phases as broken lines, and gravitational
contraction phases as dotted lines. The thin lines display equal intervals of time since $\log T_{eff}=4.0$. The observational PN positions of
Pottasch (1983) (dots) and of Kaler (1983) (open circles) are also shown,
as are the luminosities of globular cluster red giant tips, 47 Tuc Miras
and typical Galactic disc Miras.

2. RESULTS

Fig. 1 shows evolutionary tracks for PN nuclei of mass 0.6, 0.7
and 0.76 $M_\odot$ for both the ejection possibilities under consideration. In
both series of tracks, mass loss at the rate of 3×10^{-5} $M_\odot$ yr^{-1} was
applied until the star's effective temperature had increased to $\log T_{eff}$
= 3.8; thereafter $\dot{M}$ was set to zero. The main effect of ejection at a
helium flash is to prolong the evolution at high luminosities, particularly for the higher mass cores. A similar result has been noted by
Iben (1983).

3. DISCUSSION

Also shown on Fig. 1 are the observationally derived luminosities and T_{eff} values for PN nuclei from Pottasch (1983) and Kaler (1983); in some cases the Kaler points represent only limiting positions for the nuclei. These results, particularly those of Pottasch, indicate that the bulk of PN nuclei should have masses ≤ 0.6 $M_\odot$ based on their positions relative to the evolutionary tracks. In fact the bulk of these observations indicate evolution away from the AGB at luminosities less than those characteristic of globular cluster giant branch tips (in particular the luminosities of the 47 Tuc Miras) and much less than those of a typical Mira variable with P=375 days. Since the bulk of the PN near the Sun come from stars more massive than stars presently on the globular cluster giant branches, and probably come from stars similar to a typical Mira variable, it is difficult to see how any meaningful conclusions can be drawn at present from a comparison of observation and theory using the HR diagram plane alone. Indeed, we would suggest that the present observational estimates of luminosities of PN Nuclei are too low by an average factor of ~3. An alternative approach is to introduce evolutionary time-scales into the comparison by means of the (M_V, time) diagram of Schönburner (1981), and we have experimented with such comparisons using our theoretical results and published observational data. In making such a comparison it has usually been assumed that the zero point of the theoretical timescale can be set at some arbitrarily chosen T_{eff} value, but it seems to us that this is of dubious validity in view of our ignorance of the duration, mass loss rate, etc., of the PN ejection process itself. Once again we have been forced to the conclusion that the (M_V, time) diagram also yields little information on the phase of the flash cycle at which ejection occurs. Full details will be published in Faulkner and Wood (1983, in preparation).

REFERENCES

Iben, I.: 1983, preprint.
Kaler, J.B.: 1983, Astrophys. J., 271, pp. 188-220.
Pottasch, S.R.: 1983, IAU Symp. No. 103, pp. 391-407.
Schönberner, D.: 1981, Astron. Astrophys., 103, pp. 119-130.
Tuchman, Y., Sack, N., and Barkat, Z.: 1979, Astrophys. J., 234, pp.
        217-227.
Wood, P.R.: 1974, Astrophys. J., 190, pp. 609-630.
Wood, P.R., and Zarro, D.M.: 1981, Astrophys. J., 247, pp. 247-256.

DISCUSSION

Cox: In constructing models for pulsation calculations, we found that there is frequently a convective shell even at 10^5 K. This is due to the high stellar material opacity. Do your models have surface convection?

Wood: Not in the high temperature parts of the evolutionary tracks.

<u>Weidemann</u>:   These questions have been extensively discussed at the Erice and London meeting by Schönberner and myself, with the result that the observational evidence, especially on the M_v distribution, favours evolution by quick hydrogen- rather than helium-burning. The production of DB white dwarfs by phasing of the shell flashes as investigated e.g. by Iben is also not convincing since the fraction of DB stars is actually smaller (10-12 %) than predicted.

<u>Schönberner</u>: Tomorrow I shall present an HR diagram for central stars which gives better agreement between theory and observations. Now my question: What are the mass loss rates used to get rid of the envelopes?

<u>Wood</u>: $\dot{M} = 1\ M_\odot\ yr^{-1}$ down to an envelope mass of 0.01 $M_\odot$. Then $\dot{M} = 10^{-5}$ $M_\odot\ yr^{-1}$ until log T_{eff} had increased to 3.8. From here on, various mass loss rates were used, $\dot{M} = 0$ being the ones discussed in detail here.

CONSTRAINTS ON TERMINAL AGB EVOLUTION FROM PROPERTIES OF MIRAS AND OH-IR
SOURCES

L. A. Willson and George H. Bowen
Iowa State University
Ames Iowa 50011 USA

Near the tip of the AGB we find the optical Mira variables and also
the OH-IR stars. Observational data on the luminosities, periods, and
population membership of the OH-IR sources are beginning to be available;
in principle this information should constrain AGB models very strongly.
In practice, current data are found to be insufficient to unambiguously
determine the relation between Miras and these sources; in fact three
very different interpretations are currently possible.

Three currently available period-luminosity relations for the OH-IR
sources and Miras are compared in Figure 1. These give very different
pictures of the relationship between Miras and OH-IR sources.

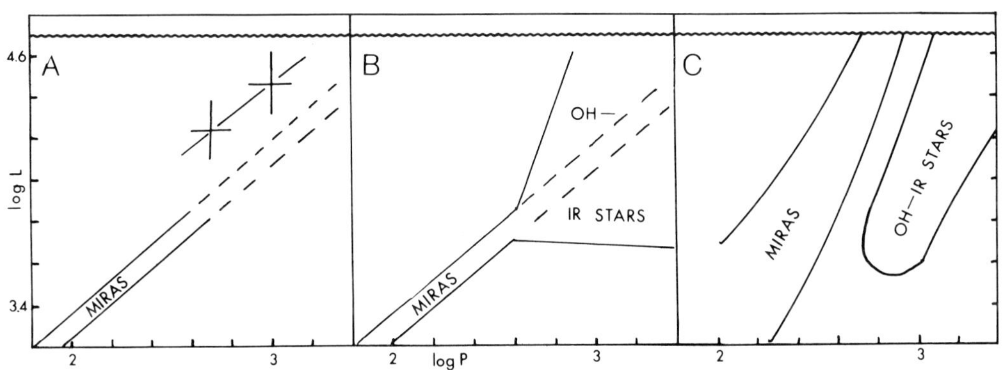

Figure 1. Period – luminosity relations for Miras and OH-IR sources based
on A: an analysis of radio luminous IR sources by deJong, (1983); B:
observations and analysis of galactic sources by Engels et. al. (1983 =
EKSS); C: observations and analysis of red stars in the LMC by Wood,
Bessell, and Fox (1981 = WBF). The AGB limit log $L/L_\odot$ = 4.7 is
indicated. In A, B the Mira period-luminosity relations are from Glass
and Lloyd-Evans (1981).

A. Maeder and A. Renzini (eds.), Observational Tests of the Stellar Evolution Theory, 183–186.
© *1984 by the IAU.*

deJong selected radio-luminous infrared sources with periods near 500, 1000^d; in Fig. 1A his derived luminosities are compared with the period-luminosity relations of Glass and Lloyd-Evans (1981) for the LMC (upper line) and galactic Miras. The OH-IR sources studied by deJong lie above the extrapolated Mira relation; a similar result is obtained if M_{ms} is plotted against P for the same data. One interpretation of these plots is that for fixed luminosity, P(OH-IR) / P(Mira) ~ 1/3 -- as would result from fundamental mode pulsation for the Miras and first overtone pulsation for the OH-IR sources.

In contrast, Engels (EKSS and this volume) finds a period-luminosity relation for the OH-IR stars which is continuous with that of the Miras (Figure 1B). The scatter in luminosity, particularly at the longer periods, is very large; this may be a reflection of observational uncertainty, or it may represent a real range of L for fixed P. The simplest interpretation of this relation is that the Miras and at least most of the OH-IR stars are pulsating in the same mode.

Finally, WBF found for the long period sources in the LMC and SMC a discontinuity between the "Miras" and the longer period sources, which they interpret as a shift from overtone "Mira" pulsation to fundamental mode pulsation for the longer period objects (Figure 1C). However their definition of "Mira" variables is less strict than that usually used. Strictly, a Mira is a variable with visual amplitude >2.5 mag, emission lines of hydrogen at least part of the time, and a light curve with constant mean magnitude which is regular in timing and amplitude from cycle to cycle to within 15%. Other red long period variable belong to various semiregular (SR) classes; it is not yet known what relation the SRs are to the Miras. If questionable "Miras" are eliminated from WBF's data, and if the same bolometric corrections and distance modulus are used, then the Glass and Lloyd-Evans relation is recovered, at least for $P < 350^d$ (Lloyd-Evans, this volume).

What can be done both observationally and theoretically to select the correct picture from among the three? First, we need to know whether there is in fact a discontinuity in the properties of Miras and OH-IR sources -- and in what properties this discontinuity can be seen. Second, sorting properties by period may not produce groups of sufficient homogeneity for the statistical analyses usually undertaken. The expected range of L at fixed P is very large even if only one mode is present, because stars of differing M reach the same P at different L. Ultimately, direct comparison of the distribution of L vs. P with theoretical predictions will be required to sort rigorously among the models.

Theoretical calculations as well as observations of red LPVs in globular clusters indicate that as a star evolves up the AGB it first pulsates in an overtone, then converts to fundamental mode pulsation at some critical (L,T,M). This has led to two evolutionary scenarios including Miras and OH-IR sources, in both of which the more evolved OH-IR sources are fundamental mode pulsators (e.g. Willson 1982, Wood 1982). Figure 1C is consistent with Wood's scenario, where OH-IR sources are the

result of a mode switch from overtone Mira pulsation to fundamental
mode. Figure 1B is consistent with the Willson scenario, where both OH-IR
sources and Miras pulsate in the fundamental mode. If Figure 1A applies,
however, that is if there is a distinct group of high-luminosity overtone
pulsating OH-IR stars, then a mechanism must be found for a second mode
switch, and/or for more massive, more luminous stars to develop rapid mass
loss while remaining overtone pulsators.

A possible mechanism is inherent in the mass loss process itself:
When the star's mass loss rate exceeds some critical value, probably
around 10^{-5} $M_{\odot}$/ yr , the star develops an optically thick dust shell.
This produces the characteristic appearance of the OH-IR source; it must
also have some effect on the temperature boundary condition at the
surface of the star. If we model the circumstellar dust shell very
simply, as a spherical "blanket" several stellar radii out, then we can
calculate the backwarming effect in two simplified cases. If the stellar
radiation is absorbed by the blanket without scattering, and re-radiated
at the equilibrium Planck temperature (near 1000K), then the effect on the
boundary temperature will be small -- less than 1% for stars above
2500K. However if scattering is important, so that perhaps 50% of the
photons are scattered back towards the star without change in energy, then
the boundary temperature will be raised by 10% -- easily enough to move
the star back across the boundary in (L,T,M) between fundamental and
overtone pulsation.

Further model calculation efforts should focus on determining (a) the
dependence of the preferred pulsation mode on $(L, T, M_{*}, M_{core}, Y, Z)$;
(b) the scaling law for pulsation enhanced mass loss from Miras; (c) the
effects of such a mass loss process on the evolutionary tracks and in
particular (d) the effects of an opaque, scattering circumstellar envelope
on the stellar characteristics and pulsation mode. At the same time,
better observational constraints are needed, particularly luminosities,
mass loss rates and current masses of Miras and OH-IR sources as a
function of at least two observable parameters -- period and probably
infrared amplitude or expansion velocity.

Bibliography

deJong, T. 1983: (to appear - Ap. J.)

Engels, D., Kreysa, E., Schultz, G. V., and Sherwood, W. A. 1983: Astron.
Astrophys. 124, p. 123.

Glass, I. S., and Lloyd-Evans, T., Nature 291, p. 303.

Willson, L. A. 1982: Proceedings of the 1982 Boulder conference on
Pulsations in Classical and Cataclysmic Variables", J. Cox and C. Hansen,
editors, p 269.

Wood, P. R. 1982: ibid., p 284.

DISCUSSION

<u>Frogel</u>: How high an albedo does the dust need to give sufficient back-scattering?

<u>Willson</u>: I assumed 50 % backscattering, that is a very high albedo; I also assumed single scattering of photons. These effects are in opposite directions. The purpose of this calculation is just to show that a 10 % effect is reasonable; this is very large in relation to the pulsation mode and the HR diagram tracks for these stars.

RADIAL VELOCITY OBSERVATIONS OF BARIUM, CH, AND R STARS

Robert D. McClure
Dominion Astrophysical Observatory
Herzberg Institute of Astrophysics

The Ba II, CH, and R0-R3 carbon stars are related in that they are giant stars having about the same position in the HR diagram, and having enhancements of carbon features. The Ba II and CH stars are also noted for enhancements of s process elements. The Ba II stars normally don't exhibit C_2 bands that are characteristic of the CH and R stars, indicating that the Ba II stars have a C/O ratio close to but less than unity (Smith 1983). The Ba II and R stars are members of the old disk population (e.g. Eggen 1972), whereas the CH stars are high velocity, population II stars (Keenan 1942).

Radial velocity measurements of Ba II stars have been obtained during the last 4 years at the Dominion Astrophysical Observatory (DAO), and the conclusion has been reached that these stars are likely all members of binary systems. These observations, as of the end of 1982, have been published by McClure et al. (1980), and McClure (1983), and will not be discussed further. The connection between multiplicity and abundance peculiarities is uncertain, but two possibilities exist. Either a companion has evolved and exchanged enriched material onto the present Ba II star, or the companion has somehow affected the internal structure of the Ba II star making it mix. In the former case, a white dwarf companion might be expected, and some have been found (Böhm-Vitense 1983; Böhm-Vitense et al. 1983; Dominy and Lambert 1983) from IUE spectra. Dominy and Lambert (1983) argue that because the majority have no observed white dwarf companions, this hypothesis is ruled out, whereas Böhm-Vitense et al. (1983) argue in favour of mass exchange, feeling that there is a magnitude limit beyond which the white dwarf companions are just not observable.

The question arises: if all Ba II stars are binaries, are the related CH and R stars also members of binary systems? One might question this suggestion in the case of the CH stars since these are population II objects, and found in several globular clusters, populations normally thought to be deficient in binary systems (e.g. Abt and Levy 1969; Crampton and Hartwick 1972; Gunn and Griffin 1979). Figure 1 shows radial velocity data for CH stars, and subgiant CH stars (Bond

187

A. Maeder and A. Renzini (eds.), Observational Tests of the Stellar Evolution Theory, 187–190.

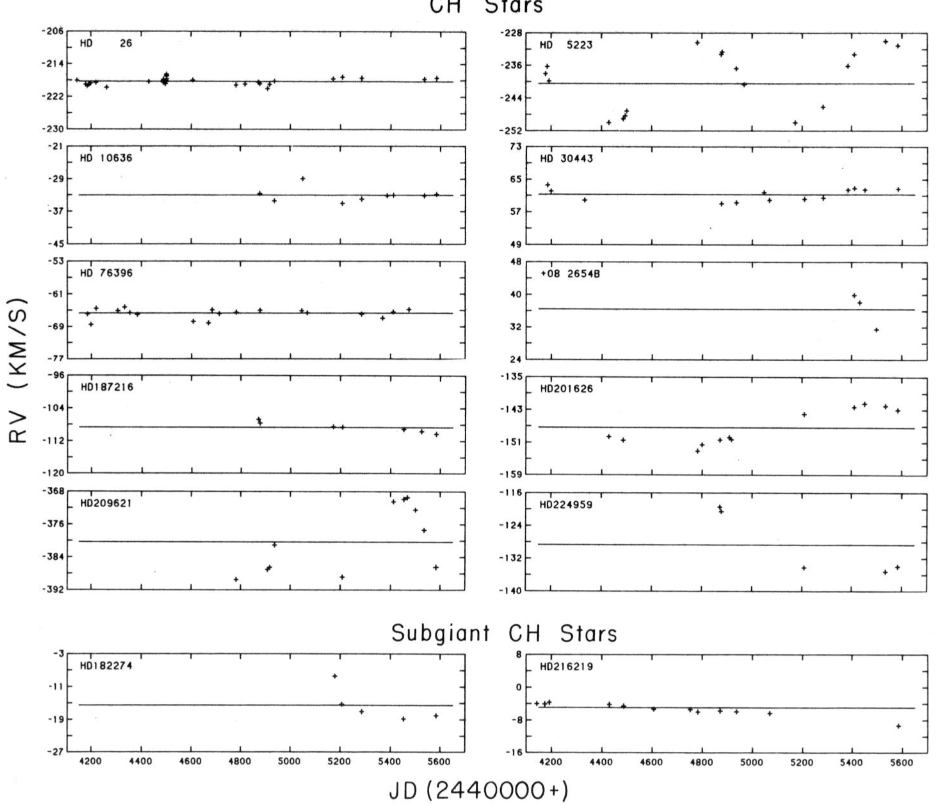

Figure 1.   Radial velocities of CH stars versus Julian data.   The solid
lines represent the mean of the velocities.

1974), obtained with the DAO radial velocity coudé spectrometer during
the last several years.   Two effects are apparent.   First, the majority
of CH stars do indeed have long term velocity variations such as
expected of a giant member of a binary system.   Second, the scatter of
the observations from night to night appears to be larger than the
expected precision of the instrument, which is normally better than 0.5
km s^{-1}.   The latter value is represented in the figure by the size of
the symbols.   Since these are population II giants, and must have very
low surface gravities, perhaps turbulence in their atmospheres is adding
scatter to the velocity curves.

Since the frequency of spectroscopic binaries among normal K giants
has been found to be 25-30% (Gunn and Griffin 1979;   Harris and McClure
1982), and since the frequency of binaries in population II is thought
to be lower than in population I, the high frequency found here so far
for CH stars (at least 65%) is highly significant.   Experience has shown
from our Ba II star observations, that increasing the number and time
span of observations tend to turn up more binaries, so that the frequency determined from the present data will likely increase.   It seems

then that all CH stars are binaries as is the case for Ba II stars. In fact it perhaps strengthens the hypothesis that the CH stars are population II counterparts of the population I Ba II stars, and raises some doubts about binary frequency in population II being low.

Observations of R stars are much less advanced, but there are a significant number of members of this class for which the DAO data show long term velocity variations indicating multiplicity. About an equal number for which multiple observations have been obtained show no velocity variations. Although the frequency of binaries perhaps looks high (~ 50%), at this point, a much larger observational base is needed for the R stars.

Three areas for further observational study are: 1. continued observations of the R stars to tell us whether these too are multiple systems, and therefore related closely to the Ba II and CH stars, or predominantly single, in which case stellar evolution theory will have to account for the mixing which has occurred. 2. a thorough study of Yamashita's (1975) "CH-like" stars, which have spectra similar to the CH stars, with enhanced carbon and s process elements, but which have low space motions. Like the Ba II and R stars, these are probably members of the old disk population. Are these stars, therefore, a link between the Ba II and R stars? They have C_2 bands indicating a C/O ratio greater than unity, but they have enhanced s process elements not found among R0–R3 stars. 3. the "mild", "marginal", or "Ba0" barium stars need to be understood. These appear to have a high binary frequency, but not all are binaries (Griffin 1983; McClure 1983). Some have slightly enhanced Sr II and Ba II but no enhancement of CN (Sneden et al. 1981). Others have enhanced Sr II and CN, but no enhancement of Ba II (Bond 1983, private communication).

REFERENCES

Abt, H.A., and Levy, S.G.: 1969, A.J. **74**, 908.
Böhm-Vitense, E.: 1980, Ap.J. Letters **239**, L79.
Böhm-Vitense, E., Nemec, J., and Proffitt, C.: 1983, this volume, p .
Bond, H.E.: 1974, Ap.J. **194**, 95.
Crampton, D., and Hartwick, F.D.A.: 1972, A.J. **77**, 590.
Dominy, J.: 1983, Ph.D. thesis, University of Texas.
Dominy, J., and Lambert, D.L.: 1983, Ap.J. **270**, 180.
Eggen, O.J.: 1972, M.N.R.A.S. **159**, 403.
Griffin, R.F.: 1982, Observatory **102**, 82.
Gunn, J.E., and Griffin, R.F.: 1979, A.J. **84**, 752.
Harris, H.C., and McClure, R.D.: 1983, Ap.J. Letters **265**, L77.
Keenan, P.C.: 1942, Ap.J. **96**, 101.
McClure, R.D.: 1983, Ap.J. **268**, 264.
McClure, R.D., Fletcher, J.M., and Nemec, J.M.: 1980, Ap.J. Letters **238**, L35.
Smith, V.V.: 1983, Ph.D. thesis, University of Washington.
Sneden, C., Lambert, D.L., and Pilachowski, C.A.: 1981, Ap.J. **247**, 1052.
Yamashita, Y.: 1975, Publ. Astron. Soc. Japan **27**, 325.

DISCUSSION

<u>Schatzman</u>: What is the connection of the observational results, showing
binarity of BaII stars with stellar evolution theory?

<u>McClure</u>: The next paper by Böhm-Vitense will discuss this. The abundance
peculiarities of these stars must be connected in some way with binarity
since this class of stars contains only binary systems. Either there was
mass exchange from a higher mass companion, or else the companion has
somehow caused the BaII star to mix. There is a great deal of debate
now as to which of these has happened, and I am not able to answer this.

<u>Richer</u>: The inferred separations of the stars must be rather large. Does
this worry you regarding the mass transfer?

<u>McClure</u>: If the companion was a reasonably massive star it could have
become quite large by the time it reached the upper part of the AGB,
large enough compared with the ~2 AU separations.

<u>Wing</u>: If all CH stars are binaries, can the fact that the CH stars in
ω Cen lie within the cluster's giant branch be used to place a limit on
the luminosities of their companions?

<u>McClure</u>: I don't think so. The companion is likely either a white dwarf
or a main sequence dwarf, and therefore several magnitudes fainter so
that it would not affect the magnitude of the CH star.

ON THE ORIGIN OF THE BARIUM STARS

Erika Böhm-Vitense
Astronomy Department, University of Washington, Seattle, WA

ABSTRACT

We show that Ba stars cannot originate from single stars, they can only be
formed in binaries by mass transfer. The companion must then have been
an evolved star with $\log L/L_\theta \geq 3.2$, requiring a radius larger than 0.5
a.u., explaining the long periods of the observed Ba star binaries. If
the companion lost half of its mass to the present Ba star we expect to
see a white dwarf companion now. Companions with $T_{eff} \approx 12000K$ have been
seen for the nearest Ba stars. For Ba stars at larger distances companions can only be detected if they have $T_{eff} \geq 20000K$.

In order to make a Ba star the slow neutron capture process has to
go on in the interior of a star, and some process has to bring the newly
formed elements to the surface of either the same star or a companion.
In order to understand how this happens we first have to know where in
the H-R diagram we find the Ba stars. Temperatures can be determined
spectroscopically, so can gravities, at least to within a factor of 2.
If we know the masses then radii and luminosities can be computed. For
different masses we find different luminosities. The luminosity increases
proportional to the assumed mass M. Another relation between mass M and
luminosity L is given by the evolution theory. From these two relations
between M and L we can determine approximate masses for the Ba stars. In
figure 1 we show the approximate positions for the Ba stars in the H-R
diagram assuming masses of 2.5 M_θ. Also shown is the evolutionary track
for a 2.5 M_θ star as extrapolated from the calculations of Becker, Iben
and Tuggle (1977). The spectroscopic luminosity for a 2.5 M_θ star fits
this track rather well. The observed luminosity, assuming a 5 M_θ star,
would not fit on the corresponding evolution track. We conclude that the
Ba stars have masses between 2.5 and 1.5 M_θ, except for ζ Cap which seems
to have a larger mass. Absolute M_V determined from the Wilson Bappu relation are generally in rough agreement with the spectroscopic M_V.

The Ba stars reside in a region of core He burning. They probably
have gone through the He flash but they have not gone through double shell

191

A. Maeder and A. Renzini (eds.), Observational Tests of the Stellar Evolution Theory, 191–194.
© *1984 by the IAU.*

source flashes. According to present stellar evolution theory they have
not yet gone through a phase where the slow neutron capture process can
have taken place, which needs $\log L/L_\odot > 3.2$ according to Iben (1981).
The only way out of this difficulty is a neutron capture process that took
place in another star, a companion, which then shed its envelope on to the
present Ba star.

The companions of the Ba stars must then all be evolved stars which
have lost their envelopes, they must be white dwarfs. They have been
confirmed for ζ Cap, ζ Cyg (Böhm-Vitense 1981, Lambert 1982), 56 Peg,
(Schindler et al. 1982), and for ζ^1 Cet, (Böhm-Vitense and Johnson 1983).

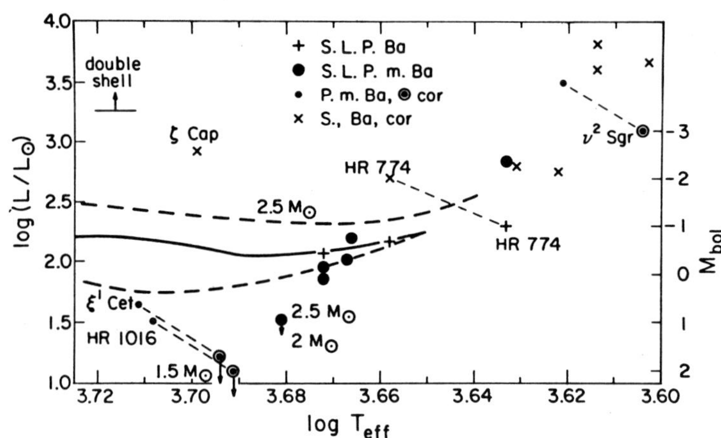

Figure 1: The position of the Ba stars in the H-R diagram
suggests a mass of about 2.5 $M_\odot$. They appear to be at a stage
of core helium burning.

Indirect evidence is seen for HR 1016 and 56 U Ma (Böhm-Vitense
1983), for which forbidden lines of OIII and NIV are observed.

The last Figure shows in a diagram the distances of the Ba stars
which have been observed. We also indicate for which stars the white
dwarf companions have been seen. Cool white dwarfs have been seen for
the nearest stars. For the Ba stars which are further away we can see
the companions only if they are still very hot. I think the observations
confirm what we expect to be able to see.

If mass transfer leads to the formation of Ba stars, why are the
binary periods so long? McClure (1982) observed periods between 80 days
and several years. Before the mass transfer the present WD companion
must have been evolved enough for the slow neutron process to have occurred. If this requires $\log L/L_\odot > \log L_{crit}/L_\odot = 3.2$ (Iben 1981), then we
calculate that $\log R/R_\odot > 2$ if $T_{eff} = 3500$ K at this time. The distance
d of the stars must then be d > 0.5 a.u. Otherwise mass transfer occurs

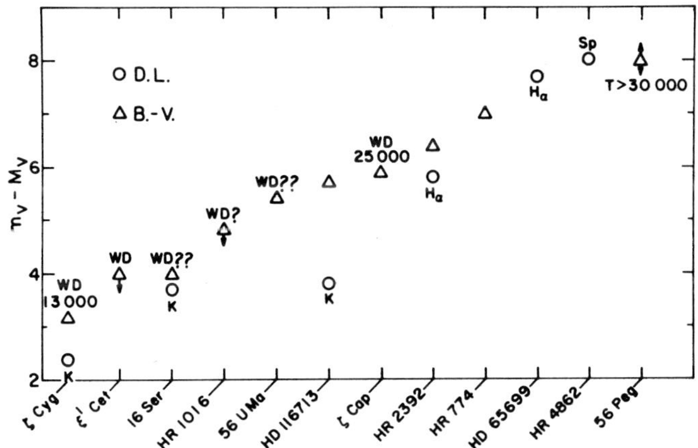

Figure 2: The distances of the Ba stars are shown. Stars for which white dwarf companions have been seen are indicated and approximate temperatures are given. Question marks indicate stars for which indirect evidence for the companions has been seen.

too early. For $M_1 + M_2 \sim 3\ M_\theta$ we find that the period must be longer than 72 days. This seems to agree rather well with the shortest observed period of 80 days for HD 77247, which could, however, turn out to be a triple system. If that should be the case, then the shortest observed period is a factor 6 larger which could mean either that the distance of the binaries increased during the mass transfer or that the critical luminosity is larger than $\log L_{crit}/L_\theta = 3.2$.

This study was supported by a NASA grant NAG 8378, which is gratefully acknowledged. We are obliged to the astronomers and staff of the IUE observatory without whose help and support with the observations and data reductions, the observations would have been impossible.

REFERENCES

Böhm-Vitense, E.: 1981, Astrophys. J.
Böhm-Vitense, E. and Johnson, H.: 1983, in preparation.
Böhm-Vitense, E.: 1983, submitted to Astrophys. J.
Becker, R., and Iben, I. and Tuggle: 1977, Astrophys. J. 218.
Iben, I.: 1981, In Physical Processes in Red Giants, p. 3, 3d. Iben and
    Renzini, Reidel Publ. Co., Dordrecht, Holland.
McClure, R.: 1982, Astrophys. J.

DISCUSSION

<u>Renzini</u>: I don't quite see the difficulty with evolutionary timescales. Accreting secondaries could well have been main sequence stars, and then one can have any delay whatsoever between the formation of the WD and the arrival of the Ba star on the red giant branch.

<u>Böhm-Vitense</u>: The problem is that main sequence Ba stars have not yet been discovered. If I did my statistics right, one would expect to see one Ba star out of 1000 main sequence stars. We may have missed the one, then this may well be the answer. But Dr. Lambert feels that we should have seen it.

<u>Renzini</u>: But there will certainly be selection effects against dis-covering MS Barium stars.

<u>Despain</u>: Does your model assume that ^{22}Ne is the neutron source, as for Iben and Truran's scenario?
Why couldn't the s-processing occur during some hydrodynamic event during the core flash?

<u>Böhm-Vitense</u>: No, the source is not specified.
The abundances of any neutron donor element is too low at that time. At least 80 neutrons have to be available per Fe atom in order to make the observed abundances of the s-process elements. It also seems to me that the times available at the flash are too short for the slow neutron capture process.

SOME OBSERVABLE INDICATORS OF s-PROCESS ENVIRONMENTS?

Keith H. Despain
Los Alamos National Laboratory

1. ABSTRACT

By definition, the chemically peculiar stars have element abundances
that are non-solar. Many of these stars show peculiarities in s-process
elements. This paper discusses observable indicators of the neutron
environment for the s-process in these stars.

2. ISOTOPIC INDICATORS?

The neutron sources usually considered for producing s-process elements
are ^{13}C and ^{22}Ne (Cameron, 1955; Burbidge, Burbidge, Fowler, and Hoyle,
1957...). Cowan and Rose (1977) have suggested that under certain conditions ^{14}C might be produced in observable quantities if ^{13}C were the
neutron source. However, they ignored neutron reactions on all elements
heavier than ^{16}O (Despain, 1977), which cause a reduction in ^{14}C production of at least a factor of about 50 (Despain, unpublished).

Scalo and Miller (1981) proposed that ^{99}Tc could be used as an indicator
of the nucleosynthesis and mixing processes in thermally-pulsing red
giants, using ^{22}Ne as a neutron source. However, they did not take into
account the enhanced beta decay of ^{99}Tc in the hot stellar interior,
which places their interpretation of its detection in doubt (Cosner,
Despain, and Truran, 1984).

3. ELEMENTAL INDICATORS

Elemental indicators have previously been discussed by Butcher (1976)
and Cowley & Downs (1981); their indicators give a measure of the total
neutron exposure, τ--defined by Clayton, Fowler, Hull, and Zimmerman
(1961)--but not of the neutron density, n. The following presents an
s-process indicator for measuring both τ and n. Assuming the peculiar
abundance observed, $\{X_*\} \equiv (X_*/Fe_*)/(X_r/Fe_r)$ is the consequence of mixing an ordinary (i.e., "solar like") envelope of mass M_e with abundances

A. Maeder and A. Renzini (eds.), Observational Tests of the Stellar Evolution Theory, 195–198.

X_e, with a shell of mass M_s, with s-process abundances, X_s, where X_r is the abundance in a reference star, $\{X_*\}$ is given, after some algebra, by

$$\{X_*\} - R = fR\,(X_s/X_e - 1)\quad , \tag{1}$$

where $f = M_s/(M_e + M_s)$ and $R = (X_e/Fe_*)/(X_r/Fe_r)$. Assuming $(X_e/Fe_e) = (X_r/Fe_r)$, $R = (Fe_e/Fe_*)$. A reasonable range for R is $1 \leq R \leq (1-f)^{-1}$, the limits being given by $Fe_s/Fe_e = 1$ and by $Fe_s/Fe_e = 0$. To eliminate the factor fR in Eq. (1) we examine ratios of elements. For the s-process, the proposed observables are

$$S_{12} \equiv \langle Sr+Y+Zr \rangle / \langle Ba+La+Ce \rangle \text{ and } S_{23} \equiv \langle Ba+La+Ce \rangle / \langle Nd+Sm+Eu+Gd \rangle, \tag{2}$$

where $\langle \Sigma X_i \rangle \equiv \Sigma(\{X_i\} - R)/N$. S_{12} is the ratio of the average enhancement of elements in the first s-process element peak, determined by the closed nuclear neutron shell at 50 to those in the second closed shell at 82. It is a measure of the neutron exposure, τ. S_{23} is the ratio of the average enhancement of elements in the second peak to a group of elements on the "plateau" after the peak. Since at a higher neutron density a nuclear shell will generally close at a lower atomic weight and (eventually) atomic number, this can serve as a measure of n.

4. COMPARISON WITH OBSERVATIONS

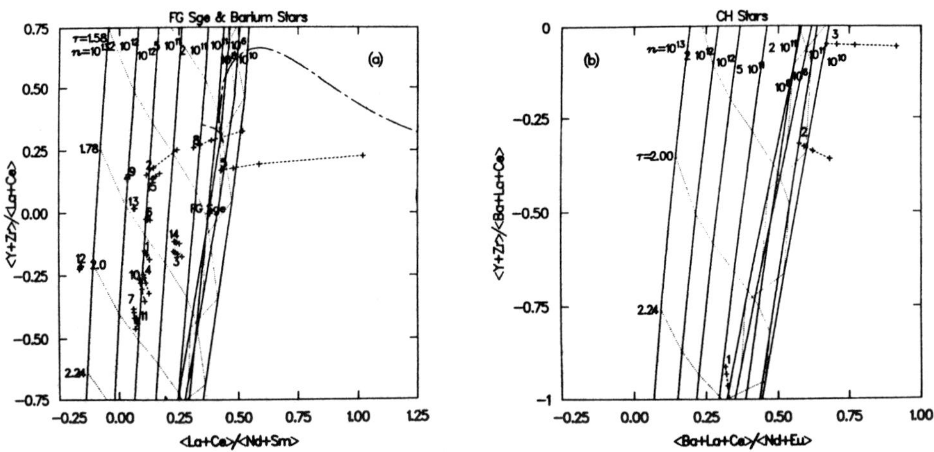

Figure 1. Theory vs Observation (see text for discussion)

Figure 1 compares observations and calculations for FG Sge (a), barium stars (a), and CH stars (b). The dashed curves are the observations with R = 1, 1.1, 1.25 and 1.5. The numbers correspond to the stars in Table I. The solid curves are calculations at constant neutron density, given across the top. The dotted curves connect the points on the curve at constant τ, as indicated on the left. These values of τ correspond to the number of neutrons per iron seed captured on all elements heavier than, and including, ^{22}Ne in the range from roughly 400 to 1000 (top to bottom). The chain dash curve is the loci of models of the thermally-pulsed ^{22}Ne source, as it operates in the AGB stars (Iben, 1976; Cosner,

1982; etc.). As can be seen, these stars appear to show a large range in neutron densities, but virtually all show neutron exposures too heavy for the AGB ^{22}Ne source. Note that a different set of elements is used in the definition of S_{12} and S_{23} than in section (4) because one (or more) abundance is missing for some of the stars.

Table I

# Star	Reference (see reference section)							
1	HD 46407	W	6	HD 178717	W	11	HD 204075	CD		FG Sge	LKA
2	HD 83548	W	7	HD 183915	W	12	HD 46407	BB	CH Stars		
3	HD 92626	W	8	HD 204075	W	13	HD 204075	T	1	HD 26	WG
4	HD 116713	W	9	HD 211594	W	14	HD 116713	D	2	HD 176201	SB
5	HD 175674	W	10	HR 774	C	15	HD 83548	D	3	HD 204613	SB

5. CONCLUSIONS

1. There _are_ some potentially useful, _observable_ s-process indicators.
2. Abundance measurements on as many of the elements in each of the sets (Sr,Y,Zr), (Ba,La,Ce), and (Nd,Sm,Eu,Gd) would be helpful in characterizing the s-process environment.
3. Barium stars, FG Sge, and CH stars have all apparently produced their s-process elements in quite rich neutron environments, with significant star to star variations.

6. ACKNOWLEDGEMENTS

I would like to acknowledge the help of Jim Truran and Ken Cosner, who made available their computer code for computing the s-process, which was modified and used in the calculations reported here. This research was supported by USDOE contract W-7405-ENG. 36.

7. REFERENCES

Burbidge, EM and Burbidge, GR 1957, ApJ,_126_,357. (BB)
Burbidge, EM, Burbidge, GR, Fowler, WA, and Hoyle, F 1957, RMP,_29_,547.
Butcher, HR 1976, ApJ, _210_, 489.
Cameron, AGW 1955, ApJ, _121_, 144.
Clayton, D, Fowler, W, Hull, T, and Zimmerman, B 1961, Ann Phys, _12_,331.
Cosner, KR 1982, PhD thesis, Univ. of Ill.
Cowan, JJ, and Rose, WK 1977, ApJ, _212_, 149.
Cowley, CR, 1968 ApJ, _153_, 169. (C)
Cowley, CR, and Downs, PL 1980, ApJ, _236_, 648. (CD)
Danziger, I 1965, MNRAS, _131_, 51. (D)
Despain, KH 1977, ApJ _212_, 774.
Iben, I 1975, ApJ, _196_, 549
Langer, G, Kraft, R, and Anderson, K 1974, ApJ, _189_, 509. (LKA)
Scalo, J. and Miller, G 1981, ApJ, _246_, _251_.
Sneden, C, and Bond, H 1976, ApJ, _204_, 810. (SB)
Tech, JL 1971, NBS Monograph 119. (T).
Wallerstein, G, and Greenstein, J 1964, ApJ, _169_, 1163. (WG)
Warner, B 1965, MNRAS, 129, 263. (W)

DISCUSSION

<u>Müller</u>: The elements for which you wish to have very good abundance
determinations are precisely some of those which are very difficult to
determine: the non-LTE plays an important role in their spectral line
formation, and for non-LTE calculations of spectral lines accurate
atomic cross-sections are needed which, however, are not available and
are very difficult to determine for these multilevel atoms.

<u>Renzini</u>: I think that Iben and Truran never suggested the ^{22}Ne source in
connection with Barium stars. Indeed, much more neutrons per iron seed
are produced in AGB stars of small core mass, where post-flash carbon/
hydrogen semiconvection leads to the activation of the ^{13}C source.

<u>Despain</u>: The indicated neutron per iron seed for the Barium stars is
400-800, while the neutrons per iron seed indicated by Iben for the
above mechanism was ~26, more than an order of magnitude too low for the
indicated S_{12} parameter.

<u>Acker</u>: Using spectra of FG Sge taken in 1979-1982, I show that this star
does not evolve toward a Barium star (as suggested), and I give new
indications about s-process element abundances.

HCN AND C_2H_2 IN CARBON STARS

K. Eriksson, B. Gustafsson, Uppsala Astronomical Observatory
U. G. Jørgensen, Å. Nordlund, Copenhagen
Astronomical Observatory

At the low temperatures of cool carbon star atmospheres polyatomic molecules form. Most abundant of these for the temperature range 2500 - 3000
are HCN and C_2H_2 (Fig. 1).

These molecules have numerous bands at wavelengths where a major part
of the stellar flux is transported. This opacity could therefore be of
great importance when constructing models for cool carbon star atmospheres,
but has not been included in earlier models.

Models without the HCN and C_2H_2 opacity show a strange transition to
"thin", high-pressure atmospheric structures when T_{eff} is decreased below
about 2900 K (Fig. 2). This is caused by the decrease in opacity when the
diatomic carbon molecules - important opacity sources - are depleted due
to formation of HCN and C_2H_2. These "thin" models predict Na D and H_2
quadrupole lines which are far too strong compared with observed lines.

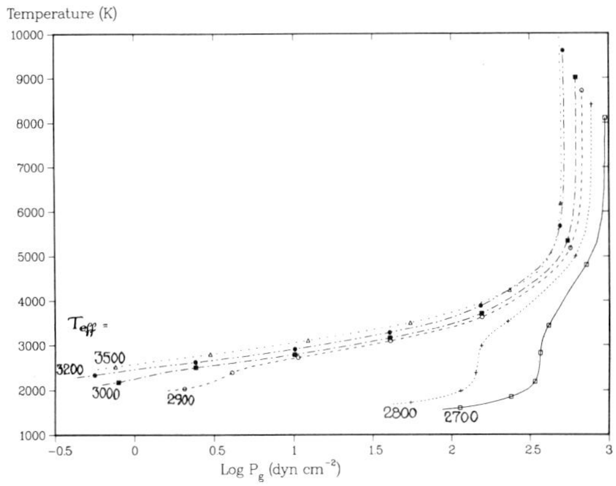

Figure 2

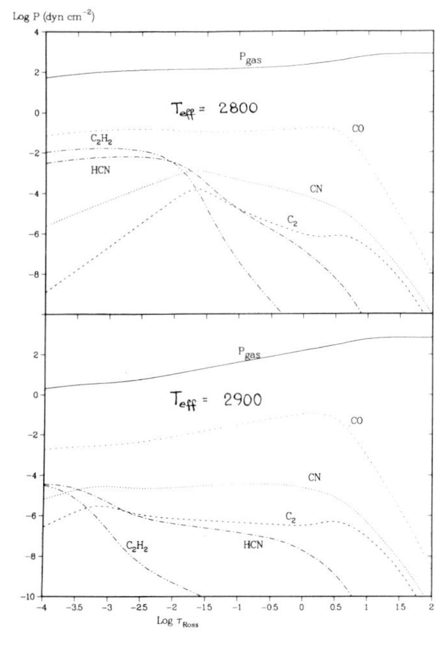

Figure 1

199

A. Maeder and A. Renzini (eds.), *Observational Tests of the Stellar Evolution Theory, 199–201.*

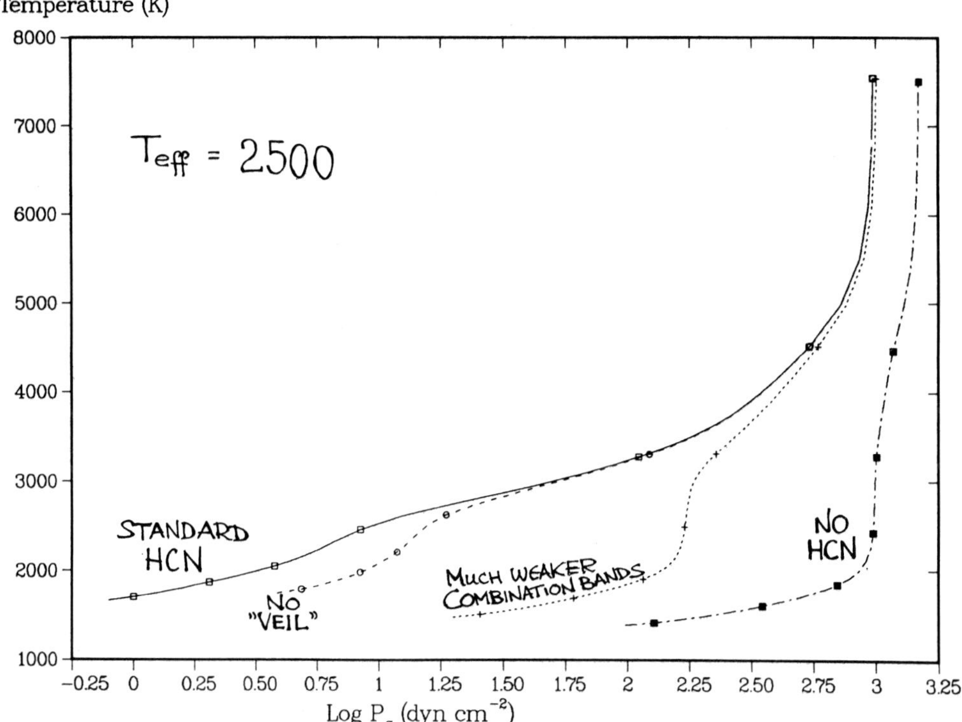

Figure 3. Note the large drop in pressure
at T_{eff} = 2500 K (1-2 orders of magnitude!).
No important effect at T_{eff} = 3000 K.

Calculation of HCN (and C_2H_2) opacity. In the laboratory mainly the
fundamental bands have been studied. No sufficiently detailed *ab initio*
quantum mechanical calculations exist (yet). We have therefore assumed
that the transition probability for a combination band relative to the
transition probability for a fundamental band scales as the corresponding
ratio for other, more well-studied molecules.

We calculated the strengths of many ($\sim 10^6$) lines and added the rest
of the total bandstrength as a "veil" of continuous opacity. We then
transformed this opacity into smooth ODF's (Opacity Distribution Functions)
in the wavelength region 1 μ to 12.5 μ.

When this opacity was added in the model atmosphere calculations,
great effects on the structure were found for the lower temperatures.

Reference:

Eriksson, Gustafsson, Jørgensen & Nordlund, 1983: Astron. Astrophys.,
    in press

DISCUSSION

<u>Wing</u>: Do the computed abundances of polyatomic molecules in carbon stars show a dependence on the luminosity? I am interested in the possibility of using the observed infrared bands as spectroscopic luminosity indicators.

<u>Eriksson</u>: We have not computed any synthetic spectra yet, but will in the near future.

<u>Tayler</u>: Do you expect to find even larger molecules such as HCCCN in these atmospheres?

<u>Eriksson</u>: When T_{eff} becomes still lower than $2500^{O}K$ you will find larger molecules but also dust formation will be important then.

EVOLUTION OF S STARS IN THE MAGELLANIC CLOUDS

T. Lloyd Evans
South African Astronomical Observatory

ABSTRACT.   Early S stars occur between M and C stars in the colour
magnitude diagrams of intermediate age globular clusters in the
Magellanic Clouds.   Most have $-4.2 \geq M_{bol} \geq -4.8$ and are probably
brighter in younger or more metal rich clusters.   The galactic globular
cluster NGC 6723 contains two marginal S stars at $M_{bol} \sim -3.3$.   The rare
CS stars have $M_{bol} \sim -6$, with no faint examples.

Rich intermediate-age clusters in the Magellanic Clouds contain S stars
in a luminosity interval between M and carbon stars, confirming that
they represent an intermediate stage in the increasing ratio of C/O in
the outer layers (Bessell, Wood & Lloyd Evans 1983).   These S stars
have spectral types equivalent to early or middle M with luminosities
in the range $-4.2 \geq M_{bol} \geq -4.8$.   The luminosity boundaries depend on
age and metal content of the cluster, being brighter in the younger or
more metal rich clusters (Lloyd Evans, in press).   This is understandable as the quantity of pre-existing oxygen and heavy elements will be
smaller if the envelope mass is small and the metal abundance low
(Renzini & Voli 1981).

Open clusters of similar age in the Galaxy are so poor that they
contain few carbon stars while no S stars are known to be members.
Two unusually red stars in the globular cluster NGC 6723 have enhanced
ZrO and extend the S star domain to $M_{bol} = -3.3$. ω Centauri contains
even fainter S stars but there is evidence that these result from a
primordial composition which is unusually rich in heavy elements (Lloyd
Evans, in press).

The field of the Magellanic Clouds contains carbon stars brighter
than the brightest found in the clusters, which have $M_{bol} \sim -5.6$.   Two
large amplitude variables with CS spectra in the SMC (Lloyd Evans 1980)
have $M_{bol} \sim -6$, at the upper end of the field carbon star luminosity
function, and AGB stars brighter than this have MS or M spectra (Wood,
Bessell & Fox 1981, and preprint (WBF)).   CS and SC spectra have not
been found among the less luminous AGB stars in clusters or the field

A. Maeder and A. Renzini (eds.), Observational Tests of the Stellar Evolution Theory, 203–204.

and in particular NGC 121/V8, a faint carbon star with colours which
place it between S and C stars (WBF), has a featureless spectrum in the
red.   This may result in part from a higher surface temperature;   the
M→S→C transitions appear to occur at lower temperature in more luminous
(massive) stars.   A second reason for the absence of faint CS stars is
that the number of shell flashes required to produce a carbon star is
quite small so that the carbon enrichment process occurs by large steps
so that any narrow range in C/O will rarely occur.   The carbon enrichment process at high luminosity may be much more gradual, because the
large envelope mass dilutes new material so that many shell flashes are
needed to produce a carbon star or because newly-formed ^{12}C is being
converted into ^{14}N at the base of the envelope (WBF).   Each star will
thus pass through the S→C transition more gradually, with a higher
probability of attaining the narrow range of C/O required for a CS or
SC spectrum.

REFERENCES

Bessell, M.S., Wood, P.R. and Lloyd Evans, T.:   1983, Mon. Not. R. astr.
   Soc. 202, pp. 59-76.
Lloyd Evans, T.:   1980, Mon. Not. R. astr. Soc. 193, pp. 333-336.
Renzini, A. and Voli, M.:   1981, Astron. Astrophys. 94, pp. 175-193.
Wood, P.R., Bessell, M.S. and Fox, M.W.:   1981, Proc. astr. Soc.
   Australia 4, pp. 203-205.

NON-LTE ANALYSIS OF CENTRAL STARS[*]

R.P. Kudritzki[4], R.H. Méndez[123] and K.P. Simon[4]

[1] Instituto de Astronomia y Fisica del Espacio, C.C. 67, Suc 28,
1428 Buenos Aires, Argentina

[2] Visiting Astronomer, Cerro Tololo Inter-American Observatory, operated
by the Association of Universities for Research in Astronomy, Inc.,
under contract with the U.S. National Science Foundation

[3] Member of the "Carrera del Investigador Cientifico", Conicet, Argentina

[4] Institut für Astronomie und Astrophysik der Universität München,
Scheinerstr. 1, D-8000 München 80, Federal Republic of Germany

[*] Based partly on observations made at the ESO, La Silla, Chile

1. INTRODUCTION

Most of our present knowledge about central stars of planetary
nebulae (CPN) is obtained by indirect methods using the emission line
spectra of the surrounding nebula. These methods, which apply the beautiful recombination theory, can provide us with information about temperature, distance and radius of the CPN. However, we know that these methods
are subject to severe problems, which then lead to strong discrepancies
in the parameters of the CPN.

Therefore, we decided to make use of the direct method, which is
normally applied to stars, namely, the quantitative analysis of the
photospheric line spectrum. This 'model atmosphere approach' has recently
become possible after extensive computations of non-LTE model atmospheres
(carried out to a large part in our group) and after the development of
efficient detectors attached at big telescopes.

The main motivations for this approach are two. The first is to be
able to discuss the evolution of CPN without all the uncertainties related to the use of nebular distances; this requires to place our objects

A. Maeder and A. Renzini (eds.), Observational Tests of the Stellar Evolution Theory, 205–208.
© *1984 by the IAU.*

Table 1

ATMOSPHERIC PARAMETERS OF CENTRAL STARS

Object	T_{eff} (10^3 K)	log g	$y = N(He) / (N(He)+N(H))$
Longmore 1	65 +10	5.7 +0.3	0.10 +0.03
NGC 1360	60 +15,-5	5.2 +0.2	0.05 +0.02
NGC 1535	50 +10,-5	4.5 +0.3	0.09 +0.03
Abell 7	75 +10	7.0 +0.5	0.01 +0.005
K1-27	100 +30	5.7 +0.5	0.60 +0.30
Abell 15	65 +10	6.0 +0.5	0.09 +0.04
Abell 33	100 +30	6.0 +0.5	0.09 +0.05
NGC 3242	70 +30,-20	4.5 +0.5	0.10 +0.03
NGC 4361	80 +10	5.5 +0.3	0.05 +0.02
Longmore 8	65 +10	5.0 +0.5	0.11 +0.03
Abell 36	65 +10	5.2 +0.3	0.13 +0.04
NGC 7293	90 +10	6.6 +0.3	0.01 +0.005

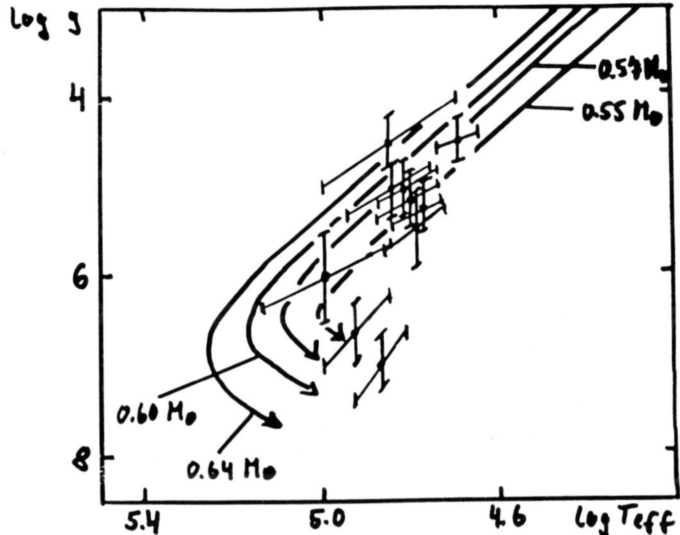

Fig. 1: The log g - log T_{eff} diagram for CPN.
Evolutionary tracks (Schönberner, 1981) descending from the AGB
and evolving towards the white dwarfs are also shown (the
numbers refer to masses in $M_\odot$). For discussion see text.

on the log g - log T_{eff} diagram instead of the H. R. diagram. The second reason is to explore the surface helium abundance of CPN.

2. METHOD

Cassegrain spectra of moderate dispersion (29 A/mm, 45 A/mm, 58 A/mm) covering the spectral range between 4000 to 5000 A were taken using the IDS at the ESO 3.6 m and the SIT - Vidicon at the CTIO 4 m. The profiles of hydrogen and helium lines were compared with detailed non-LTE model atmosphere and line formation calculations in order to determine simultaneously T_{eff}, log g and helium abundance. Details of the analysis technique are described in Méndez et al. (1981, 1983 a, b) and Kudritzki et al. (1981).

3. RESULTS AND DISCUSSION

Table 1 contains the atmospheric parameters of our program stars, as obtained from our non-LTE analysis. Figure 1 shows their position on the log g, log T_{eff} - plane. A comparison with evolutionary tracks descending from the AGB (Schönberner, 1979, 1981) reveals that our objects form an evolutionary sequence of hot objects with constant luminosity (upper diagonal of the tracks in the log g, log T_{eff} - plane) towards the white dwarf stage. The masses are in a narrow range between 0.5 to 0.6 $M_{\odot}$, which fits nicely with the mass distribution of DA white dwarfs (Koester et al., 1979). Interestingly, the two objects with the highest gravities (NGC 7293 and Abell 7), which are more advanced in their evolution towards the white dwarf stage, have surface helium abundances significantly (a factor of ten) smaller than solar. This implies the onset of gravitational settling and provides a very strong observational connection between these central stars and DA white dwarfs.

REFERENCES

Koester, D., Schulz, H., Weidemann, V.: 1979, Astron. Astrophys. $\underline{76}$, 653
Kudritzki, R.P., Méndez, R.H., Simon, K.P.: 1981, Astron.Astrophys.$\underline{99}$,L15
Méndez, R.H., Kudritzki, R.P., Gruschinske, J., Simon, K.P.: 1981,
      Astron. Astrophys. 101, 323
Méndez, R.H., Kudritzki, R.P., Simon, K.P.: 1983a, IAU Symp. 103 on
      Planetary Nebulae, p. 343, ed. D.R. Flower
Méndez, R.H., Kudritzki, R.P., Simon, K.P.: 1983b, submitted to Astron.
      Astrophys.
Schönberner, D.: 1979, Astron. Astrophys. $\underline{79}$, 108
Schönberner, D.: 1981, Astron. Astrophys. $\underline{103}$, 119

DISCUSSION

Cox: Do any of your stars show strong carbon features in its spectrum?

Kudritzki: Yes, NGC 246 (as already noted by Heap, 1977), Longmore 3 and Longmore 4 show very strong CIV lines at $\lambda 4441$ and $\lambda 4646$, respectively. In addition, they are extremely helium-rich. The analysis of these objects, which from the atmospheric models is extremely difficult, is subject to our future work.

Spruit: Can one compare the distribution of number density of the objects along the evolutionary track with predictions?

Kudritzki: For our sample of objects I would guess: no, because it is not statistically complete. However, it appears to be worthwhile to complete the sample in our future work and to carry out an analogous comparison, as has been done by Schönberner with his different method, which does depend on distance scale but not on T_{eff}, whereas our results are independent of distance but use T_{eff}.

THE H. R.- DIAGRAM OF CENTRAL STARS OF PLANETARY NEBULAE

D. Schönberner
Institut für Theoretische Physik und Sternwarte
Universität Kiel
Olshausenstraße 40
2300   Kiel, F.R.G.

During the last years some progress has been made in the determination of temperatures of central stars of planetary nebulae (CPN). The main reasons are the successful deployment of space crafts which made the more temperature sensitive spectral region in the UV accessible, and the application of NLTE-spectral analyses (e.g.Mendez et al.,1981). Therefore, a re-examination of the H.R.-diagram of CPN's seems to be in order.

We collected informations about the temperatures of central stars from different sources (see references in the Table) and accepted only objects with at least two independent determinations. We discarded all objects with too discrepant temperatures. WC-type nuclei have also been omitted. The selected objects are listed in the Table, and the quoted temperatures represent mean values, the error of which seems to be in most cases much less than $\pm$ 25%. For the distances, we used the recent calibration of Daub (1982), upscaled to the (revised) distances of Cahn and Kaler (1971). We did not consider objects with $R_{Neb} < 0.05$ pc because of the uncertainty of their distances. The resulting absolute magnitude are also given in the Table (interstellar absorption has been taken into account). Fig. 1a shows all objects of the Table in a H.R.-diagram, i.e. M_V vs. T_{eff}. They occupy a well defined strip, ranging from M_V = -1 to 9 and log T_{eff} from 4.5 to 5.2. Superimposed are the evolutionary tracks of post-AGB models with different masses as they follow from Paczynski (1971) and Schönberner (1979, 1983). The latter are burning hydrogen in a shell under quiet conditions. Within the errors ($\Delta M_V \approx \pm$ 0.5, Δ log $T_{eff} \approx \pm$ 0.1), we find an excellent agreement between the observed loci of CPN and those of post-AGB models with $M \lesssim 0.65$ $M_\odot$. Only the two faintest CPN ($M_V \gtrsim 8$) seem to have higher masses ($\sim 0.9 - 1.0$ $M_\odot$).

Fig. 1b shows the (M_V, age)-diagram for the same objects and the same evolutionary tracks. We have used a nebular expansion velocity of 20 km/s in those cases where an individual velocity is not available. Both figures are consistent with each other,in that the CPN studied are confined (with only a few exceptions already mentioned above) within the 0.55 and 0.65 $M_\odot$ post-AGB tracks, and the theoretical predicted temperatures (read off from Fig.1b) are in reasonable agreement with the ob-

A. Maeder and A. Renzini (eds.), Observational Tests of the Stellar Evolution Theory, 209–212.
© *1984 by the IAU.*

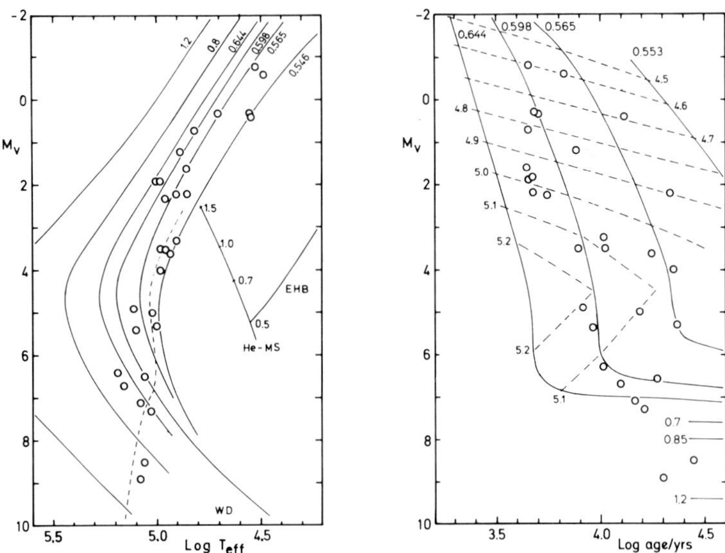

1a: H.R.-diagram for CPN and post-AGB evolutionary tracks from Paczynski (1.2, 0.8 $M_\odot$) and Schönberner (0.644, 0.598, 0.565, 0.546 $M_\odot$). On the dashed line 40 000 years are spent since the models left the AGB (zero age is at $T_{eff} = 10^{3.7}$ K).

1b: M_V vs. age of the models (zero point defined as in Fig. 1a) and CPN. The 0.553 $M_\odot$ track has been interpolated. Dashed lines connect points of equal effective temperatures as predicted by these models. The ages of observed CPN are given by nebular radius devided by the expansion verlocity.

Object	$T_{eff}/10^3$K	M_V	T_{eff}(pred)/T_{eff}	Object	$T_{eff}/10^3$K	M_V	T_{eff}(pred)/T_{eff}
NGC 650	150 (3,7)	6.4	0.79	NGC 6891	35 (2,3,6)	0.4	1.38
1535	50 (2,4)	0.3	1.13	7293	110 (1,2,3,4,7)	7.3	~1
2022	100 (6,8)	1.8	0.94	IC 2448	75 (2,6,8)	1.6	1.17
2392	30 (2,5)	-0.6	1.33	4593	33 (3,8)	-0.8	1.21
2610	90 (3,9)	3.5	1.22	A 15	82 (2,3,4,7)	3.3	1.45
3242	70 (2,6)	2.2	1.50	20	105 (2,3)	5.0	1.22
3587	120 (2,3,6,7)	7.1	~1	31	115 (3,7,9)	8.5	-
4361	95 (2,3,4,6)	3.5	1.26	33	95 (3,4,7)	5.3	~1
6058	75 (3,7,8)	1.2	0.92	36	85 (2,3,4,7)	3.6	1.32
6326	90 (3,8)	2.3	1.14	43	85 (2,3)	2.2	0.93
6720	130 (3,6,7)	4.9	1.20	84	115 (3,7)	6.5	~1
6772	125 (6,7)	5.4	1.13	K 1-16	93 (3,9)	4.0	1.21
6818	100 (3,8)	1.9	1.03	J 320	65 (3,8)	0.7	1.00
6826	37 (2,3,5,6,8)	0.3	1.60	YM 29	120 (3,7,9)	8.9	-
6853	135 (1,3,6)	6.7	~1				

1) Bohlin et al. (1982)
2) Clegg and Seaton (1983 b)
3) Cahn (1983)
4) Mendéz et al. (1983)
5) Natta et al. (1983)

6) Pottasch (1983)
7) Kaler (1983)
8) Martin (1981)
9) Schönberner (unpublished)

served ones (see Table). We found, however, a systematic difference between the two sets of temperatures because Fig. 1b predicts CPN-tempera-
tures which are, on the average, 15% larger than those which follow from
the observations (see Table).

This difference can be accounted for by an average distance increase of,
say, 15% (see also Schönberner, 1981). This would be in line with other
arguments favouring an increase of the Cahn/Kaler distances (Weidemann,
1977; Schneider et al., 1983). Alternatively, this discrepancy might
also indicate a faster fading of real nuclei by a factor of about two
compared to the models. This would imply that actual mass loss rates
are larger than those used in the computations of Schönberner (1979,
1983).

In conclusion we can say that post-AGB models with quiescent hydrogen
burning as shown in Fig 1a give in all respects a very satisfactory
description of the evolution of at least those central stars which have
been studied in this work.

References

Bohlin, R.C., Harrington, J.P., Stecher, I.P.: 1982, Astrophys. J.
          252, 635
Cahn, J.H., Kaler, J.B.: 1971, Astrophys. J. Suppl. 22, 319
Cahn, J.H.: 1983, preprint
Clegg, R.E.S., Seaton, M.S.: 1983a, IAU-Sypos. No. 103, Planetary
          Nebulae, ed. D.R.Flower (Dordrecht:D.Reidel), p. 536
Clegg, R.E.S., Seaton, M.S.: 1983b, private communication
Daub, C.T.: 1982, Astrophys. J. 260, 612
Kaler, J.: 1983, Astrophys. J. 271, 188
Martin, W.: 1981, private communication
Méndez, R.H., Kudritzki, R.P., Gruschinske, J., Simon, K.P.: 1981,
          Astron. Astrophys. 101, 323
Méndez. R.H., Kudritzki, R.P., Simon, K.P.: 1983, IAU-Sympos. No. 103,
          Planetary Nebulae, ed. D.R.Flower (Dordrecht: D.Reidel),
          p. 343
Natta, A., Pottasch, S.R., Preite-Martinez, A.: 1981, Astron. Astrophys.
          84, 284
Paczynski, B.: 1971, Acta Astron. 21, 417
Pottasch, S.R.: 1983, IAU-Sympos. No. 103, Planetary Nebulae, ed.
          D.R.Flower (Dordrecht: D. Reidel), p. 391
Schneider, S.E., Terzian, Y., Purgathofer, A., Perionotto, M.: 1983,
          IAU-Sympos. No. 103, Planetary Nebulae, ed. D.R. Flower
          (Dordrecht: D. Reidel), p. 411
Schönberner, D.: 1979, Astron. Astrophys. 79, 108
Schönberner, D.: 1981, Astron. Astrophys. 103, 119
Schönberner, D.: 1983, Astrophys. J. 272, 708

DISCUSSION

Conti: Are any of your central stars WC-type and if so, where do they sit in the HR-diagram?

Schönberner: No.

Mould: Looking at your final diagram, one gets the impression that the selection criterion on nebula size may have depopulated the $M > 0.6$ side of the diagram. Could this selection effect influence the derived mass distribution?

Schönberner: Theoretically we don't expect CPN on the left side of the $M = 0.644\ M_\odot$ track because the fading times are too short. Rather, they should show up at $M_V \gtrsim 7$. Of course this ensemble is too highly biased by selection as to make any statement about a general CPN mass distribution.

SPECTROSCOPIC OBSERVATIONS OF NUCLEI OF PLANETARY NEBULAE

Agnès Acker
Observatoire de Strasbourg, France

Up to date, 1518 true, possible, and probable planetary nebulae (PN) are known in our Galaxy ; 463 nuclei (NPN) were observed. For about 150 stars, a spectral study has been done ; only about 30 NPN have a well studied spectrum. (Acker, Gleizes et al, 1982 "Catalogue of the central stars of true and possible PN")
It must be remembered that this is a difficult study, as the PN are very **distant objects (the nearest, Helix, is** at about 120 pc), and that their central stars are very faint (the 35 brightest have a magnitude of from 9 to 12), are generally of a particular spectral type, and have a spectrum in which nebular lines interfere with stellar features.

SPECTRAL CLASSIFICATION

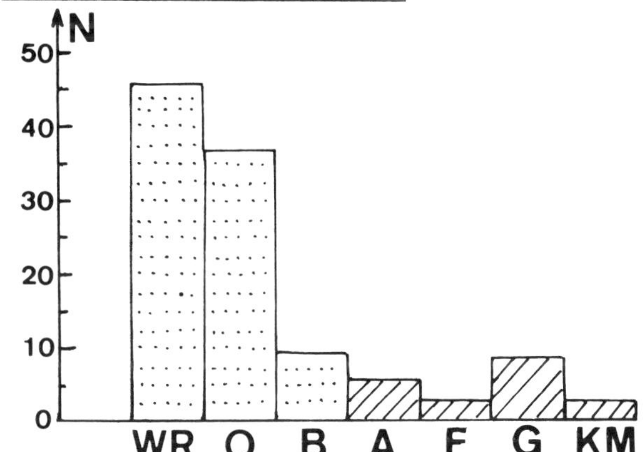

Number N of NPN for which a spectral type has been given in the "catalogue of the central stars of PN" (Acker, Gleizes et al, 1982).

*A spectral class is given for 127 NPN. 14 % have a spectrum classified as continuous, on the basis on observations made in the visible ;

35 % are of Wolf-Rayet type (18 % WC, 10 % WN)

29 % are O stars (9 % Of) ;

7 % are B stars ; in addition, the BQ [ ] type, i.e. early type stars with forbidden emission lines, is attributed to 5 stars called "proto-PN" ;

14 % of the nuclei have a cold type, which implies the presence of a warmer companion, or an instability of the star (case of FG Sge).

* New lines have been discovered in the U.V. (Heap, 1983). 39 NPN show strong P-Cygni type lines ; all WR-and Of-type NPN have high-velocity winds ; the mass loss rate does not exceed 10^{-6} $M_\odot \cdot yr^{-1}$ (Perinotto, 1983).

A. Maeder and A. Renzini (eds.), Observational Tests of the Stellar Evolution Theory, 213–214.

TEMPERATURES

The temperatures of the NPN have been calculated by classic methods, using photometric and spectral data of the nebulae and of the stars, in the visible and in the U.V. The different temperature values obtained vary greatly in height, because of the different magnitudes and distance scales used by the authors. The regrouping of NPN in the H – R diagram into one single evolutive sequence, or into various sequences according to mass, is not yet clearly defined.

TEMPERATURE SCALES OF NUCLEI OF PLANETARY NEBULAE	
Extreme values of the Temperature	Authors (objects)
100 000 < T < 700 000	POTTASCH 1981 (29 very faint nuclei of brightest PN)
33 000 350 000	PREITE-MARTINEZ, POTTASCH 1982 (14 NPN)
70 000 200 000	KALER, FEIBELMAN 1982 (23 nuclei of large PN)
40 000 140 000	CLEGG, SEATON 1982 (20 NPN)
120 000	KOHOUTEK, MARTIN 1982 (62 NPN)

SPECTROSCOPIC VARIATIONS ; BINARIES

For 13 NPN, spectroscopic variations are due to an instability of the star (FG Sge ; proto-PN such as HBV 475, V 1016 Cygni, ...). For other NPN, variations are due to a binarity ; 5 nuclei are single lined (SB1) spectroscopic binaries :nuclei of IC 418 (P≈0.2 d.), NGC 1360 (30d.?), NGC 2346 (16.2 d.), NGC 6826 (0,238 d.), and A 36 (several d.); the nucleus of NGC 1514 has a composite (Ao+sd 0) spectrum.
Other NPN are known to be binaries : 9 astrometric, 4 eclipsing. In addition, about 20 objects with an advanced spectral type, and about 10 % of the WR–nuclei (4 to 5 objects) are possible binaries. On the other hand, Peimbert's type I PN with bipolar structure seem to have binary nuclei (Acker, 1983).
The occurence of binary and peculiar stars must be considered in working out the evolutionary study for PN.

NEW OBSERVATIONS

A new spectroscopic study is made by F. GLEIZES, for about 60 NPN, using a spectrograph with image-tube (dispersion ≈ 90 $\overset{o}{A}$ /mm). A new spectral classification is given, and new values of the temperature were calculated ; the results were published in 1984.

REFERENCES
ACKER A., 1983, 5e Journée de Strasbourg : Les étoiles binaires dans le diagramme H.R., p. 103
HEAP S.R., 1983, IAU Sympos. 103 (London), Reidel, p. 375
PERINOTTO M., 1983, IAU Sympos. 103 (London), Reidel, p. 323

Other references in ACKER et al (1982) Catalogue of the Central Stars of PN (Strasbourg).

NLTE ANALYSIS OF SDO STARS

U.Heber[1], K. Hunger[1], R.P. Kudritzki[2], K. P. Simon[2]
1) Institut für Theor. Physik und Sternwarte, Kiel
2) Institut für Astronomie und Astrophysik, München

Among the faint blue halo stars the sdO's form a group whose evolutionary status is still not fully understood. Phenomenologically they appear
to be related to the horizontal branch. The knowledge of stellar parameters and abundances may give a clue for understanding their evolutionary status. As the sdO's are hot ($T_{eff} \geq 35000K$), the assumption of LTE
leads to substantial errors. Hence detailed NLTE analyses are inevitable.

THE METHOD OF ANALYSIS

High (or medium) resolution visual and UV spectra are analysed by means
of complete NLTE techniques in the following way:
In the first step a grid of NLTE model atmospheres is constructed which
is then used for additional line formation calculations for hydrogen,
He II and He I. A fit of the observed profiles of hydrogen and helium
lines then yields simultaneously effective temperature, gravity and
helium abundance. From the finally adopted model multi-level (or two-
level) calculations are carried out for different metal ions to determine metal abundances. An illustrative example for the procedure of
model fitting is given by Kudritzki and Simon (1978).

ATMOSPHERIC PARAMETERS

26 sdO's up to now have been analyzed for atmospheric parameters. The
results are summarized in Table 1 and plotted in Fig.1. Most strikingly
there is a subdivision of the group with respect to the helium content.
We find three groups:
1) 5 stars are extremely helium rich. Their photospheres consist almost
   entirely of helium (helium number fraction y = 1.0).
2) 10 stars have intermediate helium rich photospheres ($y \approx 0.5$).
3) 11 stars are helium poor ($0.0007 \lesssim y \leq 0.12$).
Moreover, there is a clear borderline which separates helium rich objects from the helium poor ones, the critical temperature being 40000K.
The helium rich sdO's are hotter, the helium poor ones cooler then this.
The helium poor sdO's are intermediate in T_{eff} to the sdB stars (see
Fig.1) which are also helium poor and are better called sdOB stars.

A. Maeder and A. Renzini (eds.), Observational Tests of the Stellar Evolution Theory, 215–218.
© *1984 by the IAU.*

Table 1: Atmospheric parameters of sdO stars (for a list of references
          see Heber et al., 1983b)

star	T_{eff}/K	log g	y	star	T_{eff}/K	log g	y
CD-31°4800	42500	5.5	1.0	BD-3°2179	>55000	4.5	0.3
BD+39°3226	45000	5.5	1.0	LB 1566	42600	5.0	0.2
BD+37° 442	55000	4.0	1.0	SB 884	40000	6.1	0.12
TONS 103	40000:	6.5:	1.0	SB 169	37000	6.3	0.04
SB 21	36300	5.4	1.0	HD 149382	35000	5.5	0.04
SB 58	38000	4.5	0.7	Feige 110	40000	5.0	0.03
HD 127493	42500	5.25	0.6	SB 38	39000	5.4	0.03
BD+75° 325	55000	5.3	0.6	EG 55	38000	5.8	0.02
HD 49798	47500	4.25	0.5	Feige 66	36000	6.0	0.02
SB 705	44700	5.8	0.5	SN 1006	38500:	6.5:	0.015
LS 630	55000	6.0	0.5	LB 3241	41000	5.7	$\leq$ 0.005
SB 933	49000	5.5	0.4	LB 3459	40000	5.3	0.003
HD 128220	{45000	4.75	0.37	SB 707	34000	6.0	$\leq$ 0.0007
	{40000	4.25	0.17				

: uncertain due to low quality observations

METAL ABUNDANCES

Abundances of C,N and Si have been obtained for 3 helium rich sdO's and
3 sdOB's. The results are listed in Table 2. For the helium rich sdO's
the data given in Table 2 indicate that the material in their atmospheres has been processed in the CNO cycle. For the sdOB's the abundances are considerably different (see Table 2). Silicon is deficient
in all analyzed stars by large factors. Nitrogen is approximately solar.
Carbon is moderately deficient in HD 149382 and Feige 66 while it is
strongly deficient in Feige 110 (by more than a factor of 300000).These
strange abundance patterns of the sdOB stars can be explained by
diffusion. Due to the high gravity helium, carbon and silicon settle
downwards out of the photosphere. The prerequisite for gravitational
settling is an atmosphere that is quiet. For effective temperatures near
40000K a helium convection zone occurs, which may impede diffusion.

Table 2: Abundances of sdO stars (log of mass fraction)

star	H	He	C	N	Si
HD 49798	-0.70	-0.10	-3.7	-1.6	-3.0
HD 127493	-0.85	-0.07	-4.1	-2.1	-3.0
BD+ 75°325	-0.85	-0.07	-3.7	-1.9	-3.1
CNO cycle	-0.85	-0.07	-4.1	-1.9	
sdOB stars:					
HD 149382	-0.07	-0.85	-4.4	-2.9	< -7.7
Feige 66	-0.03	-1.12	-4.5	-2.7	< -7.8
Feige 100	-0.05	-0.96	<-7.9	-3.1	< -6.8
sun	-0.152	-0.55	-2.4	-3.0	-3.16

(with $T_6 = 2 \cdot 10^7 K$)

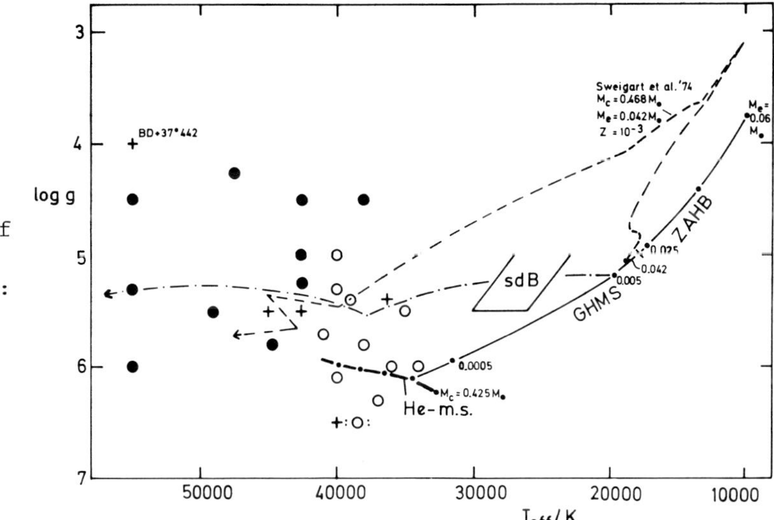

Fig.1:Position of
sdO stars in the
log g-T_eff-plane:
+ = extremely
    helium rich
● = intermediate
    helium rich
o = sdOB

THE EVOLUTIONARÝ STATUS OF THE SDO STARS

In Fig.1(the log g-T_eff-plane) zero age horizontal branch models (Gross,
1973) are plotted.The models are labelled with their envelope masses M_e.
For $M_e < 0.02\ M_☉$ the hydrogen rich envelope is inert.These models may be
called generalized helium main sequence models (GHMS, see Heber et al.,
1983a). The true helium main sequence is also shown. The sdO's lie to
the left of the ZAHB and GHMS but above the helium main sequence. We can
reach the position of the subdwarfs in the log-T_{eff}-plane when evolution is considered. Two relevant evolutionary tracks are shown in
Fig.1. The dashed curve represents evolution of a ZAHB-star (Sweigart
et al., 1974), while a GHMS star proᵇably evolves similar to a pure helium star along the dashed dotted curve. From Fig.1 we conclude that
there are essentially two ways sdO's can form: either by evolution from
horizontal branch B stars and/or from sdB stars. The latter are slightly evolved GHMS stars ($M_e < 0.02\ M_☉$). Some of the high gravity sdOB's
are located at the helium main sequence, which means that their hydrogen
rich envelope masses are even smaller ($M_e < 10^{-3}M_☉$). This situation is
reminiscent of DA white dwarfs which are known to have very little hydrogen floating atop the helium rich layers. Some (low gravity) stars
(e.g. BD+37°442) are probably remnants of the asymptotic giant branch.

REFERENCES:

Gross,P.G.: 1973, Monthly Notices Roy. Astron. Soc. <u>164</u>, 65
Heber,U., Hunger,K., Jonas,G., Kudritzki,R.P.: 1983a, Astron.
        Astrophys., in press
Heber,U., Hamann,W.-R., Hunger,K., Kudritzki,R.P., Simon,K.P.
        Mendez,R.H.: 1983b, Astron. Astrophys., submitted
Kudritzki,R.P., Simon, K.P.: 1978, Astron. Astrophys. <u>70</u>, 653
Sweigart,A.V., Mengel,J.G., Demarque,P.: 1974, Astron. Astrophys.
        <u>30</u>, 13

DISCUSSION

Renzini: How can an HB star be turned into an extreme He star?

Heber: No idea.

Cox: What are the luminosities above which there is convection to prevent the He settling?

Heber: The onset of convection in the photosphere is not very sensitive to luminosity. Instead it is more sensitive to the effective temperature of the star.

Mould: Isn't it possible that an extremely large mass loss could have completely stripped the hydrogen envelope from these stars?

Heber: Yes. In order to account for the very low envelope masses M_e of the subdwarfs ($M_e \lesssim 0.05\ M_\odot$) we have to consider mass loss both on the red giant branch and during the core helium flash. At present nobody is able to tell you what is going on during the core helium flash.

Schatzman: Is there any information on the rotational velocities of these objects?

Heber: From **our** spectra we conclude that they must be very low (below the detection limit).

THE H-R DIAGRAM OF O-TYPE SUBDWARFS

J. S. Drilling
Louisiana State University Observatory
Baton Rouge, LA   70803-4001

and

D. Schönberner
Institut für Theoretische Physik und Sternwarte
Universität Kiel

A spectroscopic survey by Drilling (1983) of non-emission OB+ stars in the Case-Hamburg survey and its extension to b = ±30° for l = ±60° has turned up 12 hot, relatively bright O-type subdwarfs (V<13). Drilling's survey covers more than 13 times the area of the south galactic polar survey of Slettebak and Brundage (1971), but has revealed only the hottest and/or hydrogen-poor objects because only the weakest-lined O- and B-type stars were observed. We have observed all of these stars in the large-aperture, low-resolution mode of the IUE satellite. A full description of their spectral appearance is given elsewhere (Schönberner and Drilling, 1984). The spectra were corrected for interstellar absorption using the Seaton (1979) reddening law to nullify the 2200 Å interstellar feature, and in all cases $E_{B-V} < 0.25$.

We determined the temperatures using R, the ratio of the total flux between 1240 and 1945 Å to that between 1945 and 3120 Å. This quantity is dependent on effective temperature and nearly independent of the color excess: $\Delta E(B-V) = 0.1$ corresponds to only a 5% change in R! In order to calibrate R against temperature, we chose four hot stars from the IUE data center with reasonably well known temperatures, gravities and compositions (see Table 1). Unfortunately two of them are helium rich, whereas the others are helium poor. Because the UV flux also depends on the composition (see Schönberner and Drilling 1984), we calibrated R against the color temperature, T_{BB}. This is the temperature of the black body which best matches the flux between 1200 and 3100 Å of the model atmosphere corresponding to the published values of temperature, gravity and composition. These black body temperatures, T_{BB}, are also listed in Table 1, together with the measured values of R.

Our subdwarf O stars have R in the range 2.5 to 3.9, indicating color temperatures from 60,000 K to over 100,000 K according to Table

A. Maeder and A. Renzini (eds.), Observational Tests of the Stellar Evolution Theory, 219–222.
© *1984 by the IAU.*

1. We estimated the compositions from visual spectra (Drilling, 1983) and assigned to each star the effective temperature of a model whose UV flux matches that of a black body with temperature T_{BB}. We derived distances from the color excesses as described by Schönberner and Drilling (1984). The resulting absolute magnitudes cover the range from 1.5 to 8.

Fig. 1 shows the loci of our objects in the H-R diagram together with those of previously analyzed sdO's. Obviously our new hot subdwarfs occupy a previously empty region in the H-R diagram, filling the luminosity range from central stars to white dwarfs. This spread in luminosity is in agreement with their spectral appearance (see Schönberner and Drilling 1984) and indicates differences in their origin. Fig. 1 suggests that the less luminous sdO's are post-EHB

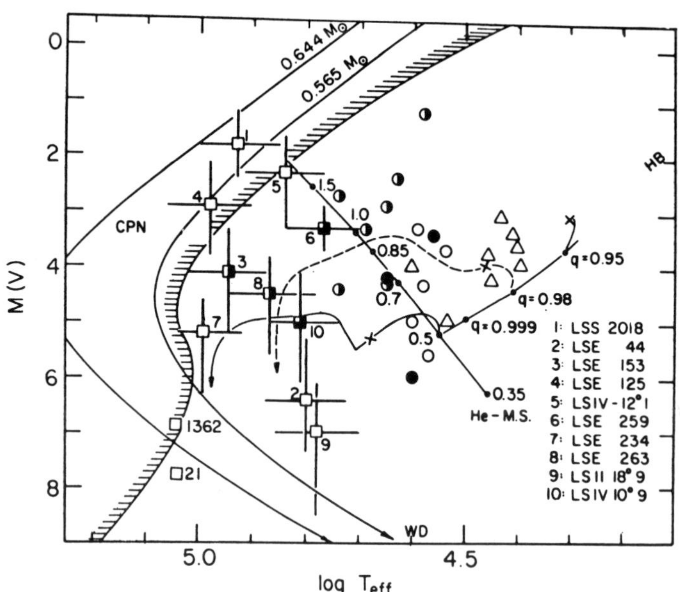

Figure 1. H-R diagram with the central star region (CPN) to the left of the shaded line, the helium main sequence (Paczynski, 1971), the extended horizontal branch (q), and the loci of our objects (squares). The evolutionary paths of post-AGB models of 0.565 and 0.644 M_O are from Schönberner (1979, 1983), and that of a 0.5 M_O pure helium star is from Paczynski (1971). The central helium-burning phase of an EHB model with M = 0.5 M_O and q = 0.95 (Sweigart and Gross, 1976) is also shown. The (estimated) evolution of a 0.5 M_O EHB star with q = 0.98 is marked by a dashed line. The crosses indicate the termination of central helium burning. The sdO's (circles) from Hunger et al. (1981) and sdB's (triangles) from Heber et al. (1983) are also displayed. The n(He)/n(H) ratios are indicated as follows: open symbols: normal or helium poor; half filled: helium rich; three-quarters filled: probably extremely helium rich; filled: extremely helium rich.

objects with 0.5 $M_\odot$ and a helium burning shell as their energy source. The evolution could then proceed as indicated in Fig. 1 by the dashed line, thereby providing the low mass tail of the white dwarf mass distribution. The peculiar helium-to-hydrogen ratios then have to be explained by diffusion and convective mixing in surface layers (Michaud et al. 1983; Wesemael et al. 1982). Another possibility is that some of these stars are descendants of HB stars which were more massive than 0.5 $M_\odot$, but not massive enough to climb the AGB ($M < 0.57\ M_\odot$). Such stars also have a hydrogen burning shell which contributes significantly to their luminosity.

This work was supported in part by the National Science Foundation (Grant Nos. AST-8018766 and INT-8219240) and the National Aeronautics and Space Administration (Grant No. NAG 5-71).

TABLE 1

Star	T_{eff}	n(He)/n(H)	T_{BB}	R	Ref.
HD 49798	47,500	1.0	58,000	2.30	Kudritzki and Simon, 1978
BD+75°325	55,000	1.5	63,000	2.50	Kudritzki, et al. 1980
HZ 43	62,500	<0.001	97,000	3.10	Auer and Shipman, 1977
NGC 7293	105,000	0.01	145,000	4.00	Bohlin, et al. 1982; Mendez, et al. 1983

REFERENCES

Auer, L. H., and Shipman, L. H. 1977, Ap. J. (Letters), 211, L103.
Bohlin, R. C., Harrington, J. P., and Stecher, T. P. 1982, Ap. J., 252, 635.
Drilling, J. S., 1983, Ap. J. (Letters), 270, L13.
Heber, U., Hunger, K., Jonas, G., and Kudritzki, R. P., 1983, Astr. Ap., in press.
Hunger, K., Gruschinske, J., Kudritzki, R. P., and Simon, K. P., 1981, Astr. Ap. 95, 244.
Kudritzki, R. P., and Simon, K. P., 1978, Astr. and Ap., 70, 653.
Kudritzki, R. P., Gruschinske, J., Hunger, K., and Simon, K., 1980, Proc. 2nd-year IUE Conference, Tubingen.
Mendez, R. H., Kudritzki, R. P., and Simon, K. P., 1983, IAU Symp. No. 103, 343.
Michaud, G., Vauclair, G., and Vauclair, S., 1983, Ap. J., 267, 256.
Paczynski, B., 1971, Acta. Astr., 21, 1.
Schönberner, D., 1979, Astr. and Ap., 79, 108.
Schönberner, D., 1983, Ap. J., in press.
Schönberner, D., and Drilling, J. S., 1984, Ap. J., in press.
Seaton, M. J., 1979, Mon. Not. Roy. Astron. Soc., 187, 73P.
Sweigert, A. V., and Gross, P. G., 1976, Ap. J. Suppl., 32, 387.
Slettebak, A., and Brundage, R. K., 1971, Astr. J., 76, 338.
Wesemael, F., Winget, D. E., and Cabot, W., 1982, Ap. J., 254, 221.

DISCUSSION

<u>Lub</u>: Would you please care to indicate the errors in effective tem-
perature and absolute magnitude in your diagram as these two are
correlated!

<u>Schönberner</u>: The quantity R is nearly <u>independent</u> of the assumed colour
excess. The reason is that the $\lambda 2200$ bump compensates for the increasing
absorption below 1900 Å. The main error in temperature stems from the
somewhat uncertain surface composition. We estimate it to be around
20 %. The error in absolute magnitude is somewhat around 1 magnitude.

<u>Cannon</u>: Could you explain how your sample of new sdO stars was selected,
i.e. what was the discovery technique?

<u>Schönberner</u>: These sdO stars are contained in the Case-Hamburg survey
of luminous stars and in its extension made by J. Drilling. They are
classified as OB+ stars.

THE   ATMOSPHERE   OF   SUBLUMINOUS   B   STARS

U. Heber[1], K. Hunger[1], G. Jonas[1], R.P. Kudritzki[2]

1) Institut f. theoretische Physik und Sternwarte, Kiel
2) Institut f. Astronomie und Astrophysik, München

8 sdB stars have been analyzed for effective temperatures, gravity and
helium content by means of LTE and NLTE model atmospheres (Heber et al.,
1983). The analyses are based on the following observations:
1.) B&C IDS spectrograms obtained with ESO's 3.6m and 1.5m telescopes
     (4250-4800$\overset{o}{A}$ at 29 $\overset{o}{A}$/mm, or, 4000-5000$\overset{o}{A}$ at 59 $\overset{o}{A}$/mm).
2.) IUE spectra in low resolution (6$\overset{o}{A}$) through the large aperture.
3.) Johnson and Strömgren colours, which are taken from the literature.
Effective temperatures are obtained via their definition from the integrated fluxes. Surface gravities then follow from the line profiles of
Hγ and (when available) Hβ. The helium abundances are derived from the
equivalent widths of He I, λ4471$\overset{o}{A}$.
The results are summarized in Table 1. The majority of the program stars
(6 out of 8) have effective temperatures around 26000K. SB 707 and
LB 3241 are distinctly hotter and thus belong to the sdOB subgroup.
Helium is depleted in all program stars, the depletion varying by large
factors. In SB 707 and LB 3241 no He I, λ4471$\overset{o}{A}$ or He II, λ4686$\overset{o}{A}$ are seen
and hence upper limits are derived. The helium content of SB 707 is the
lowest known in any subdwarf. These low helium abundances (far below
primordial) cannot be explained by canonical stellar evolution. It is
brought about by diffusion (gravitational   settling of helium).

Table 1: Atmospheric parameters of sdB stars

star	T_{eff}/K	log g	n_{He}/n_{H}
SB 459	25000	5.3	0.0016
LB 1559	25200	5.2	0.0150
SB 410	26700	5.1	< 0.0012
SB 485	27100	5.0	0.0040
SB 290	28200	5.5	0.0044
SB 815	28800	5.4	0.0035
SB 707	34000	6.0	< 0.0007
LB 3241	41000	5.7	< 0.0050

A. Maeder and A. Renzini (eds.), Observational Tests of the Stellar Evolution Theory, 223–224.

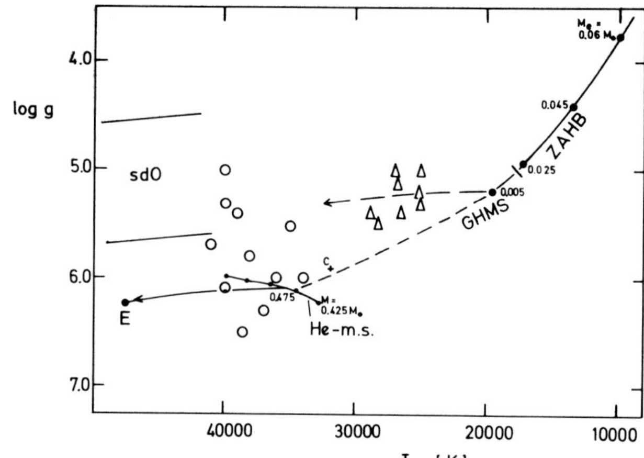

Fig. 1: Position of
analyzed sdB (Δ) and
sdOB (o) stars in the
(log g, T_{eff})-plane.
See text.

The Evolutionary Status of the sdB Stars

In Fig. 1 the position of all sdB's and sdOB's sofar analyzed are shown
in the (log g, T_{eff})-plane. Also shown are Zero Age Horizontal Branch
models (ZAHB, Gross, 1973) with core masses M_C=0.475 $M_\odot$, helium mass
fraction Y=.25 and metallicity Z=0.01. The models are labelled with
their envelope masses (in solar units). The model from Caloi (1972)
labelled with an C in Fig. 1 has a very low envelope mass M_e=0.0005 $M_\odot$.
When the core mass fraction q exceeds 0.95, hydrogen burning in the
envelope is negligible ($L_H/L_{He} \leq$ 0.2): a helium main sequence is reached
which has an inert hydrogen rich envelope (henceforth called generalized
helium main sequence GHMS). The true helium main sequence is also
plotted in Fig. 1. The dashed line is an interpolated GHMS line for
envelope masses less than M_e=0.005 $M_\odot$. The position of q=0.95 which
separates hydrogen shell burning models (ZAHB) from GHMS models is indicated by a tick mark. The observed sdB's and sdOB's (except two
objects) lie to the left and above the GHMS. Evolutionary tracks for
GHMS models are not yet available. Since the envelope is inert the
evolution.   proceeds presumably similar to that of a pure helium star.
In Fig. 1 the evolutionary track of a pure helium star of 0.5 $M_\odot$
(Paczynski, 1971) until helium core exhaustion (point E) is shown. If
we apply the same (horizontal) evolutionary vector to the GHMS stars
we reach the domain of the observed subdwarfs. This implies that the
sdB's are slightly evolved GHMS stars and have very low envelope masses
($M_e \leq$ 0.02 $M_\odot$). The envelope masses of the majority of the sdOB's are
further reduced to almost nothing ($M_e \leq 10^{-3}$ $M_\odot$).

References:

Caloi, V.: 1972, Astron. Astrophys. <u>62</u>, 411
Gross, P.G.: 1973, Monthly Notices Roy. Astron. Soc. <u>164</u>, 65
Heber, U., Hunger, K., Jonas, G., Kudritzki, R.P.: 1983, Astron.
        Astrophys., in press
Paczynski, B.: 1971, Acta Astron. <u>21</u>, 1

MASS LOSS FROM HIGHLY EVOLVED STARS

Wolf-Rainer Hamann
Institut für Theoretische Physik und Sternwarte
Universität Kiel

1. INTRODUCTION

This is a summarizing discussion about detailed studies of stellar
winds from hot,subluminous stars of different classes: two sdO's (Hamann et al.,1981), three extreme helium stars (Hamann et al.,1982) and
one CPN (Hamann et al.,1983). All program stars have been analyzed previously in Kiel by non-LTE techniques with respect to effective temperature, gravity and abundances. For comparison, we have also included
the Of-supergiant ζ Pup and the B-main sequence star τ Sco, because their
winds have been investigated in detail by the same methods as applied
in this paper (Hamann, 1980, 1981).

2. THE METHOD

The observational material for the mass-loss studies consists of IUE
spectra in the high-dispersion mode. Those of the UV-resonance lines
which indicate stellar winds are fitted with theoretical profiles.The
line formation is calculated with the comoving-frame technique,assuming
pure scattering by two-level atoms. The overlap of the doublet components is taken into account, and the photospheric contribution is also
included if necessary.An example of a line fit is given in Fig. 1.

Only a lower limit of the mass loss rate can be deduced from the
line fit, unless the ionization fraction of the considered ion is known.
We also calculated theoretical ionization fractions,but because of the
"superionization" of these stellar winds the calculations have free
parameters.Only in some cases restrictions of the ionization fractions
are obtained from theory.Often an upper limit of the mass loss rate
is set only by the fact that the photosphere appears to be static.
Generally, all mass loss rates derived from UV lines suffer from the
great uncertainties in the ionization fractions.

A. Maeder and A. Renzini (eds.), Observational Tests of the Stellar Evolution Theory, 225–228.
© *1984 by the IAU.*

3. RESULTS AND DISCUSSION

Principally, we find that the wind parameters derived for these subluminous stars fit well into the general picture of stellar winds from
early-type stars.Is there a general formula, predicting the mass loss
rate of a star when certain parameters are given?Our program stars,to-
gether with the (commonly studied) OB supergiants,provide a wide range
of stellar parameters and hence an improved basis to discuss this
question. Fig.2a gives the empirical mass loss rates versus luminosity.
ζ Pup and the subluminous stars can be related by $\dot{M} \propto L^{4.5}$ .However, the
scatter around this relation is probably larger than the errors in the
data: τ Sco lies below it, and the three extreme helium stars have
similar L,but very different mass loss. A better alignment is achieved
in Fig.2b with the Reimers formula $\dot{M} \propto LR/M$, which was established
originally for red giants and calibrated, e.g.,with Antares(Kudritzki
and Reimers,1978). A similar relation was confirmed empirically for
massive OB stars (Lamers,1981). ζ Pup lies above the predictions of this
formula by one order-of-magnitude, but τ Sco fits well. For our three
helium stars, the R/M-dependance goes into the wrong direction.

The radiation-driven wind theory (Castor et al.,1975) predicts the
mass flux (per surface unit) in terms of effective temperature and gravity. Comparison with our program stars reveals poor agreement, which
cannot be demonstrated here for sake of brevity.

Towards luminosities below 100 $L_\odot$, a strong constraint on the mass
loss comes from the existence of helium-deficient OB subdwarfs. Even
under optimum conditions, helium depletion by gravitational settling
can only be efficient if the mass loss rate is smaller than predicted
from extrapolation of Reimers' formula by at least a factor of 1000.

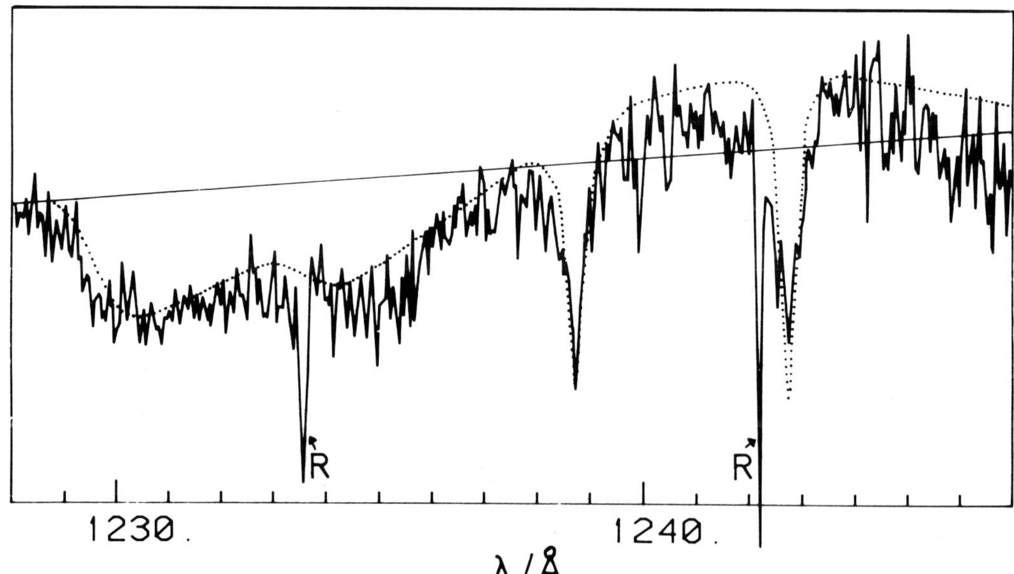

Fig.1: N V resonance doublet of NGC 3242 (CPN): IUE tracing and
theoretical profile (dotted), R = reseau mark

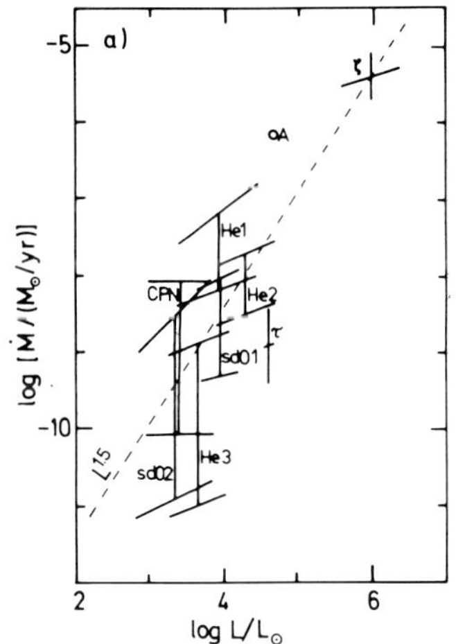
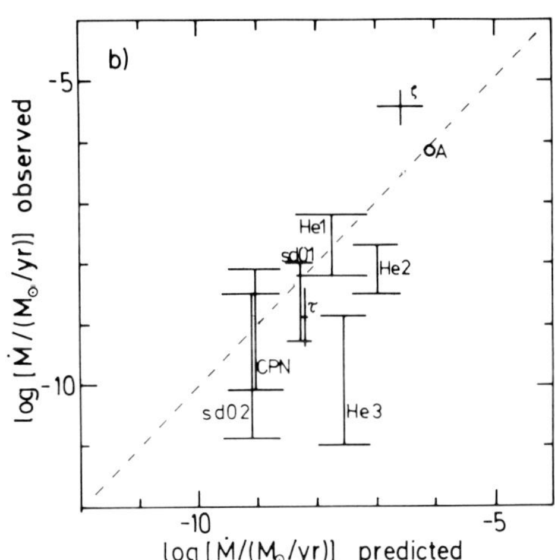

Fig.2:Empirical mass loss rates versus stellar parameters.
O-subdwarfs:sdO1 = HD 49798, sdO2 = HD 128220;
extreme helium stars: He1 = HD 160641, He2 = BD -9°4395,
He3 = BD +10°2179;
Central Star: CPN = NGC 3242;
comparison stars: ζ = ζ Pup (OfI), τ = τ Sco (BOV), A = Antares
a) $\dot{M}$ versus L   b) $\dot{M}$ versus prediction of Reimers' formula

4. CONCLUSIONS

The Reimers-formula agrees with most observed mass loss rates within one
order-of-magnitude. This holds for the subluminous stars studied here,
as well as for the luminous (massive) OB stars and the red giants.Some
stars behave peculiar, e.g. Of stars, Wolf-Rayet stars, extreme helium
stars. An extrapolation of the mass loss prediction down to 100 $L_\odot$ is
possibly prohibited.

5. REFERENCES

Castor,J.I., Abbott,D.G., Klein,R.I.: 1975, Astrophys.J. 195, p. 157
Hamann,W.-R.: 1980, Astron. Astrophys. 84, p. 342
Hamann,W.-R.: 1981, Astron. Astrophys. 100, p. 169
Hamann,W.-R., Gruschinske,J., Kudritzki,R.P., Simon,K.P.: 1981,
        Astron. Astrophys. 104, p. 249
Hamann,W.-R., Schönberner,D., Heber,U.: 1982, Astron. Astrophys.
        116, p. 273
Hamann,W.-R., Kudritzki,R.P., Méndez, R.H.: 1983, Astron. Astrophys.
        (in preparation)
Kudritzki,R.P., Reimers,D.: 1978, Astron. Astrophys. 70, p. 227
Lamers,H.J.G.L.M.: 1981, Astrophys. J. 245, p. 593

DISCUSSION

<u>Walborn</u>: τ Sco has a very peculiar UV spectrum showing an <u>enhanced</u> stellar wind relative to normal stars of similar spectral <u>types</u>. A stronger mass or temperature dependence of mass loss along the main sequence than the relation you showed may be indicated.

<u>Hamann</u>: Yes, possibly. The problem is that only τ Sco has been studied in detail. From an only qualitative comparison of the spectral wind-features, one cannot distinguish whether the mass loss rates are different or the ionization conditions are not the same.

<u>Richer</u>: Just a comment. The few red giants that do have well determined mass loss rates (e.g. α Sco) have mass loss rates that are higher than Reimers by up to 3 orders of magnitude.

<u>Conti</u>: I'd like to comment on your presentation of the relation between $\dot{M}$ and L, along with such other parameters as M, R, T_{eff}, gravity. Dr. Garmany and I have just completed an analysis of some 50 O-type stars with systematic spectroscopy. We too find that $\dot{M}$ depends on L with an exponent of order 1.6. There remains considerable scatter. The scatter is reduced by parameterizing the M, R, T_{eff}, gravity in various combinations, but we cannot unequivocally determine the correct parameterization - some scatter still remains. It will be interesting to now compare our data since your objects have much smaller masses and radii.

CONSTRAINTS ON STELLAR EVOLUTION BY OBSERVATION OF WHITE DWARFS

V. Weidemann
Institut Theor. Physik und Sternwarte
Universität Kiel, W.-Germany

The recent observation of white dwarfs in the open cluster NGC 2516 and the determination of their surface gravity and effective temperatures (Reimers and Koester, 1982) has enabled the establishment of the initial-final mass relation for low and intermediate mass stars which was published a few months ago and is presented here (Fig. 1 of Weidemann and Koester, 1983a). The most important conclusions drawn are:

1. The limiting mass for white dwarf progenitors is 8-9 $M_\odot$ rather than 5-6 $M_\odot$, with supernova production beyond;
2. The rather flat run of the initial-final mass relation in the main range of star production, 1-5 $M_\odot$, explains the observed narrow mass distribution of white dwarfs and central stars of planetary nebulae around 0.6 $M_\odot$;
3. High mass white dwarfs exist, as shown in the rich, young cluster NGC 2516, but are extremely rare in general.

A second constraint is provided by the fraction of white dwarfs which do not go through the PN stage. From Schönberner's evolutionary tracks (Schönberner, 1983) it can be seen that stars leaving the AGB below M = 0.55 $M_\odot$ do not show planetary nebulae. They evolve through the sdO-region towards the white dwarf stage, as demonstrated in the contri-butions of Kudritzki et al, Heber et al and Drilling and Schönberner at this meeting.

I present here the most recent mass distribution of DA white dwarfs, based on evaluation of Palomar multichannel photometry (Weidemann and Koester, 1983b) which not only demonstrates the steep increase at about 0.45 $M_\odot$, the He-flash core mass, as expected from stellar evolution theory, but also shows the flatter tail towards higher masses. Cutting this distribution at 0.55 $M_\odot$ we can estimate the fraction of white dwarfs produced without planetary nebulae to be 44 or 28 %, respectively, de-pending on the calibration of the multichannel photometry (Hayes and Latham, 1975, or Oke and Gunn, 1983). For details I refer to the full publications.

A. Maeder and A. Renzini (eds.), Observational Tests of the Stellar Evolution Theory, 229–230.
© *1984 by the IAU.*

REFERENCES

Drilling, J.S., Schönberner, D.: 1984, this volume, p. 219.
Hayes, D.S., Latham, D.W.: 1975, Ap.J. 197, 593.
Heber, U., Hunger, K., Kudritzki, R.P., Simon, K.P.: 1984, this volume.
Kudritzki, R.P., Mendez, R.H., Simon, K.P.: 1984, this volume, p. 205
Oke, J.B., Gunn, J.E.: 1983, Ap.J. 266, 713.
Reimers, P., Koester, D.: 1982, Astron. Astrophys. 116, 341.
Schönberner, D.: 1983, Astrophys. J., in press.
Weidemann, V., Koester, D.: 1983a, Astron. Astrophys. 121, 77.
Weidemann, V., Koester, D.: 1983b, Astron. Astrophys., submitted.

These remarks have been made as a conclusion of session I on low and
intermediate mass stars.

II. EVOLUTION OF MASSIVE STARS

EVOLUTION AND MODELS

BASIC OBSERVATIONAL CONSTRAINTS ON THE EVOLUTION OF MASSIVE STARS

Peter S. Conti
Joint Institute for Laboratory Astrophysics, University
of Colorado and National Bureau of Standards, Boulder,
Colorado 80309

ABSTRACT

The sources of information and uncertainties in the intrinsic
stellar parameters of luminosity, effective temperature, mass, composition and mass loss rates are discussed. These are used to compare
the observed positions of massive stars in the Hertzsprung-Russell Diagram (HRD) with evolutionary tracks. The current status of this effort
is briefly reviewed. A short summary of the kinematic properties of
massive stars is made. A preliminary but fairly extensive discussion
of the distributions and numbers of O-type and Wolf-Rayet stars in the
galaxy and other members of the local group is then given.

I. INTRODUCTION

I have been asked to review the basic observational features
of "massive stars." I shall take as a lower mass limit a value $M \gtrsim$
10 $M_\odot$ as an arbitrary dividing line between these stars and those of
"intermediate" or "low" mass discussed elsewhere in this symposium.
This corresponds to a lower luminosity limit $L \gtrsim 10^{3 \cdot 5} L_\odot$. This mass
is roughly the lower mass limit for supernova progenitors; similarly
it is near the upper limit of mass for stars with an asymptotic giant
branch. In spectral-type terms, "massive stars" include the O to B3
main sequence, and luminous supergiants of all spectral types.

From work over the past decade or so, we know that for most of
these massive stars, mass loss in the form of stellar winds plays an
important, in some cases, even dominant, role in their evolution (Conti
and McCray 1980). The Wolf-Rayet (W-R) stars, and other luminous hot
variable stars, such as η Car, P Cyg, R71, R122, the S Doradus stars
and the Hubble-Sandage variables, are clearly related to massive stars
and undoubtedly represent advanced evolutionary stages (e.g. Maeder
1984). I shall refer to the non W-R or "other," hot stars as "luminous
blue variables," or LBV, in my talk. LBV are thus nicely contrasted
with LPV, long-period variables, which are luminous red supergiants.

A. Maeder and A. Renzini (eds.), Observational Tests of the Stellar Evolution Theory, 233–254.
© *1984 by the IAU.*

W-R stars come in two types, WN and WC. Subtypes, based mostly on
line ratios also are identified: WN subtypes range from WN2 to WN9; WC
subtypes from WC4 to WC9. In both cases, the higher ionization stages
have smaller numbers analogous to MK subtypes. Unlike MK types, these
W-R subtypes may not be an effective temperature sequence.

As is well known, the intrinsic properties of a star which most
clearly define its evolutionary state are the mass and composition
(throughout). Recently, we have needed to include mass loss, $\dot{M}$, as an
input to stellar evolution. The angular momentum and magnetic fields
may also play a role but I will not discuss them here. The observable
quantities, in addition to M,$\dot{M}$ and composition are luminosity, L, and
effective temperature, here T.

The _interplay_ among these parameters, both observationally and
theoretically, forms the current drama for stellar evolution. The
theoreticians write the script; the observers provide the stage set
and costumes. We are the actors in this play, which is going onward
amongst continual script and set revisions, role exchanges, and general commotion. The audience is, for the most part, members of the
astronomical community, who provide applause at important junctures;
granting agencies provide the critical financial backing. Drama
critics, better known as paper referees, provide comments on the
actors and their roles. Unlike real-life productions, actors and
critics frequently play both roles at once. I do not know whether
or not our stellar evolution play is a comedy or a tragedy, but I do
know, as a participant, that it is fun!

II. INTRINSIC PARAMETERS

A. Luminosities

Stellar distances are intimately related to the determinations of
luminosity. Massive stars, which are quite rare, are not close enough
to the Sun to have measurable parallaxes. Cluster and association membership with distances provided by the B-type stars (Walborn 1972; Conti
and Alschuler 1971) provides the spectroscopic calibration for nearly
all O-type stars. A recent compilation is given by Conti, Garmany, de
Loore and Vanbeveren (1983). These authors also discuss the M_V calibration of W-R stars, based mostly upon membership in the Large Magellanic Cloud. These M_V magnitudes ultimately depend on cluster fitting
methods extending back to the Hyades and the absolute values depend on
that distance. The LBV are even more rare than W-R stars but their luminous properties (e.g. Humphreys and Davidson 1979) make them readily
visible at large distances and hence they were among the first objects
detected in nearby galaxies (see also Lamers, de Groot and Cassatella
1983; Wolf, Appenzeller and Cassatella 1980; Wolf, Appenzeller and Stahl
1981). Surprisingly, the distance scale for local group galaxies
such as M33 seems to be uncertain to a factor two (Sandage and Carlson
1983) so moduli for many LBV are not well determined. P Cyg and η Car

can reasonably be assigned to associations in their vicinity; some LBV are known in the Magellanic Clouds.

In addition to distances, one needs apparent magnitudes, m_V, and colors to determine the interstellar reddening of distant stars. These observables are reasonably well in hand for massive stars, particularly with the recent absolute spectrophotometry of W-R stars by Massey (1983). With M_V determined, it is then necessary to estimate the bolometric correction, e.g., which can be quite large for hot stars in which most of the radiation is emitted below visible wavelengths. This determination depends critically on the T and the stellar model, which is not well known in some cases. My estimate of the current status of the uncertainty on luminosity as deduced from the spectrum is as follows:

	M_V	log L	No. of Stars
MK types	$\pm 0^{m}_{.}5$	± 0.3	$\sim 10^3$
W-R	$\pm 1^{m}_{.}0$	± 0.5	$\sim 10^2$
LBV	$\pm 1^{m}_{.}2$	± 0.5	~ 10

B. Effective Temperatures

Basically, three independent measures have been used to determine T: the absorption line spectrum; the overall continuum, and the Zanstra procedure. These cannot always be used together: the latter depends on the detection of the H II region surrounding the star and an estimate of the radiation escaping below 912 Å. All methods are very model dependent, and the models, particularly for W-R stars, are not well determined.

For O stars, and LBV with absorption lines, the spectroscopic method has been heavily utilized. The non-LTE models of Auer and Mihalas (1972) were utilized by Conti (1973) in determining a temperature scale for O stars. These were recently revised downward for the (mostly) Of stars of highest temperatures by Kudritzki, Simon and Hamann (1983) and Simon, Jonas, Kudritzki and Rahe (1983) by use of lower gravity models. Isolated estimates of individual temperatures of LBV have been made by, e.g., Lamers et al. (1983), Wolf et al. (1980, 1981). For W-R stars which generally do not show absorption lines, spectroscopic methods give only constraints of the ionization temperatures from the emission lines in the wind. These are typically a few $\times 10^4$ to 10^5 K.

Considerable recent work on continua of massive stars has been published by Underhill (1980, 1981). Among O stars, she finds little

correlation between temperatures found by fitting plane parallel models to
the observed ultraviolet data and those from spectra. Among W-R stars the
temperatures are in the range 20,000-40,000 K. These efforts have recently
been criticized by Garmany, Massey and Conti (1984). These authors make
use of absolute spectrophotometry from 1200 Å - 7000 Å to find the
observed continua of a number of W-R stars. They point out that
Underhill's adoption of a standard reddening law, or more exactly a
UV extinction law for all parts of the galaxy leads to inconsistent
results. Massa, Savage and Fitzpatrick (1983) have presented similar
arguments based upon a different sample of stars. I conclude that, at
least for now, intrinsic continua of hot stars in the far UV cannot
readily be determined until the extinction is established
independently. Hence model continua fitting cannot be trusted since
most radiation is emitted in this region or below 912 Å.

My estimate of the current status of the uncertainties of effective temperature are as follows:

	T	No. of Stars
MK types	±10%	~100
WR	± 50%	~10
LBV	± 50%	A few

C. Masses

For this important parameter, one must depend on binary systems.
In particular, direct mass estimates can come only for double-lined
spectroscopic binaries (there are no suitable visual binary candidates)
which are also eclipsing, so that the inclination can be found. The
spectroscopic analysis provides the velocity amplitude while a separate photometric analysis provides the period, orbital eccentricity
and inclination.

Catalogues of O-type stars (Cruz-Gonzales et al. 1974) and of W-R
stars (van der Hucht, Conti, Lundstrom and Stenholm 1981) provide much
of the background material on binary systems. Reviews of the masses
determined have been provided by Popper (1980) and Massey (1982).
There is little direct information on masses of LBV since none are
known to be double-lined eclipsing systems.

Indirect information on stellar masses comes from the location of
the stars in a Hertzsprung-Russell diagram and evolutionary scenarios.
These are good to perhaps a factor two. My estimate of the current
status of uncertainties in the mass is as follows:

	Masses	No. of Stars
MK types	± 30%	~10
WR	± 50%	~10
LBV	Factor 2	A few

D. Composition

This parameter is obtained from the spectrum; a good atmospheric
model is absolutely essential. With the advent of UV spectroscopy,
particularly the IUE satellite, the important information on resonance
lines of carbon, nitrogen and silicon ions is now becoming available.
With the exception of the analysis of the B stars τ Sco (Hunger 1955)
and γ Peg (Aller and Jugaku 1959) many years ago, detailed studies
of massive main sequence stars have generally not been done. Coarse
analysis of many O-type stars (Conti 1973) suggests compositions which
are solar to a factor two. A small subset of OB stars has strong lines
of nitrogen or carbon (e.g. Walborn 1970) suggesting abundance anomalies. The O4f star ζ Pup appears to be slightly helium rich according
to a modern analysis of Kudritzki et al. (1983). The important issue
of possible nitrogen enhancement in this star is not yet settled.

Among W-R stars the available information indicates helium and
nitrogen enhancement among WN stars and helium, carbon and oxygen enhancements among WC stars (Smith and Willis 1983; Willis 1982; Nugis
1982; Conti, Leep and Perry 1983). The anomalies are consistent with
what would be expected from observing the products of H-burning in CNO
equilibrium (WN stars) and He-burning (WC stars). We are thus observing the "cores" of highly evolved massive stars.

There has been little work to date on the compositions of LBV
specifically, although the analysis of Luud (1967) on P Cygni should
be noted.

The largest uncertainties in all these composition analyses is
the stellar models. These are reasonably good for OB and Of stars but
not too reliable for W-R stars and LBV. The recent analysis of the
ejecta of η Car by Davidson, Walborn and Gull (1982) is very important. The material which left the star in the last century has a
density state which can best be described by the physical relationships of interstellar material, rather than stellar atmospheres. They
find large N/C ratios, similar to highly processed material. In this
one LBV object, we see evidence for a highly evolved stage.

My estimate of the current status of the compositions of massive
stars is as follows:

	Composition	Uncertainties	No. of Stars
MK types	"Normal"	Factor 2	~10
WN	CNO Equilibrium	Factor 2	~10
WC	He Burn Products	Factor 2	~10
LBV	?	Factor 2	A few

E. Mass Loss Rates

There has been considerable progress on this topic in the past few years. I provided a review recently (Conti 1981), but additional efforts over the past few years have given us many new numbers. In particular, Abbott, Bieging and Churchwell (1982) have provided $\dot{M}$ for 21 W-R stars using free-free emission measures and Garmany, Olson, Conti, and Van Steenberg (1981) and Garmany and Conti (1984) have given mass loss estimates for 50 O stars based on UV spectra. Other data have been provided by Olson and Castor (1981).

My estimate of the current status concerning $\dot{M}$ is as follows:

	$\dot{M}$ ($M_\odot$ yr^{-1})	Uncertainties	No. of Stars
MK Types	$\lesssim 10^{-7}$ to a few $\times\ 10^{-5}$	Factor 2	$\gtrsim 50$
W-R	Few $\times 10^{-5}$	Factor 2	~20
LBV	Few $\times\ 10^{-5}$ to 10^{-7}	Factor 2	A few

III. LOCATION OF MASSIVE STARS IN THE HRD

It will be recalled that Humphreys (1978) and Humphreys and Davidson (1979) provided the first modern compilation and discussion of the location of luminous stars of the galaxy and Large Magellanic Cloud in the HRD. The central features of those figures were:

1) The existence of luminous blue supergiants with inferred masses $\gtrsim 60$ $M_\odot$ <u>and</u>

2) the absence of similarly luminous (and massive) red supergiants with M_{BOL} brighter than $-9^m.8$.

3) The presence of considerable numbers of hot supergiants redward of the termination of hydrogen core burning tracks, compared to those stars still burning hydrogen.

This latter has been described as "main sequence widening" by Stothers and Chin (1977); Chiosi, Nasi and Sreenivasan (1978); Cloutman and Whitaker (1980); Bressan, Bertelli and Chiosi (1981). Meylan and Maeder (1982) noticed a similar effect in an analysis of young cluster magnitude diagrams.

I should like to present a comparison of stars of the galaxy, LMC and SMC with a current theoretical HRD of Maeder (1983). Similar diagrams will be shown by Humphreys (1984) later in this symposium based on a different sample of stars. My data for the galaxy are limited to those stars within 2.5 kpc of the Sun. Garmany, Conti and Chiosi (1982) showed that counts of O stars were reasonably complete to this distance; I assume this is also the case for the later-type supergiants. A volume limited sample has an advantage in comparing populations in different parts of the HRD.

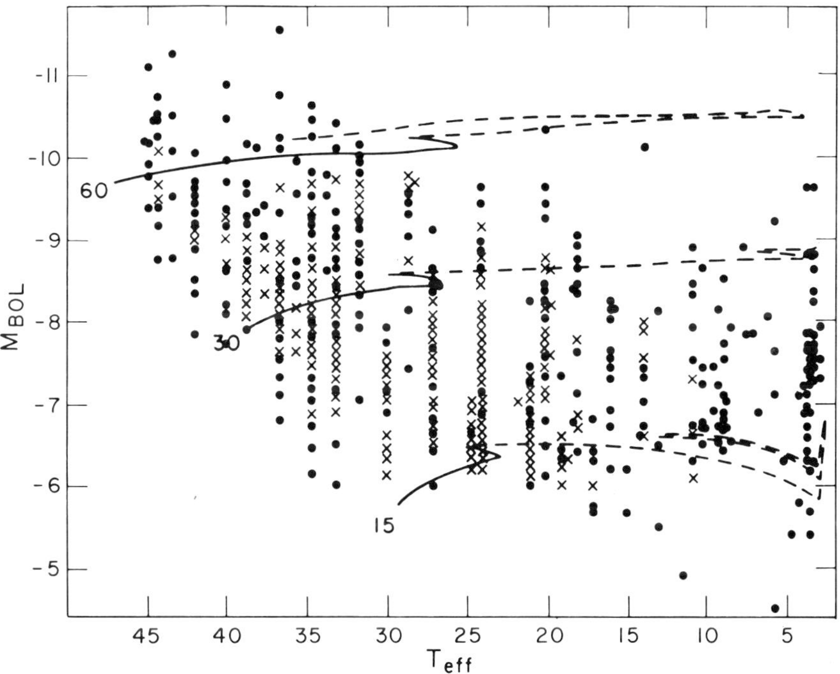

Fig. 1. Observational/theoretical HRD for luminous stars in our galaxy to 2.5 kpc from the Sun. The filled circles are single stars; the crosses represent more than one star at that position, often up to ten or so. The evolutionary tracks are those of Maeder (1983).

The central features of the Humphreys-Davidson HRD are illuminated: Luminous blue supergiants with inferred masses to ~120 M☉, the absence of comparable initial mass red supergiants, and still a strong

effect of main sequence widening. These tracks incorporate mass loss
and mixing but even so too many stars exist to the right of the termination of core HB. This is most clearly indicated for the 30 $M_\odot$ track;
that for 60 $M_\odot$ may be wide enough. Other evolution tracks have been
calculated by Doom (1982). Like the Maeder (1983) tracks, they all
fail to traverse far enough to lower temperatures to explain the large
numbers of B and A supergiants, even though various prescriptions of
mass loss and mixing are incorporated. Completeness of the sample is
probably not a problem above 30 $M_\odot$; it may begin to be an issue below
there so we cannot draw conclusions about the 15 $M_\odot$ tracks.

The absence of luminous red supergiants above some mass limit,
here $\sim$50 $M_\odot$, is almost certainly related to the appearance of W-R
stars as the core helium burning counterparts. Maeder (1982) has
suggested various channels for production of W-R stars from their
O-type predecessors. I shall return to this issue later.

In Fig. 2, I show an incomplete HRD for stars of the LMC. Photometry
of hot stars is insufficient to indicate their temperatures so we have
plotted only those O-stars with classification spectra. Garmany, Massey
and I have a major classification project under way to remedy the incompleteness of this figure for the hot stars. I will just use this figure to indicate that the same problem with the tracks exists here!

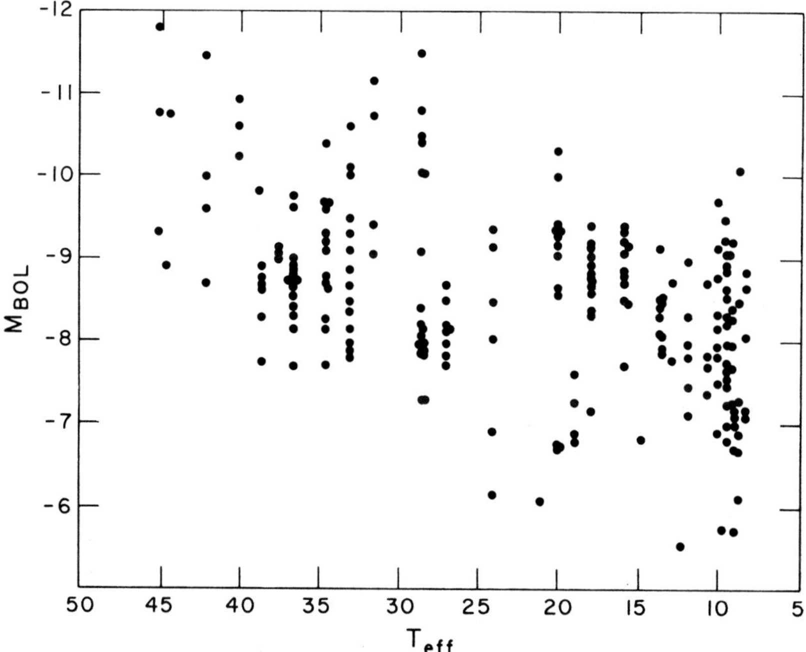

Fig. 2. Observational/theoretical HRD for luminous stars in the LMC.
Only those hot stars with photometry and classification spectra are plotted. Symbols as in Fig. 1. Sample is not complete, see text.

Too many stars are to the right of the core hydrogen burning termination point. In particular, there is a group of middle B type supergiants (between 15,000-10,000 K and M_{BOL} near -9^m) which have few
counterparts in the galaxy. Humphreys (1984) believes this grouping
to be a selection effect, but I am not so sure.

In Fig. 3, I show an <u>incomplete</u> HRD for stars of the SMC. As for
the LMC, only those O-type stars with spectral classifications have been
plotted. Along with Garmany, Humphreys and Massey, we have a major
classification program under way for the hot stars of the SMC. Again,
too many stars exist to the right of the core hydrogen burning tracks.
Perhaps this statement will need to be modified when the classifications are complete but I doubt it, given the number of candidate O
stars, of order one hundred.

In summary, I have demonstrated that the current evolutionary
tracks which incorporate mass loss and mixing do not yet sufficiently
account for the "main sequence widening" of the solar vicinity, or
Magellanic Clouds. Some physical ingredient(s) are not yet included
in the models. Stars more massive than ~50 $M_\odot$, with these tracks, do
not become red supergiants, or exist there only a very short time.

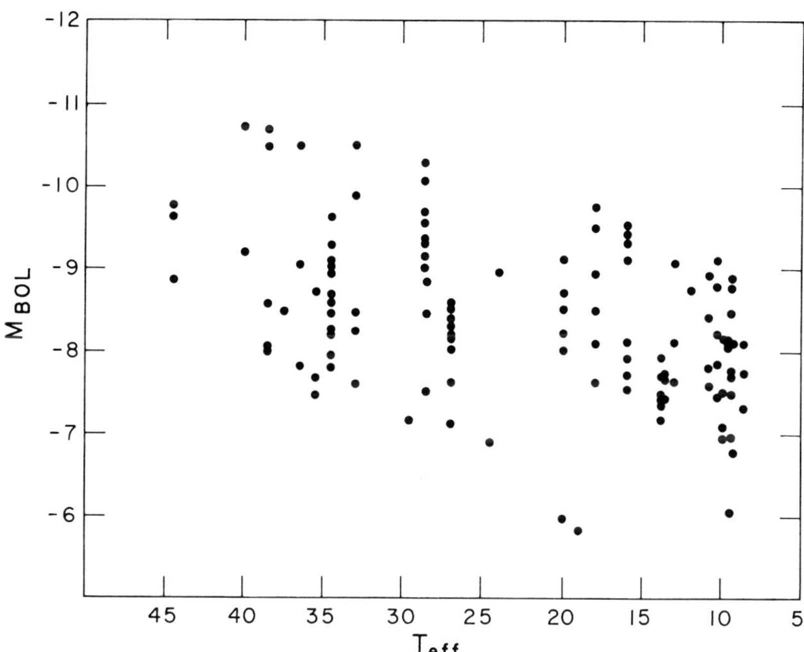

Fig. 3. Observational/theoretical HRD for luminous stars in the SMC.
Only those hot stars with photometry and classification spectra are plotted. Symbols as in Fig. 1. Sample is not complete, see text.

IV. EXTRINSIC PARAMETERS

A. Kinematics

Radial velocities are provided in various catalogues for O stars (Cruz-Gonzalez et al. 1974) and W-R stars (van der Hucht et al. 1981). The latter are not well determined, being based upon measures of broad emission lines, often differing from ion to ion. In some spectroscopic binaries, γ velocities from the O components may give good estimates of real kinematic radial velocities. Recently, Torres and Conti (1984) attempted to derive meaningful radial velocities of some narrow lined WC9 stars. The results were not too satisfying, having fairly large uncertainties. There is no catalogue of properties of LBV but scattered data are found in the literature.

My estimate of the uncertainties in radial velocity measures are as follows:

	Uncertainties	No. of Stars
MK Types	~ 10 km s^{-1}	Few 100
W-R	30-40 km s^{-1}	~ 20
LBV	10 km s^{-1}	A few

Scattered proper motion data exist in the literature but as most stars are well beyond 1 kpc these numbers do not provide much independent kinematic data aside from proving or disproving cluster membership.

B. Binary Fraction

Some 40% of the MK stars, at least along the main sequence, are binaries (Abt and Levy 1978; Garmany, Conti and Massey 1980). This is similar to the fraction of W-R type binaries (Massey 1982), at least those with massive companions. There is an entire scenario of massive binary evolution which suggests some W-R systems will be found with neutron star companions. Moffat (1982a) sums up the evidence for the existence of these objects. Although at one time W-R systems were all supposed to be binaries (e.g. Paczynski 1973), more recent arguments (e.g. Vanbeveren and Conti 1980; Massey 1982) suggests this is not the case in some systems. Among LBV, P Cyg has been suspected of being a binary (e.g. De Groot 1969) but the case has not been proven. While the binary nature of some massive stars clearly will affect their subsequent evolution, the physical importance of mass loss and mixing in these objects also plays a key role.

There has been little systematic study of duplicity for stars
outside our galaxy but Moffat (1982b) has begun such an effort for W-R
stars in the Magellanic Clouds.

C. Galactic Distribution of Massive Stars

Garmany, Conti and Chiosi (1982) have studied the longitude distribution of a nearly complete sample of O stars to 3 kpc. In Fig. 4,
I show their distribution, projected onto the plane of the galaxy. The
field star distances are derived from the O spectral type calibration;
the cluster stars are at distances given by the B star calibration.

Three "spiral arms" are seen: Scorpio-Sagittarius-Carina interior
to the Sun; the Cygnus arm near ℓ = 90°; and the Perseus-Auriga arm
outwards. These features are ill defined and certainly cannot de-
lineate a "spiral" nature by themselves. We presume them to be so by
reference to the appearance of other neighboring galaxies. The large
"width" of the arms is perhaps surprising but one wonders whether or
not this is caused by the individual stellar distance uncertainties.

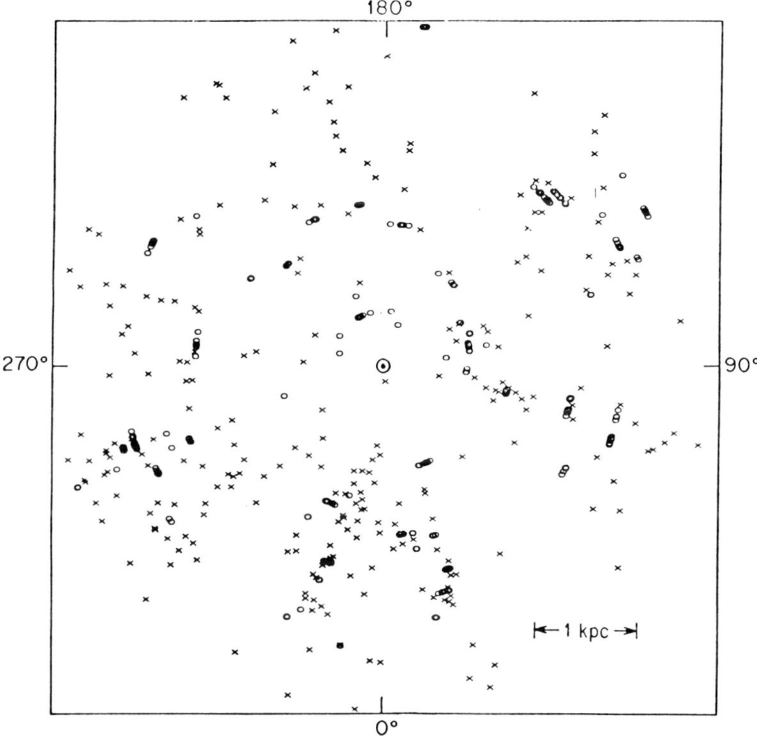

Fig. 4. O stars within 3 kpc of the Sun, projected onto the plane of
the galaxy. Open circles are cluster members; crosses are field stars
(adapted from Garmany, Conti and Chiosi 1982).

I will return to this issue later. The "arms" in Fig. 4 are about as
well defined as any other stellar features (e.g. Cepheids).

In Fig. 5, I show the galactic distribution of W-R stars, from
the work of Conti, Garmany, de Loore and Vanbeveren (1983). These
distances are based upon the W-R spectral type distribution of these
authors and are more uncertain than that for O stars. We see some
features like that of Fig. 4 (stars in the inner and solar arms) but
few outward from the Sun. The complete absence of stars in the anti-
center has been known for some time (e.g. Roberts 1962).

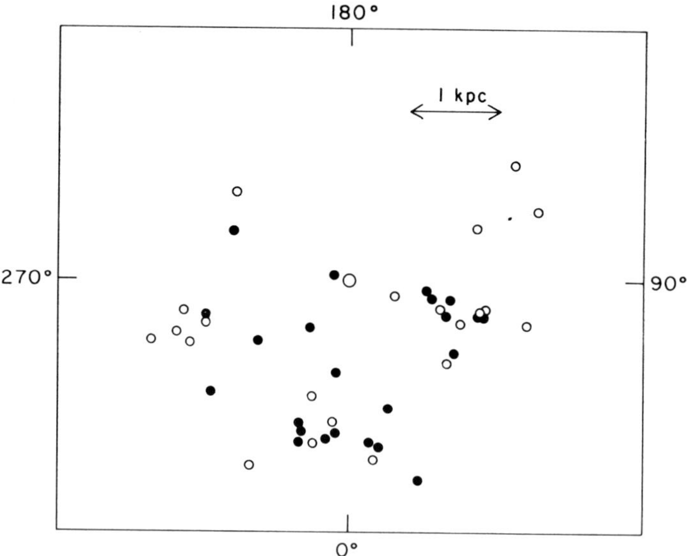

Fig. 5. W-R stars within 2.5 kpc, projected onto the plane of the
galaxy (adapted from Conti, Garmany, de Loore and Vanbeveren 1983).
WN stars are o; WC types have ● symbols.

There is clearly a gradient in the W-R star distribution outward
from the galactic center which is not obviously present in the overall
O-star distribution. It has been suggested that this is due to different evolutionary histories of the progenitor stars in terms of their
initial chemical composition (e.g. Smith 1973; Maeder 1981). However,
as Conti, Garmany, de Loore and Vanbeveren note, another factor may be
important. These authors divided the O stars of Fig. 4 into two mass
ranges, those between 20–40 $M_\odot$ and those above 40 $M_\odot$. The division
was somewhat arbitrary initially, in that models of this mass value
are published. Figures 6 and 7, adapted from their paper, show these
two mass distributions.

We see that the lower mass O stars (Fig. 6) do not show much of
a gradient with galactocentric distance, while such a gradient is
present in the massive O star distribution (Fig. 7). Garmany et al.

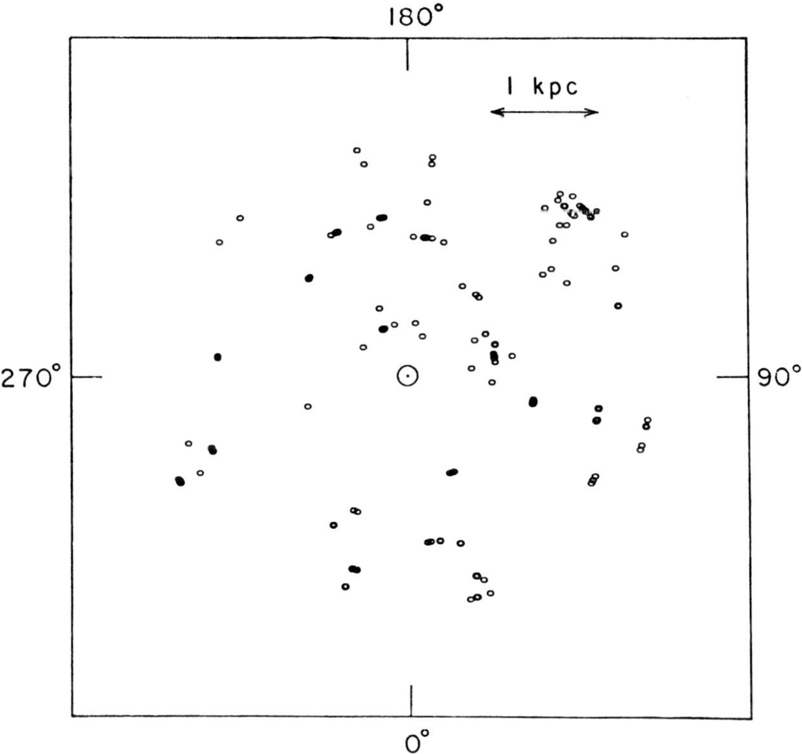

Fig. 6. Galactic distribution of O stars within 2.5 kpc with initial masses between 20-40 $M_\odot$ (from Conti, Garmany, de Loore and Vanbeveren 1983).

(1982) and Conti, Garmany, de Loore and Vanbeveren (1983) suggested that the observed difference in the O star mass population -- a difference in the slope of the massive end of the initial mass function (IMF) -- could be a major contributor to the W-R star gradient. A natural consequence would then be that W-R stars are mostly descendant from massive O stars, and not from all O stars.

In retrospect, I regret that in Conti, Garmany, de Loore and Vanbeveren, we did not sufficiently stress that the 40 $M_\odot$ dividing line was not absolute: Plots of 35 $M_\odot$ and 45 $M_\odot$ analogous to Fig. 7 were not very different from one another in the appearance of the gradient. A 30 $M_\odot$ did show less of a galactocentric gradient. I believe that it is safe to conclude that most W-R stars appear to be related to the most massive O stars, the lower limit being near 40 $M_\odot$ but not as small as 30 $M_\odot$. Of course, exceptions (e.g. binary condition) may allow lower initial mass stars to become W-R. Also, this sample is near the Sun where the chemical gradient is small (factor two across the sample). As Maeder (1981) has stressed, the initial composition, if very different,

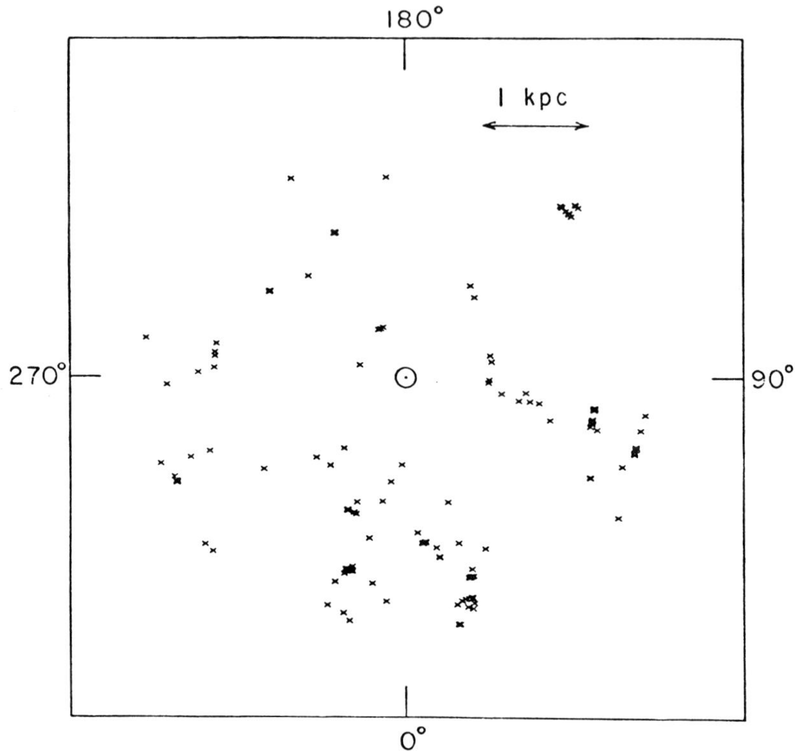

Fig. 7. Galactic distribution of O stars within 2.5 kpc, with initial masses greater than 40 $M_\odot$ (from Conti, Garmany, de Loore and Vanbeveren 1983).

may also have an important function. In this regard, the massive star population of, say, the SMC, with a "metal" composition perhaps 20% of the Sun, may give us useful evidence.

In Fig. 5, I have plotted the galactic WN and WC stars with different symbols. A careful perusal will indicate that these objects do not have quite the same galactic distribution: there are relatively more WC vis-à-vis WN stars toward the galactic center and conversely toward the anti-center. Of course, with such a small sample, statistics are uncertain and I do not expect skeptics to be persuaded by this figure. I will return to the variation in the WN/WC ratio later.

D. Distribution of Massive Stars in the Magellanic Clouds

There has been no comparable study of a complete sample of O stars in the Magellanic Clouds although a number of us are at work on this problem. A nearly complete sample of W-R stars is available for the Clouds: Breysacher (1981) lists 100 stars for the LMC and Azzopardi and Breysacher (1979) list eight stars for the SMC. While the distribution in these galaxies of the WN + WC types has not been studied, it is known from these listings that the WN/WC ratio is about 4 for the LMC and 7 for the SMC. Furthermore, whereas nearly all WN and WC subtypes are present in the solar vicinity sample, and nearly all WN subtypes in the LMC, only early WC subtypes are found there. Also in the SMC the seven WN stars are of early subtype, as is the lone known WC star. What are we to make of this? Frankly, whereas one might devise evolutionary scenarios that produce early or late WN or WC, these are purely speculative since we have not linked the subtypes to either an effective temperature scale, or to a mass or to a luminosity scale. The curious distribution and presence and/or absence of WN and WC subtypes is going to give us information on massive star evolution eventually.

E. Distribution of Massive Stars in Local Group Galaxies

The landmark survey of bright supergiants in M33 (Humphreys and Sandage 1980 -- HS) will be a starting point for every modern study of luminous stellar populations. Spectral types exist for only a handful of the M33 stars and the photometry is very incomplete, so we are unable to discuss individual stellar distributions. HS did assign association boundaries to obvious groupings of bright blue stars. Their paper clearly shows the spiral arm distribution of the luminous stellar population. LBV are, of course, also known in M33.

What of the W-R stars in this galaxy? Phil Massey (mostly) and I have been studying candidate W-R objects found from survey plates taken "on-line (λ4686 Å)/off-line" with narrow band filters. With data already in the literature we have discussed a sample of 80 W-R stars in M33 (Massey and Conti 1983). This number is, at most, 50% incomplete and better data may become available this fall. Even so, the distribution of WN and WC stars within this galaxy is interesting: the WN/WC ratio changes with galactocentric distance within M33, with increasing values away from the center analogous to the sketchy information available for our galaxy (Fig. 5). The subtype distribution is also curious: Nearly all WN stars are of early subtype, as are all the identified WC stars. This is unlike the galaxy, or the Magellanic Clouds. What are these subtype distributions telling us? We don't know yet, but surely they are somehow related to initial masses, or initial compositions, or combinations of these parameters or possibly others.

A few W-R stars are known in the small irregular galaxies NGC 6822 and IC 1613 (Westerlund et al. 1983; D'Odorico and Rosa 1982). A survey of the massive spiral M31 (Shara and Moffatt 1982) revealed some 21 W-R stars. It seems hard to believe this sample is complete, as claimed by these authors, but I must confess my surprise at the smallness of the "total," given the mass of M31 compared to, say, M33. Unless Moffat and Shara are incomplete by a factor 100, the apparently small W-R population of M31 suggests an exceedingly small number of massive stars in this galaxy.

The distribution of W-R stars in M33 is indicated in Fig. 8, adapted from Massey and Conti (1983). The spiral arm structure is a little difficult to make out! A careful examination of this figure reveals that most W-R stars are in arms, but not all arms have W-R stars! In particular, the arm going NE, ending at NGC 604 has no W-R stars (or candidates). We do not understand this yet. I should also draw your attention to the "width" of the arms, like that in Figs. 4-7. By stellar appearance, we would conclude that our galaxy, in the solar vicinity, is much like M33 in type.

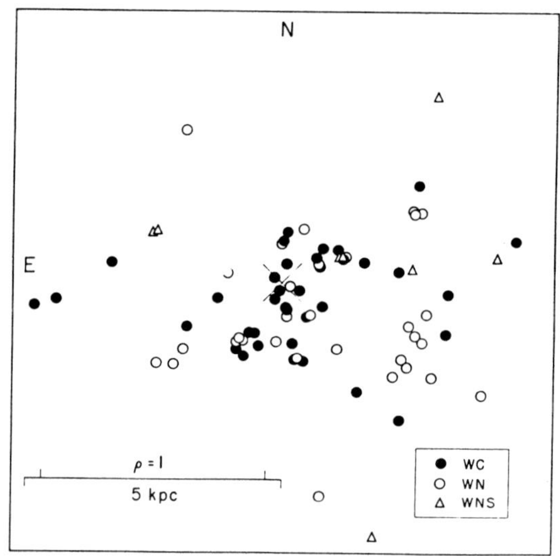

Fig. 8. Distribution of known W-R stars in M33, projected on the plane of the galaxy which has been tilted to remove the inclination (from Massey and Conti 1983).

V. SUMMARY AND CONCLUSIONS

Massive stars are important stellar constituents of our galaxy and other spiral and irregular types. They provide the supernova progenitors, most of the ionizing radiation and greatly modify their environments by their stellar winds. Their light dominates the spiral arm structure. The evolution of massive stars, discussed in detail later in this symposium by Maeder, is a very exciting topic. The general outlines are understood to involve luminous blue supergiants, LBV, and W-R subtypes, in order of increasing evolution. The details are still controversial but their resolution will help in our understanding of

how spiral galaxies evolve. W-R stars and their subtype distributions
may provide clues to the progenitor initial masses and/or compositions.
Luminous stars are among the first objects individually studied outside
our own galaxy. They can just barely be investigated spectroscopically
in galaxies outside the local group. Problems of the distance scale
perhaps can be addressed with these objects.

ACKNOWLEDGMENTS

     I am deeply appreciative of the collaborative efforts of my JILA
associates Drs. Katy Garmany and Phil Massey on many aspects of the
work reported here. I particularly thank Katy for providing Figs. 1-3
and other help on the manuscript. I have benefitted from discussions
with Dr. Andre Maeder. This paper was written during visits to institutes in Brussels and Liége, Belgium, and I am grateful for the hospi-
tality provided by Drs. de Loore and Vreux. The National Science
Foundation has provided support under grant AST81-17357 through the
University of Colorado.

REFERENCES

Abbott, D., Bieging, J. H. and Churchwell, E. B.: 1982, Astrophys. J.
     263, p. 207.
Abt, H. and Levy, S.: 1978, Astrophys. J. Suppl. 36, p. 241.
Aller, L. and Jugaku, J.: 1959, Astrophys. J. Suppl. 4, p. 109.
Auer, L. and Mihalas, D.: 1972, Astrophys. J. Suppl. 24, p. 193.
Azzopardi, M. and Breysacher, J.: 1979, Astron. Astrophys. 75, p. 120.
Bressan, A. G., Bertelli, G. and Chiosi, C.: 1981, Astron. Astrophys.
     102, p. 25.
Breysacher, J.: 1981, Astron. Astrophys. Suppl. 43, p. 203.
Chiosi, C., Nasi, E. and Sreenivasan, S. R.: 1978, Astron. Astrophys.
     63, p. 103.
Cloutman, L. D. and Whitaker, R. W.: 1980, Astrophys. J. 237, p. 900.
Conti, P. S.: 1973, Astrophys. J. 179, p. 181.
Conti, P. S.: 1981, IAU Colloq. #59, "Effects of Mass Loss on Stellar
     Evolution" (Reidel, Dordrecht), p. 1.
Conti, P. S. and Alschuler, W.: 1971, Astrophys. J. 170, p. 325.
Conti, P. S., Garmany, C. D., de Loore, C. and Vanbeveren, D.: 1983,
     Astrophys. J. 274, in press.
Conti, P. S., Leep, E. M. and Perry, D.: 1983, Astrophys. J. 268, p.
     228.
Conti, P. S. and McCray, R.: 1980, Science 208, p. 9.
Cruz-Gonzalez, C., Recillas-Cruz, E., Costero, R., Peimbert, M.,
     Torres-Peimbert, S.: 1974, Rev. Mex. Astr. Astrof. 1, p. 211.
Davidson, K., Walborn, N. R. and Gull, T.: 1982, Astrophys. J. 254, p.
     L47.
de Groot, M.: 1969, Bull. Astron. Inst. Netherlands 20, p. 225.
D'Odorico, S. and Rosa, M.: 1982, Astron. Astrophys. 105, p. 410.
Doom, C.: 1982, Astron. Astrophys. 116, p. 303.

Garmany, C. D. and Conti, P. S.: 1984, Astrophys. J., in press.
Garmany, C. D., Conti, P. S. and Chiosi, C.: 1982, Astrophys. J. 263, p. 777.
Garmany, C. D., Conti, P. S. and Massey, P.: 1980, Astrophys. J. 242, p. 1063.
Garmany, C. D., Massey, P. and Conti, P. S.: 1984, Astrophys. J., in press (Mar. 15).
Garmany, C. D., Olson, G., Conti, P. S. and Van Steenburg, M. E.: 1981, Astrophys. J. 250, p. 660.
Humphreys, R.: 1978, Astrophys. J. Suppl. 38, p. 309.
Humphreys, R.: 1984, this symposium.
Humphreys, R. and Davidson, K.: 1979, Astrophys. J. 232, p. 409.
Humphreys, R. and Sandage, A.: 1980, Astrophys. J. Suppl. 44, p. 319.
Hunger, K.: 1955, Z. Astrophys. 36, p. 42.
Kudritzki, R.-P., Simon, K. P. and Hamann, W. R.: 1983, Astron. Astrophys. 118, p. 245.
Lamers, H. J. G. L. M., de Groot, M. and Cassatella, A.: 1983, Astrons. Astrophys., in press.
Luud, L. S.: 1967, Sov. Astron. 11, p. 211.
Maeder, A.: 1981, Astron. Astrophys. 101, p. 385.
Maeder, A.: 1982, IAU Symp. #99 "Wolf-Rayet Stars: Observations, Physics and Evolution" (Reidel, Dordrecht), p. 405.
Maeder, A.: 1983, Astron. Astrophys., in press.
Maeder, A.: 1984, this symposium.
Massa, D., Savage, B. D. and Fitzpatrick, E. L.: 1983, Astrophys. J. 266, p. 662.
Massey, P.: 1982, IAU Symp. #99 "Wolf-Rayet Stars: Observations, Physics and Evolution" (Reidel, Dordrecht), p. 251.
Massey, P.: 1983, Astrophys. J., submitted.
Massey, P. and Conti, P. S.: 1983, Astrophys. J. 264, p. 126.
Meylan, G. and Maeder, A.: 1982, Astron. Astrophys. 108, p. 148.
Moffat, A. F.: 1982a, IAU Symp. #99 "Wolf-Rayet Stars: Observations, Physics and Evolution" (Reidel, Dordrecht), p. 263.
Moffat, A. F.: 1982b, IAU Symp. #99 "Wolf-Rayet Stars: Observations, Physics and Evolution" (Reidel, Dordrecht), p. 515.
Nugis, T.: 1982, IAU Symp. #99 "Wolf-Rayet Stars: Observations, Physics and Evolution" (Reidel, Dordrecht), p. 131.
Olson, G. and Castor, J. I.: 1981, Astrophys. J. 244, p. 179.
Paczynski, B.: 1973, IAU Symp. #49 "Wolf-Rayet and High Temperature Stars" (Reidel, Dordrecht), p. 143.
Popper, D.: 1980, Ann. Rev. Astron. Astrophys. 18, p. 115.
Roberts, M.: 1962, Astron. J. 67, p. 79.
Sandage, A. and Carlson, G.: 1983, Astrophys. J. (Letters) 267, p. L25.
Shara, M. M. and Moffat, A. F. J.: 1982, IAU Symp. #99 "Wolf-Rayet Stars: Observations, Physics and Evolution" (Reidel, Dordrecht), p. 531.
Simon, K. P., Jonas, G., Kudritzki, R. P. and Rahe, J.: 1983, Astron. Astrophys., in press.
Smith, L. F.: 1973, IAU Symp. #49 "Wolf-Rayet and High Temperature Stars" (Reidel, Dordrecht), p. 15.

Smith, L. and Willis, A.: 1983, Monthly Notices Roy. Astron. Soc., in press.
Stothers, R. and Chin, C. W.: 1977, Astrophys. J. 211, p. 189.
Torres, A. and Conti, P. S.: 1984, Astrophys. J., in press.
Underhill, A. B.: 1980, Astrophys. J. 239, p. 220.
Underhill, A. B.: 1981, Astrophys. J. 244, p. 963.
Vanbeveren, D. and Conti, P. S.: 1980, Astron. Astrophys. 88, p. 230.
van der Hucht, K., Conti, P. S., Lundstrom, I. and Stenholm, B.: 1981, Space Sci. Rev. 28, p. 227.
Walborn, N.: 1970, Astrophys. J. 161, p. L149.
Walborn, N.: 1972, Astron. J. 77, p. 312.
Westerlund, B. E., Azzopardi, M., Breysacher, J. and Lequeux, J.: 1983, Astron. Astrophys. 123, p. 159.
Willis, A.: 1982, IAU Symp. #99, "Wolf-Rayet Stars: Observations, Physics and Evolution" (Reidel, Dordrecht), p. 87.
Wolf, B., Appenzeller, I. and Cassatella, A.: 1980, Astron. Astrophys. 88, p. 15.
Wolf, B., Appenzeller, I. and Stahl, O.: 1981, Astron. Astrophys. 103, p. 94.

DISCUSSION

Iben: You have established a galactic gradient in both the O-star and the WR-star distributions, but it does not then follow that all WR progenitors must be O-stars. Here are other galactic gradients such as those in average "metallicity" and in the WC/WN ratio. If, for example, one were to assume that the average mass loss rate from massive stars decreases with decreasing metallicity one can immediately account for both the WR and the WC/WN gradients. This interpretation is reinforced by the evidence that all of the WR-stars in the SMC (whose metallicity is down by about a factor of 4 from that of the Sun) may be binaries, suggesting that when Z (and $\dot{M}$) is too small, one needs the help of Roche-lobe overflow to expose highly processed layers. The gradient in the frequency of WR-stars is larger than that of O-stars because O-stars can of course become WR-stars, but as their frequency decreases, obviously this source of WR-stars decreases, leaving only less massive stars (with smaller $\dot{M}$'s) and binaries.

de Loore: 1. According to our computations for massive single stars (mass loss, overshooting modelling Humphreys' diagram, or by adopting Roxburgh's criterion, see Doom) the initial mass producing the most luminous red supergiant is 33 $M_\odot$.
2. On one hand I am coresponsible for the lower limit determination of WR-stars (Conti et al, Ap.J., 1983), on the other hand we made new computations for massive close binaries, especially for the accreting components. Comparison of these accreting models with observations allows to determine the inclination, hence to determine the mass of the O-component, and hence also the WR mass. This reveals that the masses of WR binaries (determined for 8 systems) are in the range 7-10 $M_\odot$, and the initial masses of the initial primaries can then be as low as 15-20 $M_\odot$.

<u>Maeder</u>: The WR-star excess you find depends clearly on the limit of 40
$M_\odot$ for initial stellar masses leading to the WR-stars you have chosen.
When looking at the data from a quantitative point of view one can check
that the galactic gradient of WR-stars is steeper than the gradient of
O-stars for any lower mass limit one may choose. In my opinion, the best
way of gathering information on the lower initial mass for WR formation
is to look at clusters and associations containing WR stars. H. Schild
has done so and obtained a value around 20 $M_\odot$.
Now, regarding another point, don't you think that, in view of the strong
dependence of the predicted number of WR-stars on metallicity Z, even a
very small dependence of mass loss rates $\dot{M}$ on Z (well inside the data
scatter) would be helpful to account for the gradient of WR stars?

<u>Conti</u>: I agree that the observed galactocentric gradient is steeper for
WR-stars than for O-stars of any other lower mass limit, thus suggesting
that other parameters, such as the initial metal content, may play a role.
On the other hand, the observed numbers of WR-stars are not large and
small number statistics may affect this result.
As to the second issue, concerning the small differences in metal content affecting $\dot{M}$: I agree this is possible. On the other hand, we have
shown (Garmany and Conti, 1984, Ap.J. in press) that among stars of the
same cluster, thus presumably of the same metal content, there is quite
a wide disparity in observed $\dot{M}$ (for a given luminosity). This scatter
presumably is due to other factors affecting $\dot{M}$ - such factors may dominate metal abundance effects.

<u>McCarthy</u>: You list 101 WR-stars in Large Magellanic Cloud. Can you tell
us if WN-stars predominate toward the center while WC-stars abound toward the outer regions? What can you say of the distribution of these
stars in the Large Cloud?

<u>Conti</u>: We haven't had a chance to look at this yet, being busy with
M33, but it sounds like a fine idea.

<u>McCarthy</u>: A comparison of these early-types and their distribution should
be most interesting. I suggest a comparison between the 30 Doradus
region and the area designated by Shapley as constellation III.

<u>Cox</u>: I have two questions. What about the Underhill very blue edge of
the main sequence? For the red side, for these upper main sequence stars,
there is a core helium and H-shell burning region just merged to the
core hydrogen burning. Why do you then not count this region, which is
also long lived, as part of the main sequence band when you compare with
the observed HRD distribution? Both these questions bear on the accuracy
of stellar opacity values.

<u>Conti</u>: The first question concerns the effective temperature scale of the
hottest main sequence O-stars. Underhill found the earliest types were

not hotter than $\sim 40000^{\circ}$K. I am sorry she is not here to take issue with my response but I will say I don't think her result is correct: she included no O3-stars among the earliest types and she fitted dereddened continua to plane parallel models. We have recently argued that a standard reddening law cannot be applied to determine ultra-violet extinction. Furthermore, the models do not match the continua over wavelength ranges from the UV to the IR, better physics is needed.
The second question bears on the issue of counting stars to the right of the _present_ evolution tracks core hydrogen burning point. As I have argued, there are too many stars rightwards of the CHB point to be considered post-CHB. Thus the tracks must be extended to cooler temperatures. The problem might be helped by opacity but as I understand it, this arises other problems. There is still some missing physics - perhaps increasing again turbulent diffusion will help.

Renzini: When superficially looking at HR diagrams for massive stars I've been always confused by the apparent virtual absence of stars close to the ZAMS. Could you comment on that?

Conti: I have noticed this too. It seems to be a clumping of stars a little away from the current ZAMS. Part of the problem may be the necessary quantization of M_V at various temperatures, since for many galactic stars we get both the luminosity and the temperature from the spectrum. Perhaps some ZAMS stars are still hidden in dense clouds. Our study of the Magellanic Clouds, when complete, will help this since we will obtain the M_V from the m_V directly.

Tayler: You commented on a shortage of stars close to the zero age main sequence. Presumably the uncertainties of 10 percent in T_{eff} and $\frac{1}{2}$ magnitude in luminosity means that the ZAMS is not the lower envelope. In addition, 40 percent binarity will move stars away from ZAMS. Do you think that a significant number of young massive stars could still be obscured in clouds?

Conti: My personal belief is very few are hidden, at least within 2.5 kpc from the sun. Infra-red workers I have questioned feel few have been missed. The IRAS satellite will have the data to answer this question.

de Groot: I want to help Peter Conti with his statistics: As far as proper motions are concerned P Cygni's has been determined at least twice; in 1967 and rather recently. There also is a binary among the LBV, because the Heidelberg people (Wolf, Zickgraf) have found that R81 in the LMC which is exactly like P Cygni has a lightcurve like an eclipsing binary. This latter may mean I can also give you a mass for an LBV, once the radial velocity curve has been determined and the system analysed more completely.

Janes: Is it possible that there is a significant population of stars that have not yet reached the zero age main sequence?

Conti: I don't think so. Scaling arguments suggest pre-main sequence
lifetimes are at most 1 % of the MS lifetime. Thus a few stars in our
sample, or not yet found, could be pre-main sequence.

Richer: Could you amplify your remarks concerning uncertainty in the
distance modulus to M33?

Conti: Actually, I'd rather not get into this controversy, but for what
it's worth, the distance to M33 is uncertain by a factor two. Sandage
now assigns a somewhat further distance based upon his analysis of
Cepheids. Madore finds a distance somewhat closer based upon IR photometry. The mean WR magnitudes, compared to galactic stars, tend to
favour the smaller distance but I would be hesitant to use this to settle
the matter.

Humphreys: You mentioned a problem with the relative numbers of Wolf-
Rayet stars in the LMC and M33. What happens if the distance to M33 is
doubled?

Conti: I think we would still be relatively complete in our M33 survey
for WR candidates. The mean magnitudes would be brighter, though.

ULTRAVIOLET SPECTRAL MORPHOLOGY OF ON AND OC STARS

Nolan R. Walborn*
Laboratory for Astronomy and Solar Physics
NASA/Goddard Space Flight Center

Robert J. Panek
Astronomy Department
Computer Sciences Corporation

We have undertaken a systematic survey of 115 O-B0 type spectra between 1200 and 1900Å, by means of International Ultraviolet Explorer high-dispersion data rebinned to a constant wavelength resolution of 0.25Å and uniformly normalized (Walborn and Panek 1983). The ultraviolet spectral features, both photospheric and stellar wind, are found to be smoothly correlated with the optical spectral types for the majority of normal stars. In particular, the C IV ($\lambda\lambda$1548,1551) and N V ($\lambda\lambda$1239,1243) resonance doublets show similar strong P Cygni profiles, which decline along the main sequence beginning at type O7 but remain strong in all higher luminosity spectra. In contrast, the stellar wind effect in the Si IV ($\lambda\lambda$1394,1403) doublet displays a pronounced luminosity dependence, being absent from the main sequence spectra at all types and developing gradually through the intermediate luminosity classes into a full P Cygni profile in the supergiants. Relative to this consistent framework provided by the normal stars, a number of peculiar spectra can be succinctly described, including several of types OBN and OBC (Walborn 1976).

For instance, the supergiants HD 105056 (ON9.7 Iae), 123008 (ON9.7 Iab), 152249 (OC9.5 Iab), and 152424 (OC9.7 Ia) show pronounced anomalies in the ultraviolet C and N features relative to the normal standard α Cam (O9.5 Ia). The saturated Si IV P Cyg profiles confirm that all of these stars are indeed supergiants. However, the two OC spectra have weak, unsaturated N V profiles, while conversely C III λ1247 is strikingly deficient in the two ON spectra. Similarly, the latter have weaker C IV P Cyg profiles, with peculiar sharp emission at λ1551 in HD 105056. An additional anomaly in HD 105056 is the presence of strong stellar wind profiles in Al III $\lambda\lambda$1855, 1863, unlike any other star at these spectral types examined so far. These spectrograms will be published shortly (Walborn and Panek 1984).

*National Research Council Senior Associate

A. Maeder and A. Renzini (eds.), Observational Tests of the Stellar Evolution Theory, 255–259.

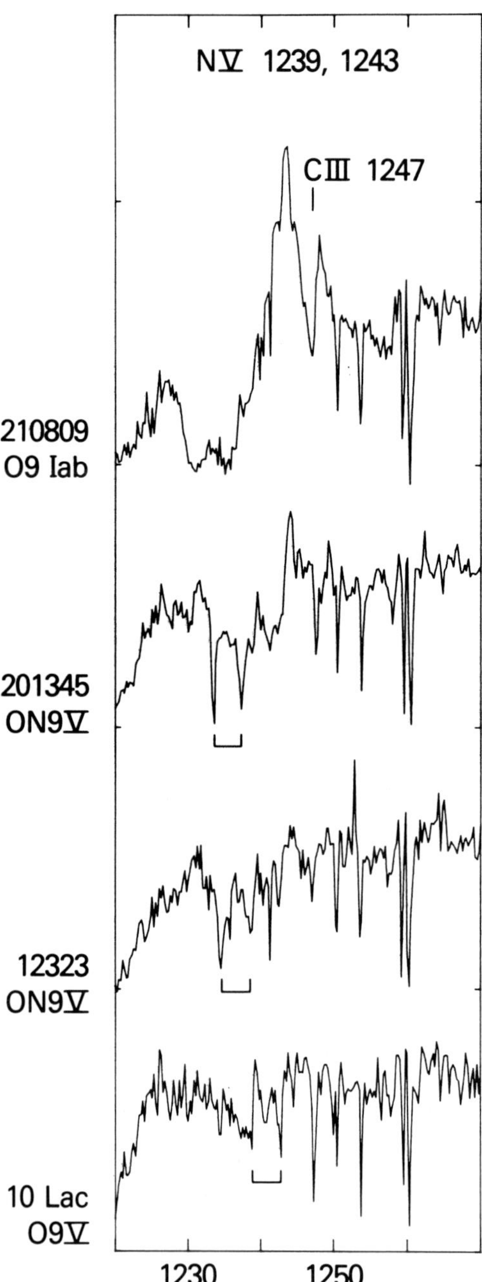

Fig. 1 - O9 spectra, λλ1220–1270.   The O9 V standard shows weak, asymmetrical but unshifted N V lines, while the supergiant has a strong P Cygni profile.   However, the two ON spectra have strong, highly blueshifted sharp wind absorptions in N V, with emission as well in HD 201345.   Note also the weaker C III line in the two ON spectra.

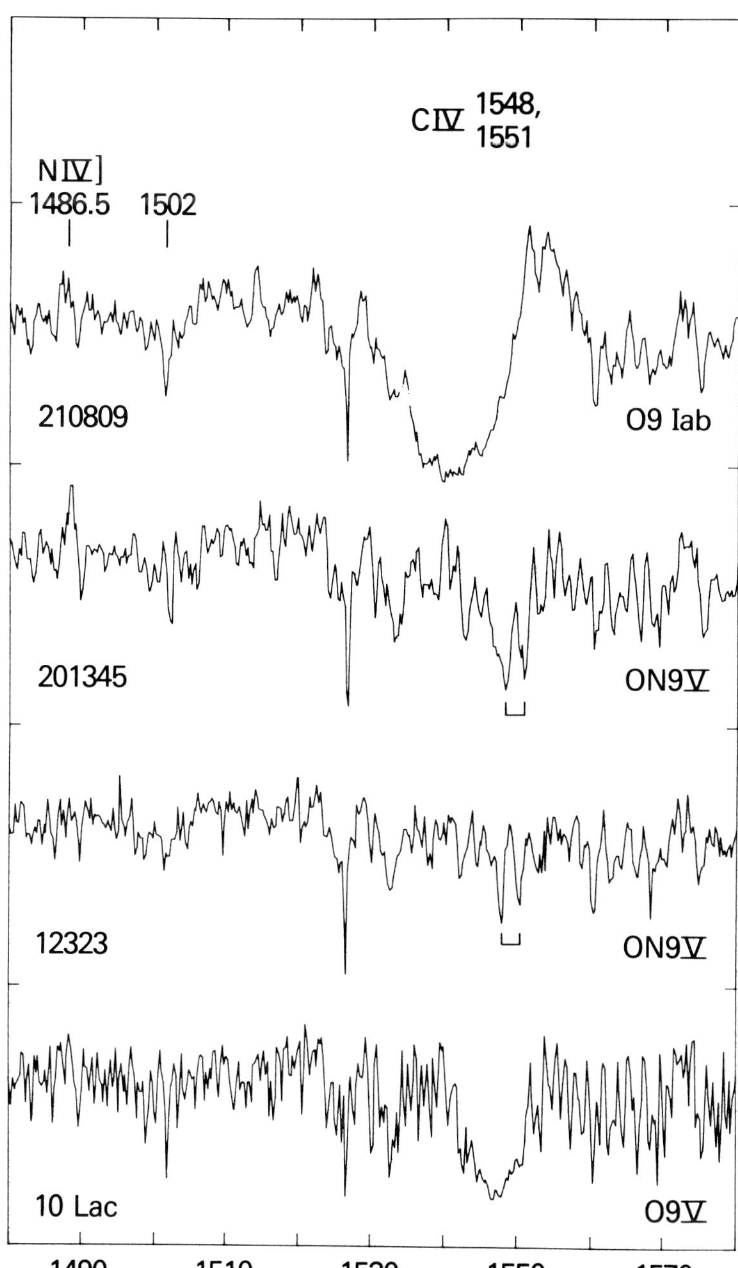

Fig. 2 - O9 spectra, $\lambda\lambda1480$–1580. The O9 V standard has a very strong C IV absorption trough but no emission, while the supergiant has a strong P Cygni profile. The weakening of the C IV in the two ON spectra is striking; the doublet may be entirely interstellar in the case of HD 12323. Note also the enhanced N IV]$\lambda1486.5$ emission in HD 201345.

Figures 1 and 2 compare ultraviolet C and N features in two ON9 V spectra to those of normal dwarf and supergiant standards at the same spectral type; blue-violet spectrograms of three of these stars are illustrated by Walborn (1970). The ultraviolet Si IV doublet shows no stellar wind effect in 10 Lac, HD 12323, or HD 201345, but a strong P Cyg profile in HD 210809, in agreement with the optical luminosity classifications. Since the C IV and N V emissions have disappeared from the normal main-sequence spectra by type O9, these features now also show a luminosity effect similar to the Si IV. The pronounced anomalies in the C and N features of the two ON9 V spectra, relative to the systematic behavior of those features in the normal spectra, strongly suggest an origin in abundance effects. This spectral type is later than the C IV and N V maxima in normal spectra, in which the two features are strongly correlated; hence, it is difficult to imagine how a change in other physical conditions could simultaneously weaken the former and strengthen the latter. The enhanced N IV] emission found in HD 201345 is also seen in the nitrogen-rich globular cluster UV star discovered by Bohlin et al. (1983).

REFERENCES

Bohlin, R. C., Cornett, R. H., Hill, J. K., Smith, A. M., Stecher,
     T. P., and Sweigart, A. V. 1983, Ap. J. Letters 267, L89.
Walborn, N. R. 1970, Ap. J. Letters 161, L149.
_______. 1976, Ap. J. 205, 419.
Walborn, N. R. and Panek, R. J. 1983, Workshop on Spectral
     Classification, University of Toronto.
_______. 1984, in preparation.

DISCUSSION

<u>Iben</u>: What fraction of all O stars are ON stars?
What is the abundance of N in ON stars?

<u>Walborn</u>: I have classified 327 galactic O stars among which I found 7 of
type ON, 5 of type OC, and 11 with more moderate CNO anomalies
(Ap.J. <u>205</u>, 419, 1976; A.J. 87, 1300, 1982). However, these fractions
are upper limits because several of the peculiar objects were selected
on the basis of suggestive remarks by authors of more extensive, lower
dispersion surveys. The main sequence ON stars may well be products of
mass transfer in binary systems. On the other hand, a possible interpretation of the anomalous supergiants is that the OC's, found predominantly in certain young associations, have normal CNO abundances,
while the morphologically "normal" majority of O-type supergiants are
moderately nitrogen-rich due to mixing and/or mass loss, and the ON's are
extreme cases. In this picture, <u>all</u> O-type supergiants would be involved
in the phenomenon.
Baschek, Kodaira and Scholz (Ap.Lett. <u>12</u>, 227, 1972) found a nitrogen
abundance >2.5 times larger in HD 188209 (morphologically normal) than
in ζ Orionis (moderately N-deficient), suggesting an admixture of 20 to
40 per cent CNO-cycled material in the atmosphere of the former star.
Lester (Ap.J. <u>185</u>, 253, 1973) found N 33 times more abundant in
HD 201345 than in 10 Lacertae. Kudritzki, Méndez, Simon, Heber and
Schönberner (ESO Workshop on Primordial Helium, p. 189, 1983) found N
overabundant by at least a factor of 10 in HD 48279 relative to τ Scorpii.

<u>Kudritzki</u>: D. Schönberner, K.P. Simon and myself have recently finished
the NLTE-analysis of some ON V-star and found an overabundance of a
factor of 10 for nitrogen accompanied by an increase of the helium number
fraction to 14 % (relative to the "normal" value of 9 %). This has been
published in the recent proceedings of the ESO-Workshop on "Primordial
Helium".

CNO-CYCLED MATERIAL EJECTED BY ETA CARINAE

Kris Davidson
University of Minnesota

R.J. Dufour
Rice University

N.R. Walborn & T.R. Gull
Goddard Space Flight Center

Eta Carinae has provided more incisive clues to the evolution of the most massive stars than any other individual object. Its luminosity is comparable to that of a very massive O3 star, but it appears to have evolved beyond the O3 stage, to the limit where its surface is rather unstable. We think that it has begun to lose mass semi-catastrophically, in outbursts which may recur at intervals of perhaps several hundred years. The suspected binary nature of Eta Carinae introduces some conceptual uncertainty but is most likely not crucial.

It is an interesting fact that our knowledge of Eta Carinae is largely independent of the methods used for most other stars discussed in this symposium. The luminosity is observed as dust-re-radiated infrared emission and does not depend upon a conventional bolometric correction; the suspected "surface temperature" of roughly 30 000 K is based on non-standard arguments (see K. Davidson, 1971, M.N.R.A.S. 154, 415); and the surface composition -- the main topic of our work -- is estimated from nebular emission lines rather than from analyses of a photosphere or a stellar wind. Eta Carinae is clearly special from more than one point of view.

The spectrum of the star itself, plus its surrounding dense halo of ejecta from the outburst observed in 1837--1858, is intractably complicated. But farther out, at distances of the order of 0.15 pc from the star, are some lower-density condensations with nebular emission-line spectra. These condensations have measurable proper motions and certainly were ejected from the star during the past two centuries, "apparent time" (see N.R. Walborn, B.M. Blanco, and A.D. Thackeray, 1978, Ap.J. 219, 498). The brightest condensation, denoted "S", is quite unusual among nebular objects in that it has prominent emission lines of 5 successive ionization stages of nitrogen (NI through NV) but scarcely any carbon or oxygen emission (see K. Davidson, N.R. Walborn, and T.R. Gull, 1982, Ap.J. Letters 254, L47). The implication is clear: CNO-cycle-processed material has been mixed to the surface of the star and then ejected.

A. Maeder and A. Renzini (eds.), Observational Tests of the Stellar Evolution Theory, 261–264.
© *1984 by the IAU.*

We have made some improved observations.   K.D., N.R.W., and T.R.G. have used several shifts of IUE time to observe additional condensations and to obtain better UV data on the "S" condensation. R.J.D. has used the Cerro Tololo 4-m telescope and SIT vidicon system to obtain quantitative visual-wavelength spectral data on "S". Here we can offer only a preliminary, somewhat qualitative account of the results.

The figure below shows tracings of the UV spectra of condensations "S", "ES", "W", and "NN". These appear to show the same sets of emission lines. In each case there is a significant UV continuum, which seems stronger (relative to the lines) in condensation "NN". Generally, though, we can take "S" to be representative, aside from its greater brightness.

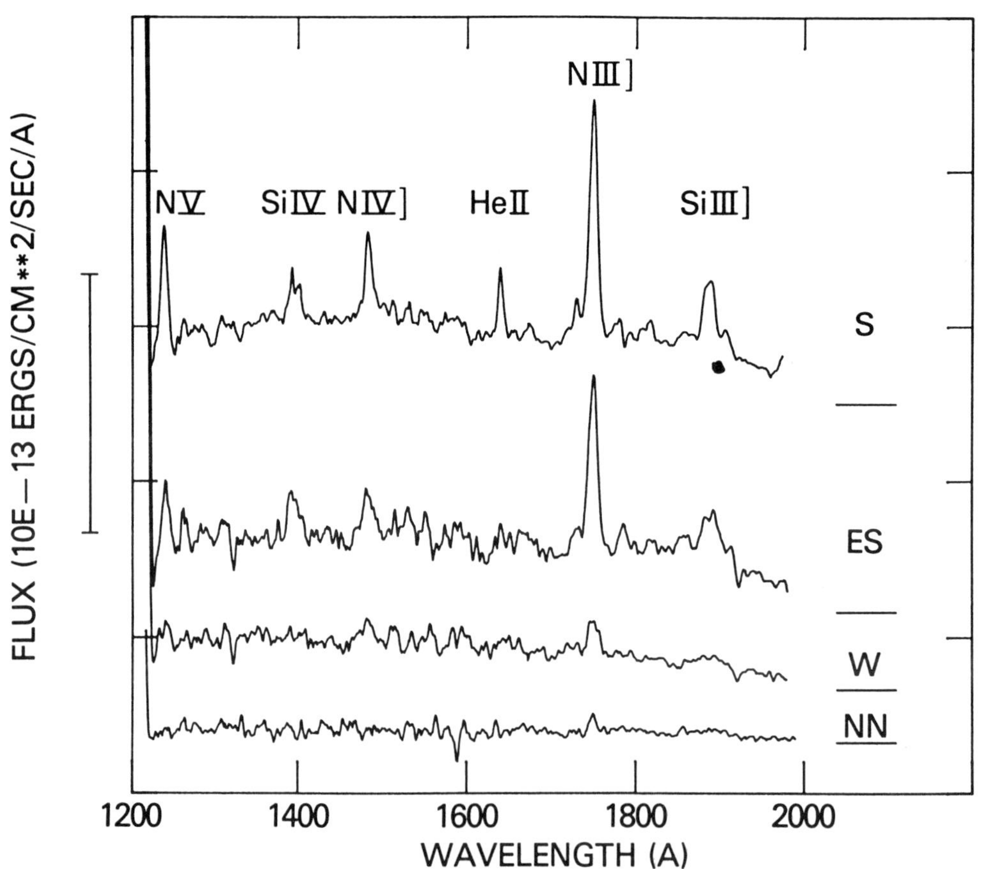

Tracings of short-wavelength IUE data on several condensations around Eta Carinae. Condensations are labelled (S, ES, W, and NN, see Ap.J. 219, 498) on the right side, with the zero level for each tracing marked by a horizontal line under each label.

There are several difficulties in analyzing the ionization structure, which we can only mention briefly here. Corrections for interstellar and circumstellar reddening are uncertain. The UV continua mentioned above may be largely due to scattering by dust grains within the condensations; therefore, internal extinction within each condensation may affect the UV emission lines. If our preliminary UV/visual-wavelength relative intensity calibration is valid, the intrinsic He II $\lambda1640/\lambda4686$ line ratio greatly exceeds the "Case B recombination" value. This probably indicates (along with other arguments) that the gas is shock-heated and has a complicated ionization structure with strong temperature gradients. Pending appropriate theoretical calculations, we tentatively think that the bulk of the gas contains moderate ionization stages (H^+, He^+, N^{++}, some He^0 and N^+) at temperatures only modestly above 13 000 K.

With the most plausible assumptions regarding ionization structure, our data suggest that the nitrogen mass fraction is roughly 0.005, of which about 70% is doubly ionized. This mass fraction, while less than the total CNO mass fraction in the Sun, is large enough to be consistent (within the uncertainties) with the idea that most of the CNO nuclei are nitrogen. Weak C IV, C III], and [O II] emission lines appear to be present; these suggest that the C/N abundance ratio is of the order of 0.01, and O/N is even smaller, in the observed gas. We suspect that most of the oxygen in each condensation is in silicate dust grains. Silicon emission lines are prominent in the spectra of the condensations and of the gas nearer the star; this may imply that some silicon was "left over" after the formation of silicate grains, which must therefore have been limited by the supply of oxygen. Thus we conjecture that the oxygen abundance (including grains as well as gas) is less than about twice the silicon abundance. This would correspond to O/N < 0.15, roughly.

With the same ionization assumptions, we estimate that the <u>helium</u> mass fraction is Y = 0.42 ± 0.04, which of course is larger than the Solar value. While the nitrogen overabundance is <u>qualitatively</u> significant because it indicates that the star has evolved and that mixing has occurred, the helium abundance may be <u>quantitatively</u> significant because it must be related to the degree of evolution.

...While working on this problem, K.D. and N.R.W. have been guest observers with the IUE (supported by NASA) and R.J.D. has been a guest observer at Cerro Tololo Inter-American Observatory (supported by NSF). N.R.W. is a National Research Council Senior Associate in the Lab. for Astronomy and Solar Physics at GSFC. The figure was designed by Nolan Walborn and kindly executed by Joy N. Heckathorn at the GSFC Regional Data Analysis Facility.

DISCUSSION

Spruit: If the oxygen is bound in dust, can the carbon be bound in dust
as well?

Davidson: In principle, yes; there may be carbon grains. Note, however,
that the O/N ratio must be low, because gaseous oxygen is thoroughly
"depleted" by silicate grain formation, with some gaseous silicon left
over. If O/N is low, we tend to expect low C/N also (unless some triple-
alpha carbon nucleosynthesis has occurred). Eventually, we hope the
2200 Å extinction feature may give information about carbon grains in
the relevant gas. A significant part of the observed reddening is due
to grains associated with Eta Carinae itself, rather than ordinary
interstellar dust; and the 2200 Å feature is only moderately strong in
the "S" condensation UV continuum. But we cannot yet promise you any
particular results.

NON – LTE ANALYSIS OF MASSIVE O-STARS

K. P. Simon   and   R. P. Kudritzki
Institut für Astronomie und Astrophysik der
Ludwig-Maximillians-Universität, München

Abstract: Non-LTE analyses of five very massive O-stars yield effective temperature, gravity and helium abundance of their photospheres. From these results together with photometric data an estimation of the stellar mass is possible. The comparison of the position in the log g - log T_{eff} diagram relative to theoretical evolutionary tracks allows three compatibility-checks: (1) masses, (2) helium-abundance and (3) age of the stars. From these checks it is inferred, that (a) the stars are in an advanced main-sequence-phase, and (b) they have suffered from considerable mass-loss. Furthermore it seems to be likely that (c) mass-loss-rates $\dot{M} = -N \cdot L/c^2$ do not scale with $N > 200$, (d) the age of the stars in the η Carinae region is less than 2 mio yrs., and (e) HD 93128 is in the background of Tr14.

1. INTRODUCTION

The integral parameters of the stars on the very hot end of the main sequence are rather uncertain due to the weakness of the most important temperature indicator, the neutral helium line 4471 $\overset{\circ}{A}$. According to Walborn (1971) the absence of this line classifies the spectral type O3. Short evolutionary time scales and the observational fact of the considerably widened upper main sequence, however, make these stars candidates for tests of stellar evolution.

2. OBSERVATIONS

The low noise emulsion IIIaJ renders it possible to detect HeI even in the spectra of O3-stars. So spectrograms of the stars HD 93128, HD93129A, HD93250 and HDE303308(12 $\overset{\circ}{A}$/mm) have been taken using the Coude spectrograph of the ESO 1.52m telescope in March 1978 and February 1980 (Kudritzki, 1979, Simon et al., 1983). Furthermore spectrograms of HD 66811 (2.6 $\overset{\circ}{A}$/mm) were secured using the holographic grating (Kudritzki et al., 1983). Calibration was performed with the ESO-ETA spectrograph. The plates were traced with a PDS microdensitometer. We

A. Maeder and A. Renzini (eds.), Observational Tests of the Stellar Evolution Theory, 265–268.

found no signs of spectrum variability, not even in the spectra of HD93129A, though this star was reported to be probably spectroscopically variable (Smith, 1968). So we superimposed several spectra of each star in order to improve the signal to noise ratio. For further details of the observations refer to the above cited papers.

3. NON-LTE ANALYSIS AND MODEL FIT

The necessity of the use of non-LTE-methods for model atmospheres of hot stars is established elsewhere(cf Auer and Mihalas,1972). Details of our computations are given in former papers, Kudritzki (1976) and Kudritzki and Simon (1978). The fit is performed by adjusting the profiles of the Balmer-lines and the equivalent width of the HeI line presuming normal helium abundance. This assumption is then verified by inspection of the lines of ionized helium. For ζ Puppis the helium abundance must be increased in order to achieve a satifying fit. H_β and HeII 4686 Å were not used, because their profiles may be affected by atmospheric motions. The masses are evaluated from gravity and radius, the latter was estimated using photometric data. Results are compiled in Table 1.

	log g $\pm$0.15	T_{eff}	X $\pm$0.06	M / M$_\odot$
HD 66811	3.50	42000 $\pm$2000	0.61	40 +40/-20
HD 93128	3.85	48000 $\pm$3000	0.70	30 +30/-15
HD 93129A	3.60	45000 $\pm$3000	0.70	70 +70/-40
HD 93250	3.95	52500 $\pm$2500	0.70	120 +80/-50
HDE 303308	3.90	45500 $\pm$3000	0.70	50 +50/-30

Table 1: Results of the non-LTE analyses.(X=mass fraction of hydrogen)

4. DISCUSSION

The analyses reveal that the massive O-stars are significantly cooler than previously assumed: Conti and Frost (1977), f.i., estimated T_{eff} of the O3-stars to be higher than 50000K. The main reason for this misjudgement was the assumption of log g = 4. According to the NLTE-calculations a supergiant of the same spectral type as a main sequence star has a much lower temperature. This is illustrated by figure 1.

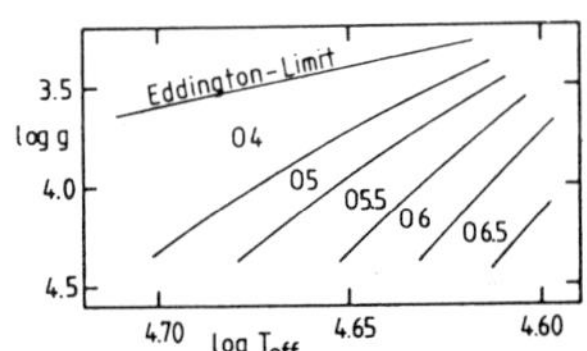

Fig. 1. The regions of the spectral types O4 to O6.5 in the log g - log T_{eff} diagram as obtained from NLTE calculations. The boundaries are given by W_λ(HeI 4471)/W_λ(HeII 4542) = 0.25,0.35,0.50,0.63, and 0.79 .

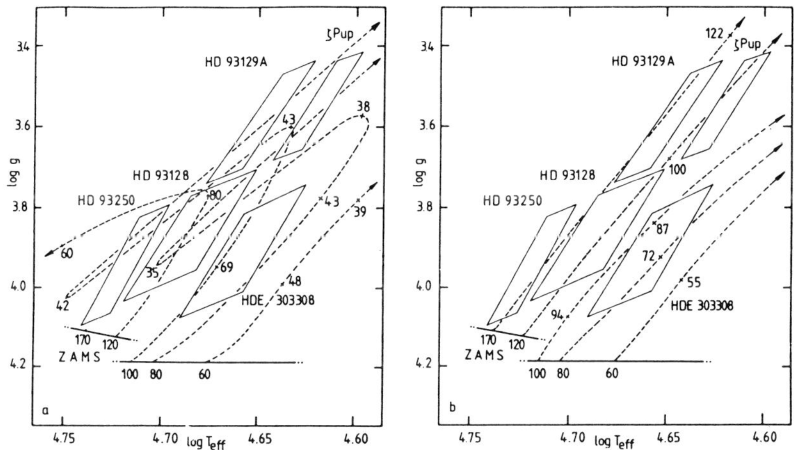

Fig. 2. The positions of the stars in the log g - log T_{eff} diagram and evolutionary tracks. Numbers denote masses in solar units. Right part: 'normal' mass loss rates - left part: 'high' mass loss rates.

The tracks in fig. 2 are adapted from De Loore et al. (1978) and Maeder (1980). The ZAMS is broken due to different input physics in the two sets. With the exception of HD93128 the stars are located on tracks of the appropriate mass in both diagrams. Perhaps HD93128 is in the backgroung of Tr14. Assuming 'normal' mass loss rates ($\dot{M}=-N\cdot L/c^2$, $N\sim100$), the age of all stars is less than 2 mio. yrs, in case of 'high' rates ($N\sim300$) the Of-stars must be older than 2 mio. years. This is compatible with estimations of the age of the η-Car-associations from the expansion velocity of the nebula. The theory of stellar evolution predicts an increased helium abundance of the outer layers long before hydrogen exhaustion. This is confirmed for ζ Puppis, where we find X=0.61 (see Tab. 1). From the observed helium abundance we suspect, that the mass loss rates do not scale with N>200. Otherwise the Of-stars should exhibit a very helium rich atmosphere. Even in the case of 'normal' mass loss rates this restricts the initial mass of HD93129A to M_i < 150 $M_\odot$.

REFERENCES

Auer, L.H., Mihalas, D.:1972, Astrophys.J.Suppl.24, pp. 193-246
Conti, P.S., Frost, S.A.:1977, Astrophys. J. 212, pp. 728-742
DeLoore,C.,DeGreve,J.P.,Vanbeveren,D.:1978,Astr.Astroph.67,pp.373-379
Kudritzki,R.P.:1976, Astron. Astrophys. 52, pp. 11-21
Kudritzki,R.P.:1980, Astron. Astrophys. 85, pp. 174-183
Kudritzki,R.P., Simon,K.P.:1978, Astron. Astrophys. 70, pp. 653-663
Kudritzki,R.P., Simon,K.P.,Hamann,W.R.:1983,Astr.Astroph.118.pp.245-254
Maeder,A.:1981,Astron.Astrophys.99,pp.97-107
Simon,K.P.,Jonas,G.,Kudritzki,R.P.,Rahe,J.:83,Astr.Astroph.125,pp.33-44
Smith, L.F.:1968, Monthly Notices Roy. Astron. Soc. 138,pp.109-124
Walborn, N.R.:1971, Astrophys. J. 167, pp. L31-L33

DISCUSSION

<u>Lynas-Gray</u>: Do you predict eventual helium abundances as high as ~ 50 % or more by numbers?

<u>Simon</u>: The abundances are for hydrogen and quoted as mass fractions. The high helium abundances are predicted by the evolutionary calculations and are exhibited by stars in a stage when hydrogen-burning is nearly finished.

<u>Schatzman</u>: Which temperature profile did you use for your non-LTE calculations?

<u>Simon</u>: The temperature stratification of the atmospheres is a result of the self-consistent non-LTE models. We assumed radiation equilibrium.

ON THE HYDROGEN ABUNDANCE IN WOLF-RAYET STARS

Werner Schmutz
Institute of Astronomy
ETH-Zentrum CH-8092 Zürich
Switzerland

ABSTRACT

With a semiempirical NLTE-model in spherical geometry we investigate the hydrogen abundance in Wolf-Rayet stars. We find that the flux contribution of the H-Balmer to the He-Pickering lines lies between the usually applied approximate formulae for the optically thin and optically thick case. This result can be understood with our finding that there are no optically thin helium and hydrogen lines. The influence of the temperature structure on the relative intensity of the hydrogen-helium blends to the helium lines is vanishingly small, even for peculiar temperature laws. The conclusion is that Wolf-Rayet stars are hydrogen deficient objects.

1. INTRODUCTION

The majority of Wolf-Rayet specialists are convinced that the elemental abundances in Wolf-Rayet stars are abnormal, especially that they are hydrogen deficient. However, up to now there have been no elaborate investigations of the formation of the helium and hydrogen lines in the extended atmospheres of Wolf-Rayet stars. To analyse the Balmer and Pickering lines the approximate formula of Castor and van Blerkom (1970, henceforth CB) for the optically thin case has usually been applied

$$\Delta \log(W_\lambda) \approx \log\left(\frac{N[\mathrm{H}^+]}{N[\mathrm{He}^{++}]}\right) \approx \log\left(\frac{N[\mathrm{H}]}{N[\mathrm{He}]}\right),$$

where $\Delta \log(W_\lambda) = \log(I(\mathrm{He+H})/I(\mathrm{He}))$ is the ratio of the intensity of a Balmer-Pickering blend to the estimated intensity of the Pickering line alone. The optically thin approximation has been justified by CB, who showed that the Pickering lines from the upper level $n > 10$ are optically thin. However, the model used by CB is a one point model. This was criticised e.g. by Underhill (1982) or by Sahade (1981), who writes: "(...) *at least we ought to be cautious and not state so firmly that WR stars are H-deficient objects.*" The approximation of Massey (1980) for the optically thick case takes into account the different emitting volumes of the lines. But the lines used to derive the hydrogen abundance are not expected to be optically thick.

The purpose of this paper is to investigate the formation of HI and HeII lines in a spherically extended atmosphere and the influence of the temperature structure on these line intensities.

269

A. Maeder and A. Renzini (eds.), Observational Tests of the Stellar Evolution Theory, 269–272.
© *1984 by the IAU.*

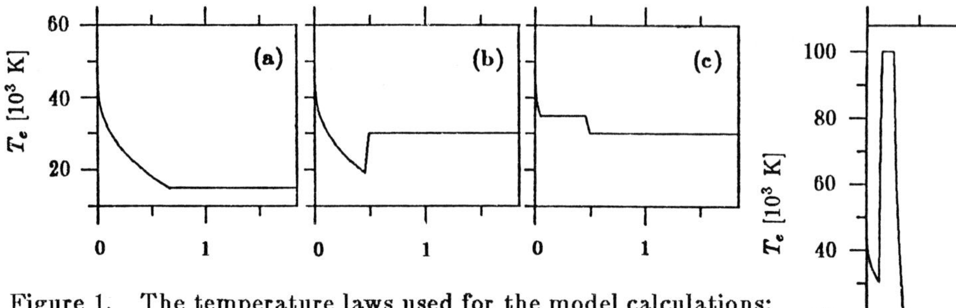

Figure 1.   The temperature laws used for the model calculations: (a) the law used in section 3a, (b) – (d) the laws used in section 3b. The continuum optical depth $\tau_{R_{oss}} = 1/3$ is at $\log(r/R_c) \approx 0.25$.

2. THE MODEL ATMOSPHERE

The model is semiempirical, i.e. the velocity law $V(r)$ and the temperature law $T(r)$ are free input parameters. Further physically important input parameters are the mass loss rate and the abundances. The elements included in the calculation are hydrogen and helium. The calculation of the continuum radiation field and the level populations are performed alternately until convergence of the level populations ('Λ-iteration'). The NLTE level populations are calculated with the Sobolev approximation of Castor (1970), modified to treat the diffuse radiation field. Interaction between the helium and hydrogen lines of the same wavelength are taken into account by using the same combined escape probability for both elements. We set the inner boundary R_c where the velocity field is $V(R_c) = V_\infty/100$; we do not include deeper layers because the validity of the Sobolev approximation would break down. At the inner boundary we assume the outgoing radiation field to be $I^{out}(\nu, R_c) = B(\nu, T(R_c))$ and LTE level population. For the models presented here the inner boundary lies at $\tau_{R_{oss}} \approx 8$, therefore the observable quantities are not very sensitive to the boundary conditions.

Following Klein and Castor (1978) we write for the equivalent widths

$$W_\lambda = \frac{h\nu}{L_\nu^c} \int 4\pi r^2 \left[\beta^{out} N_u A_{u,l} - (N_l B_{l,u} - N_u B_{u,l})\overline{J}^{out}\right] e^{-r_\nu^c(r)}\, dr = \int E(r) + A(r)\, dr,$$

where β^{out} is the escape probability of a line photon to leave the atmosphere and $\overline{J}^{out}$ is the mean outgoing continuum radiation seen by the line at the position r.

For this investigation only the abundances and the temperature law (given in figure 1) are varied. The assumed mass loss rate is $\partial M/\partial t = 4 \cdot 10^{-5}\ M_\odot/year$ and the velocity law is

$$V(r) = 3000 \left(1 - \frac{R_c}{r}\right)^1\ [km/sec],$$

with $R_c = 5.5 R_\odot$. Though we have no particular star in mind we notice that the resulting flux and line strengths resemble the LMC WN4 star FD13 (Smith and Willis, 1983).

	$N[H]/N[He]$	opt. thin	opt. thick	Model
	2.0	0.47	0.32	0.41
Table 1.	1.0	0.30	0.20	0.24
Intensity ratio $\Delta \log(W_\lambda) =$	0.5	0.18	0.12	0.15
$\log(I(He + H)/I(He))$ of the Pickering-Balmer blends to the Pickering line intensity.	0.1	0.040	0.027	0.030

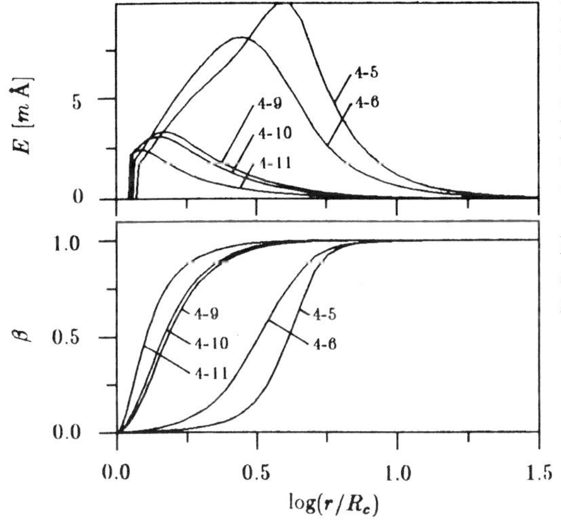

Figure 2a. He II and combined H I and He II line emission $E(r)$ (see text) as a function of radius for a model with $N[\mathrm{H}]/N[\mathrm{He}] = 1$. The labels on the curves refer to the He II transitions. The curves 4-6 and 4-10 include the contribution from H I 2 3 respectively 2-5.

Figure 2b. The escape probability $\beta(r)$ as a function of radius for the same transitions as in figure 2a.

3. RESULTS OF THE MODEL CALCULATIONS

(a) The Balmer-Pickering line strengths for different abundances

In Table 1 we give the calculated contribution of the Balmer to the Pickering lines for four different hydrogen abundances. Also given are the corresponding values of the two approximate formulae. It is interesting to note that our value lies between the two approximations. To understand this result we present figure 2a, where the quantity $E(r)$ from the formula above is plotted as a function of radius. In figure 2b we give the escape probability $\beta(r)$. A comparison of these two graphs shows that the main contribution to the line strength arises from the region where $\beta \approx 0.5$. This is true for all lines (except for optically thick ones), even for those from higher n. Hence, *there are no optically thin He lines*; the result of CB was an effect of the one point approximation.

The figures of Conti et al. (1983) give an idea of the quality of the observations: the detection limit is $N[\mathrm{H}^+]/N[\mathrm{He}^{++}] \approx 0.3$.

(b) The influence of the temperature structure

For the calculations reported in section 3a the temperature law of figure 1a was used. To analyse the influence of the temperature law, we ran the three different temperature structures given in figures 1b, 1c and 1d. Though the absolute line strengths are dependent on the laws *we do not find any measurable change of* $\Delta \log W_\lambda$. This is due to H^+ and He^{++} being isoelectronic. If both hydrogen and helium are fully ionized, line emissivities of H I and He II react practically in the same way to variations in N_e and T_e. This implies also that an inhomogenious atmosphere, as suggested by Underhill (1983), would have no influence on the relative line strengths of hydrogen to helium.

4. CONCLUSION

We found that there is no mechanism to suppress the H I/He II line ratio. Neither optical thickness, i.e. density structure, nor a peculiar temperature law changes the relative emission. An upper and a lower limit on the $N[\mathrm{H}^+]/N[\mathrm{He}^{++}]$ ratio can be set by the approximate formulae. As under all conditions He^{++} will recombine before H^+, the Pickering and Balmer line intensities will give an upper limit for the hydrogen abundances.

ACKNOWLEDGEMENTS

This research has been supported by the Swiss Science Foundation.

REFERENCES

Castor,J.I.: 1970, *Monthly Notices Roy. Astron. Soc.* **149**,111.
Castor,J.I. und van Blerkom,D.: 1970, *Astrophys. J.* **161**,485.
Conti,P.S.,Leep,E.M. und Perry,D.N.: 1983 *Astrophys. J.* **268**,228.
Klein,R.I. und Castor,J.I.: 1978, *Astrophys. J.* **220**,902.
Massey,P.: 1980, *Astrophys. J.* **236**,526.
Sahade,J.: 1981, Rev. Mexicana Astron. Astrof. **6**,189.
Smith,L.J. und Willis,A.J.: 1983, *Astron. Astrophys. suppl.*, in press.
Underhill,A.B.: 1982, 'Wolf-Rayet Stars: Observations, Physics, Evolution', W.H.de Loore
        und A.J.Willis (eds.), *IAU Symp.* **99**,p19.
Underhill,A.B.: 1983, *Astrophys. J.* **265**,933.

DISCUSSION

Kudritzki: Did you include the HeII-resonance lines in your rate equations
and did you encounter then convergence difficulties within the Λ-iteration
scheme?

Schmutz: Yes, I include the HeII-resonance lines and I got indeed problems
namely that the convergence is very slow. On the basis of a few tests
I think that it is a real convergence.

CIRCUMSTELLAR DUST SHELLS AROUND WN10-11 and WC8-10 STARS:
AN EVOLUTIONARY SEQUENCE?[*]

K.A. van der Hucht
Space Research Laboratory, Utrecht, The Netherlands
P.M. Williams
The Royal Observatory Edinburgh, UK and
United Kingdom Infrared Telescope Unit, Hilo, Hawaii, USA
P.S. Thé
Astronomical Institute 'Anton Pannekoek', Amsterdam,
The Netherlands

ABSTRACT

In a recent IR photometric survey of late-type WC and WN stars, it was
discovered that not only most WC8-10 stars have circumstellar dust
shells, but that two extreme late-type WN stars also have strong IR
excesses from circumstellar dust. The latter shells appear to have
significantly different density distributions. In this paper the
possibility of an evolutionary sequence is suggested.

INTRODUCTION

Late-type WC stars are known to differ from other Wolf-Rayet subtypes
in the sense that they exhibit strong thermal infrared excesses emitted
by heated dust particles. The circumstellar (CS) dust shell radii are
typically of the order of 100 AU (Allen et al., 1981). At similar radii
FeIII UV absorption lines were observed (Van der Hucht et al., 1982),
indicating gas in the shells as well. Since the dust shells are located
well within the stellar wind spheres, they must be continuously replenished.

In order to investigate individual dust shells around different
WR stars, the authors conducted an IR photometric survey of most of the
24 WC8-9 stars in the 6th WR catalog (Van der Hucht et al., 1981, henceforth HCLS). Recently, some of these stars have been reclassified by
Massey and Conti (1983) to extreme late-type WC subclasses. They also
classified the faint star WR122 as WN10, which represents a natural
extrapolation of the WN sequence. Following this trend, Lundström and
Stenholm (1983a) classified the faint star LSS4005 (Stephenson and
Sanduleak, 1971) as WN11. However, on the basis of their line identification study, they argue that perhaps both WR122 and LSS4005 are possible pre-WN stars. We observed also these two objects in the IR.

OBSERVATIONS

The observations were conducted in June 1982 and March 1983 with the
ESO 1 m and 3.6 m telescopes (J-Q) and in June 1982 and July 1983 with
UKIRT (J-[19.5]).

A. Maeder and A. Renzini (eds.), Observational Tests of the Stellar Evolution Theory, 273–277.

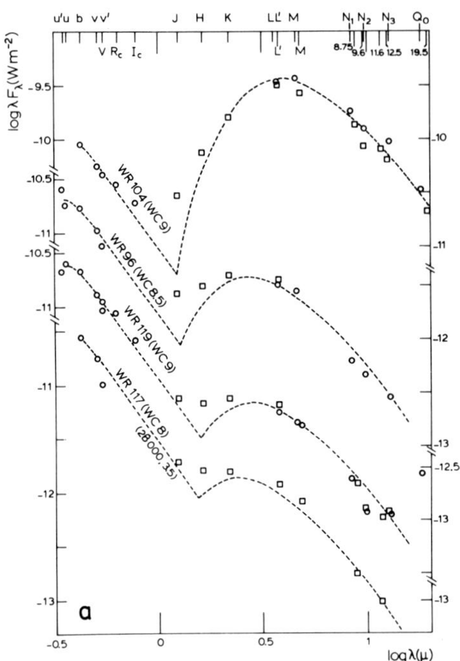

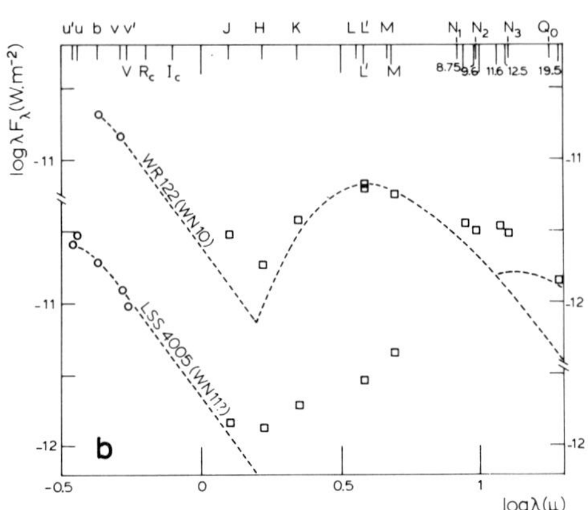

Figure 1a.
Dereddened energy distributions of WC8-9 stars. ○ : observed from ESO. □ : observed from UKIRT. Optical data points fit with model atmosphere flux from Kurucz (1979).
Figure 1b.
Dereddened energy distributions of WN10-11 stars. Same format as Fig. 1a.

Thermal emission by heated dust was found for most of the objects, even among some of the WC8 stars. A detailed account of all the observational results will be published elsewhere (Williams, van der Hucht, and Thé, in preparation).

The observations are corrected for extinction using Van de Hulst curve no. 15. Visual extinction values are based on $(b-v)_o = -0.42$ for all WC8-10 stars, $(b-v)_o = -0.33$ for the WN10-11 stars, and narrow-band **bv photometry** from HCLS, with improvements by Lundström and Stenholm (1983b). Additional data points are from Lundström and Stenholm (1979) and VRI fluxes are from one of us (PST). Extinction free energy distributions for a sample of our program stars are presented in Figure 1a,b.

THE 9.7 μm SILICATE ABSORPTION FEATURE

With the available narrow-band filters around 9.7 μm we observed silicate absorption in many of our program stars. We define a silicate index S.I. as:

$$\text{S.I. (ESO filters)} = N_2 - 0.5\,N_1 - 0.5\,N_3 \quad \text{and}$$
$$\text{S.I. (UKIRT filters)} = [9.6] - 0.5[8.7] - 0.5[12.5].$$

In Figure 2 we present A_V vs. S.I.. Recently Roche and Aitken (1983) observed a few late-type WC stars. From their 8-13 μm grating spectra they show that the silicate absorption feature at 9.7 μm is correlated with the visual extinction by $A_V = 18.5\,\tau_{9.7}$. With our definition of S.I. this means $A_V = 27\,(\text{S.I.})$. This line is marked in Figure 2. One star, having indices measured on both systems, WR119, shows more silicate

absorption than expected. This extra
silicate absorption can not come from
the CS shell responsible for the
general emission (otherwise we would
see a silicate emission feature), but
could originate in a much cooler shell,
at least 10.000 times further away
from the WC star than the carbon
shell. The star with the most negative
index, WR121, may also be variable and
deserves further study.

Figure 2.

*Visual extinction A_v vs. silicate
index S.I. for WC8-10 stars.
○ : observed from ESO;
□ : observed from UKIRT.
Connected data points refer to
the same star.*

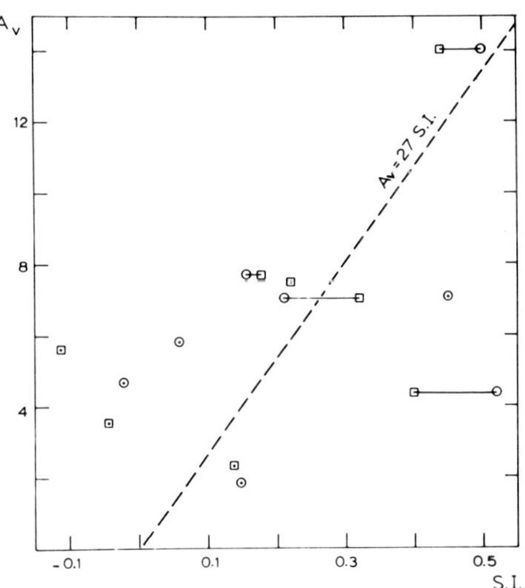

CIRCUMSTELLAR DUST SHELLS

The energy distributions in Figure 1a are close to black bodies having
temperatures ranging from 950 K for WR104 to 1500 K for WR117. We do
not believe the shells to be optically thick, because we observe much
less IR flux than this predicts. The smoothness of the energy distributions is compatible with carbon as the grain material. We have constructed model shells wherein the grain density falls of as $\rho \sim r^{-2}$ from
an inner radius determined in the fitting process. The dust temperatures
are given by thermal equilibrium on the grains. Our models fit the WC8-9
star observations better than the black bodies, and yield dust shell
masses ranging from $5 \times 10^{-9}\ M_\odot$ for WR117 to $10^{-6}\ M_\odot$ for WR104, and shell
inner radii ranging from 1300 R_* for WR119 to 2200 R_* for WR104. Attempts
to fit shells having density distributions differing from $\rho \sim r^{-2}$ were
not successful, confirming that the grains are replenished in the stellar
wind.

The energy distributions of the WN10-11 stars WR122 and LSS4005 are
shown in Figure 1b. It is obvious that also for these stars the IR excess
is caused by thermal emission by heated dust. That of WR122 might even
be explained in terms of two CS dust shells, one comparable to those
around the WC8-10 stars and the second much cooler ($T_C \approx 280$ K), more
distant and more massive. Alternatively, the observations are compatible
with a CS shell wherein the density is constant or increases slightly
with radius. This is obviously quite different from the late-type WC dust
envelopes. The JHKLM data for LSS4005 are not sufficient for modelling,
but they do show that the star is surrounded by an substantial amount of
heated dust. It is remarkable, that, except for the possible multiplicity
of their dust shells, these two late-type WN stars have a characteristic
in common with the late-type WC stars, namely CS dust. Thus, the low-
ionization end of both the WN sequence and the WC sequence are characterized by CS dust shells.

HYPOTHESIS

The observations presented above show that not just WC8-10 stars have
CS dust, but also the recently classified WN10-11 stars. Dust formation
is likely to occur in regions of relatively high density, perhaps caused
by a drop in the stellar wind velocity (van der Hucht et al., 1982).
All WN and WC stars appear to have, at least within a factor 2, the same
high value for their mass loss rates: 2.8×10^{-5} $M_{\odot}$ /yr (Abbott, 1982).
We expect WC stars to evolve from WN stars. Let us assume that stellar
wind parameters for low-ionization WC stars are not different from those
of low-ionization WN stars. Before these stars were in the WR phase,
e.g. during their post-RSG phase (Maeder, 1982), their stellar wind
parameters and mass loss rate may have been quite different. Notably
the mass loss rate will have been smaller, initially. Wind parameter
variations during the pre-WN phase, could very well have caused high
density regions and subsequent CS dust formation, pertaining from the
WN phase into the WC phase. On the basis of this argumentation we would
like to hypothesize that WC8-10 stars with CS dust have evolved from
WN10-11 stars like WR122 and LSS4005. A supporting factor is, that
their galactocentric distances (5.9 and 3.5 kpc respectively
Hidayat et al., 1984) are within the sphere where the late-type WC stars
are found $(r \lesssim 9$ kpc, viz., van der Hucht, 1982, Fig. 3). Because of
their small number the WN10-11 lifetime may be about 10x shorter than
the WC8-10 lifetime.

ACKNOWLEDGEMENT

We would like to thank Dr. Chris Impey for timely observations of
LSS4005, and Pieter Mulder (SRL) for conducting part of the ESO observations.

LITERATURE

Abbott, D.C.: 1982, Astrophys. J. 263, 723.
Allen, D.A., Barton, J.R., and Wallace, P.T.: 1981, Monthly Notices
      Roy. Astron. Soc. 196, 797.
Hidayat, B., van der Hucht, K.A., Supelli, K.R., and Admiranto, A.G.:
      1984, to be submitted to Astron. Astrophys.
van der Hucht, K.A.: 1982, in: S. D'Odorico, D. Baade and K. Kjär (eds),
      The Most Massive Stars, Proc. ESO Workshop, Garching, 23-25 Nov.
      1981, p. 157.
van der Hucht, K.A., Conti, P.S., Lundström, I., and Stenholm, B.: 1981,
      Space Science Reviews 28, 227.
van der Hucht, K.A., Conti, P.S., and Willis, A.J.: 1982, in: C. de
      Loore & A.J. Willis (eds), Wolf-Rayet Stars, Observations, Physics,
      Evolution, Proc. IAU Symp. No. 99 (Dordrecht, Reidel), p. 277.
Kurucz, R.L.: 1979, Astrophys. J. Suppl. 40, 1.
Lundström, I., and Stenholm, B.: 1979, Astron. Astrophys. Suppl. 35, 303.
Lundström, I., and Stenholm, B.: 1983a, in: Proc. Nordic Astronomical
      Meeting, Oslo, August 15-17, 1983.
Lundström, I., and Stenholm, B.: 1983b, in preparation.
Maeder, A.: 1982, Astron. Astrophys. 105, 149.

Massey, P., and Conti, P.S.: 1983, Publ. Astron. Soc. Pacific 95, 440.
Roche, P.F., and Aitken, D.: 1983, Monthly Notices Roy. Astron. Soc.,
    in press.
Stephenson, C.B., and Sanduleak, N.: 1971, Publ. Warner and Swasey
    Observatory 1, No. 1.

---

*
Based on observations collected at the European Southern Observatory
(La Silla, Chile) and at the United Kingdom Infrared Telescope (Mauna
Kea, Hawaii, USA).

DISCUSSION

de Loore: Are these circumstellar dust shells also observed around
Of-stars? The presence or absence could be an important clue for the
investigation of the mass loss mechanism in Wolf-Rayet stars.

van der Hucht: To my knowledge nothing about CS dust shells around Of-
stars has been published. Maybe the search for this has not been deep
enough.

MASSIVE STAR EVOLUTION IN NEARBY GALAXIES

Roberta M. Humphreys
University of Minnesota

In this review I will primarily be discussing the observational data relevant to understanding the process of stellar evolution in galaxies of different types. This discussion will focus on the stellar content of the nearer galaxies; those galaxies in which the brightest individual stars are resolved and can be observed.

The most luminous stars are also the most massive stars. Because of their intrinsic brightness, they are our first probes for observational studies of stellar evolution in other galaxies. The most massive stars provide our first hints that the progress of stellar evolution may have been different in a particular galaxy.

Basically we want to know if stellar evolution as we understand it for our region of the Milky Way has produced similar massive star populations in different galaxies. In the past decade we have already learned in our own galaxy that the observed HR diagram for massive stars cannot be explained adequately by conservative, non-mass loss evolution. Phenomena such as stellar winds and mass loss alter the evolution of the most massive stars and physical processes such as turbulence and mixing in the stellar interiors may ultimately control their short but often flamboyant lives.

The study of massive star evolution in other galaxies offers many advantages. In our own Milky Way our observations are limited to a small fraction of the galaxy by interstellar dust. We are uncertain how our current ideas may be influenced by incompleteness and whether massive star evolution may differ in different regions. Observations in other galaxies will eventually permit us to determine if the rate of star formation and the initial mass function for massive stars vary with location in a galaxy and with galaxy type. How is stellar evolution influenced by environmental variations such as possible chemical composition gradients in galactic disks? I will not be able to answer these questions in this review paper, but I will show that there are great similarities among the massive star populations in different galaxies and some important differences. I will try to untangle some of the

A. Maeder and A. Renzini (eds.), Observational Tests of the Stellar Evolution Theory, 279–297.
© *1984 by the IAU.*

possible causes of the observed differences such as IMF variations and chemical composition.

1.   THE HR DIAGRAM

The HR diagrams (M_v vs. spectral type or M_{Bol} vs. $\log T_{eff}$) provide an efficient overall perspective for comparison of the massive star populations in difference galaxies. The M_{Bol} vs. $\log T_{eff}$ diagram can also be compared directly with stellar models and evolutionary tracks.

The observational data for the HR diagrams include photometry and spectral types for the individual stars. Their luminosities, corrected for interstellar extinction, are then derived from the true distance modulus of the galaxy. For this paper I have used the bolometric correction-effective temperature scale from Flower (1977) with somewhat lower temperatures for the early O-type stars from Kudritzki (1981). The extensive observational data required for hundreds of stars in a galaxy means that reasonably complete or representative HR diagrams are available for the massive star populations in only three nearby galaxies: the Milky Way and the Large and Small Magellanic Clouds.

Figure 1 is the M_{Bol} vs. $\log T_{eff}$ diagram for the solar region of our galaxy. There are 2354 luminous stars in this diagram and all are members of 91 stellar associations and young clusters (see Humphreys 1978) whose distances are reasonably known. Even though the stars are essentially restricted to a region only 6 kpc across, centered on the Sun, this diagram is representative of the stellar population in the spiral arms in the outer parts of our galaxy. The most significant feature of this HR diagram for massive stars is the observed upper envelope to their luminosities. The upper luminosity boundary declines with decreasing temperature for the hottest stars, but becomes eesentially constant for the cooler stars (<10000 K) at $M_{Bol} \simeq -9.5$ to -10 mag. There are many very luminous, hot stars with $M_{Bol} = -10$ to -11 mag but they have no cooler counterparts. The evolutionary tracks are from models with mass loss from Maeder (1981, 1983).

The HR diagram for the Large Cloud (Figure 2) looks very much like that for the luminous stars in the solar region of our galaxy, revealing very similar distributions of stellar temperatures and luminosities in the two galaxies (Humphreys and Davidson 1979). The upper envelope of stellar luminosities is the same – the decreasing luminosity with decreasing temperature for the hot stars and the upper boundary for the later-type supergiants. When making any comparisons of this type, possible incompleteness of the data and the effects of observational selection must be considered. One noticeable difference between the LMC and galactić supergiants is the greater relative number of high luminosity late B and early-A type stars in the Large Cloud. This is most likely due to observational selection.

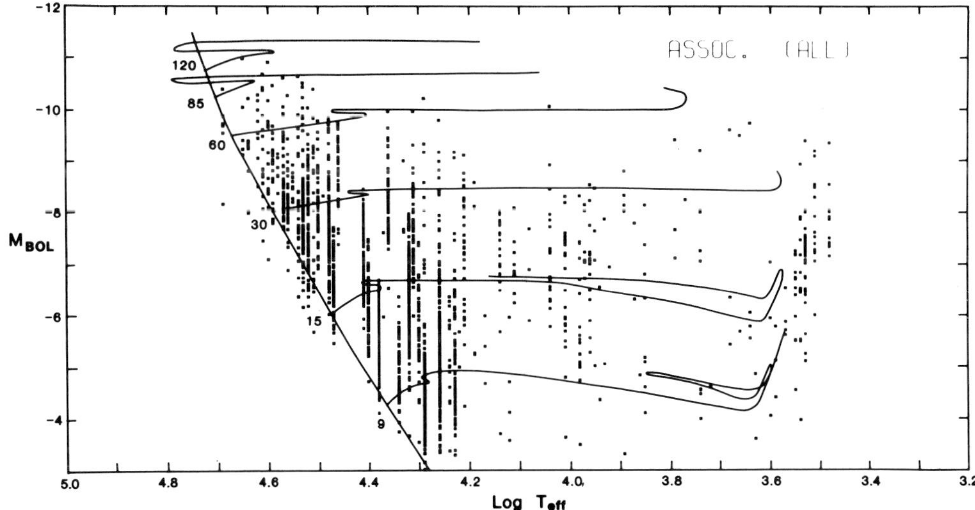

Figure 1 – The HR diagram, M_{Bol} vs. log T_{eff}, for O-type stars, supergiants, and less luminous early-type stars in 91 stellar associations and clusters in the solar region of our Galaxy.

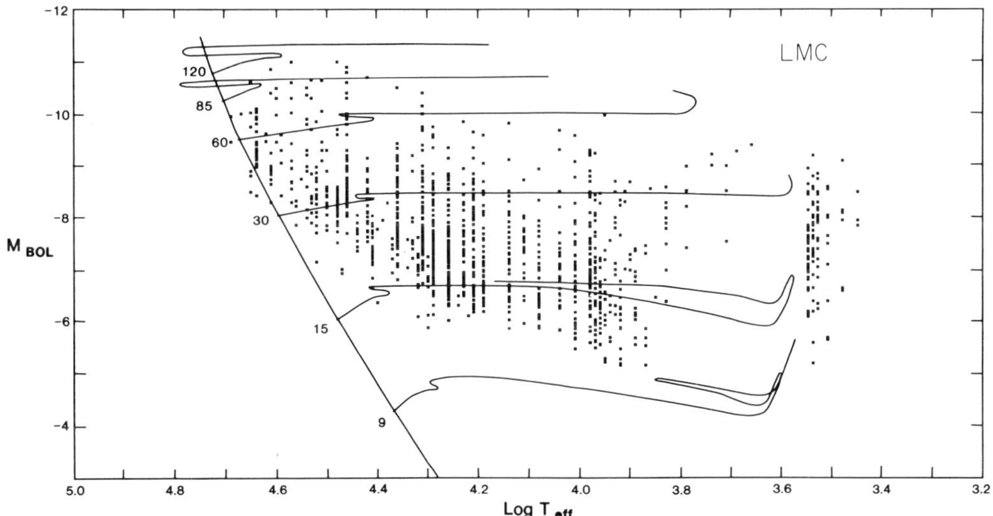

Figure 2 – The M_{Bol} vs. log T_{eff} diagram for confirmed and suspected luminous stars in the Large Magellanic Cloud.

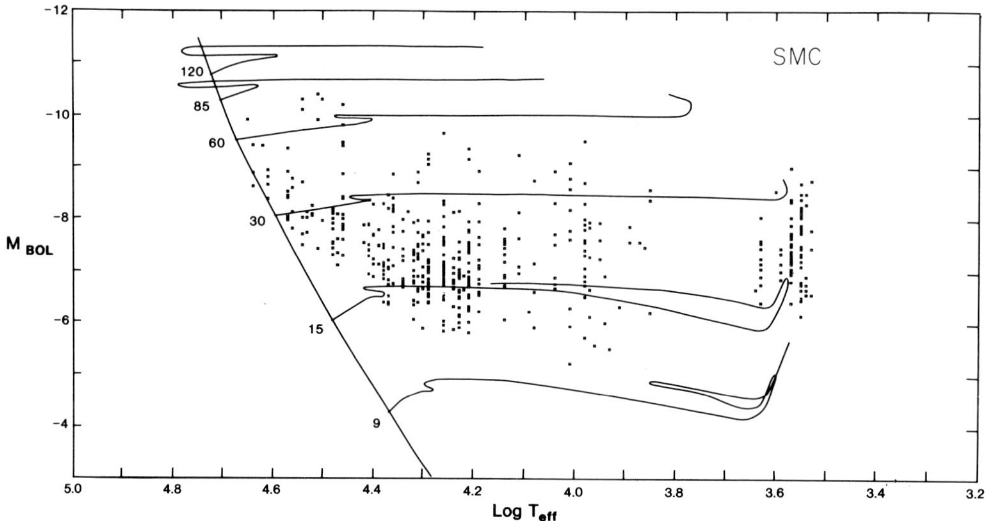

Figure 3 – The M_{Bol} vs. log T_{eff} diagram for confirmed and suspected luminous stars in the Small Magellanic Cloud.

Overall this comparison reveals very similar massive star populations and presumably similar evolutionary histories in our galaxy and the Large Cloud.

In the Small Cloud (Figure 3) there are obvious and significant differences. The hottest stars in the SMC are both less luminous and fewer in number (Humphreys 1983a). On the basis of presently available spectra and photometric data there appear to be no stars in the SMC with initial masses much above 80 $M_\odot$. There are no known O3 stars and only one star has been classified O4 III by Walborn (1978), but there is no photometry for it. It is shown on the HR diagram using the luminosity calibration. Conti, Garmany, Massey, and I have a program to classify the candidate O stars in the SMC to determine if the lack of highly luminous hot stars is real. Nevertheless, there are fewer of the brightest stars at all of the early spectral types in the SMC. This may be a size of sample effect in the Small Cloud which is a much smaller, less luminous galaxy, although it is also possible that star formation produces a different statistical distribution of initial masses in the SMC. It is not due to chemical composition differences.

Even with these differences in mind, the large scale features of the HR diagram for the Small Cloud are essentially the same as for the Galaxy and the Large Cloud. It is especially important to note that the upper luminosity boundary for the late-type supergiants is similar.

It is well known that the Small Cloud has a significantly lower heavy element abundance than either the solar region of our galaxy or the LMC. The effects of the lower metallicity show up very strongly in the spectral type distribution of the M supergiants in the Small Cloud. Histograms of the M supergiant spectral types for our solar region, the LMC and SMC (Figure 2 in Humphreys 1979a and Figure 1 in Elias et al. 1983) show a shift to earlier spectral types from our galaxy to the LMC and SMC. This is most likely due to the shift of the Hayashi track, the nearly vertical slope of the red giant branch, to warmer temperatures with lower metallicity. As the heavy element abundance decreases the surface temperature increases because of the lowered opacity. This effect is also observed for the red giant branches of globular clusters, and will very likely explain the Blanco/McCarthy (Blanco, McCarthy, and Blanco 1980) results for the ratio of M giants to carbon stars between our galaxy and the Clouds. If the M giants have also shifted to earlier spectral types then the early M giants should be searched for in the Clouds and compared with the carbon star population.

Another consequence of the lower metallicity in the SMC is smaller mass loss rates. Hutchings (1980, 1982) has found from IUE observations of hot supergiants in the Magellanic Clouds that the stellar wind phenomena are either lacking or very weak in the SMC stars, and while stronger in the LMC stars, they are weaker than in galactic stars. He concludes that the mass loss rates may be two or more times lower than in their galactic equivalents. This is attributed to the lower metal abundances. Interestingly recent infrared observations at 10µm by Elias, Frogel, and Humphreys (1983) of the M supergiants in the Clouds, reveal that the strength of the 10µm circumstellar silicate feature depends on metallicity. The 10µm excess is very weak or nonexistent in the SMC M supergiants. It is somewhat larger in the LMC red supergiants, but smaller than for the galactic M supergiants of the same type. This may be due either to lower mass loss rates or less dust formation around the Cloud M supergiants. These results may be important for evolutionary models for the Cloud stars because as discussed later mass loss may influence massive star evolution.

A few evolutionary tracks for massive stars have been computed with the chemical abundances of the LMC (Maeder 1980, Brunish and Truran 1982b) and SMC (Hellings and Vanbeveren 1981, Brunish and Truran 1982b). The principal difference is that the models are bluer and slightly more luminous at comparable evolutionary stages.

Our knowledge of massive star evolution in the more distant members of the Local Group is more limited. For observational reasons only the visually brightest stars have been observed, so we are lacking information on the individual hottest, most luminous stars. We are naturally very interested in stellar evolution in our two neighboring spirals M31 and M33.

We know very little about the stellar content of M31. Because of its large size, high tilt angle and tightly wound arms there are no extensive surveys for the blue and red stars in the most massive galaxy in our Local Group. Our information is essentially limited to Baade's Field IV (Baade and Swope 1963), one of four fields in M31 selected by Baade along the major axis south of the nucleus. It is the only field for which a color-magnitude diagram exists, and I have taken spectra of several of the brightest stars in this field. Figure 4 in Humphreys (1979b) shows the color-magnitude diagram for the four associations in Field IV. Field IV is in the outermost part of M31, 18.5 kpc from the nucleus, which when scaled to the Milky Way corresponds to about 14 kpc from the center. The color-magnitude diagrams, ages and dimensions of the associations and even the blue-to-red supergiant ratio are all very similar to what we observe in the spiral features in the outer parts of our galaxy. The evolution of the Population I members of M31 is presumably similar to that in our solar region, although the results apply to only one small region in one spiral arm. Detailed studies in other regions of M31 are definitely needed.

M33 is a much easier galaxy to study for this purpose because of its more open spiral arms and nearly face-on appearance. Consequently, Sandage and I (Humphreys and Sandage 1980) completed a survey for its brightest stars, and spectra have been taken (Humphreys 1980b) of some of the visually brightest blue and red supergiants. We also identified 143 stellar associations. Figures 18 and 19 in HS show the color-magnitude diagrams for the stars inside and outside the associations. There are clear differences; the visually brightest blue stars inside the association boundaries are more than two magnitudes brighter than those outside the stellar groups. Nevertheless, it is also clear that non-association OB stars exist as in our galaxy. Color-magnitude diagrams were also measured for a few of the individual associations and ages were estimated from the main sequence turnoff. All six associations have essentially the same age of $\approx 5 \times 10^6$ yrs.

The M_{Bol} vs. log T_{eff} diagram has been determined for the stars in M33 with spectra (Figure 4 in Humphreys 1980b). The lack of hot stars on the diagram is because only the visually brightest stars were observed. Comparison with the observed color-magnitude diagram shows that many candidate OB stars are present. By analogy with our galaxy we would expect the brightest O stars to appear at $M_V \simeq -7$ mag, and they do in M33. Most of these stars will occupy the region between $M_{Bol} = -9$ mag to -11 mag. There is one star with a spectrum as early as BO with a total luminosity $M_{Bol} \simeq -11$ mag and an initial mass near 100 $M_\odot$. Using the HS stellar content survey, Berkhuijsen (1982) finds similar IMF's for M33 and the Galaxy. The fraction of high mass stars is similar in the two galaxies, but the rate of star formation per unit HI mass is smaller in M33.

The upper luminosity boundary based on the HR diagrams for our solar region and the LMC is also appropriate for M33. On the basis of this first look at the brightest stars in M33 there is no reason to suspect any significant differences in the evolution of the Population I stars in our galaxy, the Large Cloud and in M33.

The brightest stars have also been observed in the two dwarf irregular galaxies NGC 6822 and IC 1613 (see Figures 3 and 4 in Humphreys 1980a). In both galaxies there are confirmed supergiants with initial masses near 80 $M_\odot$, but very massive stars near 100 $M_\odot$ appear to be missing as in the SMC. Again comparison with the observed HR diagrams show that there are candidate OB stars in each galaxy, but the brightest candidates appear to be visually less luminous than in the spirals and the LMC ($M_V \simeq -6.5$ in NGC 6822 and IC 1613). Therefore we do not expect any stars much more massive than 80 $M_\odot$ in these two galaxies. In contrast the red supergiants in both these small galaxies have upper luminosities comparable to those in the more luminous galaxies. Interestingly, in IC 1613 there are no known M supergiants classified later than M1 similar to the SMC, but in NGC 6822 I have recently classi- fied one as M3-M4. These results correlate with the known metallicities of the two galaxies. IC 1613 has lower metallicity even than the SMC (Davidson and Kinman 1982) while NGC 6822 is intermediate between the LMC and SMC in heavy element abundances (Pagel and Edmonds 1981). IC 1613 also has a different luminosity function (Lequeux <u>et al</u>. 1980) from the other dwarf irregulars (SMC and NGC 6822) which means either a low star formation rate or a very different IMF (Hoessel <u>et al</u>. 1983).

Among the six Local Group galaxies for which M_{Bol} vs. log T_{eff} dia- grams are available, the primary difference is that the smaller galaxies have fewer of the most luminous, hot stars. Otherwise, the HR diagrams for each galaxy show the same upper envelope of declining luminosity with temperature for the hot stars and the upper luminosity limit for the cool supergiants.

Observational HR diagrams (M_V vs. color) and photometry of the brightest stars have recently been observed for a few of the even smaller, less luminous dwarf irregulars in our Local Group; Sextans A (Sandage and Carlson 1982, Hoessel, Schommer, and Danileson 1983), Pegasus (Hoessel and Mould 1982, Christian and Tully 1983), GR 8 (Hoessel and Danielson 1983) and LGS 3 (Christian and Tully 1983). Hoessel and his associates conclude that the luminosity function for Sextans A is similar to that for our galaxy and the LMC, that recent star formation (>15 $M_\odot$) in Pegasus has been very subdued, and that GR 8, the lowest luminosity galaxy studied, has a flat luminosity function which may mean an unusual IMF. Using the distances adopted by Hoessel, the visual luminosities for both the brightest blue and red stars show a sharp decline among these three galaxies.

The dependence of the visual luminosity of the brightest blue star on galaxy luminosity is well known (Sandage and Tammann 1974, Sandage and Tammann 1982, Figure 1; Humphreys 1983b, Figure 1). The visually brightest stars are all late B or A-type supergiants. As we have already seen, the most massive, most luminous stars in the fainter, smaller galaxies are both fainter and fewer in number. As their numbers decrease in a galaxy, the probability of finding A-type supergiants at a certain luminosity should also decrease. When there are fewer of the most massive star progenitors, the visually brightest star should be less luminous; consequently, there should be a similar relation between M_{Bol} for the most luminous stars and galaxy luminosity (Figure 2 in Humphreys 1983b). This correlation for the most luminous stars shows that the relation for the visually brightest stars is due to differences in the massive star populations (>50-60 $M_\odot$). It is not caused by metallicity differences, but may be due to statistical effects because of smaller population samples in smaller galaxies or possibly to real IMF variations. Very likely both play a role.

The situation for the brightest red supergiants is very different. We have already emphasized the observed upper limit to the luminosities of the M supergiants. Figure 4 in Humphreys (1983b) shows the visual luminosities for the brightest M supergiants in six Local Group galaxies covering a range of nearly six magnitudes in galaxy luminosity. This very tight upper limit is not fortuitous. It is a consequence of massive star evolution discussed at the end of this paper.

2. THE INITIAL MASS FUNCTION AND BLUE TO RED SUPERGIANT RATIOS

Variations in the observed luminosity function and the initial mass function have been mentioned above as possible causes of some of the observed differences among the massive star populations in different galaxies. Lequeux and his associates (1979a,b, 1980, Vangioni-Flamm et al. 1980) have made several studies of this type, and conclude that there are real variations in the rate of massive star formation per unit gas mass from galaxy to galaxy with some tendency for it to be smaller in less evolved galaxies. They also found no reason to suspect that the IMF significantly differs in any galaxy from that of the solar region. Recently, however, Garmany, Conti, and Chiosi (1982) reported a significant variation in the slope of the IMF for O-type stars in our galaxy with galactocentric distance. Their data set was restricted to O-type stars and was volume-limited to 2.5 kpc from the Sun.

Their result is sufficiently important and interesting that it is worth repeating with a more complete data set. We compiled a list of all known supergiants, O-type stars and less luminous B-type stars from the catalogues of stars with Mk types by Buscombe (Kennedy and Buscombe 1974, Buscombe 1977, 1980, 1981). Our list includes 5044 stars; 2354 of these are in 91 stellar associations and clusters, and the sample is complete to 3 kpc from the Sun for which there are 4058 stars. Our IMF is defined as the number of stars per kpc^2 per year per log mass ($M_\odot$) and is determined for mass intervals defined by Maeder's evolutionary tracks (see Figure 1).

, The IMF for the massive stars within 3 kpc is shown in Figure 4. The IMF for >30 $M_\odot$ clearly deviates from the Miller-Scalo IMF (Miller and Scalo 1979) for the solar region (r < 1 kpc). For the least squares solution the points were weighted by the number of stars in each mass interval. The data is incomplete for the 9-15 $M_\odot$ interval and likely for the 15-30 $M_\odot$ interval, as well. Also notice that the data point is very low for the highest mass interval. The IMF calculation has been repeated for the same data set divided between inside and outside the solar circle (Figure 4). The significant difference with galactocentric distance (Garmany et al. 1982) is confirmed. The IMF interior to the sun's orbit is flatter meaning that more of the most massive stars are formed. It is too early to conclude that there is an IMF gradient in the galactic disk, because these results depend strongly on the stars in a few associations.

For comparison we have also determined the IMF for the Magellanic Clouds. There are nearly 1300 stars in the Large Cloud and 500 stars in the Small Cloud for which spectral types or two-color photometry is available. The results are shown in Figure 4. The IMF's for the LMC and very likely the SMC, as well, are both more like the outer galactic region and their slopes are consistent with the Miller-Scalo IMF for the solar neighborhood. Because there are no stars above 80 $M_\odot$ in the SMC and the lower mass intervals are incomplete, this solution for the SMC is based only on two points. The results for the IMF's are summarized in Table 1.

Table 1

The Initial Mass Function for Massive Stars

Milky Way $(M \geq 15\ M_\odot)$

$\qquad r \leq 3$ kpc $\psi = 1.2 \times 10^{-3} (M/M_\odot)^{-1.38}$

$\qquad\qquad$ Inner $\quad R \ominus 10$ kpc

$\qquad\qquad\qquad \psi = 3 \times 10^{-4} (M/M_\odot)^{-0.96}$

$\qquad\qquad$ Outer $\quad R \geq 10$ kpc

$\qquad\qquad\qquad \psi = 7.2 \times 10^{-3} (M/M_\odot)^{-1.98}$

LMC $(M \geq 15\ M_\odot)$

$\qquad\qquad \psi = 5.5 \times 10^{-3} (M/M_\odot)^{-1.90}$

SMC $(M \geq 30\ M_\odot)$

$\qquad\qquad \psi = 3.0 \times 10^{-3} (M/M_\odot)^{-1.50}$

The IMF's for the Large and Small Cloud and the outer parts of the solar region are essentially the same. Why then is the inner part of the solar region so different? Is there an IMF gradient? This question can best be answered by observations of luminous stars in other spiral galaxies. Interestingly, all four regions have very few if any stars in the highest mass interval. Of course these are small number statistics, but in addition, it is very difficult to produce the most massive stars. Perhaps the shape of the IMF for massive stars may be more accurately represented by a power law plus an exponential.

The IMF for the SMC is especially important because of suggestions that the IMF's for some dwarf irregulars may be different. This IMF

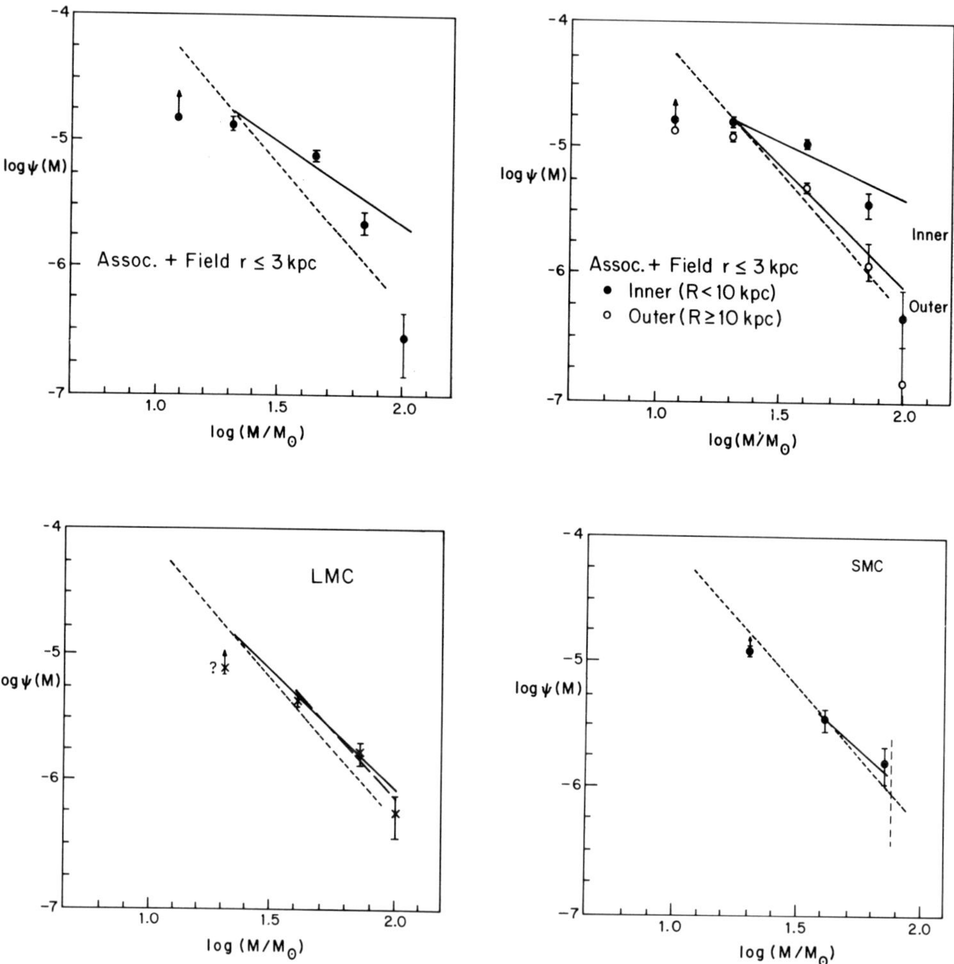

Figure 4 – The initial mass function $\psi(M)$, the number of stars $kpc^{-2}\ yr^{-1}\ (\log M/M_\oplus)^{-1}$ for the solar region ($r \leq 3$ kpc), the inner and outer sections of the solar region, the LMC, and the SMC. The solid lines are the least squares solutions to the observed points. The dashed line is the Miller-Scalo IMF for the solar neighborhood ($r < 1$ kpc).

would predict 5 stars between 85 and 120 $M_\odot$ in the SMC, but none are observed. This difference is significant with a probability of less than 1%, but the same is true for the galactic and LMC IMF's, because of the sharp turn down in the observed numbers of the most massive stars in all three galaxies. Final conclusions on the IMF for the Small Cloud hinge on observations in progress.

Blue to red supergiant ratios are often used as diagnostics to test models of stellar evolution, and counts of blue and red stars in other galaxies are considered possible indicators of metallicity variations. But interpretation of the B/R ratio is not entirely straightforward. They may be affected by other factors such as variations in the IMF. A much more complete data set is now available for the massive stars in our galaxy and the Clouds. Their blue to red ratios are summarized in Table 2.

Table 2
The Blue to Red Supergiant Ratio in the
Galaxy and the Magellanic Clouds

M_{Bol} (mag)	Mass Range ($M_\odot$)	Milky Way (8–10 kpc)	(10–12 kpc)	LMC	SMC
−10.5 to −11.5	100–200	–	–	–	–
− 9.5 to −10.5	60–85	–	–	–	–
− 8.5 to − 9.5	40–60	46	22	11	4.6
− 7.5 to − 8.5	25–40	30	20	7(10)	3.8
− 6.5 to − 7.5	15–25	15(31)	7(16)	11(13)	7(6)

The B/R ratio shows two phenomena: 1) little or no variation with luminosity when the data is corrected for incompleteness, and 2) a gradient with distance in our galaxy and between us and the Magellanic Clouds which is attributed to metallicity variations. Comparison with the expected ratios from evolutionary tracks is not especially good, but models with moderate mass loss give much better agreement with the observed ratios than do the older, non-mass loss calculations. Suggestions (Brunish and Truran 1982a,b) that the number of red supergiants have been overestimated by contamination from intermediate mass stars are incorrect. Red intermediate mass stars near $M_{Bol} \simeq -7$ mag are at the tip of the assymptotic giant branch. They are very rare; they are long period variables and have very late spectral types. They will not be mistaken for M supergiants. The smaller B/R ratios in the LMC and SMC imply that the lifetimes of the M supergiants are longer in these galaxies with lower metallicity and presumably lower mass loss rates. Unfortunately the models by Brunish and Truran (1982b) for Magellanic Cloud abundances predict the opposite.

The gradient in the blue to red ratio is dependent on the luminosity interval used. At the lower luminosities ($M_{Bol} \leq -8.5$ mag) where the effects of possible IMF variations is less important, the B/R gradient is less steep. The B/R ratios in the $M_{Bol} \simeq -6.5$ to -8.5 mag interval are probably the best indicators of the metallicity gradient between our galaxy and the Clouds. Unfortunately in other galaxies we usually do not know M_{Bol} for the stars, but instead counts of blue and red stars are available only to some limiting visual magnitude (M_V). If the B/R ratio is used to infer an abundance gradient in a galaxy, the results will depend strongly on the brightness of the stars being used. The effects of metallicity and IMF variations will be mixed. For a distant galaxy, only the visually brightest stars would be resolved, and one might conclude that a steep abundance gradient exists when one may also be observing the effects of IMF differences. I do not recommend use of B/R ratios as probes of stellar evolution in a galaxy unless additional information is available.

3. EVOLUTION OF THE MOST MASSIVE STARS

How well do the current models for massive star evolution explain the observed upper envelope of luminosities for normal stars on the HR diagram? Figure 5 is a schematic HR diagram for the most massive stars ($L > 5 \times 10\ L_\odot$, $M_{Bol} \simeq -9$ mag) based on the composite of the M_{Bol} vs. log T_{eff} diagrams for the six Local Group galaxies. The empirical upper boundary of stellar luminosities for normal stars shows the envelope of declining luminosity with decreasing temperature for the hottest stars and the upper limit to the luminosities of the cooler supergiants observed in all six galaxies. The temperature dependence of the luminosity boundary suggests that this defines a critical location on the HR diagram for massive star evolution. Stars with initial masses less than 60 $M_\odot$ can apparently evolve all the way across the diagram to the red supergiant region. Why then are the hottest, most luminous stars restricted to the left side of the HR diagram? What prevents a very massive star from evolving into a highly luminous cooler supergiant?

A group of very luminous, hot stars with peculiar spectra that are also known to be unstable and undergoing unsteady mass loss provide important clues to the evolution of the most massive stars and their eventual fate. The famous star η Car is probably the most massive and most extreme member of this group. These stars have spectra with emission lines of hydrogen, HeI, HeII, FeII and [FeII], often with P Cygni profiles. Many of them are also known irregular variables with extended maxima and minima of several years. Their observed mass loss rates range from 10^{-5} to 10^{-3} $M_\odot$/yr with a possible high of 10^{-1} $M_\odot$/yr for η Car. Other well known examples are P Cyg and S Dor. As a group they are known as the S Dor variables and as the Hubble-Sandage variables in M31 and M33. Many of these stars are known in our galaxy, the Large Cloud and in M31 and M33; however, only two are recognized in the Small Cloud, and there are no candidates in IC 1613 and NGC 6822. This is a consequence of the significant difference in the distribution of stellar masses between the less massive galaxies and the spirals and LMC.

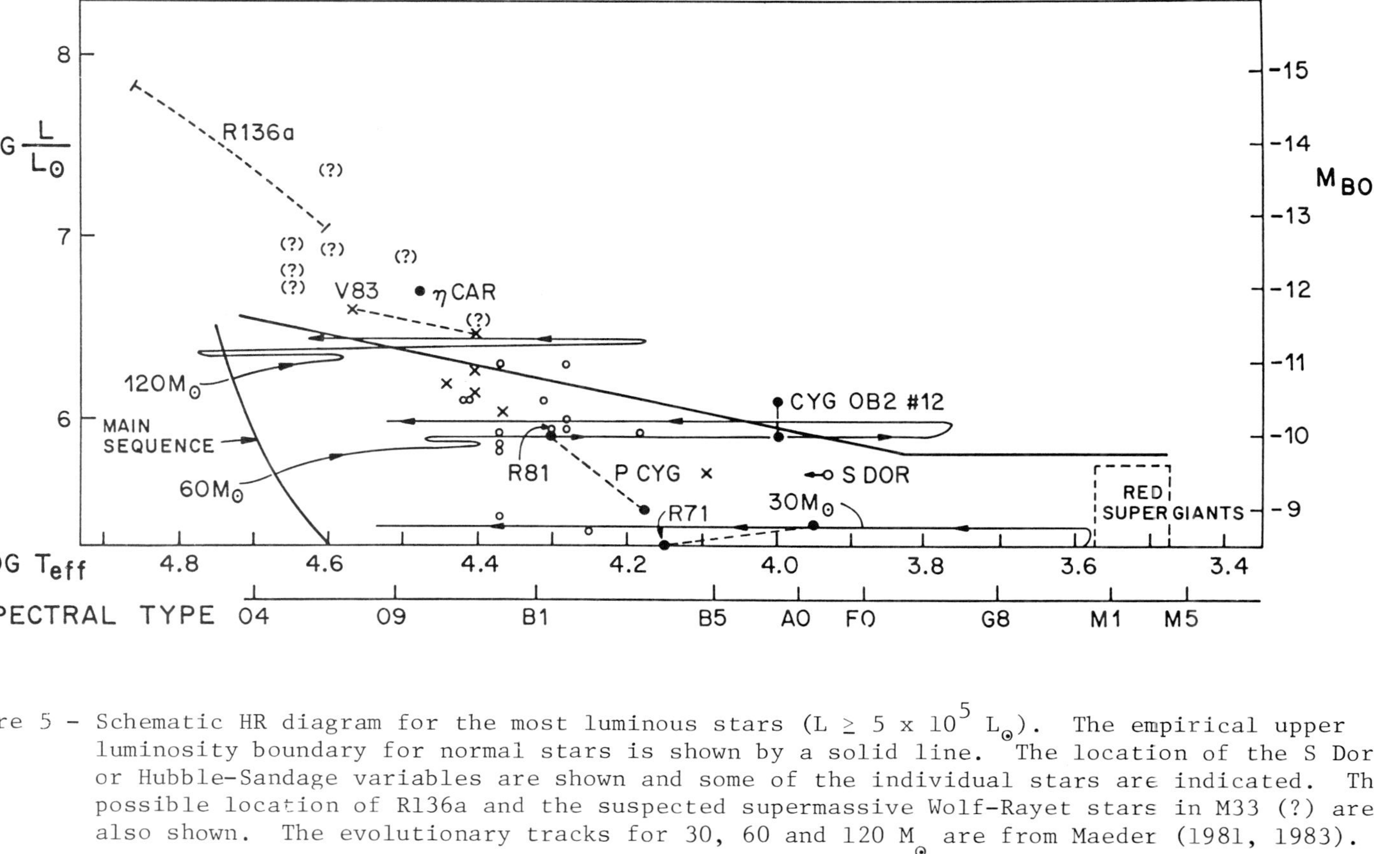

Figure 5 – Schematic HR diagram for the most luminous stars ($L \geq 5 \times 10^5\ L_\odot$). The empirical upper luminosity boundary for normal stars is shown by a solid line. The location of the S Dor or Hubble–Sandage variables are shown and some of the individual stars are indicated. The possible location of R136a and the suspected supermassive Wolf–Rayet stars in M33 (?) are also shown. The evolutionary tracks for 30, 60 and 120 $M_\odot$ are from Maeder (1981, 1983).

Information on temperatures, luminosities and mass loss rates have recently been determined for many of these stars primarily as a result of ultraviolet spectroscopy with the IUE (references are given in a review by Humphreys 1984). Their location on the HR diagram is shown in Figure 5. η Car, P Cyg, and some of the H-S variables (Var A and Var 83 in M33), thought to suffer spectacular episodes of mass ejection, are near or even above the critical upper luminosity boundary. Walborn (1983) and Stahl et al. (1983) have just reported that another of this class, R127 in the LMC, is presently undergoing an outburst.

These stars inspired us (Humphreys and Davidson 1979) to propose a scenario in which the most massive stars are prevented from evolution to cooler temperatures by an instability limit accompanied by sudden mass loss. (The line drawn in Figure 5 is an empirical envelope for 'normal' stars. The true instability boundary may be above this; perhaps η Car marks its location.) A star may evolve to this critical limit several times (Davidson 1983). The sudden instability may cause an η Car-like outburst which ejects a fraction of a percent of the star's mass. This moves the star slightly away from the critical line and temporarily relieves the instability, but then in a few centuries or decades the star evolves back to the limit and suffers another outburst; and so on until it is reduced to a Wolf-Rayet star, unless it becomes a supernova first.

This sequence of events is supported by data for η Car. We know that it has undergone more than one episode of large mass ejection (Walborn et al. 1978, Ringuelet 1958). Davidson, Walborn, and Gull (1982) have found that η Car is nitrogen-rich but carbon and oxygen-poor; thus confirming that it is indeed an evolved very massive star. Many of the S Dor or H-S variables are also known to be nitrogen-rich (Shore and Sanduleak 1984, Walborn 1982). This suggests that these hot, luminous and often unstable stars are on their way to becoming Wolf-Rayet stars of the late WN type. Several years ago, Conti (1975) proposed a scenario for the production of WR stars progressing from O to Of type stars to late WN stars and eventually to classical WN and WC stars, but the observed mass loss rates for the O stars were insufficient (see review by Maeder 1981). If we add the H-S variable stage, with its episodes of enhanced mass loss, a possible evolutionary sequence for stars $\gtrsim 60\ M_{\odot}$ is O → Of → HS var → WN7-9.

The exact cause of the proposed instability is not known, but there are several alternatives (Stothers and Chin 1983). An internal vibrational instability is a possibility, but more likely surface radiation pressure is involved, because many of these stars are near the Eddington limit for their temperatures at a given L/M ratio.

Maeder (1983) has recently proposed a very promising explanation for the instability boundary – the de Jager limit (1980). A deep external convective zone develops in the outer layers of very massive stars as they evolve to cooler temperatures and gives rise to a turbulent pressure gradient which can oppose gravity if the star's luminosity is sufficiently high. The de Jager limit is reached when the turbulent pressure

gradient equals the acceleration due to gravity. The instability occurs
when turbulence exceeds gravity. The de Jager limit halts further evolution to the right and is accompanied by enhanced mass loss. Maeder's
models show that after the enhanced mass ejection the star evolves blueward on short time scales at essentially constant luminosity but its
visual luminosity decreases. This corresponds to the observations of
R71 and S Dor (Appenzeller and Wolf 1981) between maximum and minimum
and Var 83 in M33 (Humphreys et al. 1984).

The onset of the instability limit very likely corresponds to the
observed boundary for the luminosities for the late-type supergiants.
The enhanced mass loss at the de Jager limit increases the mass fraction
occupied by the core. Previous models (Chiosi et al. 1978, Maeder
(1981a) show that when the core becomes larger than a critical value the
evolutionary tracks reverse to warmer temperatures. This result is
independent of chemical composition in agreement with the observations
for the late type supergiants. The upper luminosity limit for cool
supergiants should be the same in all star forming regions, and stars
with initial masses greater than ≈ 50 $M_\odot$ simply do not become M supergiants.

Conspicuous in the upper left of the HR diagram in Figure 5 is
R136a, the possible supermassive star in the LMC. It is shown with a
plausible range of temperatures and luminosities if it is a single star
(Savage et al. 1983). But R136a is definitely a binary (Innis 1927,
Worley 1983, Chue and Wolfire 1983), and very likely a multiple system
(Weigelt 1983) similar to NGC 3603 (Walborn 1973). These considerations
will reduce the mass estimates for the primary to a few hundred solar
masses. This is still a very high mass, comparable to or greater than
that of η Car. The suspected supermassive Wolf-Rayet stars in M33
(Massey and Hutchings 1983) are also shown in Figure 5.

The formation of these individual very massive stars and especially
the large aggregates of very massive stars represented by the η Car
group (Tr 14, 15, 16) and NGC 3603 in our galaxy and the R136 complex
in the LMC is an important problem for the future.

In conclusion I want to emphasize the similarities of massive star
evolution in the solar region of our galaxy, in the Large Cloud, in M33,
and very likely in M31 as evidenced by their HR diagrams and the properties of their brightest stars. Most of the differences we observe in the
HR diagrams for the less massive irregular galaxies, specifically the
Small Cloud, are most likely due to small variations at the high mass
end of the IMF and to statistical effects from the smaller sample size.
The lower metallicity in the SMC produces a shift to earlier spectral
types in the M supergiants and smaller blue to red star ratios which may
mean longer lifetimes as M supergiants. The lower metallicity may also
cause lower loss rates which should be included in evolutionary tracks
for the Magellanic Cloud stars.

     Overall, stellar evolution has produced similar massive star popu-
lations in the spirals and Magellanic-type irregular galaxies in our
Local Group. Studies of the fainter stars in less massive members of
our Local Group and in more distant galaxies will soon be possible with
larger ground-based telescopes and the Space Telescope.

REFERENCES

Appenzeller, I. and Wolf, B.: 1981, in "The Most Massive Stars," ed.
   D'Odorico, Baade, and Kjar (Garching: European Southern Obs.), p. 131.
Baade, W. and Swope, H.H.: 1963, Astron. J. 68, p. 435.
Berkhuijsen, E.M.: 1982, Astron. Astrophys. 112, p. 369.
Blanco, V.M., McCarthy, M.F. and Blanco, B.M.: 1980, Astrophys. J. 242,
   p. 938.
Brunish, W.M. and Truran, J.W.: 1982a, Astrophys. J. 256, p. 247.
Brunish, W.M. and Truran, J.W.: 1982b, Astrophys. J. Suppl. 49, p. 447.
Buscombe, W.: 1977, "MK Spectral Classifications" (Northwestern Univ.,
   Evanston).
Buscombe, W.: 1980, "MK Spectral Classifications" (Northwestern Univ.,
   Evanston).
Buscombe, W.: 1981, "MK Spectral Classifications" (Northwestern Univ.,
   Evanston).
Chiosi, C., Nasi, E. and Sreenivasan, S.R.: 1978, Astron. Astrophys. 63,
   p. 103.
Christian, C.A. and Tully, R.B.: 1983, Astron. J. 88, p. 934.
Chu, Y.H. and Wolfire, M.: 1983, Bull. Am. Astron. Soc. 15, p. 644.
Conti, P.S.: 1975, Mém. Soc. Roy. Sci. Liège, 6^3 serie, tome 1K, p. 193.
Davidson, K.: 1983, private communication.
Davidson, K. and Kinman, T.D.: 1982, Publ. Astr. Soc. Pac. 94, p. 634.
Davidson, K., Walborn, N.R. and Gull, T.R.: 1982, Astrophys. J. Letters
   254, p. L47.
de Jager, C.: 1980, "The Brightest Stars" (Reidel, Dordrecht), pp. 11-14.
Elias, J.H., Frogel, J.A. and Humphreys, R.M.: 1984, Astrophys. J., in
   press.
Flower, P.J.: 1977, Astron. Astrophys. 54, p. 31.
Garmany, C.D., Conti, P.S. and Chiosi, C.: 1982, Astrophys. J. 263, p.
   777.
Hellings, P. and Vanbeveran, D.: 1981, Astron. Astrophys. 95, p. 14.
Hoessel, J.G. and Danielson, G.E.: 1983, Astrophys. J., in press.
Hoessel, J.G. and Mould, J.R.: 1982, Astrophys. J. 254, p. 38.
Hoessel, J.G., Schommer, R.A. and Danielson, G.E.: 1983, Astrophys. J.,
   in press.
Humphreys, R.M.: 1978, Astrophys. J. Suppl. 38, p. 389.
Humphreys, R.M.: 1979a, Astrophys. J. 231, p. 384.
Humphreys, R.M.: 1979b, Astrophys. J. 234, p. 854.
Humphreys, R.M.: 1980a, Astrophys. J. 238, p. 65.
Humphreys, R.M.: 1980b, Astrophys. J. 241, p. 587.
Humphreys, R.M.: 1983a, Astrophys. J. 265, p. 176.
Humphreys, R.M.: 1983b, Astrophys. J. 269, p. 335.
Humphreys, R.M.: 1984, in "Structure and Evolution of the Magellanic
   Clouds," IAU Symposium No. 108 (Reidel, Dordrecht), p. 145.

Humphreys, R.M. and Davidson, K.: 1979, Astrophys. J. 232, p. 409.
Humphreys, R.M. and Sandage, A.: 1980, Astrophys. J. Suppl. 44, p. 319.
Humphreys, R.M., Blaha, C., D'Odorico, S., Gull, T.R. and Benvenuti, P.:
    1984, Astrophys. J., in press.
Hutchings, J.B.: 1980, Astrophys. J. 237, p. 285.
Hutchings, J.B.: 1982, Astrophys. J. 255, p. 70.
Innis, R.T.A.: 1927, "Catalogue of Double Stars."
Kennedy, P.M. and Buscombe, W.: 1974, "MK Spectral Classifications"
    (Northwestern Univ., Evanston).
Kudritzki, R.P.: 1981, in "The Most Massive Stars," ed. D'Odorico, Baade,
    and Kjar (Garching: European Southern Obs.), p. 49.
Lequeux, J.: 1979a, Astron. Astrophys. 71, p. 1.
Lequeux, J.: 1979b, Astron. Astrophys. 80, p. 35.
Lequeux, J., Martin, N., Prevot, L., Prevot-Burnichon, M.L., Rebeirot,
    E. and Rousseau, J.: 1980, Astron. Astrophys. 85, p. 305.
Maeder, A.: 1980, Astron. Astrophys. 92, p. 101.
Maeder, A.: 1981a, Astron. Astrophys. 99, p. 97.
Maeder, A.: 1981b, Astron. Astrophys. 102, p. 401.
Maeder, A.: 1981c, in "The Most Massive Stars," ed. D'Odorico, Baade and
    Kjar (Garching: European Southern Obs.), p. 173.
Maeder, A.: 1983, Astron. Astrophys. 120, p. 113.
Massey, P. and Hutchings, J.B.: 1983, Astrophys. J., in press.
Miller, G.E. and Scalo, J.M.: 1979, Astrophys. J. Suppl. 41, p. 533.
Pagel, B.E.S. and Edmonds, M.G.: 1981, Ann. Rev. Astron. Astrophys.
    (Ann. Reviews. Palo Alto), 19, p. 77.
Ringuelet, A.E.: 1958, Z.f. Astrophys. 46, p. 276.
Sandage, A. and Carlson, G.: 1982, Astrophys. J. 258,p. 439.
Sandage, A. and Tammann, G.A.: 1974, Astrophys. J. 191, p. 603.
Sandage, A. and Tammann, G.A.: 1982, Astrophys. J. 256, p. 339.
Savage, B.D., Fitzpatrick, E.E., Casinelli, J.P. and Ebbets, D.C.: 1983,
    Astrophys. J., in press.
Shore, S.N. and Sanduleak, N.: 1984, Astrophys. J., in press.
Stahl, O., Wolf, B., Klare, G., Cassatella, A., Krautter, J., Persi, P.
    and Ferrari-Toniolo, M.: 1983, Astron. Astrophys., in press.
Stothers, R. and Chin, C.-W.: 1983, Astrophys. J. 264, p. 583.
Vangioni-Flam, E., Lequeux, J., Maucherat-Joubert, M. and Rocca-
    Volmeranga, B.: 1980, Astron. Astrophys. 90, p. 73.
Walborn, N.R.: 1973, Astrophys. J. 182, p. 121.
Walborn, N.R.: 1978, Astrophys. J. Letters 224, p. L133.
Walborn, N.R.: 1982, Astrophys. J. 256, p. 452.
Walborn, N.R.: 1983, in "Workshop on MK Spectral Classification:
    Criteria and Applications," University of Toronto.
Walborn, N.R., Blanco, B.M. and Thackeray, A.D.: 1978, Astrophys. J.
    219, p. 498.
Weigelt, G.: 1983, private communication.
Worley, C.E.: 1983, unpublished.

DISCUSSION

Alcaino: Following the study of the very luminous galaxies M33 and M101,
Sandage has recently suggested a dependence of the absolute visual
magnitude of the brightest red giants with the luminosity of the parent
galaxy from a mean of $M_V \doteq -8$ up to $M_V = -9$, hence jeopardizing the use
of these stars as useful distance indicators. Perhaps you would like to
place a comment whether you believe this to be a true effect, or just
an overestimation of the distance to these galaxies, due to an under-
estimation of absorption?

Humphreys: I think that Sandage has probably overestimated the distance
to M33 because of reddening of the Cepheids. For M33 I have used
Madore's preliminary distance modulus of 24.6-24.7 mag. from infrared
observations of the Cepheids. The difference with Sandage of 0.6-0.7 mag
is probably due to reddening. Our very recent JHK photometry of the
known M supergiants in M33 shows that they are definitely reddened, some
by a magnitude in A_V. For M101 Sandage and Tammann had not found any
M supergiants down to $m_V \sim 21$ mag. Strom and I (1983) subsequently
identified candidate M supergiants in the arms of M101. The brightest
are $V \sim 20.7$ mag. Sandage then re-examined his plates and also found
candidate M supergiants even brighter than those found by Strom and
myself. On the basis of these stars he derived an M_V of $\simeq -9$ mag. for
the brightest red stars in M101 with his former distance modulus of
29.3 mag. Strom and I used the luminosity calibration of the M supergiants from the local group galaxies ($M_V \simeq -8$ mag.) to derive a somewhat
closer distance. Aaronson, Strom, Copps, Lebofsky and I have obtained
JHK photometry of six of Sandage's 7 brightest red candidates. Four are
foreground M dwarfs, one is definitely an M supergiant and one is uncertain. The M supergiant has $V \simeq 20.4$ mag. comparable to the brightest
Humphreys and Strom M supergiants. The luminosities of these stars depend on the distance to M101 which in my opinion is not known yet.

Frogel: Is number density of supergiants in HR diagram consistent with
evolutionary rates? Would a model maker care to comment on disagreement?

Humphreys: The observed HR diagrams for the massive stars in the Milky
Way, LMC and SMC all show evidence for a broader main sequence than
predicted by the models.

Maeder: As frequently mentioned there are too many stars outside the
main-sequence band. I shall come back to this topic in my talk. However,
note that it was first necessary to make models to see whether standard
evolution fits or not.

de Groot: In view of the difficulties of determining accurate parameters
for the hottest stars, do you consider it possible that the upper
luminosity boundary is a straight line sloping slightly across the top

of the HRD, instead of a line which is sloping for the hotter stars and horizontal for the cooler stars? Would such a straight line make it easier for the theoreticians to tell us why it is there?

Humphreys: I think the bolometric corrections are fairly well known now even for the very hottest stars thanks to the work of Kudritzki and his associates. For the coolest supergiants the bolometric corrections are now well determined from infrared photometry. Consequently I think the upper envelope of stellar luminosities is nearly correct.

Westerlund: I was surprised to hear yesterday and again today that observers believe that all O stars inside 3.5 kpc from the Sun are known now. I believe that there may still be a fair number not discovered because of heavy absorption.

Humphreys: The total data set of O-type stars, supergiants and less luminous B stars is complete to 3 kpc. The O-type stars alone may only be complete to a smaller distance ($\approx$ 2.5 kpc, Garmany et al, 1982). Of course the distribution of all of these stars is highly irregular and is determined by the local spiral structure. Consequently, any of these completeness estimates are somewhat misleading.

EVOLUTION OF MASSIVE STARS, SUPERGIANTS AND WOLF-RAYET STARS

André Maeder
Geneva Observatory
Switzerland

1. INTRODUCTION

Massive stars (say $M \gtrsim 9\ M_\odot$) play a major role in Astrophysics. They are the main agents of nucleosynthesis and galactic enrichment and present many interesting properties, like high stellar winds, chemical surface enhancements, instabilities etc. which deserve further investigation. Furthermore, massive stars also are the powering sources of giant H II regions and have important radiative, mechanical and chemical interactions with the interstellar medium.

2. SOME BASIC OBSERVATIONAL CONSTRAINTS

Let us firstly review some of the basic observational facts, which put some severe constraints on the evolutionary models and need to be accounted for.

a) The HR diagram and basic parameters

First of all, we have to mention the properties of the upper luminosity boundary of supergiants in the HR diagrams by Humphreys and Davidson (1979). 1) Culminating at $M_{bol} = -12$ for O3 stars, the envelope of O-stars shows a decreasing luminosity for later spectral types. 2) From late B- to M-type supergiants, the envelope is rather flat. The consequence is that the brightest O-stars are about 2 magn. (in M_{bol}) more luminous than the brightest red supergiants. A rather similar behaviour is exhibited by supergiants in the Galaxy, the LMC and SMC.

The distribution of massive stars shows two other points of interest. 3) When comparing their models of constant mass evolution with the observations, Lamb et al (1976), Stothers and Chin (1976) found that the observed ratio of red to blue supergiants in the solar neighbourhood is too high by a factor of 4 with respect to model predictions.

A. Maeder and A. Renzini (eds.), Observational Tests of the Stellar Evolution Theory, 299–319.
© *1984 by the IAU.*

Thus, Lamb et al (see also Brunish and Truran, 1982 ab) concluded that most red supergiants (RSG) arise from the evolution of lower mass stars, a conclusion which was not supported by Maeder (1981 b, see also § 3a below). 4) There is too large a number of late B- and A-type supergiants (cf. Stothers and Chin, 1977; Chiosi et al 1978). On the basis of a large sample of well studied young clusters in the Galaxy and the Magellanic Clouds, Meylan and Maeder (1982) found a substantial MS widening up to spectral type A, i.e. much larger than the one present in older clusters (cf. Maeder, Mermilliod, 1981). This large MS widening for massive stars remains a very hard constraint to satisfy, and its solution may have further implications on stellar evolution.

b) Statistics of massive stars at different galactic locations

The models must evidently fit the basic stellar parameters as given by the various available calibrations. In addition, they must account for the number ratios of stars of different types. The problem is, however, complicated by the fact that many of these number ratios are strongly changing with galactic location. This is certainly a major and stimulating challenge which we have to face.

As shown by van den Bergh (1968), Humphreys (1978), Humphreys and Davidson (1979), the number ratio of red to blue supergiants varies with galactocentric location, in the sense that there are relatively more RSG towards the anticenter. The Wolf-Rayet (WR) stars also have a strongly non uniform distribution, being much more numerous in galactic interior regions (cf. Smith, 1973; see also Hidayat et al, 1981). As shown by Maeder et al (1980), the number ratio of RSG to WR stars has the strongest galactic variations, being much larger in external galactic zones. This steep galactic gradient was interpreted in terms of the effect of metallicity on mass loss rates.

Table 1 from Lequeux (1983) shows the variations of different number ratios with galactic locations. We notice that, for the range of metallicities considered, the ratio of red to blue supergiants spans a range of 5, the ratio of RSG to WR stars a range of 45, while the last column presents a variation smaller than a factor of 2. These results confirm the earlier findings by Maeder et al (1980) who emphasized that the RSG/WR number ratio may be a very useful indicator of metallicity and chemical gradient in external spiral and irregular galaxies. As noted by Humphreys, Davidson (1979) and Bertelli, Chiosi (1981), the steepness of the gradients also depends on the adopted luminosity limit. It is however verified (cf. Meylan, Maeder, 1983) that for the limits M_{bol} = -6 and -7.5 (which rather correspond to the lower luminosity of WR stars) the results are essentially similar.

Table I

Galactic gradients of luminous red and yellow-to-blue stars and
red to Wolf-Rayet stars, and data for other galaxies. The
metallicity Z is indicated. The galactocentric distance of the
Sun is taken as 10 kpc., from Meylan and Maeder (1983) for the
Galaxy and Lequeux (1983) for the other data.
The figures refer to $M_{bol} < -6$.

Galaxy	$Z^{(1)}$	$\dfrac{N(KM)}{N(OBAFG)}$	$\dfrac{N(KM)}{N(WR)}$	$\dfrac{N(KM)+N(WR)}{N(OBAFG)}$
The Galaxy, 7.5-9kpc	0.03	0.04	0.53	0.11
The Galaxy, 9-11kpc	0.02	0.07	1.4	0.12
M33	$0.02^{(2)}$	0.07	< 3	> 0.09
The Galaxy, 11-12.5kpc	0.014	0.17	$13.5^{(3)}$	0.19
LMC	0.01	0.12	8	0.14
SMC	0.003	0.20	24	0.21

(1) A homogeneous scale referring to all objects has still to
    be discussed; the present scale indicates only a trend
(2) Ignoring the small metallicity gradient in this galaxy
(3) Very uncertain: only 2 WRs in this region

In the case of WR stars, the frequency of WR binaries and the ratio of
WN to WC stars also change with galactocentric distance (cf. Hidayat
et al, 1983), as does the distribution of the WR subtypes.

c) Mass loss by stellar winds

This is a general property of all luminous stars. No unique para-
metrization of mass loss rates exists for stars over the whole HR
diagram. Generally, the mass loss rates $\dot{M}$ increase with the luminosity
(except for WR stars?), with a rather large scatter around the mean
relation. Different estimates and parametrizations have been performed
for OB stars, BAF supergiants, red supergiants and WR stars (cf.
Garmany et al, 1981; Barlow and Cohen, 1977; Reimers, 1975, Bernat, 1979;
Barlow et al, 1981; Conti, 1982; Bieging et al, 1982).

Such observed rates or some fitted parametrizations have been used
by most authors of computations with mass loss, as well for massive
stars (cf. § 3) as for asymptotic branch giants.

d) The case of WR stars: a flavour of controversy

The general properties of WR stars have been reviewed in the
recent IAU Symposium 99 on WR stars (Ed. De Loore and Willis, 1982; see
also van der Hucht, 1981; Conti, this meeting). The WR stars, which are
supposed to be bare stellar cores, represent a strong constraint over
the whole evolutionary scheme of massive stars.

In particular, mechanisms such as mass loss, binary transfer, mixing etc. appear necessary for the formation of WR stars.

The scenario of massive O-stars evolving to WR stars as peeled off stars in the advanced stages has repeatedly been criticized by Underhill (e.g. 1983 ab) on the basic reason that the observed mass loss rate for WR stars "requires that much more momentum be deposited in the wind of a WR star than is available from the radiative field of a star". It is certainly true that the kinetic power of the winds of WR stars represents a sizeable fraction (2 - 10%) of the stellar luminosity (cf. Bieging et al, 1982) and it seems difficult to explain such a strong wind by the radiation field only. A possible way to solve this problem might well come from the finding (cf. § 5b) that most WR stars are vibrationally unstable. It is, however, important to reassert that the peeling-off which leads to the formation of WR stars is not due to the wind of WR stars, but rather to the winds acting during the previous evolutionary stages. These winds are not subject to controversy and lead the initially most massive stars to the status of bare cores.

The contention that the absence or weakness of H-lines in WR stars is due to ionization effects was not supported by the recent analysis by Willis (1982) and Conti et al (1983). In particular, there is no one-to-one relation between H/He ratios and WN subtypes which, as pointed out by Willis (1982) among other arguments, rules out ionization effects as the cause of the lack of observed H-lines. Moreover, the models of magnetically supported loops (cf. Underhill, 1983 ab), which are not mathematically formulated, do not seem to have accounted for some of the basic properties of WR stars, such as their overluminosity with respect to their actual mass. Also, the question arizes how these extended magnetic loops could survive in close binary WR stars.

Conversely, the model of bare core for WR stars well accounts for their observational properties, such as the range of masses and luminosity, where these stars occur (cf. Conti, 1982; Massey, 1982); their T_{eff} has been interpreted by de Loore et al (1982) in terms of extended atmosphere. This model also accounts for the overluminosity of WR stars with respect to their actual mass (cf. Maeder, 1983 ab) and for the observed chemical enhancements (cf. Smith, Willis, 1982; Willis, 1982). Their number statistics is also in agreement with the above picture (Maeder, 1982). Besides, the high mass loss rates for WR stars (in the order of $2 \cdot 10^{-5}$ $M_\odot y^{-1}$ and not 10^{-7} $M_\odot y^{-1}$ as suggested by Underhill, 1983 b) are in better agreement with the mass present (cf. de Jager, 1980) in the gaseous shells surrounding some WR stars.

Thus, it may be advisable to keep to the scenarios of massive stars evolving to WR stars, as the result of mass loss and possibly of some mixing processes.

For the purpose of completeness, let us also mention the constraints put by chemical abundances at the stellar surface (cf. § 4) as well as the potential information which could be given by a study of the instabilities in supergiants.

3. RESULTS FROM EVOLUTIONARY MODELS WITH MASS LOSS

a) Results in the HR diagram

The effects of mass loss in the HR diagram have been discussed intensely by several authors (cf. Chiosi et al, 1979; de Loore et al, 1978; Stothers, 1979; Maeder, 1980, 1981 ab, 1983 a; de Loore, de Greve, 1981; Chiosi, 1981; Falk and Mitalas, 1981, 1983; Sreenivasan and Wilson, 1982; Brunish and Truran, 1982 ab). The details of the effects of mass loss on the MS evolution have been discussed by most of the above authors and we only briefly mention these effects here: decrease of luminosity but the star is overluminous for its actual mass; decrease of the core mass, but the core mass fraction is larger; increase of the MS lifetime; MS widening for low and intermediate mass loss rates, MS narrowing for very large rates; decline of semi-convection; possible change of surface abundances.

Turning to post-MS evolution let us recall that in the case of constant mass evolution, the massive stars leave the main-sequence and end their life, whatever may be the exact nature of the core collapse, as red supergiants (cf. Lamb et al, 1976). When the standard treatment of convection with Schwarzschild's criterion is made, massive stars in the range of 15 to 50 $M_\odot$ burn helium as B-supergiants. This location in the blue is due to a large fully convective zone which homogenizes the intermediate stellar layers (effect nb. 3 below) and thus keeps the star to the blue (cf. Stothers and Chin, 1976; Lamb et al, 1976). Thus stars with constant mass spend no time or only a negligible time in the RSG stage. This is why the above authors concluded that there were too many observed RSG's with respect to what had been predicted.

In order to properly understand the evolutionary tracks of massive stars with mass loss, we must realize that four main competitive effects determine the bluewards and redwards motions in the HRD (cf. Stothers and Chin, 1979): 1) The classical envelope expansion which responds to core contraction (cf. Renzini, this meeting). 2) The increasing size of the helium core favours bluewards evolution. 3) The homogenization of the envelope by a large FCZ tends to maintain a blue location in the HRD. 4) Large luminosity/mass ratios favour the expansion of the envelope. This last effect makes the He-burning zone move to lower T_{eff} for very massive stars (M $\gtrsim$ 40 $M_\odot$; see Fig. 1).

The major result about the post-MS evolution of massive stars that has emerged from recent studies (cf. Maeder, 1981 ab; Chiosi, 1981) is that according to initial mass and mass loss rates, the stars may go through very different evolutionary sequences, as a result of the combination of the four above-mentioned effects. Three different evolutionary sequences can be distinguished (cf. Fig. 1):

For $M \gtrsim 60\ M_\odot$

O star - Of - BSG and Hubble-Sandage variables - WN - WC - (WO) - SN

For $25\ M_\odot \lesssim M \lesssim 60\ M_\odot$

O star - BSG - yellow and RSG - (BSG) - WN - (WC) - SN

For $M \lesssim 25\ M_\odot$

O star - (BSG) - RSG - yellow supergiant and Cepheid - RSG - SN

Overshooting or mixing will probably not affect this general scheme, but such processes might contribute to an increase of the indicated mass limits (like an increase of $\dot{M}$ would do, due to the effect nb. 2 above).

Let us now comment these three kinds of evolution.

1. For mass loss at the currently observed rates, stars with an initial mass above 50 - 60 $M_\odot$ never reach the red supergiant (RSG) stage. For these stars, the peeling off by the winds during the main-sequence and the blue supergiant (BSG) phases is high enough to remove all the outer stellar layers. Zones which initially were in the core are revealed at the stellar surface. These stars, especially the extremely massive ones, remain quasi-homogeneous (cf. Maeder, 1980) and always stay at the blue side of the HR diagram, due to effect nb. 2 above. They firstly are BSG and Hubble-Sandage variables before becoming WR stars, probably of late WN types.

2. For initial stellar masses between about 50 and 25 $M_\odot$, mass loss at the observed rates on the main-sequence and in the BSG phase is not sufficient to remove all the outer envelope and the star rapidly reaches the RSG stage. Mass loss, even small, reduces the extension of the FCZ (effect nb. 3) and thus favours the redwards motion. This allows the stars to reach the RSG stage early enough during the He-burning phase and to experience lifetimes in the RSG longer by a factor 4 to 5 with respect to the case of constant mass, thus bringing about agreement with the observations (see Fig. 6 in Maeder, 1981 b). Then the high stellar winds in the RSG stage progressively remove the outer envelope.

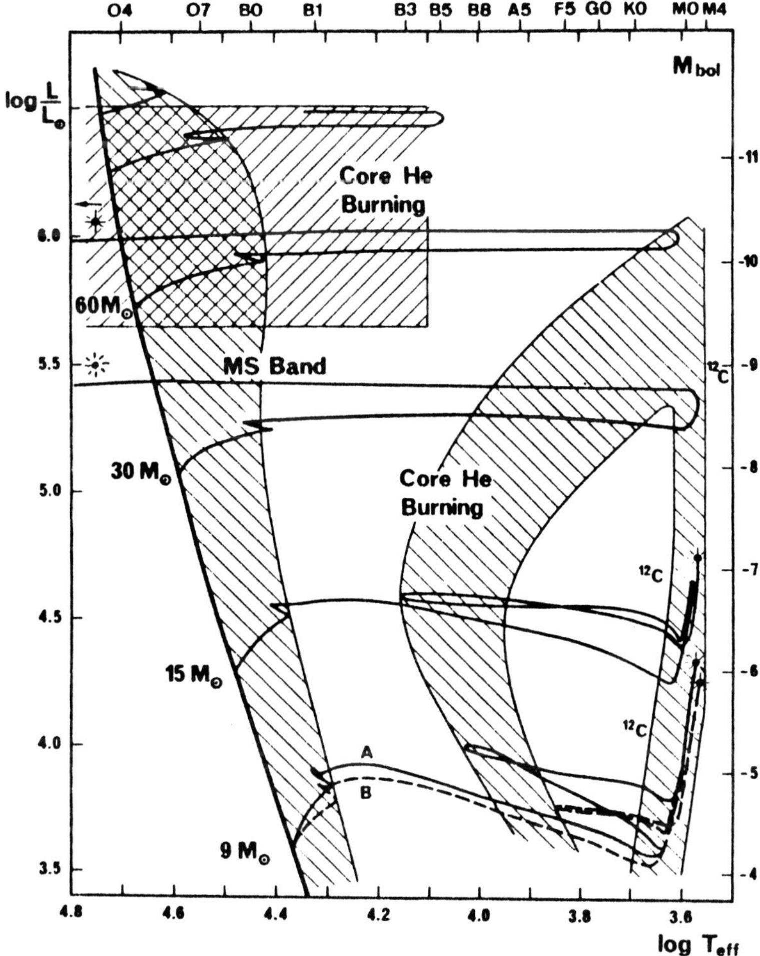

Fig. 1
*The HR diagram for evolution with mass loss up to
the end of central carbon burning (models B, 30 M_☉;
case C, Maeder, 1981 b; 1983 a, for 120 M_☉).
One well notices the 3 different cases of evolution
for massive stars. Case 1 is illustrated by the 2
higher models of 120 and 170 M_☉, case 2 by the
models of 60 and 30 M_☉, case 3 by 15 and 9 M_☉.*

When the mass fraction of the helium burning core, as a result of the
decreasing total stellar mass, becomes larger than some critical value
q_c (effect nb. 2), the star moves back to the blue in the HR diagram
and then may become a WR star (cf. Chiosi et al, 1978; Maeder, 1981 b,
q_c = 0.67 for 60 M_☉, 0.77 for 30 M_☉).

Thus, there may exist WR stars which are in a post-RSG phase. Clearly, high mass loss rates during this RSG phase and during previous phases favour the formation of WR stars, since then the critical q_c is reached earlier.

Thus, we have the complex situation that mass loss (at the observed rates) during the MS favours the formation of RSG in the range of masses 25 and 50 $M_\odot$, while a heavy mass loss in the red phase reduces the lifetime as RSG, since the star is then turned into a WR object. However, in all relevant cases the lifetime in the RSG stage remains much longer than in the case of constant mass evolution (cf. Maeder, 1981 ab).

The formation of WR star as post-RSG has been studied by Chiosi et al (1979), and Maeder (1981 ab) who examined the conditions on initial masses and mass loss rates $\dot{M}$ for this kind of evolution to occur. Two main facts are worth emphasizing. a) The fraction of stellar lifetimes spent as WR stars strongly increases with initial mass and $\dot{M}$-rates in the previous stages. Since the total duration t_{He} of the helium burning phase does not change significantly, a substantial increase of the WR phase implies a corresponding reduction of the RSG phase, as $t_{He} \simeq t_{RSG} + t_{WR}$. b) The luminosity limit above which WR stars appear is very sensitive to the mass loss in previous stages. The combination of these two facts leads to a very strong increase of the expected number of WR stars with mass loss rates. As an example, it was found that an increase of $\dot{M}$ by a factor of 2 during MS and RSG phases (which is well inside the observed scatter) gives a number of WR stars larger by a factor 17. Thus, it is clear that the expected number ratio N_{WR}/N_{RSG} is extremely sensitive to change of mass loss rates and accordingly there is thus no need of strong variations of $\dot{M}$ with metallicity in order to produce large changes in the frequencies of RSG and WR stars.

3. For initial masses below about 25 $M_\odot$, the mass loss both in the blue and in the red is never large enough to remove the outer layers. After the main-sequence, the star normally becomes a RSG, then it undergoes blue loops during which a Cepheid phase may occur; later the star again becomes a RSG. We can notice that the blue extension of the loops is significantly reduced by mass loss (cf. Maeder, 1981 b); in case of high mass loss, the loops may even be suppressed.

Different authors have presented different pictures for the advanced evolution. In particular, there are sizeable differences between the above results and those by Brunish and Truran (1982 ab). For example, these authors predict that stars with initial masses smaller or equal to 30 $M_\odot$ spend less than 1% of their total lifetime in the RSG stage. Indeed, the absence of RSG from 15 - 25 $M_\odot$ progenitors predicted by Brunish and Truran is not confirmed by the observations of red supergiants in young clusters and associations (cf. Humphreys, 1978;

Humphreys and Davidson, 1979). These observations show that the major fraction of red supergiants originates from stars initially less massive than 30 $M_\odot$. The predicted absence of RSG probably results from too small mass loss rates, so that their tracks are close to those of constant mass evolution.

The models with mass loss have been compared to the observations in several studies (cf. Chiosi et al, 1978, 1979; Maeder, 1981 ab; Chiosi, 1951; Meylan, Maeder, 1982). Most observed properties, such as the shape of the cluster sequences, the formation and expected number of RSG and WR stars seem to be accounted for by the models, as well as the surface abundances in Hubble-Sandage variables and in WR stars (cf. § 4). The most accurate comparisons in the HR diagram are those made with M_v vs. $(U-B)_o$ sequences of well analysed clusters (e.g. Mermilliod, Maeder, this meeting). However, there seems to remain a stone in the garden of the model makers, it is the so-called MS widening which will be further discussed in § 5.

  b) Some noticeable points about internal structure and
     further evolution

It is quite noteworthy that, for a wide range of mass loss rates, the He-phase ends with almost similar He + C/O core (cf. Maeder, Lequeux, 1982; Brunish, Truran, 1982 ab). For example, a model with an initial mass of 60 $M_\odot$ and which is left due to mass loss with only 25 $M_\odot$ at the end of the C-burning phase has exactly the same central conditions and the same He + C/O core as its constant mass counterpart. The comparison of the distribution of chemical elements, of the runs of temperature and density shows identical results in both cases (the physical reasons are discussed in the above ref.).

Thus, as well as a red giant may be considered as a white dwarf surrounded by an extended envelope, as much can a red supergiant be considered a WR star surrounded by a very large envelope. In the case of the white dwarf the He + C/O core is degenerate and smaller than the Chandrasekhar mass, while the WR star is not degenerate and has a larger mass.

This absence of sensitivity of central conditions to mass loss is the general rule as long as mass loss does not lead to a significant reduction in the mass of the He + C/O convective core; such a reduction occurs only when the evolution goes far enough through the WC stage, which seems to be the case mainly for stars with very large initial masses (say $M \gtrsim 50\ M_\odot$).

Thus, as a whole we can say that mass loss at the currently observed rates produces large effects in the HR diagram, while <u>the course of the evolution of the central conditions is generally little affected.</u>

This implies that most WR stars end their life as supernovae (cf. Maeder, Lequeux, 1982). Thus, from Fig. 1 we can note that for stars initially more massive than 25 $M_\odot$ the progenitors of supernovae generally are WR stars, while below this mass limit the progenitors are RSG stars with extended envelope likely giving rise to type II supernovae.

4. EVOLUTIONARY CHANGES OF CHEMICAL ABUNDANCES AT THE STELLAR SURFACES.

Changes of surface abundances also offer a very strong test of stellar evolution. Comparisons bearing on isotopic ratios and elemental abundances have been mainly performed for red giants and supergiants (cf. Lambert, this meeting). For hot stars, the analyses are very difficult and thus less numerous, despite their great interest. In particular, Wolf-Rayet stars offer us the most valuable possibility to directly observe stellar nucleosynthesis in the products of some nuclear reactions revealed at stellar surfaces as a result of mass loss and maybe of some mixing processes. The nucleosynthetic products visible in WR stars, unlike in red giants, are not highly diluted within a large envelope. [In WN stars one observes the results of CNO processing and in WC stars those of partial helium burning.] The WC stars are the only kind of stars in which the products of the 3α reaction and other He-burning reactions prominently manifest themselves at the stellar surfaces. This provides a very useful comparison basis with nucleo-synthetic predictions.

The variations of the abundances of various elements during the evolution of massive stars have been calculated recently by Noels and Gabriel (1981), Maeder (1983 ab) and Greggio (this meeting). These authors have obtained the changes of chemical abundances at stellar surface due to the fact that matter originally in the convective core is revealed by the removal of the outer layers by mass loss. At the appearance of CNO products, important changes occur. The elements ^{3}He, ^{15}N and ^{18}O disappear. As the CN-cycle rapidly reaches equilibrium, the C/N ratio almost abruptly changes from about 4.1 to 0.03 (in mass). The O/N ratio slowly changes from 9.1 to less than 0.1 because the ON-cycle takes a longer time to reach equilibrium. The abundance of ^{13}C usually keeps a factor of 3.3 below that of ^{12}C. It is to be noted that the isotopes ^{16}O, ^{17}O and ^{18}O have very different behaviours (cf. Fig. 2).

A very large discontinuity marks the appearance of the various products of He-burning at the surface (cf. Maeder, 1983 a); this is likely to correspond to the beginning of the WC stage. The physical reason for this discontinuity is that the convective core in the He-burning phase does not continually decrease (which would leave a smooth transition at the border), but the core increases in mass during most of its evolution and a chemical discontinuity is thus built at the border of the core at its largest extension.

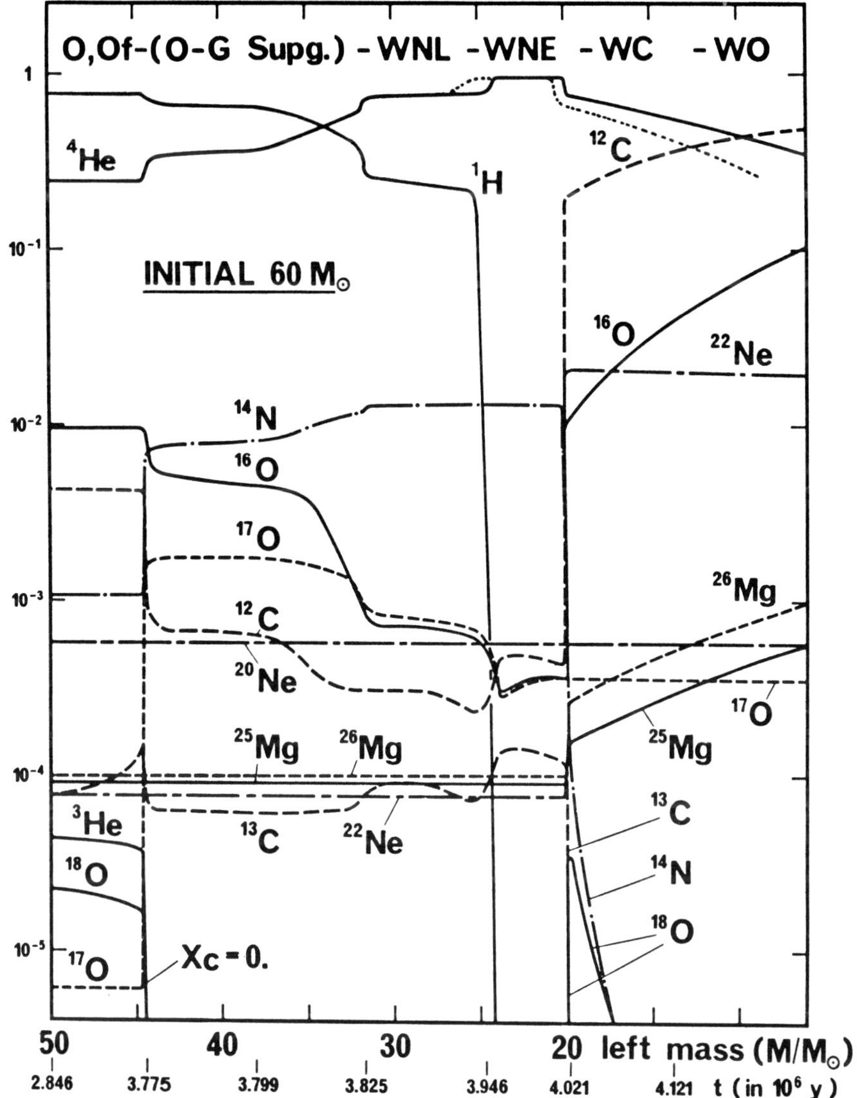

Fig. 2

Changes of the surface abundances (in mass fraction) in terms of the remaining mass for the model with initial mass of 60 $M_☉$. The ages are also indicated on the lower axis. At the top of the figure the corresponding evolutionary status is given. The dotted line for ^{4}He in the WN and WC stage corresponds to a lower mass loss rate in the WR stage. $X_c = 0$ indicates that the central hydrogen content is zero; it is just a coincidence that the central H-depletion occurs for this model just when CNO-products appear at the surface.

We note then a very steep disappearance of ^{13}C and ^{14}N, a very temporary peak of ^{18}O and above all a vertiginous rise by more than 2 orders of magnitude of ^{12}C, ^{16}O and ^{22}Ne. The abundances of ^{25}Mg and ^{26}Mg also rise strongly, particularly in the most massive WC stars, where s-elements are therefore to be expected.

The most prominent cases of peculiar abundances occur for the so-called Hubble-Sandage variables and for the WR stars. Let us review the comparisons made in these two cases.

The evolutionary status of the Hubble-Sandage variables and in particular of η Carinae is an old problem (cf. Stothers, Chin, 1983). The location of these stars in the HR diagram well corresponds to that of post MS supergiants; however, the argument cannot be regarded as sufficient in itself. In the case of η Carinae a very decisive information is coming from the recent observations by Davidson et al (1982) who determine the following values for some of the outer condensations surrounding the central object: C/N < 0.05 and O/N < 0.15 or 0.5 (see also Walborn et al., this meeting, who also find X = .57, Y = .42). These ratios differ very much from the solar ones: C/N ≃ 4, O/N ≃ 9, and are in full agreement with the ratios obtained in the blue super-giant phase of an initial 120 M$_\odot$ star evolving with heavy mass loss (at the observed rates): C/N = 0.025, O/N = 0.17 - 0.9 (cf. Maeder, 1983 a). This provides a strong constraint on the evolutionary status of this star; it is a first necessary step in order to be able to understand the origin of their strong instabilities. Further observations of O/N and C/N ratios in other blue supergiants, particularly in clusters and associations, seem a very promising programme to be undertaken.

In the conspicuous case of Wolf-Rayet stars, observations of chemical abundances have been made and analysed recently by Smith and Willis (1982), by Willis (1982), by Nugis (1982) and by Conti et al (1983). Generally speaking, the observed abundances for WNL, WNE and WC and WO stars are consistent with a progression in the exposition of nuclearly processed material from the CNO-cycles and then from He-burning reactions. The C/He, N/He and C/N ratios obtained by Smith and Willis or Nugis for WNL, WNE and WC stars have been compared in detail with the model predictions (Maeder, 1983 a). The general agreement between the observed ratios, which differ very much (10^2 for C/N) from the solar ratios, cannot be accidental, and it is a strong confirmation of the advanced evolutionary stage of WR stars as left over cores resulting from the peeling of massive stars by stellar winds. The models also suggest that the determinations of abundances for neon and magnesium in WC stars would be highly useful.

Finally, let us note that the connection between the anomalies of ^{12}C, ^{22}Ne, ^{25}Mg and ^{26}Mg in the galactic cosmic rays and the predicted excesses of these elements in WC stars has been examined (cf. Cassé, Paul, 1982; Maeder, 1983 b) with positive conclusions.

5. TOPICAL PROBLEMS AND PROGRESS NEEDED: MS WIDENING, FORMATION AND GALACTIC DISTRIBUTION OF WR STARS

Here we examine some problems arising in comparing models and observations and which potentially are lines for further progress in years to come.

a) MS widening

Standard models with mass loss do not account for this effect. The theoretically predicted gap following MS evolution extends in the HR diagram from log Teff = 4.38 to 4.15, which indeed corresponds to a narrow range of spectral type, i.e. from O9.5 to B2.5. Massive close binaries which, due to mass transfer, would be prevented from evolving to RSG could lie in this location. However, in view of their expected number (cf. Burki, Mayor, 1983) and of the large excess of stars outside the MS band (cf. Meylan, Maeder, 1982), the MS widening seems to be a real effect. Claims about the effect of a spread in the formation times of stars in a cluster have also been made (cf. Doom et al, this meeting). However, this should not affect diagrams such as those by Humphreys and Davidson (1979), which anyhow contain clusters of various ages. Moreover, it could perhaps be useful, before assigning different ages to the upper MS stars of a given cluster, to correct the colour magnitude diagrams for the well known effects of differential interstellar reddening, large rotational velocities and unsolved binaries.

Bressan et al (1981), Doom (1982 ab) have constructed models of massive stars with convective overshoot, a process which is known to extend the MS band (cf. Maeder, 1975). However, it is clear (cf. Bertelli et al, 1983) that this process is insufficient to explain alone the observed effect. More recently, Bertelli et al have shown that a suitable combination of mass loss, overshooting and opacity bumps is able to bring a satisfactory agreement. The central question relates obviously to the question of the existence or non-existence of the opacity bumps. Finally, we must note that other mixing processes may also be very efficient in massive stars, in particularly turbulent diffusion (cf. Maeder, 1982).

For the time being, the MS widening has received no satisfactory answer and it certainly remains a major line of research for the next years.

b) Formation of WR stars

This problem has been the object of several reviews in recent years (cf. IAU Symposium 99, Ed. de Loore and Willis, 1982). The question of the minimum value of the initial stellar mass for forming WR stars is a very important one in this context.

From number statistics, Firmani (1982) suggested a value of 20 $M_\odot$,
similar to that proposed by Chiosi (1981) and Maeder (1981 b, 1982).
Recently, Conti et al (1983) concluded from the analysis of the galactic
distributions of O and WR stars that most WR stars originate from stars
initially more massive than 40 $M_\odot$. If this limit is accepted as such, it
implies (cf. Conti et al) that there is a substantial excess of WR stars
with respect to theoretical predictions. More recently, Schild and
Maeder (1983), from the analysis of WR stars in clusters, supported a
value of 20 $M_\odot$, noting however that the majority of WR stars may
originate from stars more massive than 40 $M_\odot$.

Over the last 40 years WR stars have successively been associated
with almost all possible evolutionary stages (cf. bibliography by Simon
and Stothers, 1969). It is not the aim here to review all these possibilities. Four main and non exclusive scenarios which may fulfil the
above observational constraints seem to emerge nowadays (cf. IAU
Symposium 99):

> - Mass transfer in binaries
> - Mass loss in MS stars (Conti's scenario, 1975)
> - Mass loss in supergiants, particularly in
>   red supergiants.
> - Mixing by various processes.

In view of their surface compositions and frequencies, il is very
likely that most WR stars are in the He-burning stage, although some of
the massive WNL stars may still be near the end of the H-burning phase.

We shall not examine here the details of the various scenarios.
The case of WR binaries has been studied by the Brussel group (cf. de
Loore, de Greve, 1981). Recently, Hidayat et al (1983) have found 34%
of binary WR stars and have suggested that their real number could be
above 50%. The original Conti's scenario (cf. Conti, 1975; Chiosi et
al, 1979) must probably be modified so as to include during post MS
evolution an excursion through the stage of the Hubble-Sandage variables
with very high mass loss rates, before the star becomes a WR star (cf.
Maeder, 1983 a). Recently, Lamers et al (1983) have suggested that
P Cygni is on its way to the WR stage, which would be an illustration
of this case. The properties of WR stars, which are post-RSG, have been
examined above. Hidayat et al (1981) have suggested that WC 8.5 and
WC 9 stars could correspond to this case.

Internal mixing may also considerably favour the formation of WR
stars, by contributing to the homogeneisation of the star (which reduces
the ratio of central to average density and favours instability) and to
the appearance of the products of nuclear burning at the stellar surface.
Such models, with overshooting and turbulent diffusion, have been performed by Maeder (1982).

Recently Bertelli et al (1983), in the line of their proposed interpretation for the MS widening, have suggested that the combination of
mass loss, overshooting and opacity bumps plays a dominant role in
accounting for the properties of WR stars; here as for the MS widening,
this interpretation is subject to the reality of the opacity bumps.

For now, it is clear that many processes may separately or simultaneously favour the formation of WR stars, but their exact relative
importance is still rather uncertain.

A major question about WR stars is also that of the origin of the
very high wind of WR stars (cf. § 2). A recent analysis of non-adiabatic,
linear radial pulsations during the evolution of massive stars (cf.
Maeder, Lovy, in prep.) has shown that many of the models which are
turned blue and which consist of bare cores identifiable with WR stars,
are vibrationally unstable. The growth time of the instability is short,
values as small as 5 days have been found for periods of 30 minutes,
which implies a strong instability. Thus, it is tempting to·suggest,
and I do it, that the mechanical energy injected by the internal pulsations into the extended atmosphere may significantly contribute to
the large wind of WR stars. As the wind is optically thick, we may in
most cases not be able to directly observe the pulsations, which would
then manifest themselves mainly through the enhancement of the wind.

It is well known (cf. Ledoux, 1941) that the limit of vibrational
instability for homogeneous stars of mass M is given by $\bar{\mu}^2 M \stackrel{\sim}{=} $ const.
(cf. also Ziebarth, 1970), where μ is the mean molecular weight. Massive
stars may become, due to mass loss, quasi-homogeneous (Maeder, 1980,
1983 a). Thus as $\bar{\mu}$ increases due to nuclear processing, M must decrease
so as to satisfy the above expression. As the instability grows if
$\bar{\mu}^2$ M becomes too large, I suggest that <u>during their evolution WR stars
keep at the edge of the vibrational instability.</u> The instabilities found
in the above mentioned calculations support these views.

   c) Distribution of WR stars in galaxies

Many observations suggest a change of various properties of
massive stars and WR stars with galactocentric distance and from one
galaxy to another (cf. § 2b). As there are several processes contributing to the formation of WR stars, it was proposed that <u>the relative
importance of the channels, or more precisely the "traffic" through each
of the various channels leading O and Of to WR stars changes very much
with initial metallicity and thus galactic location</u> (Maeder, 1981).
In certain cases some channels are absent, while they may be very much
run through in other cases. For example, in the SMC only the binary
channel seems just capable of succeeding in WR star formation while the
possibilities of the other channels appear too small. The absence of
these other evolutionary channels also accounts for why the overall WR
frequency is so weak in the SMC.

Inversely, in the zone located in the direction of the galactic centre, the non-binary channels are strongly operating and this may explain why the overall WR frequency is so much higher there than in the SMC, and simultaneously why the number ratio (binary WR/all WR) is small there.

The dependence of the mass loss rates on the initial metallicity was considered to strongly influence the formation of WR stars as a result of MS or RSG mass loss and therefore to be responsible for the observed gradients of WR stars in the Galaxy (cf. Maeder et al, 1980; Maeder, 1981 c; Lequeux, 1983). In the case of OB stars, the theoretical models of the stellar winds by Abbott (1982) indicate an almost linear increase of $\dot{M}$ with initial metallicity. For RSG, Kwok (1980) has also suggested an increase of $\dot{M}$ with metallicity, although the dependence might be weaker than for OB stars. Thus, a higher Z involves a higher $\dot{M}$, and the models of § 3a show that this produces in turn an increase of the expected number of WR stars and a corresponding decline in the number of RSG. It was emphasized that even a very mild dependence of $\dot{M}$ on Z is sufficient to explain a dependence such as that shown in Table 1 due to the very high sensitivity of N_{RSG}/N_{WR} vs. $\dot{M}$.

An alternative interpretation was proposed by Bertelli and Chiosi (1981), Garmany et al (1982) who suggested that the galactic gradient of WR stars just reflects the galactic gradient of O-stars. Subsequently, Meylan and Maeder (1983) showed that the galactic gradient of WR stars is steeper than that of O-stars, and that the gradient of RSG is of the opposite sign; consequently, they cannot only be the reflect of the IMF. The more recent episode of this story is due to Bertelli et al (1983) who suggested that the dependence of the opacity bumps on metallicity plays the dominant role in accounting for the properties of WR stars, which still rests on the hypothetical existence of opacity bumps, which were not quite supported by A.N. Cox in this meeting. Finally, although their origin is still subject to discussions, let us emphasize that these various gradients of massive O-stars, of WR and RSG stars must have considerable effects on galactic properties, such as colours, distribution of HII regions, supernovae and nucleosynthetic yields, an important subject which has not been considered in this review.

REFERENCES

Abbott, D.C., 1982, Astrophys. J. 259, 282
Barlow, M.J., Cohen, M., 1977, Astrophys. J. 213, 737
Barlow, M.J., Smith, L.J., Willis, A.J., 1981, Monthly Notices
   Roy. Astron. Soc. 196, 101
van den Bergh, S., 1968, Journ. Royal Astron. Soc. Canada 62, no. 4
Bernat, A.P., 1977, Astrophys. J. 213, 756
Bertelli, P., Chiosi, C., 1981, in The Most Massive Stars,
   ESO Workshop, p. 211
Bertelli, G., Bressan, A.G., Chiosi, C., 1983, Astron. Astrophys.
   in press

Bieging, J.H., Abbott, D.C., Churchwell, E.B., 1982,
    Astrophys. J. 263, 207
Bressan, A.G., Bertelli, P., Chiosi, C., 1981, Astron. Astrophys. 102,25
Brunish, W.M., Truran, J.W., 1982a, Astrophys. J. 256, 247
Brunish, W.M., Truran, J.W., 1982b, Astrophys. J. suppl.ser. 49, 447
Burki, M., Mayor, M., 1983, Astron. Astrophys. 124, 256
Cassé, M., Paul, J.A., 1982, Astrophys. J. 258, 860
Chiosi, C., Nasi, E., Sreenivasan, S.R., 1978, Astron. Astrophys. 63,103
Chiosi, C., Nasi, E., Bertelli, G., 1979, Astron. Astrophys. 74, 62
Chiosi, C., 1981, in The Most Massive Stars, ESO Workshop p. 27
Conti, P.S., 1975, Mém. Soc. Roy. Sci. Liège, 6e série, tome IX, p. 193
Conti, P.S., 1982, in WR stars: observations, physics, evolution;
    IAU Symp. 99, eds. C. de Loore and A. Willis, p. 3
Conti, P.S., Leep, E.M., Perry, D.N., 1983, Astrophys. J. 268, 228
Conti, P.S., Garmany, C.D., de Loore, C., Vanbeveren, D., 1983,
    Astrophys. J. in press
Davidson, K., Walborn, N.R., Gull, T.R., 1982, Astrophys. J. 254, L47
Doom, C., 1982 ab, Astron. Astrophys. 116, 303, 308
Falk, H.J., Mitalas, R., 1981, Monthly Notices Roy. Astron. Soc. 196,225
Falk, H.J., Mitalas, R., 1983, Monthly Notices Roy. Astron. Soc. 202, 19
Firmani, C., 1982, in WR stars: observations, physics and evolution;
    IAU Symp. 99, eds. C. de Loore and A.J. Willis, p. 499
Garmany, C., Olson, G.L., Conti, P.S., 1981, Astrophys. J. 250, 660
Garmany, C.D., Conti, P.S., Chiosi, C., 1982, Astrophys. J. 263, 777
Hidayat, B., Supelli, K., van der Hucht, K.A., 1981, IAU Symp. 99,
    Eds. C. de Loore and A.J. Willis, p. 27
Hidayat, B., Admiranto, A.G., van der Hucht, K., 1983,
    Astrophysics and Space Science, in press
van der Hucht, 1981, in The Most Massive Stars, ESO Workshop p. 157
Humphreys, R.M., 1978, Astrophys. J. suppl. 38, 309
Humphreys, R.M., Davidson, M.K., 1979, Astrophys. J. 232, 409
de Jager, C., 1980, The Brightest Stars, ed. Reidel, Dordrecht
Kwok, S., 1980, Journ. Royal Astron. Soc. Canada 47, 216
Lamb, S.A., Iben, I., Howard, W.M., 1976, Astrophys. J. 207, 209
Ledoux, P., 1941, Astrophys. J. 94, 537
Lequeux, J., 1983, in Ecole de Physique, Les Houches, on Star Formation,
    in press
de Loore, C., de Greve, J.P., Vanbeveren, D., 1978,
    Astron. Astrophys. 67, 373
de Loore, C., de Greve, J.P., 1981, in The Most Massive Stars, ESO
    Workshop p. 85
de Loore, C., Hellings, P., Lamers, H., 1982, in Wolf-Rayet Stars,
    IAU Symp. 99, eds. C. de Loore and A.J. Willis, p. 53
de Loore, C., Willis, A., 1982, Wolf-Rayet Stars: observations, physics
    and evolution, IAU Symp. 99, ed. Reidel, Dordrecht
Maeder, A., 1980, Astron. Astrophys. 92, 101
Maeder, A., 1981 ab, Astron. Astrophys. 99, 97;   102, 401
Maeder, A., 1982, Astron. Astrophys. 105, 149

Maeder, A., 1983 ab, Astron. Astrophys. 120, 113; 120, 130
Maeder, A., Lequeux, J., Azzopardi, M., 1980, Astron. Astrophys. 90, L17
Maeder, A., Mermilliod, J.C., 1981, Astron. Astrophys. 93, 136
Maeder, A., Lequeux, J., 1982, Astron. Astrophys. 114, 409
Massey, P., 1981, in Wolf-Rayet Stars, IAU Symp. 99, eds. C. de Loore
    and A.J. Willis, p. 251
Meylan, G., Maeder, A., 1982, Astron. Astrophys. 108, 148
Meylan, G., Maeder, A., 1983, Astron. Astrophys. 124, 84
Noels, A., Gabriel, M., 1981, Astron. Astrophys. 101, 215
Nugis, T., 1982, in Wolf-Rayet Stars: observations, physics and
    evolution, IAU Symp. 99, eds. C. de Loore and A.J. Willis, 127, 131
Reimers, D., 1975, in Problems in Stellar Atmospheres and Envelopes,
    ed. B. Baschek et al, New York, Springer Verlag
Schild, R., Maeder, A., 1983, Astron. Astrophys., submitted
Simon, N.R., Stothers, R., 1969, Astrophys. J. 155, 247
Smith, L.J., Willis, A.J., 1982, Monthly Notices Roy. Astron. Soc. 201, 451
Smith, L., 1973, in Wolf-Rayet and High Temperature Stars, IAU Symp.
    no. 49, ed. Reidel, p. 15
Sreenivasan, S.R., Wilson, W.J.F., 1982, Astrophys. J. 254, 287
Stothers, R., Chin, C.W., 1976, Astrophys. J. 204, 472
Stothers, R., Chin, C.W., 1977, Astrophys. J. 211, 189
Stothers, R., Chin, C.W., 1979, Astrophys. J. 233, 267
Stothers, R., Chin, C.W., 1983, Astrophys. J. 264, 583
Underhill, A.B., 1983 ab, Astrophys. J. 265, 933; 266, 718
Willis, A.J., 1982, in Wolf-Rayet Stars: observations, physics and
    evolution, IAU Symp. 99, eds. C. de Loore and A.J. Willis, Reidel p.87
Ziebarth, K., 1970, Astrophys. J. 162, 947

DISCUSSION

Tayler: About 10 years ago a variety of authors (Appenzeller, Papaloizou, Talbot, Ziebarth) independently showed that vibrational instability would be stabilized at finite amplitude so that such stars would live longer than previously believed. Do you have any new ideas on this point?

Maeder: The point I emphasized is in fact different. I meant that even if the massive star is vibrationally stable during MS evolution, it may become vibrationally unstable when, due to mass loss by stellar winds, the star becomes an almost homogeneous bare He or He + C/O core with a real large mass. Now I think that massive bare cores are probably also limited to finite amplitudes by non linear effects as you mention, and this is why I suggested that they are evolving keeping at the edge of vibrational instability.

Kudritzki: What are the amplitudes of the theoretical WR-model vibrational instabilities? Could they be observed?

Maeder: The analysis we made is a non-adiabatic linear analysis, and therefore tells us the values of the periods and whether the models are unstable or not, but does not give the amplitude size. Some WR stars are known to be unstable but it is not clear what kind of instability we observe. However, it has to be noted that the wind of WR stars is thick and the optical depth $\tau \simeq 1$ from where the spectrum is supposed to come is still in the wind. This very strong wind might probably be the external manifestation of the internal instability.

Schatzman: Just a reminder. The instability is due to the kappa-mechanism or to the epsilon-mechanism?

Maeder: The vibrational instability I mentioned is due to the kappa-mechanism. The destabilizing role of radiation pressure is also important in this context.

Cox: I understand that you get much shorter growth rates now than were obtained by previous investigators. Is this because the surface material is no longer H + He, but now is carbon-rich in these Wolf-Rayet stars?

Maeder: Yes, a strong vibrational instability has been found, particularly for WC stars. For example, a 24 $M_\odot$ in the WC stage, i.e. a star with He + C + O envelope, shows a growth time of the pulsation amplitude of about 5 days, which is a very short time.

Bertelli: The observations seem to indicate that massive stars spend more than the core He-burning lifetime in the WR + RS phases. On the other hand only a small fraction of the central He-burning lifetime is available for models in the mass range 20 to 40 $M_\odot$ to appear as RS and, eventually, WR stars, if traditional models undergoing a moderate mass loss in the blue phase are considered. Can you comment on this point?

Maeder: It is quite clear that there is an excess of stars outside the MS band with respect to what is predicted by all published sets of stellar models. The question of an excess of WR stars is much more uncertain and it mainly remains on the lower mass limit assumed for their progenitors. If you take 40 $M_\odot$ for this limit, there is excess; for 20 $M_\odot$ there is none.

Humphreys: Although I believe that gradients in the B/R and WR/R ratios exist, I want to caution about the statistics especially for the red supergiants interior to the sun, e.g. 7-9 kpc galactocentric distance. Our knowledge of the M supergiants is very incomplete in this distance range, and in the 11-12 kpc range the numbers are highly influenced by one association, h and χ Per and the surrounding association. It could be just a perturbation.

<u>Cassé</u>: As far as nucleosynthesis is concerned I would like to point out
that WR stars, or more particularly WC stars at the end of their evolution, could be generous sources of s-process element (at least up to
mass 90, see e.g. Lamb et al, 1976). WC stars can be considered as open-
cast mines of ^{25}Mg, ^{26}Mg, ^{58}Fe and s-process element at the end of their
career and the extraction is automatically done by the wind.

<u>Maeder</u>: I fully agree with you and we have considered the possibility
of excesses of ^{12}C, ^{22}Ne, ^{25}Mg and ^{26}Mg in the cosmic rays originating
from WC stars and the results are very encouraging.

<u>Vanbeveren</u>: The fact that viscosity in the interior of massive stars is
so high, doesn't that prevent somehow large scale overshooting?

<u>Maeder</u>: The high radiative viscosity certainly contributes to damp
overshooting, however the buoyancy forces are also much stronger in
massive stars and hence the balance of both intervenes. In this context
we must remember that large viscosity would strongly favour turbulent
diffusion.

<u>de Groot</u>: There may be additional observational evidence for vibrational
instability: Just recently Lamers and his students have found from the
UV lines in P Cygni that an absorption component drifts in velocity
with a time scale of about 1 year. Can you say from your calculations
whether this time scale is about right for vibrational instability?

<u>Maeder</u>: Before being able to say whether this is in agreement or not,
the oscillation modes of the given star have to be identified. The
vibrational instabilities have shorter periods than the one you quote.
Thermal readjustments, shell ejections, for example, may perhaps also
be considered in this context.

<u>Gurm</u>: Are supergiants intrinsic variables or just present a surface
phenomenon of the rising convection cells which would be large size to
the order of the radius?
If these are intrinsic variables, how do you explain the multiplicity
of the periods of the variables amongst the supergiants?

<u>Maeder</u>: I would suggest supergiant stars are intrinsic variables, at
least as far as intermediate time scales are considered (from a few
days to one year). Very short term variations (shorter than one day)
could be associated to surface phenomena.
Most periods satisfy a period-luminosity-colour relation. However,
differences and scatter could be introduced by the fact that some supergiants are on blueward and some others on redward tracks with differences
in the internal structure.

Audouze: According to you, what is the minimum initial mass for a star to become a Wolf-Rayet star , (taking into account current rates of mass loss)?

Maeder: The minimum initial mass leading to WR stars is around 20 or 25 $M_\odot$, on the basis of WR stars present in clusters. The average observed mass loss rates lead to a slightly higher value (~30 $M_\odot$). However, one has to account that the scatter and uncertainty of mass loss rates are large.

Conti: André Maeder has been discussing a lower mass limit for W-R stars of 25-30 $M_\odot$, while our number is closer to 35-40 $M_\odot$. In a spirit of compromise I would suggest 33 $M_\odot$. Seriously, though, our numbers are not inconsistent with one another, given the uncertainties. I do feel the value is higher than 20 $M_\odot$, however.

Tayler: We were told yesterday that there are a few masses for OB and WR stars. Do these masses agree at all with the masses suggested by theoretical evolutionary tracks?

Conti: There are a few masses for doubled OB systems not too far up the main sequence (to type O8V). These agree with theoretical tracks. However, the predicted mass losses by stellar winds are insufficiently large in these stars to be readily discernable in the binary data. There is some additional X-ray binary pulsar information that gives masses of the neutron star companions. These stars are invariably undermassive for their luminosity, suggestive of previous mass loss. The wind rates required are a little high but the deductions are complicated by previous evolutionary history of these close binary systems. In other words, we don't have any direct, that is binary, evidence of that correctness of the wind mass loss formulations in a quantitative sense.

THE INTERACTION OF THE SUPERMASSIVE OBJECT R136a WITH THE INTERSTELLAR ENVIRONMENT

J.V. Feitzinger, Th. Schmidt-Kaler
Astronomisches Institut, Ruhr-Universität, 463o Bochum, FRG

R136a is the luminous object in the core of the 3o Dor nebula in the LMC, embedded in the extremely young star cluster NGC 2o7o with the most recent star forming event less 10^6 yr ago. R136a-1 (V = 11.o8) lies in the center of this cluster. The companion or foreground star R136a-2 (ca. 1.2^m fainter than R 136a-1) o".5 apart does not impair the dominating role of R136a-1. The light distribution is strongly peaked to the center of R136a and disagrees completely with Moffat and Seggewiss's (1983) description and their interpretation as a dense core of a cluster (Chu and Wolfire, 1983). This is strenghtened by the UV observations (Cassinelli et al. 1981, Feitzinger et al. 1983, Savage et al. 1983): the spatial spread function of the IUE satellite is dominated by an object of extremely small ultraviolet extent, less than that of a gaussian with 1".5. Speckle observations (Meaburn, 1982; Weigelt, 1981), are discordant in the interpretation of the faint background, but concordant in revealing the existence of a small bright component, with an upper limit of the diameter $<$ o.o2 pc. This is also found, with somewhat lower resolution, by Chu (1984).
The properties of this unprecedented object are listed in Table 1.

Table 1

	Multiple System or Cluster Core	Bochum	Bologna	Wisconsin
Luminosity	$1o^7 \, L_\odot$ (Moffat, Seggewiss 1983)	$7 \cdot 1o^7 \, L_\odot$	$6 \cdot 1o^7 \, L_\odot$	$6 \cdot 1o^7 \, L_\odot$
Temperature	-	$6.2 \cdot 1o^4$ K	$5.2 \cdot 1o^4$ K	$7.5 \cdot 1o^4$ K
Radius	-	65 $R_\odot$	84 $R_\odot$	5o $R_\odot$
Mass	several stars with $<$ 22o - 28o $M_\odot$ (Walborn 1933)	2500 $M_\odot$	$>$2ooo $M_\odot$	21oo $M_\odot$ ($>$8oo $M_\odot$)
Mass-Loss	-	$4 \cdot 1o^{-4} M_\odot$/a	$5.2 \cdot 1o^{-4} M_\odot$/a	$5 \cdot 1o^{-4} M_\odot$/a

The influence of the nebulae ($\sim$2o % pollution) and the neighbouring stars ($\sim$3o % pollution) is carefully taken into account. The luminosity is derived from the requirement that the star produces $5 \cdot 1o^{51}$ hydrogen

A. Maeder and A. Renzini (eds.), Observational Tests of the Stellar Evolution Theory, 321–324.
© 1984 by the IAU.

ionizations/sec in the 3o Dor nebula. The calculated mass loss of
3 1o^{-4} M$_{\odot}$/yr, derived from the line profiles of CIV and NIV, fits well
into the domain of the model parameters describing R136a as a supermassive object. A further realistic estimate of the stellar mass can be derived from the empirical proportionality between terminal and escape
speed (Abbott 1978), $v_{\infty} = 3v_{esc}$; thus we obtain M $\sim$ 26oo M$_{\odot}$. Further,
R136a fits with its temperature (Table 1) extremely well into the rela-
tion between T_{eff} and v_{∞} (Underhill 1983).
Ly-continuum radiation and stellar winds are structuring the 3o Dor nebula. Using the tabulation of return of mass and energy to the interstellar medium by winds from early type stars (Abbott 1982), the following energy input can be set up:

Source	Wind Energy	Comment
R136a	2-5 1o^{39} erg/s	The uncertainty is essen-tially due to the error of the mass loss
Nine central stars including R136b,c	2-4 1o^{38} erg/s	Uncertainty due to the fact that the 3o Dor objects are more luminous than normal WR stars
Stars within 9o" from R136	1-2 1o^{38} erg/s	

The typical energy necessary to build up a stellar wind bubble is approximately 2 1o^{38} erg/s (MacCray and Snow 1979, Dyson and Williams 198o,
Meaburn 1981) for a bubble with a radius of 5o pc and initial electron
density of 8 cm^{-3}. This means that R136a (together with the whole stellar population of 3o Dor) is just sufficient to energize the multitude
of shells and to structure the whole nebula. Walborn's (1984) O and B
stars in 3o Dor can account for at most 2/3 of the excitation (assuming
the earliest spectral classifications),at least 8 % (assuming the latest
types) with a median of 2o %. The dominating role of R136a is corrobora-
ted.
The observed velocity components (Fig. 1) (optical and UV absorptions
and emissions) can be combined to give an overall picture of the velocity structure in that particular region (Feitzinger et al. 1984). Different shells and the central bubble, powered by R136, determine the
morphology of the 3o Dor region.
The expansion velocity of 25 km/s can be used to derive the age of the
innermost shell (see Fig. 4, 5 of Feitzinger and Schmidt-Kaler 198o).
The radius of this shell is $\sim$1o pc, giving an expansion age of 4 1o^{5}yr.
By stellar evolutionary arguments R136a cannot be much older than 1o^{6}yr,
in excellent agreement with the age of the giant population deduced by
IR measurements (McGregor and Hyland 1981). Finally, the lifetime of
supermassive objects should be smaller than 1o^{6} yr to overcome the inherent instability. The high mass loss rate may stabilizes such objects.

References

Abbott, D.C., 1978, Ap.J. 225, 893
Abbott, D.C., 1982, Ap.J. 263, 723
Cassinelli, J.P., Mathis, J.S., Savage, B.D., 1981, Science 212, 1497
Chu, Y.H., Wolfire, M., 1983, BAAS 15, 644
Chu, Y.H., 1984, IAU Sympos. 108, Magellanic Clouds, p. 239.
Feitzinger, J.V., Schmidt-Kaler, Th., 1980, in Two Dimensional Photometry, ed. P. Crane, K. Kjär, ESO, Munich, p. 249
Feitzinger, J.V., Hanuschik, R., Schmidt-Kaler, Th., 1983, Astron. Astrophys. 120, 269
Feitzinger, J.V., Hanuschik, R., Schmidt-Kaler, Th., 1984, Astron. Astrophys., in press
McCray, R., Snow, T.P., 1979, Ann. Rev. Astron. Astrophys. 17, 213
McGregor, P.J., Hyland, A.R., 1981, Ap.J. 250, 116
Meaburn, J., 1981, MN 196, 19 P
Meaburn, J., Hebden, J.C., Morgan, B.L., Vine, H., 1982, MN 200, 1 P
Moffat, A.F.J., Seggewiss, W., 1983, Astron. Astrophys. 125, 83
Panagia, N., Tanzi, E.G., Tarenghi, M., 1983, Ap.J. 272, 123, (Bologna)
Savage, B.D., Fitzpatrick, E.L., Cassinelli, J.P., Ebbets, D.C., 1983, Wisconsin preprint No. 176
Underhill, A., 1983, Ap.J. 266, 718
Walborn, N.R., 1984, IAU Sympos. 108, Magellanic Clouds, p. 259.
Weigelt, G., 1981, in High Angular Resolution, ESO-Workshop, ed. M.H. Ulrich, K. Kjär, Munich, p. 95

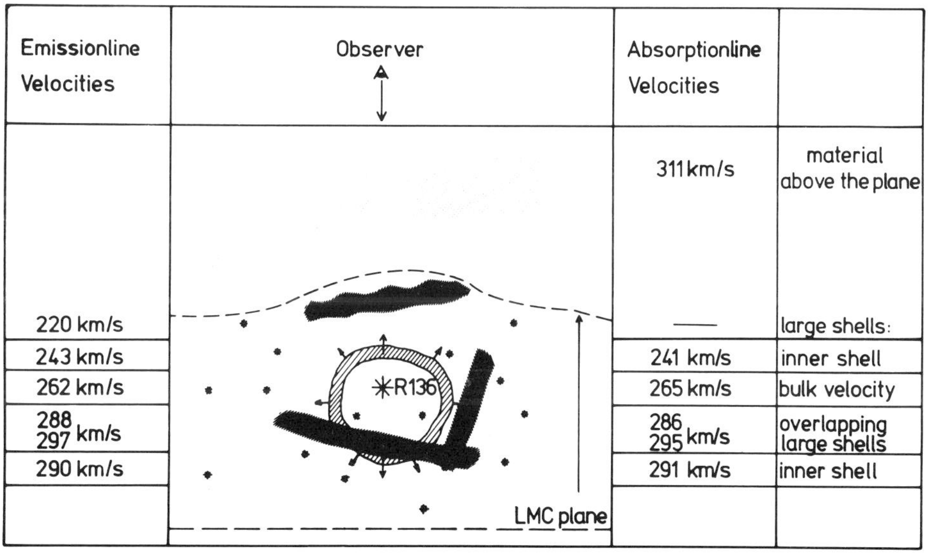

Emissionline Velocities	Observer	Absorptionline Velocities	
		311 km/s	material above the plane
220 km/s		—	large shells:
243 km/s		241 km/s	inner shell
262 km/s	*R136	265 km/s	bulk velocity
288 297 km/s		286 295 km/s	overlapping large shells
290 km/s		291 km/s	inner shell
	LMC plane		

Fig. 1 A schematic model attempting to explain the observed velocity components in the innermost part of 3o Dor.

DISCUSSION

<u>Walborn</u>: At IAU Symposium 108 in Tübingen last week 2 hours of review
and panel discussion plus 4 posters were devoted to R136. Many new
observations were presented, most of which support the interpretation
of R136 as a compact, massive multiple system. The three principal
points are (1) more than half of the ionization of the 30 Doradus Nebula
probably is provided by over 100 O-stars in the central cluster other
than R136; (2) detailed comparisons between R136 and HD 97950 in the
supergiant galactic HII region NGC 3603 show that they are very similar
O+WN multiple systems and (3) the multiple structure of R136 is observed
directly by 4 independent and concordant visual micrometer, photographic
and speckle interferometry techniques. A weak upper limit for the most
massive single object in R136 is now a few hundred, and not a few
thousand, solar masses. The interpretation of R136 is of course vital
for that of the more distant objects.

FURTHER CANDIDATES OF SUPERMASSIVE STARS IN OTHER GALAXIES

Th. Schmidt-Kaler, J.V. Feitzinger
Astronomisches Institut der Ruhr-Universität, Bochum,
Bundesrepublik Deutschland

To look for other supermassive stars in other galaxies means to answer the question: what is the spectroscopic and morphological signature of such objects? The interpretation of WR sources found in other supergiant HII complexes depends basically on our picture of 30 Dor and in particular of R136a.
Guided by the experience with 30 Dor we look for supergiant HII complexes with WR features in the spectrum, esp. near 4650 Å. Supergiant HII regions are defined by $S_5D^2 \gtrsim 5000$ Jy kpc^2 (emission measure exceeding 50 times that of the Orion Nebula; Smith et al., 1978) with D the distance in kpc and S_5 the flux at 5 GHz. The characteristic morphology of the 30 Dor nebula (and of NGC 604 as a similar object) serves as a guideline to find prospective objects.

Supergiant HII regions can often be decomposed into a cluster or a chain or a conglomerate of smaller HII complexes, which may themselves be of complex structure i.e. into cells typical of star formation. Typical examples are Mkn 325 (Coupinot et al., 1982) and NGC 5461 (Israel et al., 1975). The maximum sphere of influence of star formation is $\sim$300 pc, if mechanical energies on the order of 10^{50} - 10^{51} ergs per HII region are being returned to the ISM from massive stars (Hunter 1982).

16 supergiant HII complexes with WR contributions have been published (D'Odorico, Rosa, 1982). Experience with that sample shows that the strong NIII bands around 4650 Å scale in strength with the underlying blue continuum, giving rise to an average equivalent width of 10 Å. Since the bands are usually 30 - 50 Å broad, excellent spectra with high signal-to-noise ratios are required.
A comparison of IUE spectra of those regions with galactic and LMC OB and WR stars, as well as R136a shows:
1) All spectra bear great resemblance to R136a.
2) Broad, PCyg like profiles of CIV, SiIV, NIV are observed.
3) The spectra do not resemble the average O or WR spectra (Massey, Hutchings, 1983, D'Odorico, Benvenuti, 1983).

A. Maeder and A. Renzini (eds.), Observational Tests of the Stellar Evolution Theory, 325–327.

Why do we expect WR features? The luminous WR stars of subtypes WN 6,7
(and their possible descendants of type WC 5,6) can be identified with
late evolutionary stages of the most massive O stars (Maeder, 1982). The
absolute flux observed in the WR bands can be interpreted by the presence
a single (or only a few) very luminous objects or by a cluster of normal
WR stars, comparable in number with the equivalent of $10^3 - 10^4$ O5 stars
needed to ionize the whole nebula (Conti, Massey, 1981; D'Odorico, Rosa
1981). If the ionizing stellar cluster is formed with the normal initial
mass function, the great excess in the number of O stars compressed into
small volumes, must be explained. One severe problem arises at this
point: The WR stars are most probably the descendants of main sequence
O stars and therefore considerably older. The observation of the earliest
O types O3-O5 in the same small volume where the WR stars are located
means that a continuing star formation process must be postulated. This
contradicts the fact that the region is extremely heated up by the first
O (plus WR) star generation, effectively inhibiting the ensuing star
formation.

Two interpretations of the physics of supergiant HII regions appear
possible:
1) Burstlike star formation in small volumes. To explain the Hβ luminosity of NGC 6764 requires 1.5 10^{40} erg/s^{-1} or 3.3 10^{53} ionizing photons. This is equivalent to 6.6 10^2 O5 or 4.5 10^3 O7 or 1.8 10^4 O8
   stars in an volume of less than 1 kpc.
   For Mrk 309 with Hβ = 1.6 10^{41} erg s^{-1} the number of ionizing photons
   per sec is 3.5 10^{54}This requires the presence of 7.1 10^3 O5 stars
   (Osterbrock and Cohen, 1982). The star burst model works very well in
   the case of dwarf galaxies (NGC 5204, 5474, 5585) and their HII regions which are less extreme than typical supergiant HII regions
   (Fabbiano and Panagia, 1983).
   The alternative is:
2) Supermassive objects, responsible for 50 - 90 % of the ionization power. The best known example is R136a (cf. Feitzinger, Schmidt-Kaler,
   1983, Table 1 this volume). It can be shown that this object must be
   responsible for the major fraction of the ionization of the 30 Dor
   nebula, irrespective whether it is a single star or an extremely compact cluster of most massive O stars. This is so because its temperature is certainly exceeding 45000 K and its (de-reddened) UV flux implies M_{Bol} < -14.0, consequently producing the larger part of the
   ionization.

We have undertaken a search of possible sites of such supermassive stars
on the basis of the morphology of 30 Dor and NGC 604. Four subtypes were
used:

Subtype	Prototype	Characteristics
1	.NGC 604	spiral galaxy shows a compact knot in the outer parts comparable to central region of galaxy
2	30 Doradus	galaxy shows dominating Hα region near end of bar or spiral arm

Subtype	Prototype	Characteristics
3	NGC 5430	one or more large knots in or near edge of central region of spiral galaxy
4	IC 3576	several large bright knots similar to the galaxy nucleus

The survey was conducted on the POSS and ESO-Blue & SRC films. The following lists of galaxies were searched:

a) Elmegreen and Elmegreen, 1980: 44 Sm, Im, with diameters $\geq 4'$

b) Wray, de Vaucouleurs, 1980: 78 galaxies with identified super-associations

c) Revised Shapley-Ames-Catalogue (Sandage and Tammann, 1981) 113 galaxies of types S, SB, I with radial velocities $\leq$ 820 km/s ($<$15 Mpc)

d) Heidmann, 1983: 8 clumpy Irregulars

e) Vorontsov-Veljaminov, 1977: $\sim$50 large HII regions

The survey is almost completed. Preliminary statistics shows that supergiant HII regions do not occur in systems of luminosity class IV and V, and show up preferentially in systems of type Sc, Sd, Sm II, III. High-luminosity early spirals do not apparently occur because of selective effects. The percentage of barred spirals seems quite normal. Some of the more interesting cases are being studied by high resolution direct photographs in UBVR $H\alpha$.

References

Conti, P.S., Massey, P., 1981, Ap.J. 249, 471

Coupinot, G., Hecquet, J., Heidmann,J., 1982, MN 199, 451

D'Odorico, S., Rosa, M., 1981, Ap.J. 248, 1015

D'Odorico, S., Rosa, M., 1982, IAU Symp., Wolf-Rayet Stars, 99, p. 557

D'Odorico, S., Benvenuti, P., 1983, in press

Elmegreen, D.M., Elmegreen, B.G., 1980, AJ 85, 1325

Fabbiano, G., Panagia, N., 1983, Ap.J. 266, 568

Heidmann,J.,1983, Highlights of Astronomy, D. Reidel, p. 611.

Hunter, D.A., 1982, Ap.J. 260, 81

Israel, F.P., Goss, W.M., Allen, R.J., 1975, AA 40, 421

Maeder, A., 1982, IAU Symp., Wolf-Rayet Stars, 99, p. 405

Massey, P., Hutchings, J.B., 1983, Ap.J. in press

Osterbrock, D.E., Cohen, R.D., 1982, Ap.J. 261, 64

Sandage, A., Tammann, G.A., 1981, Revised Shapley-Ames Catalogue, Washington

Vorontsov-Veljaminov, B.A., 1977, AA Supp. 28, 1

Wray, J.D., de Vaucouleurs, G., 1980, AJ 85, 1

EVOLUTION OF MASSIVE STARS WITH MASS LOSS: SURFACE ABUNDANCES

Laura Greggio
Dipartimento di Astronomia
CP 596, I-40100 Bologna, Italy

The location of theoretical stellar models in the upper part of the
HR Diagram (HRD) depends on a variety of poorly understood physical proc
esses which may occur during the evolution of massive stars. Among the
most important we find: mass loss, convective overshoot and opacity enhancement for temperatures around 10^6 °K. The effects of these phenomena
have been recently investigated by several authors (Bertelli et al. 1983
and references therein), but, due to the large uncertainties still present in the theoretical formulation of these processes, only parametrized
formulae are avaliable for model computations. Since the theoretical dis
tribution of massive stars in the HRD turns out to be fairly sensitive
to these parameters, the evolutionary scenario for massive stars is still
far from being satisfactorily settled. On the other hand, due to mass
loss, nuclear processed material is expected to be exposed at the surface
of massive stars and then, the comparison between theoretical predictions
and observations of the surface chemical composition of these objects can
help in understanding their evolution and to set more stringent limits to
the mentioned parameters.

To this end, evolutionary sequences corresponding to 20, 40 and 60 $M_\odot$
have been computed up to core He exhaustion, following in detail the a-
bundance variations of CNO, Ne and Mg isotopes. Overshooting from convec
tive cores was taken into account during both H and He burning stages,
following the procedure described in Bressan et al. (1981). According to
Chiosi and Olson (1981), a mass loss rate given by
$\dot{M} = 3.47 \times 10^{-16} \, L^{1.75} \times \{RM/(1-\Gamma)\}^{1.03}$ $M_\odot/yr$ was applied during the MS
phase, while in the red supergiant stage a constant value of $\dot{M} = 10^{-5} M_\odot/yr$
was adopted.

Both 20 and 40 $M_\odot$ models do not present variations in their surface a-
bundances during the MS phase. When the models enter the red supergiant
region, the inward penetration of the convective envelope causes the

A. Maeder and A. Renzini (eds.), Observational Tests of the Stellar Evolution Theory, 329–332.
© *1984 by the IAU.*

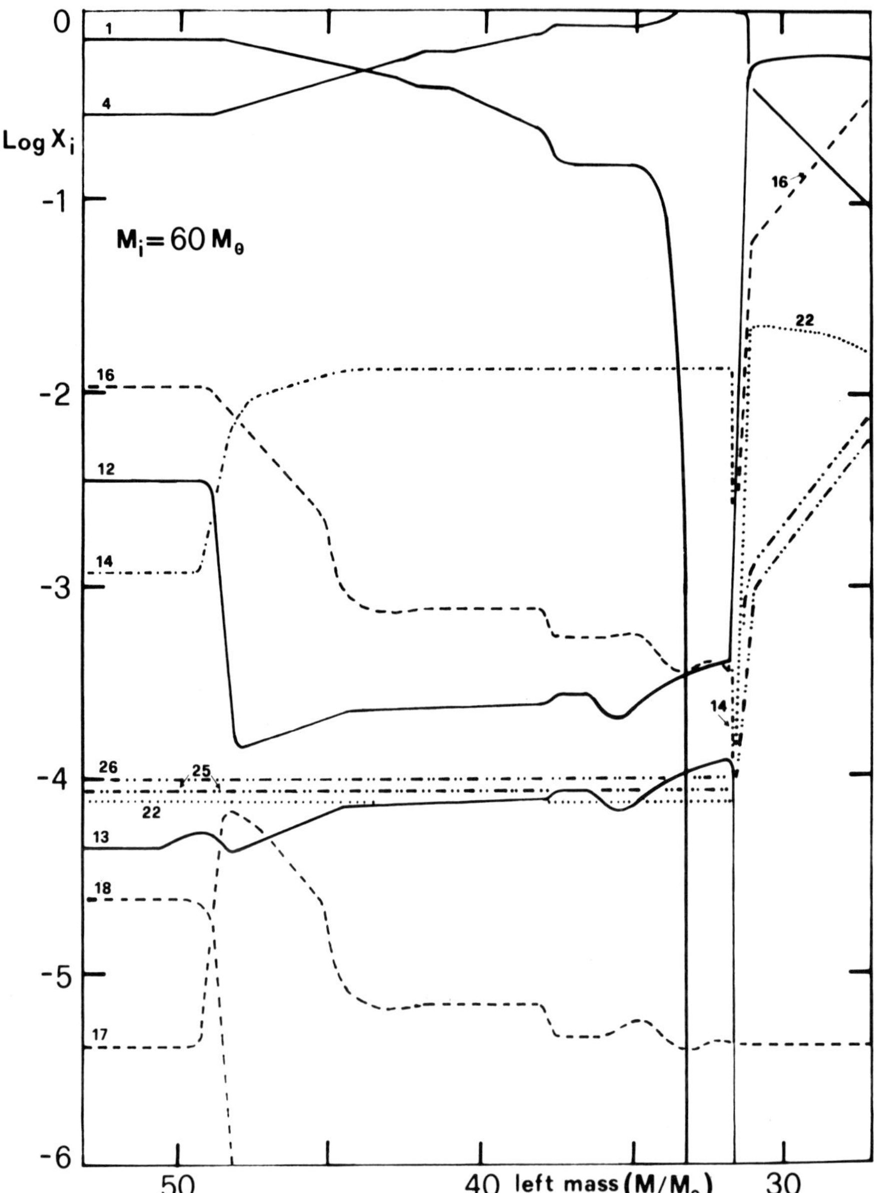

Fig.1 Evolution of the chemical abundances in mass fraction as functions of the remaining stellar mass for the 60 M_⊙ model. The mass number of the various species is indicated.

dredge up of CNO processed material: the surface ^{14}N abundance gets enhanced, while ^{12}C and ^{16}O are depleted. The surface $^{12}C/^{13}C$ ratio is considerably lowered and, to a minor extent, even the $^{16}O/^{17}O$ and $^{17}O/^{18}O$ ratios are affected. The predicted abundances are:

$C/N \cong 0.87$ $O/N \cong 2.11$ $^{12}C/^{13}C \cong 18.31$ $^{16}O/^{17}O \cong 1014$ $^{17}O/^{18}O \cong 0.87$

$C/N \cong 0.27$ $O/N \cong 0.70$ $^{12}C/^{13}C \cong 23.08$ $^{16}O/^{17}O \cong 610.4$ $^{17}O/^{18}O \cong 1.42$
for the 20 and 40 $M_\odot$ models respectively. The subsequent evolution is
spent in the red part of the HRD and the model is expected to give rise
to a type II SN event. Due to the huge mass loss and to the large size of
the convective core, the 60 $M_\odot$ model starts exhibiting CNO processed mate
rial at the surface while still on the MS. Following the representation a
dopted by Maeder (1983), Fig.1 shows the behaviour of the surface abundan
ces as functions of the remaining stellar mass. Since the dilution due to
convection is not present, the surface composition of the model during the
late MS evolutionary stage corresponds to equilibrium CNO abundances:
$C/N \cong 0.02$ and O/N steadily decreasing down to 0.05. When H is ignited in
a shell, the envelope expansion causes a redward excursion, which is
stopped and reversed by the mass loss. Eventually, the model enters the
blue region of the HRD with a low surface H abundance and is taken to rep
resent a WNL star. Further removing of the outer layers brings complete
CNO processed material at the surface and the H abundance drops to zero.
In this stage, the model may represent a WNE object. WN stars are then
predicted to exhibit CNO equilibrium abundances, that is $X_N/X_{He} = 1.5 \times 10^{-2}$
and $X_C/X_{He} = (3.8 \div 5.5) \times 10^{-4}$. When He burning products appear at the surface, the abundances suffer an abrupt change: ^{12}C, ^{22}Ne and ^{16}O are enhanced by factors of ~ 500, 200 and 100 respectively, while ^{14}N and ^{13}C
virtually disappear. In this stage the model is taken to represent a WC
star and may eventually turn into a WO star, as the surface ^{16}O abundance
keeps steadily growing.
In order to investigate upon the effect of an opacity enhancement for
$T = 10^6$ °K, a 20 $M_\odot$ evolutionary sequence has been computed adopting the
Bertelli et al. (1983) algorithm for the opacity. In this case, the model
spends $\cong$ 15% of its total core H burning lifetime as a red supergiant,
suffering a huge mass loss. As a consequence, the H rich envelope is removed and the model returns to the high temperature region of the HRD.
The dredge up of CNO processed material modifies the surface abundances
during the red supergiant stage and the predicted values are:
$C/N \cong 0.99$ $O/N \cong 2.52$ $^{12}C/^{13}C \cong 19.65$ $^{16}O/^{17}O \cong 685.4$ $^{17}O/^{18}O \cong 1.14$.
Further removal of the outer layers eventually exposes H poor material at
the surface and the subsequent evolution of the surface chemical abundances is very similar to that already described for the 60 $M_\odot$ model.

These results support the evolutionary scheme according to which single
WR stars originate from massive progenitors and the WC objects are bare
He burning cores. In particular, strong Ne and Mg surface overabundances
are predicted for WC stars. The predicted relative number of WR stars
with respect to their massive progenitors can be greatly increased if an
opacity bump around 10^6 °K is present, so that the star suffers a high
mass loss rate for an appreciable fraction of its core H burning lifetime.

However,the results are fairly dependent on the rate of mass loss which
is assumed in the red supergiant stage. Finally, a non negligible fraction of massive O-type stars is predicted to show a surface N overabundance, due to the effect of overshooting and mass loss. The same process
es are extremely important in determining the surface composition of red
supergiant. stars.
A full account of these calculations will appear elsewhere (Greggio 1984).

REFERENCES

Bertelli,G., Bressan,A.G., Chiosi,C. 1983, preprint
Bressan,A.G., Bertelli,G., Chiosi,C. 1981, Astron. Astrophys. 102, 25
Chiosi,C., Olson,G.L. 1981, private comunication
Greggio,L. 1984, in preparation
Maeder,A. 1983, Astron. Astrophys. 120, 113

DISCUSSION

Audouze: I would like you to comment on the many discontinuities of the
^{17}O abundance curves calculated in your 60 $M_\odot$ and 20 $M_\odot$ models. Moreover, how do you explain the fact that there is no ^{18}O present in the
inner zones? Are these two features connected with the uncertainties
on the nuclear reaction rates?

Greggio: The discontinuities shown by the abundance curves reflect the
distribution of the X_{17} with radius inside the star and the mixing episodes occurring during the evolution. For the 60 $M_\odot$ model, for example,
the first rise in X_{17} is due to its production up to the equilibrium
value, while the following trend simply reflects the ^{16}O behaviour, since
layers in which the two O isotopes exhibit equilibrium abundances appear
at the surface. The first plateau is due to the surface convection, developing during the redward excursion of the evolutionary model, while
the second plateau is due to the development of an intermediate fully
convective region, above the H-burning shell. Other details depend on
the evolution of the temperature structure in the hottest radiative
layers and on the sensitivity of the thermonuclear reaction rates to
temperature variations.
The lack of ^{18}O which you refer to is due to the fact that, when the
He-burning convective core has attained its maximum size, all the ^{18}O
has already gone into ^{22}Ne. The result is however dependent on the overshooting (see Maeder, this conference).

OBSERVATIONAL TESTS FOR OVERSHOOT AND OPACITY ENHANCEMENT BASED ON
MASSIVE STARS

G. Bertelli[1,3], A. Bressan[2], C. Chiosi[1,2]

1) Institute of Astronomy, Padua, Italy
2) International School for Advanced Studies, Trieste, Italy
3) Fellow of the National Council of Research (C.N.R.), Italy

SUMMARY
It is shown that a moderate increase in the opacity due to heavy elements
in models of massive stars, incorporating convective overshoot and mass
loss by stellar wind, can remove the well known discrepancy between theoretical expectation and observed frequencies of luminous stars in the HR
diagram. These models in fact extend their main sequence band far beyond
the classical limit.

1. INTRODUCTION
Meylan and Maeder (1982) comparing evolutionary models with observational
data for young clusters and associations find that too many stars are
observed outside the theoretical main sequence band. The HR diagram of
Humphreys (1978) for luminous stars in young galactic clusters and associations shows this same characteristics. In fact the comparison of the
relative number of stars in different regions of the HR diagram with the
expectation from standard theoretical models reveals that too many stars
fall outside the MS band. Unless severe selection effects are present in
the star counts for the earliest spectral types (O and B), everything
occurs as if the MS band were much wider than predicted, apparently extending up to the spectral type AO. In this paper we aim to show that an increase in the opacity due to heavy elements by a factor 2 to 4 in models
with overshoot may significantly improve upon the disagreement. The proposed increase can be physically justified by heavy element opacities in
the region around 10^7 K$^\circ$ having been plausibly underestimated by a factor
of 2 to 3 (Simon, 1982). Finally it is worth recalling that Carson's
(1976) opacities already possessed a similar increase even if it has not
received widespread acceptance.

2. MODELS WITH ENHANCED OPACITY
Previous models of massive stars with opacity by heavy elements much higher than the standard value (Carson, 1976) are by Stothers and Chin (
1977, 1978). These models however are not useful to our purpose at least
for the following reasons: i) the models do not take into account convective overshoot, which on the contrary appears to be very important. ii)
When mass loss by stellar wind during the MS phase is considered, too

A. Maeder and A. Renzini (eds.), Observational Tests of the Stellar Evolution Theory, 333–336.
© 1984 by the IAU.

high rates of mass loss have been used, which are not supported by current
observational information. In the light of this we thought worth recom-
puting models for a 20 $M_\odot$ star (assumed here as representative of the evolution in the luminosity range $- 9 \leqslant M_b \leqslant - 7$), in which the opacity enhancement is considered as a free parameter, and in which overshoot is trea-
ted according to the formulation of Bressan et al. (1981) with $\lambda = 1$. The
opacity of Cox and Steward (1970) has been modified according to the relation

$$\kappa = \kappa_{CS} (1 + \chi \exp (- 10 (5.8 - \mathrm{Log}\ T)^4)) \qquad , \quad (1)$$

where κ_{CS} stands for the Cox-Stewart opacity and χ is an adjustable parameter.   From the comparison with the opacity table of Carson the following indicative relation between the metal content Z and the parameter
χ can be established, $\chi = Z / 0.005$. It turns out from this that the
value $\chi = 1$ is compatible with $Z = 0.005$ (appropriate for SMC), whereas the
value $\chi = 4$ is about right for the galaxy. Evolutionary models computed
with the above opacity reveal that: a) the MS band becomes broader and
broader at increasing χ , amplitute of the opacity bump, untill when a
significant fraction of the core H-burning phase is found to occur in the
red supergiant region. b) Indepently of the value for χ, all the models
cross the area of the HR diagram comprised between 3.5 < Log Te < 3.9 to
4.0 on a very short time scale. Models which incorporate overshoot, opacity bump and mass loss by stellar wind (the last one being important
only during the red supergiant phase and at a rate of about 10^{-5} $M_\odot$/yr)
may account not only for a substancial broadening of the MS band in the
range of luminosity that we are considering, but also for the i) remarkable similarity of the HR diagrams of supergiant stars in galaxies with
different metal content, such as SMC, LMC and galaxy , and ii) the observed variation in the relative percentage of WR stars across the galactic
disk and from galaxy to galaxy (Maeder et al., 1981).

3. Z-INDEPENDENT MORPHOLOGY OF HR DIAGRAMS AND THE EFFECT OF Z ON WR
   FORMATION AND STATISTICS

The HR diagrams of supergiant stars in our own galaxy and two Magellanic
Clouds show nearly identical morphologies in spite of the diverse metal
content. In all the three galaxies there is a continuous distribution of
stars up to Log Te = 4.0 to 3.9, a few objects in the range 4.0 to 3.9
> Log Te > 3.7 and a clump of red supergiants. Evolutionary computations
for a 20 $M_\odot$ star with Z = 0.005 and X = 0.765 (pertinent to the case of
SMC), moderate mass loss rates during the early phases, overshoot with
λ = 1 and opacity bump with $\chi = 1$ can explain this important feature. In
fact: i) the core H-burning band extends also in this case up to Log Te
= 4.08, therefore the observed distribution of O to A type stars in SMC
 is reproduced; ii) the phase of core He-burning takes place in the red
region of the HR diagram. If mass loss by stellar wind occurs at a rate
similar to that of galactic red supergiants (about 10^{-5} $M_\odot$/yr) these models cannot lose the whole H-rich envelope before completing core He -
burning and will likely terminate their evolution as red supergiants. In
this scheme the formation of single WR stars is unlikely. On the contrary, stars with normal metal content, say about 0.02, may reach the red
supergiant stage while still burning H in the core, due to the high bump

in the opacity as a consequence of the high metallicity. Owing to the much longer lifetime now available in the red supergiant region, the entire H-rich envelope can be lost even before central He-burning ignition. These stars go back to higher effective temperatures and appear as WR 's. In this scheme, the gradient in the number of WR with respect of red supergiant stars acroos the galactic plane should be attributed to both the gradient in progenitors and to the gradient in metallicity, which by increasing the opacity would determine longer lifetimes for the post red supergiant stages, and in consequence favour the number of WR stars with respect to supergiants at increasing Z.

4. COMPARISON OF THE MODELS WITH STAR COUNTS AND CONCLUSIONS

The Table below shows the relative number of stars in each spectral type falling in the luminosity range $- 9 \leqslant M_b \leqslant - 7$ and within 2.5 Kpc from the sun. To cope with the incompleteness of the Humphreys catalogue that we have used to derive the data of the Table, we have multiplied the number of stars derived from the Humphreys catalogue by a filling factor estimated from the ratio of the number of O type stars in the catalogue of Garmany et al. (1982), which is known to be complete, to the number of the same stars in the list of Humphreys. In addition to this, when performing star counts we have also included the number of WR stars falling within the same are. The first column of the Table gives the number of stars falling within the main sequence band of models with overshoot and no opacity bump. It is evident that about 30% of stars fall beyond the limit of the MS band, contrary to the expectation of less than 10%.

Table

SP	O-B1	B2-B9	A	F-G	K-M	OB*	WR	Note
N/N_t	0.71	0.12	0.02	0.01	0.07		0.05	Humphreys
t/t_{tot}	0.69	0.10	0.01	0.01	0.09	0.04	0.06	Models

The second row of the Table gives the relative lifetime spent in the various spectral types by the 20 $M_\odot$ star computed with the following prescription: $\dot{M}_b = 0$, $\dot{M}_r = 10^{-5}$, $\dot{M}_{WR} = 0.5 \cdot 10^{-5}$ $M_\odot/yr$, $\lambda = 1$ and opacity with $\chi = 4$. It is soon evident that these new models almost entirely eliminate the disagreement between theory and observation. They spent in fact 69% and 10% of their life in the spectral range O to B1 and B2 to B9 respectively as it is required by the star counts. Another interesting feature is that about 9% and 4% of the core H-burning lifetime is spent by the models while appearing as red red supergiants and O type stars with low H content at the surface (indicated as OB* in the Table). Furthermore the whole core He-burning lifetime is now available to the WR phase, thus easily accounting for the relative percentage of WR stars in the solar vicinity. More details on this problem are reported in Bertelli et al. (1983). *As a major conclusion, we argue that opacity changes of the order we have used in models with overshoot and mass loss by stellar wind can reconcile theory and observations, and explain the major properties of supergiant and WR stars at the same time.*

This work has been financially supported by the National Council of
Research of Italy (C.N.R.-G.N.A.).

REFERENCES
Bertelli, G., Bressan, A., Chiosi, C., 1983, Astron. Astrophys. in press
Bressan, A., Bertelli, G., Chiosi, C., 1981, Astron. Astrophys. 102, 25
Carson, T.R., 1976, Ann. Rev. Astron. Astrophys. 14, 95
Cox, A.N., Stewart, J.N., 1970, Astrophys. J. Suppl. 19, 243
Garmany, C.D., Conti, P.S., Chiosi, C., 1982, Astrophys. J. 263, 777
Humphreys, R.M., 1978, Astrophys. J. Suppl. 38, 309
Maeder, A., Lequeux, J., Azzopardi, M., 1980, Astron. Astrophys. 90, L17
Meylan, G., Maeder, A., 1982, Astron. Astrophys. 108, 148
Simon, N.R., 1982, Astrophys. J. 260, L87
Stothers, R., Chin, C.W., 1977, Astrophys. J. 211, 189
Stothers, R., Chin, C.W., 1978, Astrophys. J. 225, 939

DISCUSSION

Brunish: What was the hydrogen-burning lifetime and how did the increased opacity affect the lifetime? (i.e. what was t_{tot} in years in
both cases)?

Bertelli: The opacity bump characterized by $\chi = 4$ produces only a small
increase in the total lifetime ($\sim 5 \cdot 10^5$ y.) if compared to the case in
which Cox-Stewart opacities are adopted. In the case of evolution with
opacity bump the total lifetime is $1.11 \cdot 10^7$ years.

Jeffery: On increasing the CNO opacity, did you find convective instability in the envelope?

Bertelli: No.

Maeder: Did you check the consequences of your assumed opacity bumps
in the solar model? In other words, are you able to get the proper
solar luminosity after 4.6 billion years for the standard solar
chemical composition?

Bertelli: In the case of the sun the problem does not exist. In fact
the opacity bump (due to the ultimate ionization of the CNO elements)
could appear only around $T \simeq 10^6$K and for low values of density, and
these conditions are never met simultaneously inside low mass stars.

Roxburgh: Of course you can make the solar model with the opacity bump.
There are enough free parameters in a solar model that a small change
in say the ^{4}He abundance will give a model with the solar luminosity at
the solar age.

P CYGNI STARS AS AN INTERMEDIATE STAGE BETWEEN RED SUPERGIANTS AND
WOLF-RAYET STARS

Henny J.G.L.M. Lamers
Astronomical Institute, Space Research Laboratory,
Beneluxlaan 21, 3527 HS Utrecht, The Netherlands.

Mart de Groot
Armagh Observatory, College Hill, Armagh BT61 9DG, N. Ireland.

Angelo Cassatella
Villafranca Satellite Tracking Station, European Space
Agency, Apartado 54065, Madrid, Spain.

P Cygni's evolutionary status is evaluated using its recently
redetermined basic parameters (Lamers et al. 1983) to plot the star
on the HRD (Figure 1).   The theoretical evolutionary track of a 60 $M_\odot$
star passes through P Cygni's error box twice.   From a comparison of
the time scales along the two parts of the track it is found to be
twenty times more probable to find a star in the He-core burning plus
H-shell burning phase evolving to higher T_{eff} than in the thick H-shell
burning phase evolving to the RSG region of the HRD.

According to Maeder (1981) such an object should have an atmosphere
enriched in N and depleted in C and O.   Luud (1967) found some evidence
for this but P Cygni does not really resemble a late WN star because a
thin layer of unprocessed H efficiently masks the underlying processed
atmosphere.   However, P Cygni's high mass-loss rate of 2×10^{-5} $M_\odot$ yr^{-1}
will soon make it lose its last surface layers after which its WN
character should be revealed.   Note also that P Cygni's luminosity
($M_{bol} = -9.9$) is in the same range as that of WNL stars ($M_{bol} = -8.0$ to
-9.5).

If the 6 GHz radio arc discovered by Wendker (1982) is interpreted
as a remnant of the 1600 outburst, the expansion velocity would be
1800 km s^{-1}.   This is allright for a B1 supergiant, but a factor 5
too high for P Cygni.   However, in the RSG phase typical expansion
velocities are 25 km s^{-1} and the most distant parts of the arc must
have been ejected 3×10^4 yr ago.   This is in good agreement with
the evolutionary time scale of 4×10^4 yr from RSG to BSG.

We therefore suggest the following evolutionary scenario for
P Cygni: A 60 $M_\odot$ star evolves from the ZAMS to the RSG region.   The

337

A. Maeder and A. Renzini (eds.), Observational Tests of the Stellar Evolution Theory, 337–339.

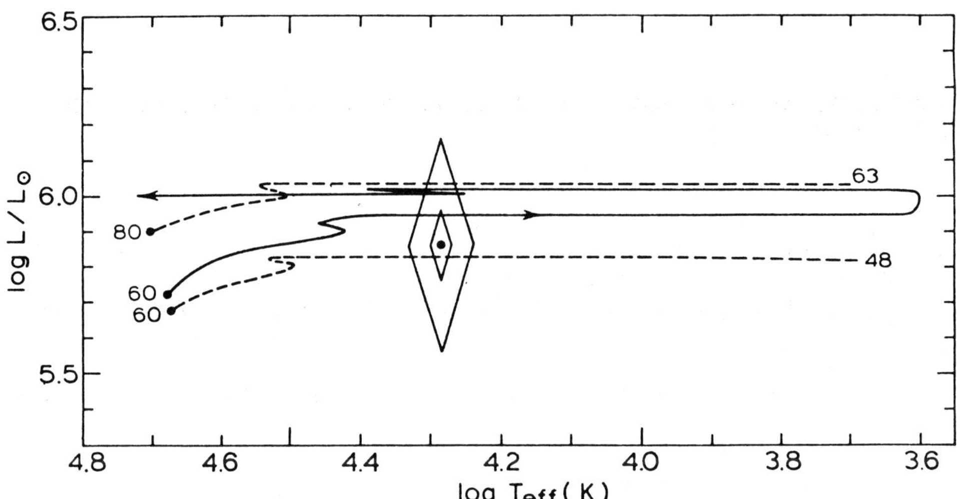

Figure 1. The location of P Cygni in the HR diagram is indicated. The
outer error box corresponds to an uncertainty of $\Delta E(B-V) = 0.05$. The
inner error box corresponds to $\Delta E(B-V) = 0.00$. The dashed curves are
De Loore et al.'s (1978) evolutionary tracks with mass loss (N = 100)
for two stars of 60 and 80 $M_\odot$. The full curve is the evolutionary track
from Maeder (1981). This last track crosses the error box of P Cygni
twice. We argue that P Cygni is in the phase of the second crossing.

RSG loses mass through a dense stellar wind and through sporadic outbursts thereafter. Remnants of this mass loss are now observed as a
radio feature and obscured the star sufficiently for it to escape naked-
eye detection before 1600 AD. In that year starlight emerged through
the dust but inhomogeneities produced the 17th century light variations
and red colour. As the dust was blown away by the stellar wind and/or
evaporated by the star's UV radiation, the star slowly brightened in the
V-band but, due to a much lower degree of re-distribution of UV radiation,
only reached $V \approx 5$ against $V \approx 3$ in the period 1600 - 1660. At present
P Cygni evolves at constant M_{bol} towards higher T_{eff}. Soon, when all
H-rich layers have been blown off, the star will be WNL.

     The complete text of this paper has appeared in Astron.Astrophys.,
123, pp.L8-L10, 1983.

References:
De Loore, C., De Grève, J.P., Vanbeveren, D.: 1978, Astron.Astrophys.
          Suppl., 34, pp.363-368.
Lamers, H.J.G.L.M., De Groot, M., Cassatella, A.: 1983, Astron.Astrophys.,
          (in press).
Luud, L.S.: 1967, Soviet Astron., 11, pp.211-219.
Maeder, A.: 1981, Astron.Astrophys., 99, pp.97-107.
Wendker, H.J.: 1982, Astron.Astrophys., 116, pp.L5-L8.

Note added after the Symposium:

From recent publications (e.g. de Jager, C: "The Brightest Stars",
D. Reidel, Dordrecht, 1980; de Jager, C.: Working Group Stars with Extended Atmospheres, Preprint No 30, 1983; Humphreys, R., Astrophys. J.
269, pp. 335-351, 1983) and further discussions at this Symposium it
seems as if stars with an initial ZAMS mass of 60 $M_{\odot}$ are about the most
massive ones that can evolve into an RSG. More massive stars encounter
an instability zone in the HRD where greatly enhanced mass loss drives
the star back to higher T_{eff}. With P Cygni so very near the upper limit
for evolution into an RSG these recent ideas might affect certain details of our scenario but leave its essential features intact.

DISCUSSION

Walborn: Roberta Humphreys' diagram shows that P Cygni lies just on the
borderline between stars which observationally become red supergiants
and those which do not.

THE "AGE" OF OB ASSOCIATIONS

C. DOOM
J.P. DE GREVE
C. de LOORE

I. INTRODUCTION.

For the determination of ages of OB associations, the traditional approach is to compare isochrones, derived from theoretical models to the observed distribution of stars in the HR diagram (see e.g. Stothers, 1972). However, even a casual look at the HR diagram of the association Per OB1 (Fig.1) shows that it is impossible to fit a unique isochrone to all stars of the group, even by allowing some scatter. We will therefore adopt a different approach and determine the initial mass and the age of each individual star in a given association. As an example, we will discuss Per OB1.

II. METHOD AND INPUT DATA.

The fundamental data for stars in OB associations are taken from the catalog of luminous galactic stars of Humphreys (1978). This catalog is complete for O and early B stars, and for stars of all spectral types, having $M_{bol} < -7$ in associations with d < 2.5 kpc. This implies that all stars in these associations, having an initial mass M_i larger than 15 $M_{\odot}$ are contained in the catalog.

In order to determine M_i and the age t of each star, we assume that :
- all O and B stars within the main sequence band are evolving from ZAMS towards Red Point;
- all Of stars are evolving from Red Point towards the end of core hydrogen burning and have a hydrogen deficient atmosphere;
- all stars outside the main sequence band are evolving towards the right in the HR diagram.

M_i and t are then interpolated between the evolutionary tracks of Doom (1982) and Doom and De Grève (1983), using log Te and log $L/L_{\odot}$ as basic quantities. The evolutionary tracks include mass loss and over-shooting. Log T_e and the bolometric corrections were determined using the calibration compiled by Vansina and De Grève (1982), extended to M5 with the calibration given by Humphreys (1978).

III. RESULTS.

Figure 1 also shows all stars of Per OB1 in a M_i vs t diagram.

A. Maeder and A. Renzini (eds.), Observational Tests of the Stellar Evolution Theory, 341–343.
© *1984 by the IAU.*

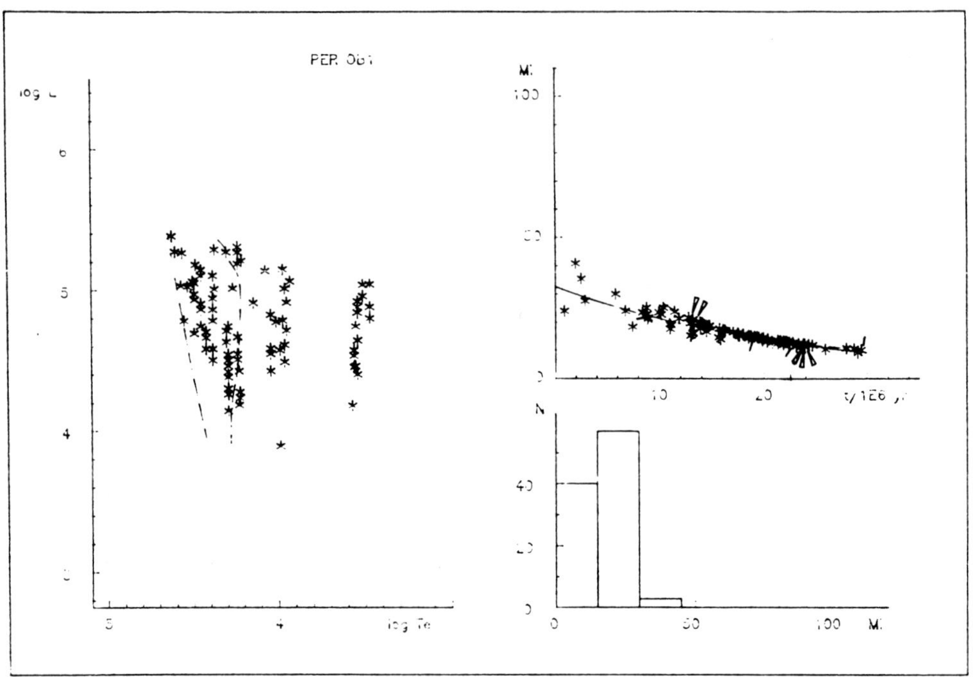

Figure Caption.

Fig. 1 : The HR diagram of Per OB1 (fundamental data are from Humphreys
(1978) (left) and the M_i vs t diagram for Per OB1 (right). The
curve gives the main sequence lifetime the lower curve is a
least squares fit: $M_i \sim \exp(-t/2.4\ 10^6\ yr)$.

This diagram shows clearly that the <u>lower mass stars</u> in this association
($M_i \simeq 15\ M_\odot$) are some 20 million year <u>older</u> than the most massive stars
($M_i \simeq 40\ M_\odot$). The same trend is visible in many other associations,
contained in the Humphreys catalog, but not shown here.

IV. INTERPRETATION.

The M_i vs t diagram of individual OB associations reveals the absence of young lower mass stars, and shows many 10-20 $M_\odot$ stars outside
the main sequence band. We interpret this as being due to the fact that
the lower mass stars in associations are formed before the more massive
stars. The same effect has been found for contracting 1-5 $M_\odot$ stars in
young associations by Iben and Talbot (1966).

The effects of binarity and rotation on the age determination will
be discussed in a forthcoming paper (Doom, De Grève and de Loore, 1984)
The results of this analysis show that binarity or rotation cannot explain the observed age spread.

The fact that the lower mass stars are formed first in many associations, contained in the Humphreys catalog explains why the same effect is observed in her complete HR diagram (a lack of lower mass ZAMS
stars, "too many" stars outside the main sequence). This HR diagram can

therefore be fully understood in terms of stellar evolution including
mass loss and overshooting.

This research is supported by the National Foundation of Collective Fundamental Research of Belgium (F.K.F.O.) under contracts nr. 2.9009.79 and 2.9002.82.

References.
Doom, C. : 1982, Astron. Astrophys. 116, 308.
Doom, C.; De Grève, J.P. : 1983, Astron. Astrophys. 120, 97.
Humphreys, R.M. : 1978, Astrophys. J. Suppl. Ser. 38, 309.
Iben, I.; Talbot, R.J. : 1966, Astrophys. J. 144, 968.
Stothers, R. : 1972, Astrophys. J. 175, 431.
Vansina, F.; De Grève, J.P. : 1982, Astrophys. Space Sci. 87, 377.

DISCUSSION

Tayler: If you had associations of all ages there would be no change in
the ratio of stars in different stages of evolution as a result of low
mass stars being formed first; there would be both very massive post-
main sequence stars left over from older associations and young
associations with only lower mass main sequence stars. Are you saying,
for example, that there are no extremely young associations and that
there is also a shortage of very massive evolved stars from older
associations?

Doom: There is indeed a certain selection effect in the presence of OB
associations with different ages. On the one hand, very young associations, containing only lower mass stars, are found (e.g. the young
Ori OB1 group). However, in this stage of evolution, the association
is more difficult to detect: it is embedded in the cloud from which it is
formed and does not yet contain very massive, luminous stars. Only when
the latter type of stars are formed, the association is seen as a "real"
OB association. On the other hand, associations being much older than
the OB associations are not found: due to the age spread, all stars will
disappear from the association (SN explosions) in a relative short time.
It is also possible that the association is dispersed by that time.

Walborn: Observationally OB associations contain subgroups of different
ages with overlapping mass ranges, e.g. the comparison between the
Orion Belt and Orion Nebula clusters. Also in h/χ Persei the two concentrated clusters should be considered separately from the extended
halo of the association.

AGES, AGE DIFFERENCES, MASSES OF STARS IN YOUNG OPEN CLUSTERS

Anne Marie Jakobsen[1]
University of Aarhus
Carlsberg Foundation Fellow, Denmark
Visiting Astronomer, Kitt Peak National Observatory[*]
P. O. Box 26732, Tucson, Arizona 85726, U.S.A.

Abstract. An extensive amount of accurate uvby Hβ data has been obtained for stars in young open clusters. Age spread earlier found within the clusters has been confirmed. Many unevolved stars have been observed. The data for these stars are positioned below ZAMS curves in HR diagrams defined by the Strömgren indices. These unevolved stars form a confined lower envelope for all the stars a little below the "normal" ZAMS curves.

A large project of obtaining accurate Strömgren photometry (uvby-Hβ) of mainly B type stars in young open clusters (example NGC 457, 663, 1960, h+χ Per) has now been completed to the distinct stage, where it is possible to derive detailed quantitative knowledge about very young stars, i.e. relationships between observed parameters, and distributions of values of stellar parameters in young open clusters. (In each of the clusters several hundred stars have been observed). These new observations, 1981–83, confirm results obtained in 1977–78, (Jakobsen, Thesis 1980, P.A.S.P. $\underset{\sim}{93}$, pp. 547, 1981), of a considerable age spread within young open clusters relative to the absolute age of the individual stars (10-100 million years).

The observed results for the individual stars were transformed to log Te, Δ Mbol, and log g values. Plots of these values for each of the example clusters NGC 457, 663, 1960 (data sample restricted as mentioned below), with Hejlesen's (1975–78) isochrones and evolutionary tracks overlaid, infer: an age, mass spread in each of the clusters from (6.5?) 7–8 in log age, 13–2.5 $M_\odot$ in mass. The masses for most stars in NGC 457:4–8 $M_\odot$, in NGC 663:5–8 $M_\odot$ in NGC 1960:2.5–5.5 $M_\odot$. This spread cannot be accounted for, by errors in observations or in the calibrations. Influence of rotation cannot be the reason either, according to a study (Jakobsen 1983) of ~ 2000 nearby B type stars with good uvby Hβ, MK, Vsini data. Unresolved binary stars can at the most effect the position of 5% of the stars in the diagrams, and will in some cases just minimize the age spread. Field star contamination is to a great extent avoided by considering distance, reddening, proper motion (pm) data. For these young clusters, the pm errors are large compared

345

A. Maeder and A. Renzini (eds.), Observational Tests of the Stellar Evolution Theory, 345–348.
© *1984 by the IAU.*

to the absolute pm values, hence the use of pm values alone for membership determination is limited. If radial velocities could be obtained for these stars, a more strict definition of the individual cluster or parts of the cluster might be possible. The age spread for a cluster can only be minimized slightly by e.g. varying the composition for the cluster as a whole or for the individual stars. Hence, we are left with a considerable age spread among the stars in the young open clusters.

However, the most striking and surprising result from this new large data sample, where an emphasis was made to observe many more fainter stars than observed before, was that a large amount of these stars were not less massive than the earlier observed brighter stars. Generally, these fainter stars have a mass of the same order as the brighter more evolved stars, but they are younger. Apparently they are so much younger that for a good part of these stars Δ Mbol turns negative, i.e. these stars are positioned below the ZAMS line. Usually, stars "below" the ZAMS curves in different diagrams have been accounted for as being influenced by observational errors, or being peculiar stars. In this case, due to the large samples, these stars cannot be accounted for as being peculiar, as it would lead to a question of what is normal and what is peculiar in an open cluster. Composite stars move the points in the opposite direction. As these data are very accurate and evenly distributed the observational or statistical errors were out of the question too (Jakobsen 1980).

To double check for systematical errors, the whole sample was reanalyzed. The presented sample consists only of data obtained with the 1 channel photometer at the KPNO No. 2, 36" telescope. The transmission curves for the filter set (uvby Hβ) are very close to the original Strömgren filters, the center of mass wavelength of the Hβ filters are very close to prevent reddening effects. Typically, 50 standards were observed per night, well covering spectral, luminosity, reddening range of the program stars. The residuals for the standard stars were $\lesssim$ 0.003 mag. Likewise, the agreement between the frequent observations of reference stars within the open clusters was excellent. There is no indication that anything due to observation or reduction could lead to data points systematically "low" in relation to ZAMS curves.

The evolutionary tracks and isochrones of Becker (1983) have been compared in detail to those of Hejlesen (1975-79). There exist some differences, but these are small compared to the questions at this stage. For given (log Te, log g, composition) the age, mass are approximately the same in the two models, as well as is the ZAMS boundary. The apparent problem does then exist in the stage from converting the observed indices to e.g. atmospheric parameters, log Te, log g. The main problem is in establishing the absolute scale, not the relative scale, for particularly the parameters: log g, Mbol, Δ Mbol; and in the definition of the ZAMS curves. Log g or Mbol, Δ Mbol are highly correlated with β, and for fixed lum. class, log Te is highly correlated with $[c_1]$, c_0 or $[u-b]$.

Therefore plots like ($[c_1]$ or $[u-b]$, β) were made, which did not involve any assumption of established ZAMS curves. Using the restricted sample of observations from NGC 457, 663, 1960 makes it possible to cover the whole B type range very well concerning "unevolved" stars, compared to the earlier samples of $\sim$ 2000 B type stars in the field and the brighter stars in open clusters, used by Crawford, Olsen, Strömgren (C.,O.,S.) for establishing ZAMS curves. For a single cluster the data points are rather evenly distributed around these ZAMS curves. NGC 1960 does not contain as many massive stars as in NGC 457, 663, but as it is closer, the observations contain many well observed unevolved late B type (including A0-2) to A-type stars. The reddening is rather constant in NGC 1960 too. Therefore determination of a lower envelope (also in the (β, M_V) diagram) for the late B type stars is now possible by using these new NGC 1960 data. According to Strömgren (1983), the sparseness of unevolved late B type stars has so far created problems in establishing e.g. accurate ΔMbol calibrations. Relatively, NGC 1960 does only contain few unevolved early B type stars, this range seems, though, to be plentily covered by unevolved stars in NGC 457, 663. The total area of rather evenly distributed points has a distinct lower envelope, which is positioned in average $\sim$ 0.2 mag in β below (i.e. higher β values) the ZAMS curves of C.,O.,S. The latter ZAMS curves differ slightly, mostly around B7 with $\sim$ 0.01 mag. (0.02 mag in β is roughly comparable to 0.5 mag in M_V or 0.25 in log g.)

Moving the ZAMS curves to the new envelopes determined by these newly observed stars in the involved calibrations, for obtaining ages, masses for the stars, will not minimize, but enhance the age spreads within the clusters, even if these unevolved stars were omitted from the age spread considerations within the clusters.

So far, calibrations for B type stars including ZAMS curves have mainly been based on field stars and the brighter parts of young clusters. These clusters are relatively far away, so it has mainly been the apparent brightest part of stars in these clusters, which has been observed and thereby formed the basis for the calibrations. By this natural selection of stars down to a limiting magnitude, or that the observed part of the total number of stars with a certain magnitude is decreasing with magnitude, it is possible that it is slightly evolved stars, which have determined ZAMS. Concerning the use of field stars as a basis, it is also plausible that they are not as unevolved as the unevolved stars in the open clusters.

This new extensive and accurate sample of data for stars in young open clusters looks very promising as a basis for dealing with the above mentioned problems. An evaluation is being made of the results from all the observed clusters. This is a description of work in progress and any comments or suggestions are highly appreciated.

[1]This research, presentation was partly supported by the K. Højgaard-, J. V. Hansen-, Danish National Science Foundations.
[*]Operated by the Association of Universities in Research for Astronomy, Inc., under contract with the National Science Foundation.

DISCUSSION

<u>Frogel</u>: Please comment on 1) field star contribution to your C-M diagram,
2) contamination in your aperture from faint companion stars.

<u>Jakobsen</u>: 1) Field star contribution is to a great extent avoided by
considering distance, reddening, proper motion data. However, radial
velocities would be very helpful.
2) The data shown are all obtained on nights with very good seeing, and
use of a small aperture. The used finding charts were enlargements from
Palomar plates, where stars 5 mag fainter than the observed ones were
easily visible, so appropriate precautions could be taken. If there was
any doubt about such a contamination, then the validity of the measurements was decided on from several observations with different centering,
or e.g. by moving the star gradually out of the aperture.

EVIDENCE FOR A MAIN SEQUENCE WIDENING FOR MASSIVE STARS

J.-C. Mermilliod[1], A. Maeder[2]

[1]Institut d'Astronomie de l'Université de Lausanne
[2]Observatoire de Genève

Theoretical isochrones, in the interval 3 to 25 10^6 y., have been obtained from the grid of stellar models with mass loss computed by Maeder (1980, 1981). They cover the main sequence evolution, the red supergiant region and the first blue loop. These isochrones have been compared to composite colour magnitude diagrams of young open clusters drawn by Mermilliod (1981).

The models correctly reproduce the main features of the observed sequences, in particular, the number of Ia supergiants and their location on the first blue loop. However, the observed number of Ib supergiants is significantly larger than the theoretically expected numbers. This phenomenon is especially noticeable in the colour magnitude diagram of clusters like NGC 884, 3293, 4755, 6871. These clusters exhibit B Ib supergiants on the first crossing trajectory, which is, according to the models, a phase of rapid evolution.

An overall comparison has been performed with a composite diagram (fig. 1) collecting all 25 open clusters considered. Two isochrones (3 and 24 10^6 y.) have been drawn, as well as the theoretical upper main sequence envelope, the binaries upper limit and the blue extremity of the loops. The supergiants whose occurrence is not predicted by the models are located between the binaries limit and the loop extremity.

We conclude that the observed main sequence of very young open clusters is subjected to some extension or widening, which is not theoretically predicted. This effect is quite similar to that found by Maeder and Mermilliod (1981) for intermediate age open clusters.

A. Maeder and A. Renzini (eds.), Observational Tests of the Stellar Evolution Theory, 349–351.

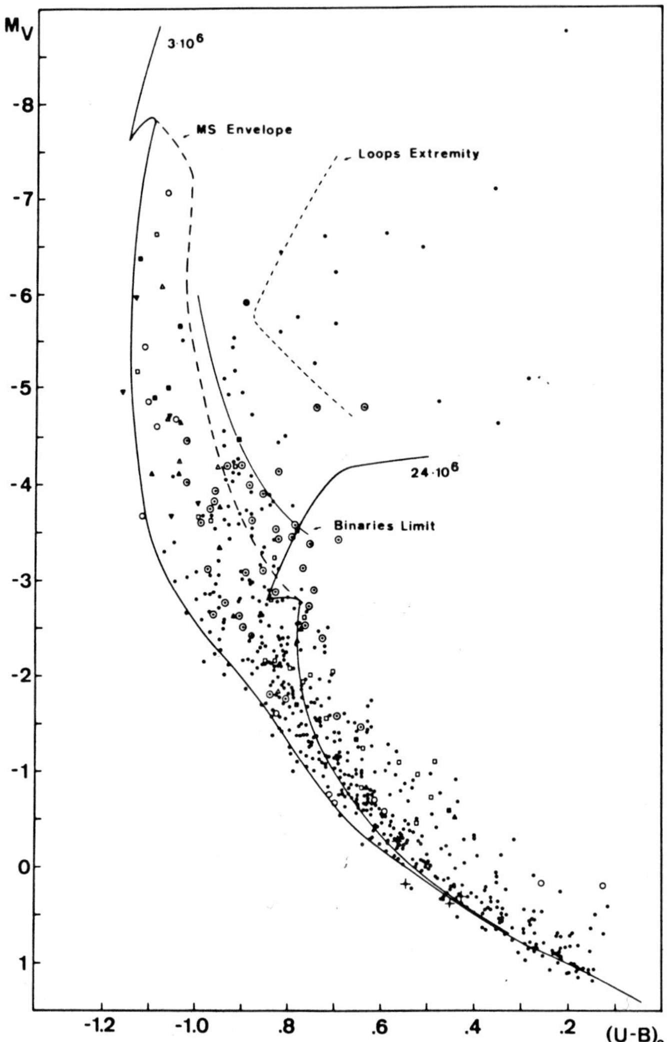

Fig. 1 Composite colour magnitude diagram for 25 young open clusters. The supergiant stars located between the binaries limit and the blue extremity of the loops are significantly more numerous than predicted by the models.

REFERENCES

Maeder, A.:   1980, Astron. Astrophys. 92, 101
Maeder, A.:   1981, Astron. Astrophys. 99, 97
Maeder, A., Mermilliod, J.-C.:   1981, Astron. Astrophys. 93, 136
Mermilliod, J.-C.:   1981, Astron. Astrophys. Suppl. 44, 467

DISCUSSION

Jakobsen: In the poster, you show evolutionary tracks in CM diagrams for different age groups, composed of stars from different clusters, assuming that all the stars in these clusters are almost coeval.
a) Where are the photoelectric data from, and what was the selection criteria for the sample of stars in each cluster?
b) Which criteria did you use in dividing the clusters into age groups?

Comment to b): Several of the clusters you have used, I have observed extensively, and depending on which "cluster age" criteria I use, I do not get the same division.

Mermilliod: The UBV data have been taken from my compilation (Astron. Astrophys. Suppl. 24, 159 (1976)) and are mostly photoelectric ones. Membership of the stars was mainly judged from the analysis of the photoelectric diagrams, in absence of other criteria.
Since the discriminating power of the U-B colour index as a temperature parameter, and hence as an age indicator decreases, I have used the published MK classifications for evolved stars to separate the clusters. This procedure has been described in Astron. Astrophys. Suppl. 44, 467 (1981).

INFRARED AND VISUAL OBSERVATIONS OF h AND χ PERSEI

M. Tapia, M. Roth, R. Costero, S. Navarro
Instituto de Astronomía
Universidad Nacional Autónoma de México

Near-infrared observations of 82 stars in h and χ Persei (NGC 869
and 884) were obtained with the Mexican National Observatory infrared
photometer/spectrometer on the 2.1 m telescope at San Pedro Mártir,
Baja California. The JHK photometry covers most of the stars in the
upper main sequence of the central parts of the clusters as well as the
giants and supergiants in a more extended region, with additional L'
and M measurements of the brightest members. In Figure 1 the (J-H) vs
(H-K) and (H-K) vs (K-L) diagrams are presented. In order to comple-
ment the available good quality UBV photometry, we carried out photoelectric measurements of 23 stars in our sample for which only low
quality photographic data has been published. These observations were
made with the pulse-counting photometer attached to the newly refur-
bished 1.5 m telescope at San Pedro Mártir.

From the infrared data alone, we find a unique value for the
extinction in the direction of both clusters and for stars of all luminosity classes, with an average value of E(J-K) = 0.28 ± .01 which, by
assuming a reddening law given by van de Hulst's curve No. 15 (Johnson,
1968) gives A_V = 1.85 ± .12. This result is in excellent agreement with
Crawford et al. (1970). Comparisons with the infrared and visual data
suggest that the observed dispersion towards large values of E(B-V) is
not due to variable reddening as stated by Wildey (1964), but it may be
intrinsic to the atmosphere of some B-type stars.

From their infrared characteristics we found two clearly distin-
guishable groups of B supergiants: those with no infrared excess up
to 4 μm which turn out to be the less luminous and five stars of luminosity class Ia with excess emission at λ > 2 μm. The shape of these
excesses is interpreted as arising due to Bremsstrahlung emission from
hot gas in strong stellar winds.

Most of the Be stars in χ Persei present significant infrared
excesses. In most cases, this appears to be dominantly Bremsstrahlung
emission from their hot envelopes while a fraction of stars can better
be modelled by a component of thermal emission from circumstellar dust.

A. Maeder and A. Renzini (eds.), Observational Tests of the Stellar Evolution Theory, 353–354.

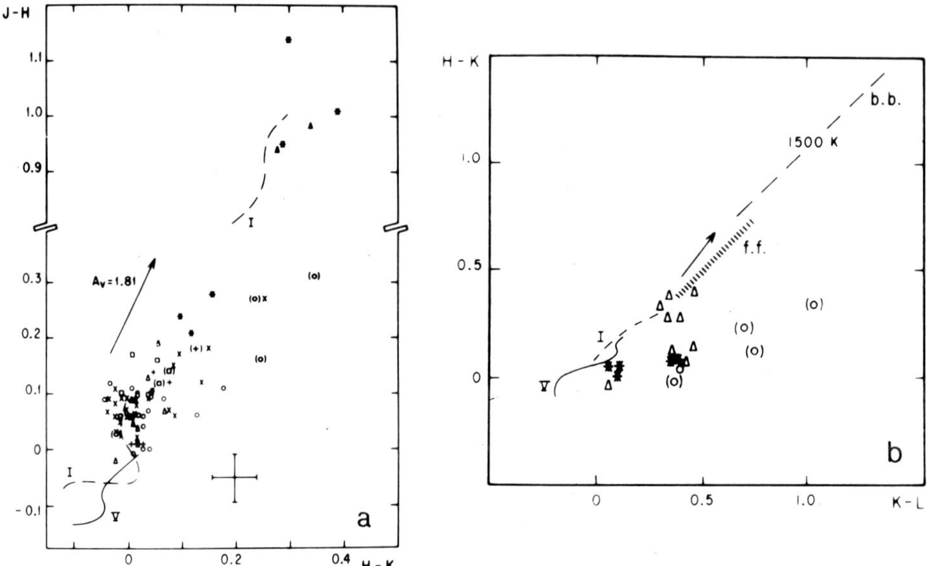

Figure 1.   (a) (J-H) vs (H-K) diagram of all stars observed.
(b) (H-K) vs (K-L) diagram of the brightest stars.
Symbols are as follows: (x) dwarfs, (+) giants, (*) supergiants in h Persei; (o) dwarfs, (□) giants, (△) supergiants
in χ Persei.   Parentheses denote emission line spectra.

The well defined turn-off points in the individual H-R diagrams of h and χ Persei from the present observations and the presence of cooler blue supergiants in χ Persei support Schild's (1967) suggestion that h Persei is younger, by a factor of two in age, than χ Persei. The presence of a number of Be and peculiar stars in the nucleus of the latter cluster, which are absent in h Persei, points in the same direction.

The details of the present work will be published in the Revista Mexicana de Astronomía y Astrofísica.

REFERENCES

Crawford, D.L., Glaspey, J.W. and Perry, C.L.: 1970, Astron. J. 75, 822.
Johnson, H.L.: 1968, in "Nebulae and Interstellar Matter", ed. B.M.
     Middlehurst and L.M. Aller (Chicago Univ. Press).
Schild, R.E.: 1967, Astrophys. J. 148, 449.
Wildey, R.L.: 1964, Astrophys. J. Suppl. Ser. 8, 439.

NEUTRINO-PAIR BREMSSTRAHLUNG IN DENSE STARS

Naoki Itoh and Yasuharu Kohyama
Department of Physics, Sophia University, Tokyo, Japan

Neutrino-pair bremsstrahlung rate is calculated in the framework
of the Weinberg-Salam theory for the dense matter relevant to the
interior of white dwarfs and neutron stars. Ionic correlations both
in the liquid state and in the bcc lattice state have been taken into
account accurately. It is found that the ionic correlation in the
liquid state reduces the neutrino-pair bremsstrahlung rate typically
by a factor 2-20 depending on the element and temperature. It is also
found that accurate inclusion of the Debye-Waller factor reduces the
neutrino-pair bremsstrahlung rate in the lattice state typically by a
factor 2.

Neutrino emission processes play crucial roles in the evolution
of dense stars, such as cooling of white dwarfs, presupernova evolution,
and cooling of neutron stars. One of the most important neutrino
emission processes in dense stars is the neutrino-pair bremsstrahlung
by a degenerate electron liquid interacting with the nuclear Coulomb
field:

$$e^- + (Z,A) \longrightarrow e^- + (Z,A) + \nu + \bar{\nu}. \tag{1}$$

Festa and Ruderman (1969) calculated the neutrino-pair bremsstrahlung
by relativistic degenerate electrons. They also included screening of
the nuclear Coulomb field and lattice effects. Their calculation,
however, was based on the V-A theory of weak interaction. Dicus, Kolb,
Schramm, and Tubbs (1976) were the first to calculate the neutrino-
pair bremsstrahlung rate using the Weinberg-Salam theory. They calculated the neutrino energy loss rate in two cases. One is the so-called
weak screening case where the Thomas-Fermi screening is taken into
account for the electron-ion interaction and no correlation between
the inos is assumed. The other is the so-called strong screening case
where a simple model of screening based on the ion-sphere model is employed.

We have calculated the neutrino-pair bremsstrahlung rate in dense
stars in the framework of the Weinberg-Salam theory by taking into
account the ionic correlations and electron screening accurately.

355

A. Maeder and A. Renzini (eds.), Observational Tests of the Stellar Evolution Theory, 355–356.

For the liquid case the ionic correlation is embodied through the use of the liquid structure factor of the classical one-component plasma which has been a subject of recent active studies. For the bcc lattice case the Debye-Waller factor has been accurately taken into account. Screening due to degenerate electrons is taken into account with the use of Jancovici's (1962) dielectric function.

It has been found that the ionic correlation in the liquid state reduces the neutrino-pair bremsstrahlung rate typically by a factor 2-20 depending on the element and temperature. It has been also found that the accurate inclusion of the Debye-Waller factor reduces the neutrino-pair bremsstrahlung rate in the lattice state typically by a factor 2. These findings are expected to have important consequences on the cooling of white dwarfs, presupernova evolution, and the cooling of neutron stars.

This research was supported in part by the Japanese Ministry of Education, Science, and Culture through Research Grant No.57340023.

REFERENCES

Dicus, D.A., Kolb, E.W., Schramm, D.N., and Tubbs, D.L. : 1976, Astrophys. J. 210, p.481.
Festa, G.G., and Ruderman, M.A. : 1969, Phys. Rev. 180, p.1227.
Jancovici, B. : 1962, Nuov. Cim. 25, p.428.

DISCUSSION

<u>Weidemann</u>: White dwarf central temperatures and densities do not go up to the high values in which your mechanism is important, however, it may be in the region between white dwarf and neutron star densities $(9 < \log \rho < 14)$.

<u>Itoh</u>: Then it refers to pre-supernova evolution.

III. BINARITY, PULSATION, ROTATION AND MIXING

BINARY EVOLUTION AND OBSERVATIONAL CONSTRAINTS

C. de Loore
Astrophysical Institute, Vrije Universiteit Brussel,
VUB, Brussels, Belgium and University of Antwerp,
RUCA, Antwerp, Belgium.

ABSTRACT.
The evolution of close binaries is discussed in connection with problems concerning mass and angular momentum losses. Theoretical and observational evidence for outflow of matter, leaving the system during evolution is given: statistics on total masses and mass ratios, effects of the accretion of the mass gaining component, the presence of streams, disks, rings, circumstellar envelopes, period changes, abundance changes in the atmosphere.
The effects of outflowing matter on the evolution is outlined, and estimates of the fraction of matter expelled by the loser, and leaving the system, are given. The various time scales involved with evolution and observation are compared. Examples of non conservative evolution are discussed.
Problems related to contact phases, on mass and energy losses, in connection with entropy changes are briefly analysed. For advanced stages the disruption probabilities for supernova explosions are examined. A global picture is given for the evolution of massive close binaries, from ZAMS, through WR phases, X-ray phases, leading to runaway pulsars or to a binary pulsar and later to a millisecond pulsar.

I. INTRODUCTION.

The aim of evolutionary computations for close binaries can be summarized as follows:
1. to discover and to describe the physical laws governing the change of:
   a. the various parameters which determine the internal structure of both components $(M,L,T,P,X,Y,...)$
   b. the parameters of the system: masses of the components, distance of the components, orbital period.
2. to determine for a given set of binaries with common characteristics the set of progenitors, hence to compute how sets of binaries (Algols, Wolf-Rayet binaries, X-ray binaries, cataclysmic binaries, bursters, Be-binaries and Be-X-ray binaries) can be formed, starting from sets of ZAMS-binaries.
3. to determine for a given binary system with known actual status the progenitor ZAMS system, to calculate the exact evolutionary sequence

359

A. Maeder and A. Renzini (eds.), Observational Tests of the Stellar Evolution Theory, 359–378.
© *1984 by the IAU.*

leading from the ZAMS system to the actual observed system.
4. to derive for a given binary system with known actual status the
   descendant.
5. as a combination of 3 and 4 complete sequences for close binaries from
   ZAMS stage to their final phases can be computed.

In the following sections we will analyse how progress in the
theory of close binaries and observations has modified the picture of
binary evolution. Mass loss changes the structure of the components,
and in an advanced mass transfer stage also the atmospheric abundance;
this is the case for the mass lossing star (the loser), as well as for
the mass accreting star (the gainer). The consequences of the mass
exchange on the two components, loser and gainer, are investigated in
section 2.

Various treatments of the Roche lobe overflow phase, conservative or not, dealing eventually with contact phases, and a discussion
of the parameters governing mass and angular momentum losses are described in section 3. The different time scales involved with binary evolution are analysed in section 4, and their relevance on the evolution is
discussed.

In section 5 is given a discussion of the connection between
groups of binaries with given characteristics and their ancestors and
descendants. Section 6 deals with different mass outflow possibilities.
In section 7 is described how theory and observations have influenced
the general ideas of binary evolution. Observed features such as gas
streams, disks, period changes are analysed and discussed critically.
Contact systems, models for the computations of such systems, and application for WUMa systems are described in section 8. In the last sections
the advanced evolution of massive close binaries, leading to supernova
explosions, is described; the effect of the explosion on the system,
disruption or not, is discussed, and the origin of binary pulsars and
runaway pulsars is explained. The millisecond pulsar can be considered
as the final evolution of a binary pulsar by coalescence.

II. CONSEQUENCES OF THE MASS EXCHANGE ON THE BINARY COMPONENTS.

From stellar evolution is known that exhaustion of a nuclear
source, or the beginning of exhaustion leads to the expansion of the
star. If this star is member of a close binary with a rotational period
equal to the orbital period, hence showing synchronous rotation, its radius can eventually equal its Roche-radius and mass transfer can occur,
at a rate $\dot{M}$ (Kuiper, 1941).

Usually for stellar evolution it is assumed that the Roche-lobe
constitutes the limiting surface for the contact component of a semi-detached system. Evolutionary computations under the assumption of conservation of mass, and angular momentum during the mass transfer stage
could explain the gross features of groups of observed binaries (a.o.
the Algol paradox). When the conservative assumption is no longer maintained the duration of the first stages of mass transfer can be drastically increased, and the final systems after the mass exchange phase
can be different from the conservative case. During the phase of rapid
mass transfer the gainer can spin up, so that synchronism it lost. It
would be very important to investigate mass accretion under non-synchro-

nous conditions.

If the gainer and/or a possible accretion disk does not expand too drastically, synchronism can be maintained, and the orbit remains circular. The conservative treatment has been justified on dynamical grounds by Lubow and Shu (1975, 1976), Prendergast and Taam (1974), Flannery, 1975 a,b).

In these circumstance (Shu and Lubow, 1981) the outflowing gas with low Mach number makes a sonic transition within a small region around L_1, falls then into the Roche lobe of the gainer and is trapped within that surface.
If the flow impinges on itself or hits the surface of the gainer, orbital energy is dissipated and the gas falls even deeper in the potential well of the gainer. No mass leaves the system. Even if a fraction of the angular momentum is stored in an accretion disk, small with respect to the masses of the two stars, for a certain time, the bulk of the angular momentum remains in the orbital motion of the two stars.

The mass gaining star accretes matter on the Kelvin-Helmholtz time scale of the loser (Paczynski, 1971), which can be different of its own Kelvin-Helmholtz time scale. The effects of rapid mass transfer on the gainer have been examined a.o. by Benson (1970),Yungelson (1973), Flannery and Ulrich (1973) Kippenhahn and Meyer-Hofmeister (1977), Neo et al. (1971), who find that the gainer drastically swells in size, fills itw own Roche lobe giving rise to contact systems, and the subsequent evolution could be different from earlier calculations.

Shu and Lubow (1981) argue however that they are not convinced by the results of Benson; they refer to the behaviour of protostars, where the accretion of matter does not lead to a drastic swelling of the receiving star. (Larson 1969, 1972, Appenzeller and Tscharnuter 1975; Winkler and Newman 1980 a,b, Stahler et al. 1980 a,b).
The reasoning is as follows: as boundary conditions used in binary evolution, an "ad hoc" thermal boundary condition is imposed on a hydrostatic star of variable mass. As boundary conditions is generally used

$$\left(\frac{\partial S}{\partial M}\right)_t = 0$$ which means that the specific entropy on the newly added shell is equal to that of the shell immediately below.

or $$\left(\frac{\partial L}{\partial M}\right)_t = 0$$ meaning that the newly added shell is in radiative equilibrium.

The two forms, which are physically equivalent for steady flow, are not completely correct, because the inflowing material has a very on short local thermal timescale and can behave in a non adiabatic way on falling down to the star, lowering the entropy as the gas radiates. Correct computations are very delicate and difficult (Stahler et al. 1980,b).
In fact the K-H-timescale is connected with the deeper, interior layers of the gainer, and is much larger than the K-H-timescale of the outer layers, where really the material is accreted. In the outer layers the material can, by radiation, lower the specific entropy of the newly accreted matter.
Hence, although the interior structure of the gainer deviates from ther-

mal equilibrium, only drastic changes of the radius occur when a very
large amount of matter (comparable to the original mass) has been accreted.

In a recent study (Hellings, 1983a) the response to accretion
of massive secondaries in close binaries was examined during their core
H burning and shell H burning stages. In this work constant accretion
rates comparable to the main loss rates of primaries during Roche lobe
overflow were used. The accreted matter in these systems is assumed to
fall gently on the surface of the secondary with no shock effects, and
with the same chemical composition as the outer layers of the accreting
star. In the binary program an instantaneous thermohaline mixing procedure (Ulrich, 1972) is used when accreted matter with a lower H abundance is added to the secondary. It is found that the accretion stars are
unable to remain in equilibrium as has been found for low mass and for
ZAMS stars but when accretion stops the secondaries restore their thermal equilibrium by a relaxation stage towards normal ZAMS structure.
This new structure containing a longer convective core and increased
core H constant is obtained by the process of semi-convection. Further
evolution does not deviate from normal main sequence evolution of single
stars.

The results of this work were used to derive the structure of
the secondary after RLOF starting from the initial parameters M_{1i}, q and
β (M_{1i} is initial mass of the primary, q is the mass ratio,secondary to
primary and β is the fraction of the mass lost by the primary and accreted by the secondary, see also section3). This work is the basis for
the mass determination of eight double lined WR + OB binaries (Hellings,
1983,b). With this method the masses of the WR components and OB components are found using the theoretical stage of the companion in the HR
diagram, and the observational results $M_{wr} \sin^3 i$ and q, the mass ratio.
The resulting masses are calculated with an accuracy of about 20%, and
are mostly in the range 7-10 $M_\odot$, which is lower than the estimated masses by Massey (1982).

III. GENERAL SURVEY OF CLOSE BINARY EVOLUTION.

Evolutionary computations for single stars show how the stellar radius changes, as a function of time. The presence of a companion
will impose restrictions on the radius increase. Theoretically this is
translated by the concept of Roche lobes, i.e. the determination of an
allowed maximum volume for the primary with increasing radius imposed
by a critical equipotential surface. This latter is calculated in a simplified way for two point masses for primary and secondary. It can also
be done a tidal lobe concept, i.e. a maximum volume imposed by the tidal interaction of the secondary on the primary.

The easiest way to calculate binary evolution is to assume
conservation of total mass and orbital angular momentum, and moreover,
to assume synchronous rotation. Constraints for conservative evolution
are that the mass ratio $q = M_2/M_1$, with M_1 mass of the primary and M_2
mass of the secondary, is not too small (larger than 0.3), that the separation A is not too small and that the envelope of the mass losing
star is in radiative equilibrium. When these conditions are satisfied
the distance A of the two components, given by

$$\frac{A}{A_o} = (\frac{M_{10}M_{20}}{M_1 M_2})^2$$

is not too much reduced by the mass transfer. The subcripts 0 refers to the initial (ZAMS) values.
If these conditions on mass ratio and distance are not satisfied the system will evolve into a contact system with a common envelope (Webbink 1979).
The effect of the mass loss on a convective envelope will be that this envelope expands, inducing a stronger mass loss rate leading to a catastrophic envelope growth. The envelope engulfes the companion, and also in this case a common envelope will be produced. The tendency of convective envelopes to expand further when mass loss occurs, is a consequence of the fact that for such envelopes the specific entropy of the gas decreases outwards, so that no energy supply for the restoration of the equilibrium is necessary, unlike for radiative envelopes.

The conditions for conservative evolution are not always satisfied, and computations of evolutionary sequences have been carried out with assumptions on mass and angular momentum losses(Yungelson,1973; Plavec et al.1973; Paczynski and Ziolkowski,1967; Vanbeveren et al.1979; Hellings et al.1982). In the case of non conservative evolution the fraction β of the mass lost by the primary and accreted by the secondary has to be specified as well as a prescription on the fraction γ of the total angular momentum leaving the system.

This can be done a.o. in the following ways.
1. arbitrary, i.e. computations are carried out for a series of values
   of β ($0<\beta<1$, with $\beta=1$)the conservative caser,and various values of γ for a set of ZAMS-binaries (Vanbeveren et al.1979). Comparison of the post Roche-lobe overflow systems with observed evolved systems (in the sense of post-mass transfer systems)allows to impose conditions on β and γ (Vanbeveren and de Loore,1980);
2. by assuming $\beta=1$ for semi-detached phases,and adopting lower values
   for contactphases: the fraction β of the mass expelled by the primary is accreted by the secondary, determined in such a way that this secondary keeps filling its Roche lobe, the rest leaving the system in a spherically symmetric way ($\beta<1$) (Plavec,1981).

After a certain time the secondary becomes detached from the contact volume, the system becomes semi-detached and the mass transfer occurs again in a conservative way ($\beta=1$).

The mass transfer comes to an end when the primary shrinks back within its Roche lobe.
3. a different way to treat the contact phase has been investigated by
   Packet (1983) in the following sense: the mass transfer rate is determined such that both stars fill a common equipotential surface, outside their Roche lobes. Until now luminosity transfer has been omitted: in the future this will be investigated.
Mass exchange is thus treated in a conservative way; only when the surface of the system corresponds with the equipotential surface through L_2, mass loss has to be considered. Test computations carried out by Packet for a binary system, with a primary of 9 M_o, and secondaries with mass ratios of 0.9 and 0.6 show that the latter situation is not reached,

and that the total binary evolution occurs in a conservative way, at
least for early B and late A-cases).

4. another possibility is that during the semidetached phase the excess
   material, or a fraction, leaves the system in a spherical symmetrical
way, in the case of massive stars as a kind of enhanced stellar wind,
driven by radiation pressure and increased by gravitational acceleration
or by the effect of the changed potential by the presence of the secondary alone.

5. a detailed analysis of the parameters determining the fractional mass
   loss (β) and the amount of the angular momentum carried along with
the escaping gas g was performed by Gianuzzi (1981), assuming β and g
functions of the mass ratio q alone.

She examined two cases : β = constant and $\beta = \dfrac{q}{1+q}$.
The change of the angular momentum depends critically on the period of
mass ejection. Idealized cases, Jeans'mode and matter leaving the system through the second Lagrangian point L_2 were investigated, leading
to the following expression for the orbital period.

$$\frac{P}{P_o} = - \left[\left(\frac{J}{J_o}\right)^3 / \left(\frac{M_t}{M_{to}}\right)^6 \right] \cdot \left(\frac{q+1}{q_o+1}\right)^6 \left(\frac{q_o}{q}\right)^3 .$$

IV. TIME SCALES.

In order to check the validity of evolutionary schemes one
can compare various parameters predicted by computations with those of
individually observed systems. Another possibility is to compute sets
of similar systems, and to compare their averaged parameters with theoretical values. These statistical methods have the advantage that the
mean values such as masses, periods and mass ratios, as well as their
scatter during or after Roche-lobe overflow can be compared with those
of a comparable group of non evolved systems (Pre-RLOF).

We have at our disposal a restricted number of binaries with
known parameters in their specific evolutionary stage, evolving, but on
timescales that are not the evolutionary timescales, for certain phases
without interaction between the two components, and phases with interactions. Successive evolutionary models have time steps, completely different from the observational time scales. We consider e.g. a primary of
$\sim$ 20 $M_\odot$. The evolutionary timescales for the various phases, i.e. the
time which enters the evolutionary code to calculate two successive models are different from the observational timescales.

The core hydrogen burning lifetime is $\sim$7 million years, the
subsequent stages are $\sim$ 10% of this time. The mass transfer occurs on
the Kelvin-Helmholtz time scale, which is in this case, of the order of
10^4 years. While the time step for successive evolutionary models during core hydrogen is of the order of 10^4 years, the time step for successive models during the mass transfer phase drops to the order of 10
years.

Let uw now turn to the observations. Two calculated successive models, model N and model (N+1) are about 5 years apart, which means
that these two models depict two phases in the life of the binary with
a time interval of 5 years, corresponding to two observations, also 5
years apart. What is happening between these two phases can be observed

but not calculated and this makes the comparison of theory and observations very difficult. This becomes even clearer when we compare the time lapse between two successive models with the orbital period. In our example about 200 orbits fill the gap between two computed successive models!

Evolutionary sequences deal with nuclear timescales, given by τ_n (the timescale related to relevant structure changes), or with thermal timescales τ_k (Kelvin-Helmholtz timescale) given by $3.10^7 \frac{M^2}{RL}$ years. Hence evolutionary models give information on transitions from a long-living situation to another one; such transitions can be described by a hydrostatic code. Evolution on a thermal timescale gives information on changes of the surface structure, changes of the dimensions of systems.

From observations can be derived information on shortliving phenomena, e.g. changes in the flow pattern, information on the impact of the flow of matter from the primary to a disk, and the generation of a hot spot.
An observing run of two weeks for a binary with a period of 10 days reveals what is happening in one period while in an evolutionary sequence for such a system, two successive computed models cover more than 100 periods.
Hence parameters derived from observations can only be used for evolutionary purposes very carefully and with suspicion. The presence of disks can be used as evidence for mass outflow but it does not mean necessarily that the disk will not disappear afterwards, carrying its mass and angular momentum in-or outwards, so that for the overal picture of the evolution with mass exchange and the final status of the system, the storage of transferred matter into a disk is perhaps not so important.

V. IMPLICATION OF OBSERVATIONS OF GROUPS ON THE COMPUTATIONS OF BINARY EVOLUTION.

Observations can give checks on the validity of evolutionary sequences, can furnish information on progenitors of binary systems, or can help to discriminate between various possible scenarios: mass ratio determinations, mass determinations from spectral variations, leading to radial velocities, spectral type determination, radius-determinations from photometric observations, spectral analysis leading to atmospheric abundances, different from ZAMS abundances.
For a review on various stellar groups and their chateristics I refer to de Loore (1983) and references therein, Popper 1980 a,b; Plavec and Koch (1978); Stencel et al. (1980) Robinson (1976), Warner (1976) Plavec (1981) Paczynski (1980), Conti (1982).

For a number of groups the charateristics of the progenitors i.e. ZAMS systems are known: mass range, mass ratios, periods. As may be seenin Table 1 this is the case for Algols, the group of the W Serpentis' Stars, Wolf-Rayet binaries. Also for certain groups links with other, more advanced groups are known, i.e. the later evolutionary scheme. This is the case for Wolf-Rayet stars evolving into massive X-ray binaries. For other groups the situation is less clear: for the progenitorsof WUMA stars there remain doubts, as well as for cataclysmic variables probably evolving into X-ray bursters, and for symbiotic stars.

<u>Table 1</u>.
Groups of close binaries with common characteristics and connected
groups of ancestors and/or descendants.

Progenitors	Group	Descendants.
Intermediate mass ZAMS binaries-short periods	Algols-periods<14d.	White dwars system.
Intermediate mass ZAMS binaries-long periods	W Serpentis-periods from 13 to 600 days	White dwarf systems.
Massive ZAMS systems	-Wolf-Rayet Stars -Be-binaries -Cataclysmic variables periods<1d. One of the components is a white dwarf-disks are present.	Massive X-ray binaries Be-X-ray binaries Low mass X-ray bina-ries.
Fission of single stars in rapid rotation or detached or semi-deta-ched binaries by angular promentum loss=magnetic braking.	WUMA Stars.	

VI. MASS MOTIONS IN BINARIES.

During the lifetime of binaries various phases are present
where the stellar material is moving. The following phases can be dis-
cerned:
1) <u>spherical outflow</u>, due to stellar wind in massive stars, during their
   hydrogen burning stage. This is not specific for binaries but occurs
also for single stars. For massive close binaries this spherical outflow of material can occur during their detached, semi-detached and contact phases.
2) <u>mass flow through L_1</u>: this occurs during Roche lobe overflow phases,
   as a flow from one of the stars towards its companion.
3) <u>mass flow through L_2 and outflow according to Jeans'mode</u>: these modes can be considered in the case of non conservative evolution during semi-detached phases or contact phases.
4) <u>mass ejection</u>. This type of heavy mass flow occurs at advanced evolutionary phases, when supernova explosions eject the outer layers
of the evolved massive star, leaving compact objects.
   Each of these types of mass flow has its repercussion on the
status of the system: changes of the structure, especially on the atmospheric layers, and changes of the orbital elements. In the case of supernova explisions the influence on the system is very large: the periods of the systems can be changed drastically, and the system can even
be disrupted.

VII. INFLUENCE OF OBSERVATIONS ON THE THEORY OF CLOSE BINARY EVOLUTION.

Theoretical as well as observational arguments have modified the schemes adopted for the computation of evolutionary sequences for close binaries. As mentioned before (see section 2) the effect of rapid mass transfer could be that the accreting star swells up, leading to overcontact and ensuing mass loss. Moreover contact systems exist:e.g. Leung & Schneider (1979) have calculated absolute dimensions of the O type binaries UW CMa, AO Cas, V729 Cyg by a combination of photometric and spectroscopic data and found for each of these systems, that they have to be in contact. Determination of the evolutionary status reveals that UW CMa and AO Cas are still burning hydrogen, but evolved from the ZAMS, hence exemples of case A mass exchange. A statistical study by Gianuzzi (1981) reveals a dependence of total mass and angular momentum on the mass ratio for semi-detached systems, pointing to a decrease of the total mass and the angular momentum as evolution proceeds.

From observations (photometry and spectroscopy) we can, apart from radial velocities, light variations, periods, eclipse durations leading to mass ratios, masses, radii, also try to get information on mass motions (streams, expanding envelopes) and mass concentrations (disks, rings, clouds, circumstellar envelopes).
Some examples will be given and discussed.

1. Gas streams in U Cep.

According to Plavec (1981) the interacting binary UCep consists of a B8V primary and a G8III secondary. Most recent mass determinations give $4.4 \, M_O$ and $2.9 \, M_O$ with radii of $2.55 \, R_O$ and $4.95 \, R_O$ respectively. The separation of the components is $15 \, R_O$. The G star is evolved and is losing mass; some returns to the G-star. The gas stream leaves the G star from the hemisphere directed towards the B star, circles around the B star; some of the gas probably falls onto this star. According to this model the rest leaves the system after orbiting 3/4 of the B star. Some of the matter can fall back onto the G star. The gas in the stream is hot enough to keep most of the Fe, Si and Mg atoms ionized. (Kondo et al. 1980). A model is shown in Figure 1. According to Plavec (1983) the line emitting region surrounds the hotter component and extends to several stellar radii. He proposed a hot tubulent circumstellar region.

2. Circumstellar cloud or disk in HR 2142.

Observations of HR 2142 (Polidan and Peters, 1980) in the UV show that most of the material that leaves the primary is not accreted immediately by the secondary. Evidence of mass ejection can be seen in CII 1335 where rather strong, sharp absorption components are present, not from interstellar origin. As most probable origin for these components a circumstellar cloud or disk is suggested. Also in HR 7085 and λ Tauri gas streams were detected.

3. Disks in W Serpentis stars.

The spectra of "Serpentides" show as common characteristics that:
a) in the UV we see strong emission lines superposed on
b) a hot continuum

The optical companions are cooler:
B8II for β Lyr, A6III for SX Cas, F5II for W Ser.
The UV is produced in a region smaller than the observed stellar surface.
In these interacting binaries emission lines of C II, C IV, N V, Si II,

III,IV, Fe III, Al II,III are present, probably related to the mass flow
and accretion; the ionization is most likely connected with a hot spot
or a hot radiative region in the interior of the thick disk. The matter
surrounding the hot component is not completely eclipsed when the star
itself disappears behind the companion, and certain components of the
shell lines remain visible against the background of continuous hydrogen
radiation. A disk model for SX Cas is shown in Figure 2, and spectra in
FIgure 3. β Lyrae has a geometrically thick, opaque disk (Huang 1963,
Wilson, 1974) β Lyrae is in the phase of rapid mass transfer (derived
from the large rate of period change).
The interval of observations is very long ∼ 65 years. In the center of
the disk sits a main sequence star of 10 to 15 M_O.
The disk is stable, otherwise the system would have brightened during
its history in the Lyrae constellation. Wilson (1979, 1981) has suggested that centrifugally limited rotation of the central star is responsible for the persistence of the disk. According to Packet (1981) a small
amount of transferred matter is already sufficient to spin up an accreting star to its critical centrifugal limit.

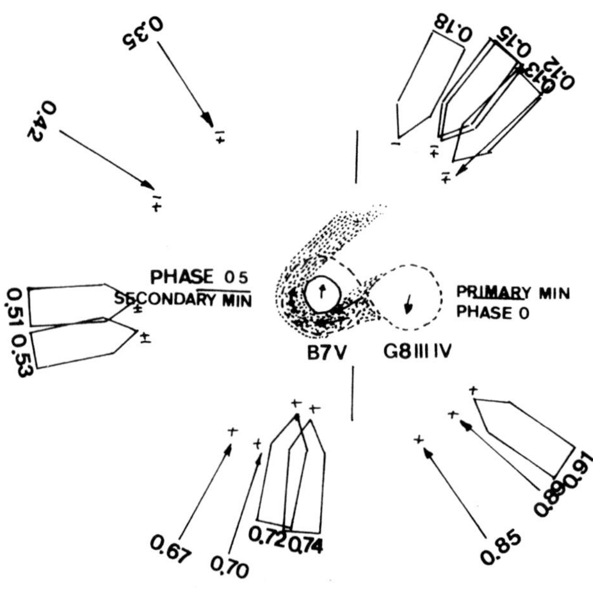

Figure 1 : Schematic representation of the gas stream as observed in
              the Fe II and Mg II resonance lines. The phases of obser-
vation are given. The + sign indicates that the radial velocity of the
gas stream projected against the B-star is away from the observer, the −
sign that the radial velocity is toward the observer; the ±sign both a-
way and toward the observer, and the ∓sign mostly toward, but with a
minor component away from the observer (From Kondo et al. 1980).

β Lyrae is, according to Wilson (1979), a contact binary.

Investigations on streams are based on particle dynamics calculations (Plavec et al. 1964; Plavec and Kriz, 1965). Prendergast 1960, Sobouti (1970), Biermann (1971), have tried to examine the hydrodynamical behaviour of circumstellar gases in binaries but they have studied velocity fields rather than trajectories. Circumstellar matter has been found around one of the components of binaries, as well as between the components, by observations of H α emission in Algol (Struve and Sahade, 1957) and the line of HeIλ4471.

From observations an estimate of the dimensions of disks can be derived. Spectra of VV Cep show emission lines of hydrogen, interpreted as generated in a structure around an invisible hotter ster, a late O or B star.

We have no observational data on the lifetime of disks. In β Lyrae the disk was probable present for the last 65 years, but this is still a small time lapse compared to the evolutionary timescale (section 4). If the disk is resolved after a certain time it has no influence on the final outcome of the system but has just acted as a storage space for matter and angular momentum not immediately transferred. If the mass of disks is small relative to the masses of the components the mass transfer can probably still be treated as conservative. The interpretation of period changes can also lead to arguments for mass loss from the system. The period of SX Cas is variable, and decreases at a rate of $\dot{P}/P = 7.6 \times 10^{-8}$ yr^{-1} (Guinan and Tomczyck, 1979). This period change is comparable to those of β Lyrae or W Ser, but the sign is opposite. Simple mass transfer from the less massive component to the other one INCREASES the period. In SX Cas the loser is certainly the less massive one. Probably strong mass loss from the system in the form of a strong stellar wind occurs. Important information on the evolutionary status and another check on the quality of calculated evolutionary models could be found in the abundance determination, e.g. the H/He ratio in the components of binaries. A non LTE analysis of the O4 star ζ Pup, carried out by Kudritzki et al. (1983), revealed a He-overabundance of NHe/(NH+ NHe) = 0.14 + 0.03. From a similar analysis for an O-type subdwarf eclipsing binary system LB 3459 a helium abundance of 0.3 % by number was found, far below the primordial He-abundance, but caused by diffusion, not by evolution (Kudritzki et al. 1982). For WR stars abnormal abundances for the He-stars, have been found (low or zero H, abnormal C or N); for the O companions, the chemical composition is not known; and neither for the O or Of components in X-ray binaries. Such an analysis would reveal information on the accretion process, and the evolutionary status. For WR binaries it is very diffucult to obtain high resolution spectra of the O components, for the X-ray binaries HD 153919 and HD 77581 however this is perfectly possible.

VIII. CONTACT SYSTEMS.

Most theories for contactsystems are developed for WUMa systems. The reason is that many WUMa systems have been observed, while for early type contact binaries only a few observations exist.
Methods which can be applied:

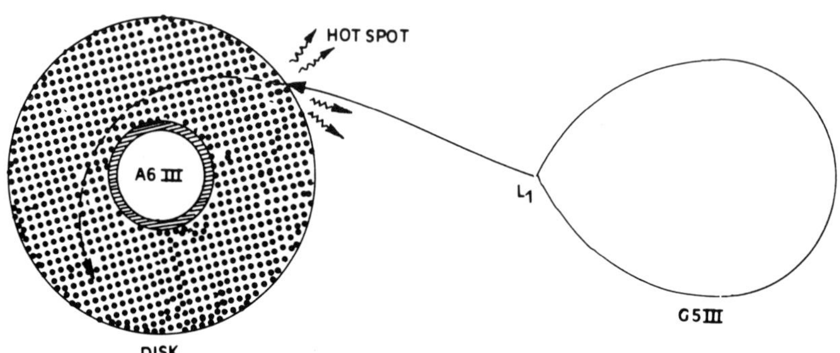

Figure 2 : Model of a close binary during the mass transfer phase, with an accretion disk and a hot spot where the flow of matter, leaving the mass losing star,impinges on the disk. Model for SX Cas by Plavec, Weiland and Koch (1981). The spectral type of the gainer is B7III, that of the loser K3III.

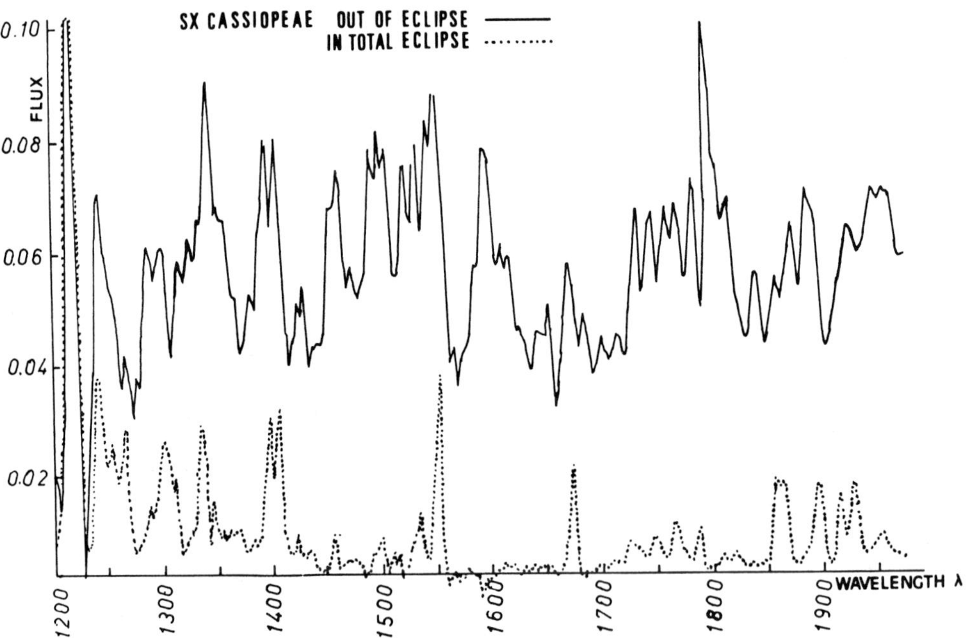

Figure 3 : IUE spectra of SX Cas in-and outside eclipse. During eclipse the continuum disappears, but the UV lines remain. (Plavec, Weiland, Koch, 1981).

1) one boundary condition on the equipotential surfaces Ω for primary
   and se condary, $\Omega_p = \Omega_s$.
We can then assume $\dot{M} = 0$ or determine $\dot{M}$ in a theoretical way. The change
in the luminosity ΔL can then be expressed as $\Delta L = K . \Delta S^m . d^n$ with ΔS
the entropy change and d the depth of the contact disontinuty (Hazlehurst
1970, Meyer and Meyer-Hofmeister, 1980). This is only valid for late
type stars with convective outer layers.
2) 2 boundary conditions :
   $\Omega_p = \Omega_s = \Omega_1$ (both stars fill their critical equipotential surfaces).
(Biermann and Thomas, 1972).
From these two conditions constraints on the mass loss rate $\dot{M}$ and the
energy loss rate $\dot{E}$ can be derived.

Contact binaries are interesting in two respects: evolving binaries can in the course of their evolution both fill their Roche lobe
(depending on their initial mass ratio and initial period), and for the
explanation of WUMa stars. These stars are probably binaries on the
main sequence, with photospheres of the same effective temperature, and
a mass ratio of $\sim$ 0.5 (Eggen, 1961; Binnendijk, 1970). These controversial facts can be explained by assuming that the two stars have an optically thick common envelope where the different luminosities of the two
stars are competely redistributed (Osaki, 1965, Lucy, 1968). Photometric and spectroscopic observations (Rucinsky, 1974, 1978; Anderson et
al. 1980) sustain this model. Real difficulties arise in modelling the
stellar interiors. The equations for stellar structure and the Roche
model are only compatible if the two stars are identical. Solutions for
this problem were presented by Lucy (1976), by a thermal relaxation model, and Shu et al (1976) by a contact discontinuity model. A discussion of these theories is given by Shu (1980).

Observations of binaries with periods below 0.5 days reveal
that the contact systems are in large majority in disagreement with the
fact that the time spent by these binaries during detached and contact
phases should be the same according to the theory of Lucy.
Lucy assumed that the specific entropies in the separate convection
zones below the inner critical surface (through L_1) are equal to the
entropy in the common envelope, condition that only can be satisfied if
hydrogen burning in one of the stars occurs via the PP-chain, and in
the other via the CNO-cycle.

In the contact discontinuity model, a nearly discontinous behaviour across the inner critical surface is allowed. In the outer
layers aboven the critical surface the closed equipotential surfaces
are common to both stars, and horizontal interchange of matter and heat
can occur. Below the critical surface, the two stars have to be de-
coupled. Pressure has to be continuous across the inner critical surface; density and temperature and their gradients are allowed to be
discontinuous. From mechanical considerations can then be derived that
a contact discontinuity allows only a discontinuity for one of the stars:
for the other star the behaviour across its Roche lobe occurs continuously. This model is valid for radiative as well as for convective conditions.
Lucy's models and Shu's model remain contraversial.

IX. SUPERNOVA EXPLOSION-PROBABILITIES FOR DISRUPTION.

Each of the two components of a massive close binary undergoes a supernova explosion. The effects of these events on the status of the system are completely different for the first or the second SN explosion. In the case of the first explosion the less massive star explodes, and consequently the probability that the system is disrupted is extremely low.

Hence most systems remain bound so that many massive X-ray binaries exist. For the second supernova explosion at a later stage, the disruption probability is very large, so that two runaway neutron stars are produced, an old one, remnant of the first explosion and a young active one, just formed.

Computations of the disruption probabilities were carried out by De Cuyper (1982), de Loore et al. (1975). Assuming an extra kick 100 kms^{-1}, a shell expansion velocity of 10000 kms^{-1} and a postsupernova remnant of 1.5 M_O these probabilities are of the order of 70-80% for ZAMS primaries between 40 and 100 M_O for all mass ratios. For lower mass ZAMS primaries (20-40 M_O) the systems always remain bound.

As well in the case of Roche lobe overflow with reverse mass transfer and mass loss as in the case of spiralling in the outer layers with the original composition leave the primary star. The mass lossing O star is "on its way" to become a He-star, hence shows more and more WR characteristics. The object SS433 is possibly an example of a binary in this stage (van den Heuvel et al. 1980), as well as Cir X-1. According to Firmani and Bisiacchi (1980) and Shlovski (1981) the mass loss rate is ~ 10^{-4} M_O yr^{-1}, with an expansion velocity of 1000 kms^{-1}, and the spectral type ranges between Of and WR (from the relative strenghts of Balmer- and Pickering lines). This agrees with the conclusions of van den Heuvel et al. (1980) about the mass loss rate. Probably the X-ray system is surrounded by an envelope of matter expelled by the non compact companion, and not by remnants of the supernova shell as is discussed by Shlovski (1981). An argument is given by the lifetime of the object according to evolutionary computations (11 10^6 years) and the large distance (3.7 - 4.7 kps). Hence the picture of SS433 by van den Heuvel et al. (1980) seems very altractive: the system is in a second X-ray stage, mass is expelled by the primary, and a large part of this matter is stored in a disk.

<u>Table 2</u>   Binary pulsars.

Name	P_{orb}	P_{pulse}(s)	Eccentricity
PSR 0656+64	24^{h}41^m	0.196	0.06
PSR 0820+02	1100d	0.865	0
PSR 1913+16	7^{h}75^m	0.059	0.617
PSR 1937+215		0.0015	

As more and more matter is expelled by the companion, more He-enriched layers will appear at the surface. Hence the next evolutionary stage should be a second Wolf-Rayel stage, i.e. a helium star with a neutron

star companion (de Loore et al., 1975)(Table 3). Since such systems
are descendants from binaries through a supernova event, they are assumed to have a large runaway velocity, and are supposed to be found at
large distance of their place of birth (Moffat and Seggewiss, 1979).
        Further evolution with further spiralling in could then lead
to ultra-short period binaries like Cyg X-3 (van den Heuvel and de Loore,
1973). Finally these systems undergo a new supernova explosion, and
are in nearly all cases disrupted, leaving two neutron stars, an old one
and a young pulsar.
In exceptional cases binary pulsars are formed.
A list of binary pulsars is given in Table 2.

X. GLOBAL PICTURE OF THE EVOLUTION OF MASSIVE CLOSE BINARIES.

        Considering typical objects such as SS433, Cyg X-3, PSR 1913 +
16 and PSR 1937 + 214, and by comparing their characteristics with those
of the standard X-ray sources, a general picture for the evolution of massive stars emerges, starting from a pair of OB ZAMS components through
successive stages of quiet and explosive mass loss, either leading to a
binary pulsar, lateron coalescing into a very short pulsar, or into two
runaway neutron stars (Henrichs and van den Heuvel, 1983). Two massive
stars with a period of ~ 10 days start their mass transfer phase during
shell hydrogen burning of the primary. The sebsequent history is sketched in table 3.
        After the mass exchange stage a Wolf-Rayet binary is formed,
later on evolving, after a SN explosion into an OB runaway, i.e. a massive star with a neutron star compagnon.
As the massive star is nearly filling its Roche lobe; the enhanced stellar wind will be sufficient to produce X-rays, hence a first X-ray stage
occurs.
The secondary evolves through the shell hydrogen burning phase, expands,
fills its Roche lobe. Matter has to leave the system since the neutron
star is not able to accrete the material expelled by its companion. The
optical star is on its way to become a Wolf-Rayet star. An accretion
disk could be formed.
This could represent a second X-ray; SS433 and Cir X-1 could be in this
phase, a phase of slow mass transfer.
At the end of this second mass loss stage, the He-remnant and the orbiting neutron star represent runaway Wolf-Rayet stars.
Due to spiral in of the neutron star in the atmoshere of the WR star, a
third X-ray stage could be produced, a WR-star and a neutron star with
a very short period, such as Cyg X-3.
The He-star continues its subsequent nuclear burning phases, and finally explodes. If the system remains bound a binary pulsar is produced,
such as PSR 1913 + 16. The two neutron stars coalesce, hence a very
short period pulsar is formed, with a period of ~ 1.5 ms. If the system
is disrupted two runaway stars are formed.

XI. CONCLUSIONS.

        Close binary evolution does not necessarily occur in a conser-
vative way. From statistical research on the total mass of close binaries and its dependence on the mass ratio of the components can be deri-

                                                                    C. DE LOORE

Table 3

Primary S_1 type	Secondary S_2 type	M_1	M_2	Age (age 10^6 yrs)	
OB	OB	20	8	0	P=4.56d
OB fills Roche lobe	OB	20	8	6.17	P=4.56d
	Direct mass exchange S1→S2				
He-star	OB	5.4	22.6	6.2	P=10.86d
	End of the first phase of mass transfer				
	Wolf-Rayet binary				
	Evolution to final stage of S1				
Neutron Star OB Runaway	OB	2	22.6	6.78	
	1st.supernova explosion				
	OB Runaway - Remnant of the Further				
	Evolved Helium star is a young neutron				
	star (Pulsar)				
Neutron Star X-ray Binary	OB	2	22.6	11.186	P=11.7d
	OB star nearly fills its roche lobe				
	enhanched stellar wind,				
	accretion on the neutron star generates				
	X-rays				
	1st X-ray phase.				
Neutron Star	OB	2	22.6	11.209	P=11.70
	Reverse Mass-transfer				
	Roche lobe overflow-mass leaves the system				
Neutron Star	Ob→He				
	S2 on its way to WR star, Slow phase of mass				
	accretion disk transfer;optical star				
	2nd X-ray phase it 'on its way' to be-				
	come a Wolf-Rayet star.				
	(Ex.SS433, Cir X-1)				
Neutron star	He	2	6.3	t=11.239 10^6 yr	
	Wolf-Rayet runaway, end of the				
	second mass loss stage.				
	He-star + neutron in common (expanding) envelope.				
Neutron star	He	2	6.3		
	Supercritical disk accretion; 3d X-ray phase(Ex.CYG X-3)				
	2n supernova explosion				
	system remains bound				
Neutron star	Neutron star				P=7.75hrs
	Ex. psr 1913 + 16 binary pulsar				
Neutron star	Neutron star				P=1.5 ms.
	Coalescence of two neutron stars. ex.PSR 1937+214.				
	Millisecond Pulsar				
	OR: system disrupted, two runaway neutron stars.				

ved that mass leaves the system during the evolution. For massive close binaries this happens already during their detached phases, as stellar wind mass losses, and for the most massive ones, only by stellar
wind, since they probably have no mass transfer phase.

Observations give us information on the flow of matter: indications for outstreaming material, disks, circumstellar material are
present, as well as period changes, pointing to outstreaming material.
However the timescales of the observations and the evolutionary timescales are different, and the parameters derived from these observations
cannot be used as general parameters for the modelling of successive
evolutionary models.
Relevant parameters, allowing discrimination among various possible op-
tions (conservative or not, fractions of outflowing matter and angular
momentum, validity of the treatment of mass transfer during contact
phases) are the mass ratios, masses, radii, determined by photometry in
eclipsing systems, periods, period changes, chemical composition, determination of the evolutionary status, lifetimes. Only for very well determined systems it can be tried to trace back the evolutionary history,
and in theory to determine the progenitor ZAMS system. But such systems
are very rare, e.g. S Cnc, an Algol system with period of ~9.5 days,
(HD 74307) studied carefully by Awadalla and Budding (1981).

Hence we are still left with the embarrassing situation that
we know more or less well how certain groups of binaries evolve into
their groups, but that we remain unable to produce individual observed
systems.
For massive stars the situation is less uncorfortable than for intermediate and low mass stars. Within certain limits we can produce Wolf-
Rayet binaries and we know that they will evolve finally into X-ray
binaries. Exact evolutionary computations reproducing well defined individual system are until now impossible. Also detailed calculations
for final stages are still lacking.
The combination of evolutionary tracks and observational data has led
in some cases to the determination of absolute dimensions of binaries.
Doom and de Loore (1984) used evolutionary tracks for massive stars, observations, and the fact that both components of a binary system have
the same age and the same sin i to determine for a number of O stars
their masses and ages.
Research on the disruption probabilities of massive binaries during supernova explosions has reasonably well advanced during the last ten
years and the existance of binary pulsars and runaway pulsars can be
explained. A plausible explanation for the millisecond pulsar by coalescence of the two neutron stars of a binary pulsar has also been published, so that we have an acceptable (although not for all evolutionary
phases completely calculated) picture of the evolution of two zero age
main sequence components of a close binary to a millisecond pulsar, or
to runaway neutron stars.

<u>ACKNOWLEDGMENTS</u>.

An important fraction of the ideas of this review are the results of discussions, suggestions and remarks of my colleagues. I would
like to acknowledge P. Hellings, C. Doom, W. Packet for allowing to use

already some of their results before publication.
I especially thank J.P. De Grève for his important remarks and suggestions concerning the content and the form of this paper.

REFERENCES.
Anderson, L., Raff, M., Shu, F.H.: 1980 in "Close Binary Stars: Observa-
            tions and Interpretation", IAU Symp. 88, eds. M.J. Plavec, D.
            M. Popper, R.K. Ulrich.
Appenzeller, I., Tscharnuter, W.: 1975, Astron. Astrophys. 40, 397.
Awadella, Budding, 1981.
Benson, R.S.: 1970, Ph. D. Thesis Univ. California, Berkeley.
Biermann, P.: 1971, Astron. Astrophys. 10, 205.
Biermann, P., Thomas, H.-C.: 1972, Astron. Astrophys. 16, 60.
Binnendijk, L.: 1970, Vistas in Astron. 12, 217.
Conti, P.S.: 1982 in "Wolf-Rayet Stars: Observations, Physics, Evolution"
            eds. C. de Loore and A. Willis.
De Cuyper, J.P.: 1982, in "Binary and Multiple Stars as Tracers of Stel-
            lar Evolution" IAU Coll. 69, eds. Z. Kopal and J. Rahe.
Doom, C., de Loore, C.: 1984, Astrophys. J. in press.
Eggen, O.J., Roy. Ols. Bull. 31, 101.
Eggen, O.J.: 1967, Mem. Roy. Astron. Soc. 70, 111.
Firmani, C.: Bisiacchi, F.: 1980, Proc. 5th. IAU Regional Meeting, Liège
            Belgium.
Flannery, B.P.: 1975a, Mon. Nat. Roy. Astron. Soc. 170, 325.
Flannery, B.P.: 1975b, Astrophys. J. 201, 661.
Flannery, B.P., Ulrich, R.K., 1977, Astrophys.J. 212, 533.
Gianuzzi, M.A.: 1981, Astron. Astrophys. 103, 111.
Guinan, E.F., Tomczyck, S.: 1979, Inform. Bull. Var. Stars, 1623.
Hazlehurst, J.: 1970, Mon. Not. Roy. Astron. Soc. 149, 129.
Hellings, P.: 1983a, Astrophys. Space Sci. in press.
Hellings, P.: 1983b, Astrophys. Space Sci. submitted.
Hellings, P., Vansina, F., Packet, W., Doom, C., De Grève, J.P., de
            Loore C.: 1982 in IAU Symp. 99 "Wolf-Rayet Stars: Observations
            Physics, Evolution" eds. C. de Loore and A. Willis, p. 397.
Henrichs, H.F., van den Heuvel, F.P.J.: 1983, Nature, vol. 303.
van den Heuvel, E.P.J., Ostriker, J.P., Petterson, J.A.: 1980, Astron.
            Astrophys. 81, L7.
van den Heuvel, E.P.J., de Loore, C.: 1973, Astron. Astrophys. 25, 387.
Huang, S.: 1963: Astrophys.J. 138, 342.
Kippenhahn, R., Meyer-Hofmeister, E.: 1977, Astron. Astrophys. 54, 539.
Kondo, Y.,Mc.Cluskey,G.E., Stencel, R.E.: 1980 in "Close Binary Stars:
            Observations and Interpretation, IAU Symp. 88 eds. M.J. Plavec,
            D.M. Popper, R.K. Ulrich, p. 237.
Kuiper, G.P.: 1941, Astrophys. J. 93, 133.
Kurdtizki, R.P., Simon, K.P., Hamann, W.R.: 1983, Astron. Astrophys.118,
            245.
Kurdtizki, R.P., Simon, P.K., Lynas-Gray, A.E., Kilkenny, D., Hill, P.W.:
            1982, Astron. Astrophys. 106, 245.
Larson, R.B.: 1969, Mon. Not. Roy, Astron. Soc. 145, 271.
Larson, R.B.: 1972, Mon. Not. Roy. Astron. Soc. 157, 121.
Leung, K.C., Schneider, D.P.: 1979, in "Mass Loss and Evolution of O-ty-

pe Stars", IAU Symp. 83, eds. P.S. Conti and C. de Loore, p. 265
de Loore, C.: 1983, in "Double Stars, Physical Properties and Genetic
          Relations", IAU Coll. 80, ed. Z. Kopal.
de Loore, C., De Grève, J.P., De Cuyper, J.P.: 1975, Mem. Soc. Astron.
          Italiana 45, 893.
Lubow, S.H., Shu, F.H.: 1975, Astrophys.J.: 198, 383.
Lubow, S.H., Shu, F.H.: 1976, Astrophys.J. Letters 207, L53.
Lucy, L.B.: 1976, Astrophys. J. 205, 208.
Lucy, L.B.: 1968, Astrophys. J. 151, 1123.
Massey, P.: 1982, in "Wolf-Rayet Stars: Observations, Physics, Evolution
          IAU Symp. 99, eds. C. de Loore and A. Willis, p. 251.
Meyer, F., Meyer-Hofmeister, E.: 1980 in "Close Binary Stars: Observa-
          tion and Interpretation" IAU Symp. 88, eds. M.J. Plavec, D.M.
          Popper, R.K. Ulrich, p. 145.
Neo, S., Miyaji, S., Nomoto, K., Sugimoto, D.: Publ. Astron. Soc., Japan
          29, 249.
Osaki, Y.: 1965, Publ. Astron. Soc. Japan, 17, 97.
Packet, W.: 1983, preprint.
Paczynski, B.: 1971, Ann. Rev. Astron. Astrophys. 9, 183.
Paczynski, B.: 1980, Astron. Astrophys. 82, 349.
Paczynski, B., Ziolkowski, J.: 1967, Acta Astron. 17, 7.
Plavec, M.J., Koch, R.H.: 1978, Inform. Bull. Var. Stars, No 1482.
Plavec, M.J.: 1981, in "Effects of Mass Loss on Stellar Evolution", eds.
          C. Chiosi and R. Stalio.
Plavec, M.J., Ulrich, R.K., Polidan, R.S.: 1973, Publ. Astron. Soc.
          Pacific, 85, 508.
Plavec, P.J.: 1983, Astrophys. J. Submitted.
Plavec, M.J., Sehnal, L., Mikulas, J.: 1964, Bull. Astron. Inst. Czeck
          15, 171.
Plavec, M.J., Weiland, J.L., Koch, R.H.: 1981, Astrophys.J.
Plavec, M.J.: Kriz, S.: 1965, Bull. Astron. Inst. Czeck. 16, 297.
Polidan, R.S., Peters, G.J.: 1980 in "Close Binaries: Observations and
          Interpretations" eds. M.J. Plavec, D.M. Popper, R.K. Ulrich,
          p. 203.
Popper, D.M.: 1980a in "Close Binaries: Observations and Interpretations"
          eds. M.J. Plavec, D.M. Popper, R.K. Ulrich, p. 203.
Popper, D.M.: 1980b, Ann. Rev. Astron. Astrophys. 18, 115.
Prendergast, K.H.: 1960, Astrophys.J. 132, 162.
Prendergast, K.H., Taam, R.E.: 1974, Astrophys.J. 189, 125.
Robinson, E.L.: 1976, Ann. Rev. Astron. Astrophys. 14, 119.
Rucinski, S.M.: 1974, Acta Astron. 24, 119.
Rucinski, S.M.: 1978 in "Non stationary Evolution of Close Binaries" ed.
          A. Zytkow, p. 117.
Schlovski, I.S.: 1981, Proc. 5th. IAU Reg. Meeting, Liège, Belgium.
Shu, F.H.: 1980 in "Close Binary Stars: Observations and Interpretations"
          IAU Symp. 88, eds. M.J. Plavec, D.M. Popper, R.K. Ulrich, p. 477.
Shu, F.H., Lubow, S.H.: 1981 in Ann. Rev. Astron. Astrophys. p. 277.
Shu, F.H., Lubow, S.H., Anderson, L.: 1976, Astrophys. J. 209, 536.
Sobouti, Y.: 1970, Astron. Astrophys. 5, 149.
Stahler, S.W., Shu, F.H., Taam, R.E.: 1980a, Astrophys.J. 241, 637.
Stahler, S.W., Shu, F.H., Taam, R.E.: 1980b, Astrophys.J. 242, 226.

Stencel, R.E., Kondo, Y., Bernat, A.P., Mc. Clusky, G.: 1980 in "Close
          Binary Stars: Observations and Interpretation", IAU Symp. No.
          88 eds, M.J. Plavec, D.M. Popper, R.K. Ulrich, p. 555.
Struve, O., Sahade, J.: 1957, Publ. Astron. Soc. Pac. 69, 41.
Vanbeveren, D., De Grève, J.P., van Dessel, E.L., de Loore, C.: 1979,
          Astron. Astrophys. 73, 19.
Vanbeveren, D., de Loore, C.: 1980, Astron. Astrophys. 86, 21.
Warner, B.: 1976, Structure and Evolution of Close Binary Systems, IAU
          Symp. 73, eds. Mitton, S. Whelan, J., p. 85.
Webbinck, R.F.: 1979, IAU Coll. 49 "Changing trends in variable star
          Research", eds. F.M. Bateson, J. Smak, I.H. Urch, p. 102.
Wilson, R.E.: 1974, Astrophys. J. 189, 319.
Wilson, R.E.: 1979, Astrophys. J. 234, 1054.
Wilson, R.E.: 1981, Astrophys. J. 251.
Winkler, K.H., Newman, M.J.: 1980a, Astrophys. J. 236, 201.
Winkler, K.H., Newman, M.J.: 1980b, Astrophys. J. 238, 311.
Yungelson, L.R.: 1973, Nauch. Inf. 27, 93.

TIDAL EFFECTS IN CLOSE BINARY STARS

Jean-Paul ZAHN
Observatoires du Pic du Midi et de Toulouse

1. INTRODUCTION

This review addresses the question of what can be learned about the stellar interiors from the study of the tidal effects in close binary stars.

The basic mechanism of tidal interaction is sufficiently well known that we need only here to recall the main results. For more detail, we refer the reader to the classical monography written by Kopal (1959).

In a binary star, each component is submitted to a tidal potential U, which can be expanded in spherical functions for its spatial dependence, and in Fourier series for its temporal behavior. Such expansion takes the form

$$U(r,\theta,\varphi,t) = \sum_{nlm} U_{nlm}(r/R)^n \, P_n^{|m|}(\cos\theta) \, \exp im\varphi \, \exp i(l\omega - m\Omega)t \ . \qquad (1)$$

For convenience, the axis of the polar coordinates r,θ,φ will be chosen perpendicular to the orbital plane. R is the mean radius of the star, ω the mean orbital angular velocity and the integers n,l,m take all values satisfying the inequalities $n \geqslant 2$, $|m| \leqslant n$.

The celestial mechanics provides expressions for the amplitudes U_{nlm} of this expansion in terms of the mass of the companion star and the parameters characterizing the orbit.

Since the star is distorted by that tidal potential, its outer gravitational field contains a non-spherical component, whose potential ϕ can likewise be expanded as :

$$\phi(r,\theta,\varphi,t) = \sum_{nlm} \phi_{nlm}(r/R)^{-n-1} P_n^{|m|}(\cos\theta) \, \exp im\varphi \, \exp i(l\omega - m\Omega)t. \qquad (2)$$

The amplitudes ϕ_{nlm} of this perturbing potential are proportional

A. Maeder and A. Renzini (eds.), Observational Tests of the Stellar Evolution Theory, 379–389.
© *1984 by the IAU.*

to the amplitudes U_{nlm} of the tidal potential, and the proportionality coefficients which link them reflect the internal structure of the star, and more generally its physical properties. Since the companion star moves in that perturbing potential ϕ, it can be used as a test particle to explore this field, and hence to determine the coefficients mentioned above. This provides one of the most powerful means of probing the interior of the stars.

Before we proceed, let us mention that the rotational distorsion can be treated in the same way, at least as long as the centrifugal force derives from a potential ; that potential then plays the same role as the tidal potential, and it also induces a perturbing field.

2. APSIDAL MOTION

To first approximation, ignoring the dynamical effects and neglecting all dissipation mechanisms, one has

$$\phi_{nlm}/U_{nlm} = 2k_n \tag{3}$$

where the k_n are constants which depend only on the mass distribution inside the star (see Kopal 1959).

By far the largest contribution to the perturbing potential comes from the quadrupolar moment of this mass distribution (n=2) ; it creates a force in r^{-4} which causes a slow uniform rotation of the semi-major axis in an eccentric orbit. For this reason, k_2 is called the __apsidal motion constant__. Its value would be 0.75 for a homogeneous star ; in real stars, k_2 drops below 10^{-2}.

Both stars contribute of course to this apsidal motion, whose period P_{aps} is related to the apsidal motion constants $(k_2)_1$ and $(k_2)_2$ of the two components by

$$P/P_{aps} = c_1 \ (k_2)_1 + c_2 \ (k_2)_2 \tag{4}$$

to that degree of approximation (n=2). The coefficients c_i are given by :

$$c_i = (R_i/a)^5 \ [15 \ (M_{3-i}/M_i) \ (1 + \tfrac{3}{2} e^2 + \tfrac{1}{8} e^4) \ (1 - e^2)^{-5}$$
$$+ \ (\Omega_i/\omega)^2 \ (1 + M_{3-i}/M_i) \ (1 - e^2)^{-2}] \ . \tag{5}$$

The first term of this expression is due to the tidal potential, and the second to the rotational distorsion ; M_i and R_i are the masses and radii of the stars, Ω_i their angular velocities (supposed uniform here, with the rotation axis being perpendicular to the orbital plane), P is the period of the orbital motion (P = $2\pi/\omega$), a the semi-major axis and e the eccentricity.

There is also a relativistic contribution to the apsidal motion, but it is in general negligible ; one exception is V889 Aql, a star studied by Gimenez and Scaltriti (1982) (and of course the Sun when considering the orbit of Mercury).

The severe a^5 dependence of P_{aps} explains why the apsidal motion can only be detected in very close binaries. For reasons that will become clear later on, it is difficult to observe in binaries which contain lower main sequence stars.

For a given binary showing this apsidal motion, the coefficients c_i can in principle be determined from the observations. The ideal case is that of an eclipsing binary, for which the fractional radii R_i/a can be measured directly. If the rotational velocities Ω_i cannot be derived from the width of the spectral lines, some assumptions must be made (corotation with the orbital motion, for instance), but they do not have a crucial influence on the result, at least for not too extreme mass ratios.

Once the coefficients c_i have been calculated, Eq. 4 provides a weighted mean of the apsidal motion contants, which can be compared with the theoretical predictions. This procedure is known as the apsidal motion test. The first of such comparisons were made by the pioneers in this field : Cowling (1938), Sterne (1939), Russel (1939), Luyten el al. (1939), and Kopal (1940) ; they found a reasonable agreement with the stellar structure theory, which was then in infancy, although they were somewhat disquietened because Eddington's standard model could not account for all the stars.

Shortly after the publication by Kushwaha ·(1957) of much improved stellar models for upper main sequence stars, a new comparison was carried out by Schwarzschild (1958). He found a trend, for the then observed stars, to have a smaller k_2 than that predicted by the models ; the discrepancy, as measured by $\Delta \log k_2 = (\log k_2)$ theory - $(\log k_2)$ observation, was of order 0.5, the reference models being those of zero age main sequence (ZAMS) stars. But since k_2 decreases as the star evolves off the main sequence, the star becoming more centrally condensed, Schwarzschild concluded that the observed discrepancy could be ascribed to evolutionary effects.

This interpretation was later questioned by Kopal (1965), and new comparisons were made by Petty (1973), Stothers (1974), Odell (1974) and Monet (1980), who used more refined models of various chemical compositions, and tried different opacities.

The most recent investigation is due to Gimenez and Garcia-Pelayo (1982) ; they rediscuss the available data, retain 55 eclipsing binaries, determine their apsidal motion period by a more refined method and compare the results with current ZAMS models. They find a very strong correlation between $\Delta \log k_2$ and $\Delta \log g$, the latter quantity refering to the gravity : $\Delta \log g = (\log g)$ ZAMS $- (\log g)$ present state. Moreover, the regression line, which has a slope of nearly + 1, passes through the origin $(\Delta \log k_2, \Delta \log g) = (0.0)$.

There is therefore little doubt left that the departures of the observed apsidal motion constants from those predicted by ZAMS models is due primarily to the evolution of the stars. However, the standard evolutionary sequences seem unable to explain the occurrence of rather large values of $\Delta \log k_2$, up to 0.8.

Two explanations are presently offered to account for these large deviations. One, already advocated by Stothers (1974), and again supported by Jeffery at this meeting, is that the Cox-Stewart opacities are inadequate and ought to be replaced by the Carson opacities, which seem to lead to a better agreement for the evolved stars. Another interesting suggestion was made by Odell (1974), namely that some mixing occurs near the convective core of the stars ; evolutionary calculations performed in allowing such mixing also yield smaller values of K_2.

At the present stade, the question can hardly be considered as settled, and one should therefore encourage both further observations and further theoretical work.

3. CIRCULARISATION OF THE ORBITS AND SYNCHRONISATION OF THE ROTATION

3.1. Stars with convective envelopes

Proceeding further with the analysis of the tidal interactions in a binary star, the next step is to consider the effects which arise from viscous friction or radiative damping. Let us do this first while still ignoring the contribution of the inertial terms ; in other words, we shall deal only, as before, with the equilibrium tide.

The dissipation mechanisms act to delay the tide by an angle α which is called the tidal lag ; its expression has been established by Darwin (1879) for a viscous stars :

$$\alpha = \frac{\Omega - \omega}{t_F}^3 \frac{R}{GM} . \tag{6}$$

The tidal lag is inversely proportional to the friction time t_F, which is given by :

$$t_F = R^2/v \tag{7}$$

v being the viscosity, and neglecting for the present purpose factors of order unity.

The same expression (6) can by used to predict the tidal lag due to radiative damping, the corresponding friction time then being the Kelvin-Helmholtz time $t_F = R^2/\kappa \sim GM/RL$ (κ : mean radiative conductivity, L :luminosity of the star).

It is that tidal lag which is responsible for the exchange of angular momentum between the rotation of the stars and their orbital motion. This has two observable consequences : the decrease of the eccentricity of the orbit, and the synchronization of the rotation of the stars wih their orbital revolution. The time scales for these two phenomena are given by :

$$t_{circ}^{-1} = - d\ln e/dt = 21/2 \; k_2/t_F \; q(1+q) \; (R/a)^8 \qquad (8)$$

for the circularization, and

$$t_{sync}^{-1} = -d\ln(\Omega-\omega)/dt = 6 \; k_2/t_F \; q^2 \; MR^2/I \; (R/a)^6 \qquad (9)$$

for the synchronization. Here q is the mass ratio of the considered star with respect to its companion, and I is its momentum of inertia. (In Eq. 8 the contribution of the second star has been omitted for simplicity sake.)

When applying these expressions to the dissipation mechanisms mentioned above, namely viscous friction and radiative damping, one obtains time scales which are much too long to explain the observations. The only mechanism which, for the equilibrium tide, yields time scales that are compatible with the observations, turns out to be the turbulent friction due to thermal convection (Zahn 1966). If the considered star has a convective envelope of sufficient depth, the friction time is $t_F \sim (MR^2/L)^{1/3}$, within the uncertainties of the stellar convection theory, and this time is extremely short (about 1 year for main sequence stars).

It is because the turbulent stresses are so efficient that most close binaries with convective envelopes have circular orbits. (For this reason, consequently, very few of such binaries can be submitted to the apsidal motion test.) The qualitative difference in behavior between upper and lower main sequence binaries is striking, and may even be used to determine the mass above which those stars no longer prossess an outer convection zone (Zahn 1966). But it is only recently that the quantitative agreement between theory and observations for late type main sequence binaries has been checked by Koch and Hrivnak (1980) : they performed computer simulations of the dynamical evolution of a sample of hypothetical binary systems, and found that their circularization proceeds on a time scale which is compatible with the observations.

Another proof of the theory is offered by the BY Dra stars. The rotation period of these K and M dwarfs can be determined accurately through their photometric variations, which are attributed to spots on the rotating surface. Bopp and Fekel (1977) have shown that the incidence of binaries among those stars is rather high, and that most BY Dra binaries are synchronized with their orbital motion. The two of them which depart significantly from synchronism (including BY Dra itself) may be interpreted as systems which are young (they also have eccentric orbits), as was discussed by Bopp at al. (1980) and by Edwards (1983).

One may thus consider as well established that the physical mechanism which is responsible for the orbital circularization in lower main sequence binaries is the turbulent friction in the convective envelope of those stars, acting on the equilibrium tide.

3.2.<u>Stars with radiative envelopes</u>

For the upper main sequence binaries, another mechanism must be invoked, namely radiative damping on the <u>dynamical tide</u> (Zahn 1975). It is well known that a star, which is a self-gravitating compressible body, can oscillate in a variety of eigenmodes, which have been classified by Cowling in p-modes (or acoustic modes) and g-modes (or gravity modes). The latter have the largest period, and they may therefore enter in resonance with the periodic tidal potential, if the considered star is a binary component. In practice, the amplitude of such resonances is severely limited by radiative damping, in the outer part of the envelope where the radiative relaxation time becomes rather short.

Since the modes which are involved in those resonances are of high order, an asymptotic theory can be applied to study them and to predict their amplitude. In the limit of a very long tidal period (i.e. close enough to synchronism), this amplitude does not depend on the details of the damping mechanism involved, but only on a series of structural parameters E_n which are much like the apsidal motion constants k_n.

For the dynamical tide, the amplitude ratio between the Fourier components of the perturbing potential and those of the tidal potential then takes the limiting form :

$$\phi_{nlm} / U_{nlm} = (3^{-1/2} + i)\, E_n\, [(l\omega - m\Omega)\, (R^3/GM)^{1/2}]^{8/3} \qquad (10)$$

It is the imaginary part of this expression which is responsible for the exchange of angular momentum ; the real part contributes, although negligibly, to the apsidal motion. The coefficients E_n are very sensitive to the structure of the star, and in particular to the size of its convective core. For zero age main sequence stars, its value ranges from $2.4\ 10^{-9}$ (for $1.6\ M_\theta$) to $3.5\ 10^{-6}$ (for $15\ M_\theta$).

We refer the reader to the original paper (Zahn 1977) for the expressions giving the relevant synchronization and circularization time scales. Here we shall repeat only the parameters which may be the most useful for the comparison with the observations. Table 1 lists the period P_c and the separation a_c below which a binary will be respectively synchronized and circularized during its main sequence life span. (The table gives the directly measurable fractional radius R/a_c rather then a_c ; the binaries are assumed to consist of two identical components.)

Several papers have been devoted recently to the verification of the predictions based on this theory, using the observational data available for detached binaries. (In semi-detached and contact binaries, the mass transfer is expected to be much more effective than the tides).

Table 1

Critical separations and periods for the synchronization and circularization in close binaries

Mass(M_o)	Synchronization		Circularization		Cause
	R/a_c	P_c(days)	R/a_c	P_c(days)	
1.0	0.022	30	0.056	5.6	(1)
1.6	0.164	1.21	0.225	0.75	(2)
2	0.142	1.59	0.200	0.95	(2)
3	0.147	1.92	0.206	1.10	(2)
5	0.153	2.19	0.214	1.33	(2)
10	0.150	3.30	0.210	2.00	(2)
15	0.142	3.98	0.200	2.38	(2)

Causes : (1) Turbulent friction acting on the equilibrium tide in the convective envelope ; (2) Radiative damping acting on the dynamical tide in the radiative envelope.

For the circularization, the observations show that most systems with $R/a \geqslant 0.25$ have circular orbits, as shown by Giuricin et al. (1984b). This limit is slightly higher than that predicted by the theory outlined above (0.200 to 0.225 depending on the mass), suggesting that the tidal coefficients E_n are probably higher than those inferred from the zero age main sequence models. A noticeable exception is α Vir, for which $e = 0.14$ whereas $R/a = 0.29$; one is therefore tempted to interpret this system as a young unevolved binary.

For the degree of synchronization, the situation is more complex. One observes a marked tendency for synchronism above R/a = 0.15, in agreement with the theory ; however, as was noticed by Rajamohan and Venkatakrishnan (1981) and by Giuricin et al. (1984a), below that limit and down to R/a = 0.05, both synchronized and non synchronized systems exist. Rajamohan and Venkatakrishnan invoke high subsurface magnetic fields to account for this stronger than expected synchronization. A more plausible explanation is that the surface layers, which experience a higher torque per unit mass, are synchronized much faster than the interior of the star. If this is confirmed, the observations can yield an upper limit of the differential rotation which is tolerated in a star.

Gimenez and Andersen (1983), and also Giuricin et al. (1984a), observe that in eccentric binaries synchronism is achieved for a velocity which is larger than the Keplerian velocity $V_K = \omega a$. Such a deviation is expected for any type of tidal friction ; its exact value, once determined, can also be used to identify the physical mechanism which is responsible for it. However, as we have seen, there is little doubt left that this mechanism is indeed the radiative damping acting on the dynamical tide.

4. CONCLUSION

The study of the tidal interactions in close binaries has already provided one of the most valuable checks of the stellar structure theory. More will certainly be learned, through better determinations of an increasing number of apsidal motion constants ; for instance they may put observational constraints on the mild mixing processes which are now suspected in radiative interiors, due to convective overshooting or hydrodynamical instabilities.

And a new field of research has opened with the observation of the degree of circularization and synchronization, which contains potentially a precious information on the age and the evolutionary state of a given system, even though some assumptions have to be made on the yet unknown initial conditions. To exploit this information completely, a new approach is probably needed, involving the simulation of the whole dynamical history of the considered binary. And the theory itself must be refined, for example to take explicitely the rotation into account, for it induces a variety of oscillation modes which have not be considered so far, namely the quasi-toroidal modes (see Provost et al. 1979).

May I conclude by expressing the hope that many more colleagues, both observers and theoreticians, will be seduced by this exciting subject !

REFERENCES

Bopp,B.W.,Fekel,F.:1977, Astron. J. **82**, 490.

Bopp,B.W.,Noah,P.V.,Klimke,A.:1980, Astron. J. **85**, 1386.

Cowling,T.G.:1938, Monthly Notices Roy. Astro. Soc. **98**, 734.

Darwin,G.II.:1079, Phil. Trans. Roy. Soc. **170**, 1.

Edwards,D.A.:1983, Astron. Astrophys. **123**, 316.

Gimenez,A.,Scaltriti,F.:1982, Astron. Astrophys. **115**, 321.

Gimenez,A.,Garcia-Pelayo,J.M.:1982,   Binary   and   Multiple   Stars   as
    Tracers   of   Stellar   Evolution,   **37**   (Z.   Kopal   and   J.   Rahe   eds.)   D.
    Reidel Publ. Co, Holland.

Gimenez,A.,Andersen,J.:1983,   Réunion   de   Strasbourg   sur   "Les   Etoiles
    Binaires dans le Diagramme HR".

Giuricin,G.,Mardirossian,F.,Mezzetti,M.:1984a,   Astron.   Astrophys. (in
    print).

Giuricin,G.,Mardirossian,F.,Mezzetti,M.:1984b,   Astron.   Astrophys. (in
    print).

Koch,R.M.,Hrivnak,B.J.:1981, Astron. J. **86**, 438.

Kopal;Z.:1940, Harvard Circ. N° 441.

Kopal,Z.:1959, Close Binary Systems, Chapman Hall, London.

Kopal,Z.:1965, Adv. Astron. Astrophys. **3**, 89.

Kushwaha,R.S.:1957, Astrophys. J. **125**, 242.

Luyten,W.J.,Struve,I.,Morgan,W.W.:1939, Publ. Yerkes Obs. **7**, 251.

Monet,D.G.:1980, Astrophys. J. **237**, 513.

Odell,A.P.:1974, Astrophys. J. **192**, 417.

Petty,A.F.:1973, Astrophys. Space Sci. **211**, 189.

Provost,J.,Berthomieu,G.,Rocca,A.:1981, Astron. Astrophys. **94**, 126.

Rajamohan,R,Venkatakrishnan,P.:1981, Bull. Astron. Soc. India **9**, 309.

Russel,H.N.:1939, Astrophys. J. **90**, 641.

Sohwarzschild,M.:1958, Structure and Evolution of the Stars, Princeton
     Univ. Press.

Shu,F.H.,Lubow,S.H.:1981, Ann. Rev. Astron. Astrophys. **19**, 277.

Sterne,T.E.:1939, Monthly Notices Roy. Astron. Soc. **99**, 662.

Stothers,R.:1974, Astrophys. J. **194**, 651.

Zahn,J.-P.:1966, Ann. Astrophys. **29**, 313, 489 and 565.

Zahn,J.-P.:1975, Astron. Astrophys. **41**, 329.

Zahn,J.-P.:1977, Astron. Astrophys. **57**, 394.

DISCUSSION

Mestel: I believe there is support for tidal synchronization from observations (especially by Abt et al) on A-stars in binary systems.
Wide binaries have normal, rapidly rotating A-stars. Intermediate
systems have A-stars with Am spectra, for which slow rotation is believed to be essential; spin-orbit synchronization will indeed produce slow rotators. In close binaries, synchronization produces rapid
rotators, so that the A stars should again be normal.

Spruit: Did you say that in early-type binaries there is agreement
with predictions as far as circularization is concerned?

Zahn: Yes.

Andersen: Two questions from the observer's point of view:
1) What we can observe spectroscopically is of course the rotation of
the surface layers (and we can say only whether it agrees with synchronism to within $\pm$ 10 % or so). If I understood correctly, you suggest
that the surface layers may well be synchronized while the interior is
still rotating more rapidly, while my recollection of a similar recent
review by Savonije (Cambridge, Aug. 1983) is the reverse?
2) Would you always expect $t_{sync} < t_{circ}$ for stars of the same type
(cf. our poster on TZ For, where $M_1 \approx M_2$, both stars are convective,
orbit is circular, and the larger star is synchronized while the
smaller one is not)?

Zahn: 1) The torque per unit mass is larger near the surface than in
the interior. One expects therefore the surface layers to be synchronized faster, if they decouple from the interior.
2) Yes, $t_{sync} < t_{circ}$ as long as the mass ratio is not too extreme.

<u>Mayor</u>: You mention a limiting period of P = 5.6 days for orbital cir-
cularization of a system with one solar mass components. After how much
time this circularization is achieved?

<u>Zahn</u>: The decay of the eccentricity is exponential in time. In a main-
sequence life span (about 10^{10} years) the eccentricity will be reduced
by a factor of $e^4 \sim 60$ for such a system.

CONSTRAINTS ON STELLAR EVOLUTION THEORY FROM PRECISE ECLIPSING
BINARY DATA

J.Andersen, J.V.Clausen, H.E.Jørgensen and B.Nordström
Copenhagen University Observatory, Denmark

Previous attempts at a detailed confrontation of eclipsing binary data with theoretical models of main-sequence evolution were faced
with the choice between data of inhomogeneous (mostly low) quality for
many systems (Kříž, 1969; Lacy, 1979) or accurate values of mass, radius, and temperature (or luminosity) for very few systems only (Popper et al., 1970). In addition, more detailed and homogeneous stellar
structure calculations for several compositions were needed. Since
1972, a coordinated photometric and spectroscopic programme at our institute contributes to building a sufficient observational basis for
such a test. Among published standard models for the range 1-10 $M_\odot$,
Hejlesen's (1980) are the most extensive, agree well with other standard models, and are presented in a format suitable for comparison
with binary data. Here we can only outline a few salient new results
from this study.

Experience in comparing with theoretical models shows that for
useful discrimination between models of different chemical composition,
mixing length, etc., a precision in the observed values of log M, log
R, and log Te of about 0.01 (0.03 in log g) or better is required.
Fig.1 displays the existing fundamental data (M and R) in this class
(Popper, 1980 and later; later published and unpublished data by our
group), and shows main sequence boundaries and isochrones for Hejlesen's models for (X,Z)=(0.70, 0.02). Although parts of it are still
sparsely populated, this diagram shows very good agreement with the
observations (same age within a system) for Z close to 0.02 in a wide
range of Y. Better coverage of strategic parts of this diagram should
permit a direct test of such effects as mixing, overshooting, etc.,
but in this mass range, standard models apparently predict correctly
the evolution in <u>radius</u>, as shown by the systems with components of
appreciably different mass.

However, when also the observed <u>temperatures</u> are compared with
the model predictions, we obtain the unexpected result that the systems do not fit one single composition, but fall into two distinct
groups: the B-Al stars fit (X,Z)=(0.70, 0.02) and the later spectral

391

A. Maeder and A. Renzini (eds.), Observational Tests of the Stellar Evolution Theory, 391–394.
© *1984 by the IAU.*

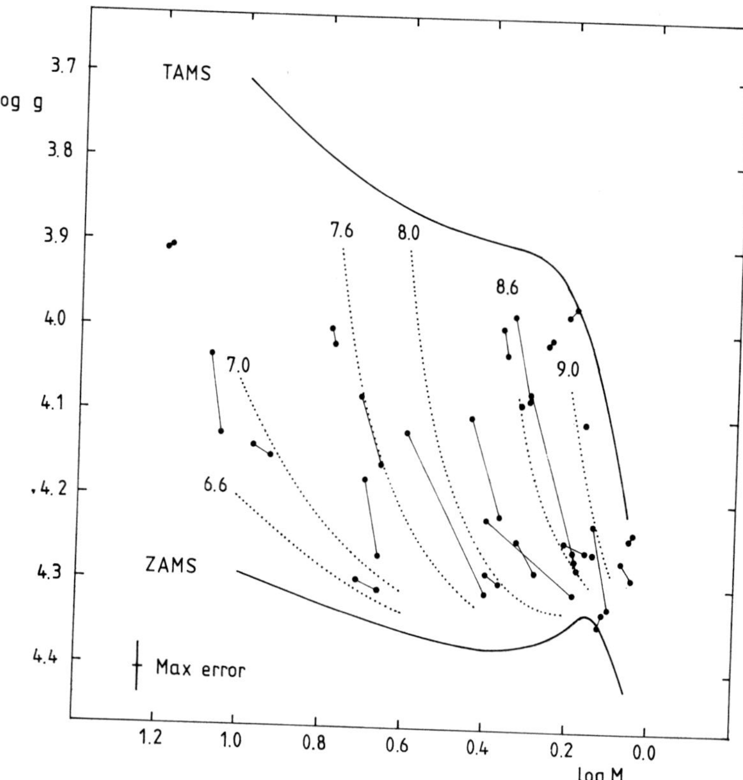

Fig.1. log M - log g diagram comparing the best available data with main-sequence boundaries and isochrones (with log age) by Hejlesen for (X,Z)=(0.70, 0.02).

types (X,Z)=(0.80, 0.02). This effect is well seen in the log Te - log g diagram, and Fig.2 shows the region of it where the stars separate into the two groups. The systems are plotted on the theoretical tracks according to their observed log M and log g values, and the error bars in log Te indicate the range in model temperatures corresponding to the observed mass limits. Arrows point to the observed log Te when the difference from the model prediction exceeds 0.01. To the left in the diagram (and higher masses) there is good agreement, to the right (and lower masses) observed log Te's are lower, by about 0.05. Choosing models of (X,Z)=(0.80, 0.02)[or (X,Z)=(0.70, 0.04) but those are excluded by other data] would reverse the situation.However, the two groups overlap in such single parameters as relative radii, orbital eccentricity, mass, temperature, gravity, surface rotation, and age, which makes it difficult to ascribe the effect to systematic errors in the models or temperature calibrations - or indeed to abundance differences. A third parameter (differential rotation, magnetic fields?) seems to be operating.

Forthcoming papers will treat these comparisons in more detail.

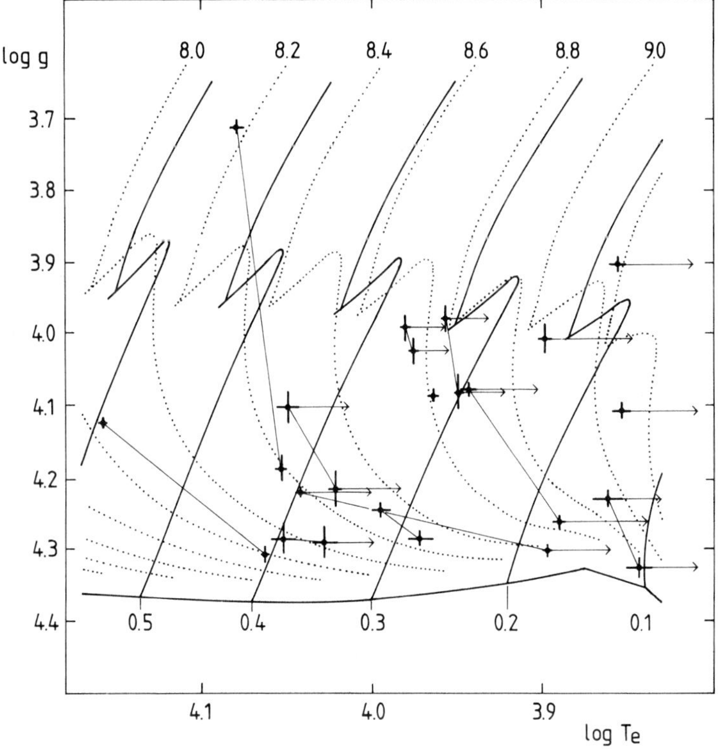

Fig.2. Evolutionary tracks (full lines, with log M), isochrones (dotted lines, with log age), and observed systems in the log Te - log g diagram (see text); (X,Z)=(0.70, 0.02).

This work is supported by ESO observing time, by the Danish Natural Science Research Council and the Danish Board for Astronomical Research. Several colleagues have contributed to individual systems and/or to discussions of the results.

REFERENCES

Hejlesen, P.M.: 1980, Astron. Astrophys. Suppl. Ser. 39, 347
Kříž, S.: 1969, Bull. Ast. Inst. Csl. 20, 202
Lacy, C.H.: 1979, Ap. J. 228, 817
Popper, D.M.: 1980, Ann.Rev. Astron. Astrophys. 18, 115
Popper, D.M., Jørgensen, H.E., Morton, D.C., Leckrone, D.S.: 1970,
        Ap. J. 161, L 57.

DISCUSSION

<u>Kudritzki</u>: How did you obtain the effective temperature of your objects?

<u>Clausen</u>: For systems analysed in the Copenhagen project we have uvby
light curves. The standard 4-colour indices for each component can therefore be calculated and the various available T_{eff} calibrations used (and
compared). See our papers in Astron. Astrophys. for references. For the
other systems included the T_{eff} values were taken from Popper's (1980)
review article and therefore based on the calibration given there.

A TEST OF STELLAR EVOLUTION THEORY BY VISUAL BINARIES

Christopher J. Corbally, S.J.
Vatican Observatory
I-00120 Vatican City State

When close visual binaries have good and homogeneous data, they can prove to be precision probes of stellar evolution theory.

This has been demonstrated by using an MK spectral classification survey of 170 visual binaries, south of -25° declination and with separations mostly between one and five arcseconds. The binaries have area scanner photometry in UBV colours (Hurly and Warner 1983; Rakos et al. 1982), which have been tested for accuracy.

The binaries with the most secure colour-magnitude data were used to test new Yale isochrones (Green et al. 1982), which include a mixing length parameter of 1.5 from the outset of the model calculations. An isochrone composition of Y=0.25, Z=0.04 proved best. Small decreases in the mixing length parameter were suggested for the coolest stars, but the good fit of the binaries to the isochrones (mean age difference between the components = 0.0 ±0.1 Gyr) showed that stellar evolution had in general been well modelled by Green et al.

A full account of this test has been submitted to The Astrophysical Journal.

REFERENCES

Green, E.M., Demarque, P., King, C.R., Light, R., and Ciardullo, R.B.:
     1982, private communication.
Hurly, P.R., and Warner, B.: 1983, M.N.R.A.S., 202, p. 761.
Rakos, K.D., et al.: 1982, Astr. Ap. Suppl., 47, p.221.

A. Maeder and A. Renzini (eds.), Observational Tests of the Stellar Evolution Theory, 395–396.
© *1984 by the IAU.*

DISCUSSION

Conti: It is interesting that you have 2 incipient blue stragglers among these binaries. These objects cannot be exchange mass systems but must be explained by other physical effects.

Andersen: 1) An additional important test of the stellar models is of course – provided the masses are accurately known – to see if both observed stars agree with the model predictions for the observed masses. 2) Although Z = 0.04 would make the observed temperatures of also our late-type binaries agree with predictions, it seems to be too high to agree with spectroscopic and other abundance determinations for Pop. I stars. And for Z = 0.04, the eclipsing binaries agree with neither the ZAMS position nor the isochrones, so this value seems uncomfortably high.

Corbally: 1) The importance of accurate masses to this test was why close visual binaries were chosen for this study. While only about 10 % have orbital elements determined, it is hoped that astrometric observations will continue for the remainder and give mass information within a reasonably small number of years.
2) I agree that Z = 0.04 seems too high in relation to abundance determinations. However, the Y = 0.25 is what would be expected, and the Y and Z abundances must change in harmony. It would certainly be interesting to test these binaries against Z = 0.02 isochrones when these are available.

A NEW TEST CASE FOR NORMAL GIANT EVOLUTION: TZ FORNACIS

J.Andersen[1], J.V.Clausen[1], B.Nordström[1], and M.Mayor[2]
1: Copenhagen University Observatory, Denmark
2: Observatoire de Genève, Switzerland

For normal giants, the only binary system known until recently with completely determined masses was Capella (α Aur, GO III+G5III). Being non-eclipsing, its masses are, however, so uncertain (20% and 8%, respectively) that the properties of the corresponding range of theoretical models are considerably more uncertain than the observationally determined temperatures and luminosities. The discovery of the eclipsing nature of TZ For = HD 20301 (Olsen, 1977) thus provides a unique opportunity to obtain absolute parameters of much higher precision.

Through concerted efforts by many observers from ESO, Marseille, Aarhus, and our own institutes, uvby light curves and radial velocity curves have been obtained with the Danish 50 cm and 1.5 m telescopes at ESO, the latter equipped with the photoelectric radial velocity scanner CORAVEL. Analysis of these, not yet quite complete data yield the following preliminary parameters for TZ For:

Star	Mass($M_\odot$)	Radius($R_\odot$)	T_e(K)	M_{bol}	vsin i(km s^{-1})
G5III	2.11$\pm$0.11	8.9:	4850	0.8:	6:
F7III	1.97$\pm$0.05	3.5:	6200	1.7:	25:
Orbital radius: 120 $R_\odot$(circular)					

Both stars are evolved well away from the main sequence. The combined uvby indices are consistent with a roughly solar metal abundance. Note that apparently the rotation of the larger, cooler primary is synchronized with the orbital motion, while the smaller, hotter secondary rotates at about ten times the synchronous rate.

The above preliminary parameters for the components of TZ For are well matched by Hejlesen's (1980 and unpublished) evolutionary tracks (Fig.1) for a single age (1.2x10^9 years) for both stars, assuming the composition (X,Z)=(0.80, 0.02) which is found to fit most main-sequence systems of similar age (Andersen et al.,this symp.). A full discussion, for which some further observations are necessary, is in preparation.

A. Maeder and A. Renzini (eds.), Observational Tests of the Stellar Evolution Theory, 397–398.
© *1984 by the IAU.*

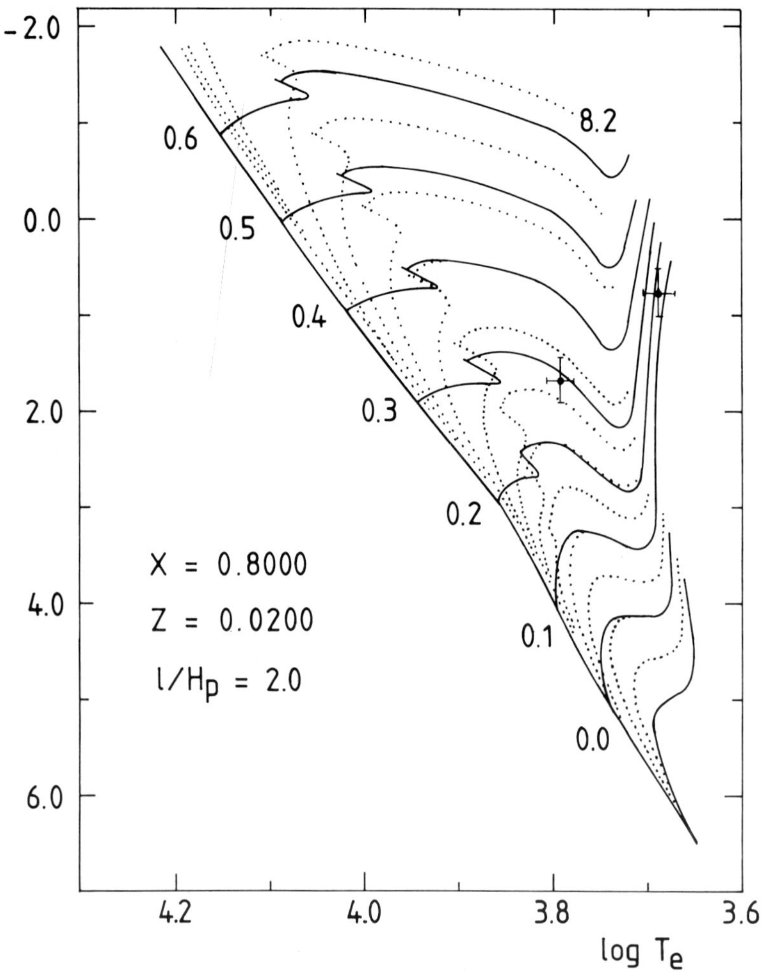

Fig.1. The components of TZ For in the theoretical HR diagram by
Hejlesen. Full lines are evolution tracks (with log M), dotted lines
are isochrones (log age steps of 0.2).

    We thank many colleagues for collaborating in the observations.
This work has been supported by the Danish Natural Science Research
Council, the Danish Board for Astronomical Research, the Swiss Natio-
nal Research Council, and by allocations of ESO observing time.

REFERENCES

Hejlesen, P.M.: 1980, Astron. Astrophys. Suppl. Ser. 39, 347
Olsen, E.H.: 1977, Astron. Astrophys. Suppl. Ser. 29, 313.

THE EVOLUTIONARY HISTORY OF BE/X-RAY BINARIES

G.M.H.J. HABETS
Astronomical Institute, University of Amsterdam,
Roetersstraat 15, 1018 WB Amsterdam

A 2.5 $M_\odot$ helium star remnant of a Case B mass transfer is evolved to off-centre neon ignition and a second mass transfer phase, Case BB, is explored. The model for the formation of Be/X-ray binaries by mass-transfer dominated evolution in an intermediate mass binary is confirmed.

INTRODUCTION

Rappaport and Van den Heuvel (1982) suggest that Be stars with neutron star companions are descendants from a mass-transfer dominated, more or less conservative evolution of systems with initial primary stars of intermediate masses ($\sim$10 to 20 $M_\odot$). The Be/X-ray binaries form an abundant class of X-rays sources with wide orbits (periods $\geq$ 15 - 20 days). The Be or Oe components tend to have masses of $\sim$10 to 20 $M_\odot$, most of them are close to the main sequence (see Rappaport and Van den Heuvel, 1982). Van der Linden (1982) calculated the evolution of binaries with primary masses $\leq$ 12 M⊙ undergoing conservative Case B mass transfer. His post-mass transfer systems have system parameters similar to those of Be binaries (cf. Križ and Harmanec, 1975). The rapid rotation of the Be star is probably due to the mass transfer. If the remnant of the original primary is a helium star of 2-4 $M_\odot$, its C-O core may become larger than the Chandrasekhar mass in the subsequent evolution and it is expected to collapse to a neutron star (see the review of Van den Heuvel, 1983). The evolution of such helium stars up to carbon ignition in the core was followed by De Grève and De Loore (1977) and Delgado and Thomas (1981). After this, the change of the radius with time is only roughly known or not at all (cf. Nomoto, 1981). A second phase of mass transfer by Roche-lobe overflow (Case BB) is possible if the original primary is not too massive ($\leq$ 14 M⊙), cf. De Grève and De Loore (1977). So far, Case BB mass-transfer after carbon exhaustion in the core is yet unexplored for helium stars with masses of 2 to 4 $M_\odot$. The final evolution of a $\sim$ 2.5 M⊙ helium core of a 10 M⊙ single hydrogen star leads to the formation of a neutron star, cf. Woosley et al. (1980) and Hillebrandt (1982). In this paper we describe the evolution of a 2.5 M⊙ helium star through helium- and carbon-burning up to off-centre neon ignition. Case BB mass-transfer is investigated with a companion of 17 M⊙. This system is the result of conservative mass-exchange from a binary with initial components of 13 + 6.5 M⊙.

A. Maeder and A. Renzini (eds.), Observational Tests of the Stellar Evolution Theory, 399–402.

MODEL CALCULATIONS AND RESULTS

The evolutionary program of Van der Linden (1982) was modified to compute the evolutionary stages of 2-4 $M_\odot$ helium stars beyond helium exhaustion in the core up to neon-burning. Simple reaction networks are included for the relevant burning phases (see Habets, 1984). The composition of the initial model (by mass) was taken to be Y = 0.97 and Z = 0.03 (i.e. .75% ^{12}C, .75% ^{16}O and 1.5% ^{14}N). The period P (= $20^d.29$) and orbital separation A (= 84.2 $R_\odot$) of the binary were chosen such that BB mass transfer sets in after carbon exhaustion in the core and such that under conservative assumptions the 13 + 6.5 $M_\odot$ progenitor system was in Case B (P = $2^d.58$). In Figures 1 and 2 the evolution of the helium stars is shown for several stages: A-B, helium main sequence and exhaustion of helium in the core, B-C, radiative shell helium-burning and gravitational contraction of the carbon core, C-D, ignition of carbon, convective core and radiative shell carbon-burning, D-E, convective shell carbon-burning and gravitational contraction of the O-Ne-Mg core, E ignition of neon and convective off-centre neon-burning. During phase D-E the core-mass (i.e. mass of the convective helium core and of the C-O core, which is defined by the region interior to the helium shell with maximal energy generation) grows larger than the Chandrasekhar limiting mass (Figure 2). The treatment of mass transfer is similar to that by Van der Linden. The total amount of mass transferred is 0.3 $M_\odot$ (in ~ 2.9×10^3 yr = 12 $\tau_{thermal}$) and the final period is 28d with A = 105 $R_\odot$. After this, two short-lasting (4-10 years) mass-transfer phases with maximal $\dot{M}$ ~ $10^{-4.7}$ $M_\odot$ /yr can not be avoided just before neon ignites. The last exchange phase has not been calculated. The evolution of the inner regions is depicted in Figures 3a and 3b. The dotted lines limit the C-O core. Notice that the extent of the C-O core is not much affected by the mass exchange.

The post-carbon compositional and thermodynamic structure of the inner regions of the 10 $M_\odot$ star as computed by Woosley et al. is similar to the one of our 2.5 $M_\odot$ helium star (even if mass loss is included). Therefore we expect that the final evolution will be almost the same, i.e. neon and oxygen ignite off-centre, but the inner 0.14 - 0.15 $M_\odot$ central region of neon and oxygen remains unburned. The central Ne-Si core contracts, the helium layer will be ejected, the remnant evolves through core collapse into a neutron star of ~ 1.44 $M_\odot$ (cf. Hillebrandt). Consequently, the proposed scenario for the formation of Be/X-ray binaries can indeed explain the observed features of these systems. The occurence of the BB mass transfer after carbon exhaustion in the core does not alter the above picture considerably. The super-nova explosions will be hydrogen-deficient (Type I?) and the amount of mass ejected will be relatively small (< 0.7-1 $M_\odot$), which will not strongly affect the orbit.

The author thanks Th. Van der Linden for the introduction to his stellar evolution code. This work is supported by the Netherlands Organization for the Advancement of Pure Research, ZWO (ASTRON).

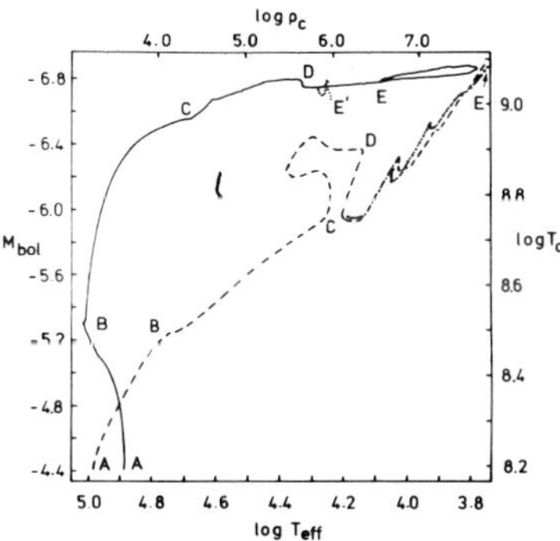

Fig.1: Fully drawn curve: the evolutionary track of the 2.5 $M_\odot$ helium star in the HR diagram. Dashed curve: plot of central temperature vs. density of the evolving star. The dotted part of these curves is with Case BB of mass-transfer.

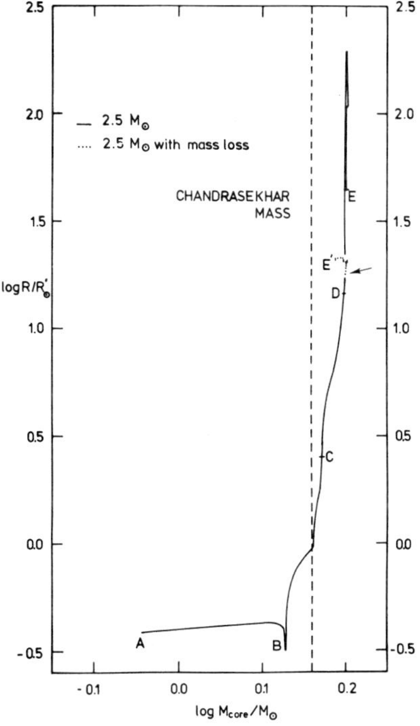

Fig.2: Radius as function of core mass. The arrow indicates the onset of mass exchange in the 2.5 + 17 $M_\odot$ system. For the letters in Figures 1 and 2 see the text.

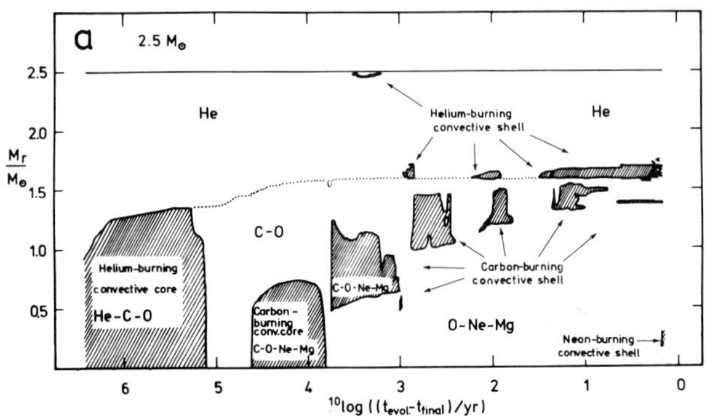

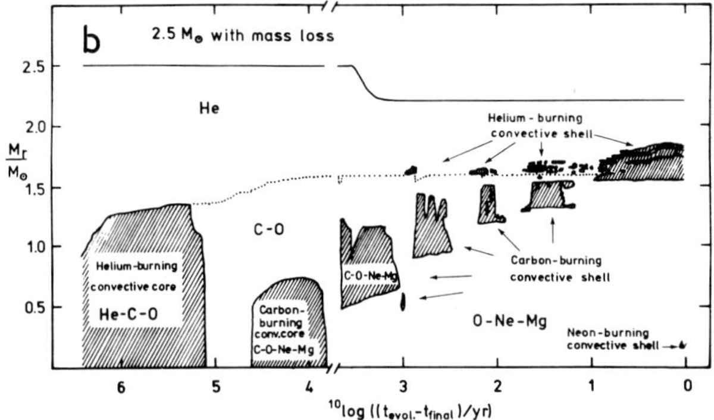

Fig. 3a: The evolution of the interior regions of the 2.5 M⊙ helium star without mass loss. 3b: Same with Case BB mass loss to the 17 M⊙ companion.

REFERENCES

De Grève, J.P. and De Loore, C.: 1977, Astrophys. Space Sci. **50**, 75.
Delgado, A.J. and Thomas, H.C.: 1981, Astrophys. **96**, 142.
Habets, G.M.H.J.: 1984, PH.D. Thesis, Univ. of Amsterdam.
Hillebrandt, W.,: 1982, Astron. Astrophys. **110**, L3.
Križ, S. and Harmanec, P.: 1975, Bull. Astron. Inst. Czech. **26**, 65.
Nomoto, K.: 1981, in "Fundamental Problems in the Theory of Stellar Evolution", D. Sugimoto, D.Q. Lamb, and D.N. Schramm (eds.), p. 295.
Rappaport, S. and Van den Heuvel, E.P.J.: 1982, in "Be Stars", M. Jaschek and H.-G. Groth (eds.), IAU Symp. **98**, p. 327.
Van den Heuvel, E.P.J.: 1983, in "Accretion-Driven Stellar X-ray Sources" (eds. W.H. Lewin and E.P.J. van den Heuvel), Cambridge Univ. Press, p. 303.
Van der Linden, Th.J.: 1982, PH.D. Thesis, Univ. of Amsterdam.
Woosley, S.E., Weaver, T.A. and Taam, R.E.: 1980, in "Type I Supernovae", J.C. Wheeler (eds.), Univ. of Texas Press, p. 96.

OBSERVATIONAL STUDY OF THE EVOLUTION OF MASSIVE BINARY STARS

G. Burki, M. Mayor
Observatoire de Genève

Six years ago an observational program on supergiant stars using CORAVEL was initiated at Geneva Observatory. About 1500 radial velocities were obtained out of a sample of 181 northern supergiants of F, G, K, M type. Nineteen new SB have been discovered and 16 others are suspected to be SB. The total rate of binary systems among northern supergiants is in the range of 31-38 % (Burki and Mayor, 1983). This value neither depends on spectral type nor luminosity class.

Note that a similar rate of binaries is found among cepheid stars: 25-35 % (Burki, 1983). This is quite normal since cepheids are simply F-G supergiants in a peculiar evolutionary phase.

The comparison of the binary rate among supergiants F0-K0Ib with that among their progenitors on the main sequence, B2-B5IV-V (Abt and Levy, 1978) is very interesting. Table 1 shows that the large fraction of binaries of period P<500 d on the main sequence is considerably reduced in the case of supergiants. Consequently, a large fraction of close binary systems never produce cool supergiants. Table 1 shows that this effect concerns about one third of the main sequence stars. Thus, this effect must be taken into account when stellar evolutionary models

	B2-B5 IV,V	F0-K0 Ib
Single stars	51 %	65 %
Binary stars		
P<500 d.	29 %	5 %
P>500 d.	20 %	30 %

Table 1

are checked by comparing the number of stars in various zones of the HR diagram.

A. Maeder and A. Renzini (eds.), Observational Tests of the Stellar Evolution Theory, 403–406.

CIRCULARIZATION OF ORBITS

First of all let us define the quantity P_{circ} for a binary system of period P and eccentricity e:

$$P_{circ} = P(1-e)^{3/2}$$

Thus, P_{circ} corresponds to the period of the circular orbit having the distance d_p between components at the periastron of the real elliptical orbit for radius. Of course, d_p is an important parameter for the mechanisms of orbital circularization.

From the catalogue of Batten et al (1978) and the results of our own survey, the orbital elements are known for 25 SB having at least one supergiant component of type F to M. In Figure 1, the eccentricity e of these systems is plotted versus log P_{circ} (P_{circ} in days). The type of symbol refers to the luminosity class. In each luminosity class all systems with P_{circ} shorter than a critical value have nearly circular orbits. The critical value is very well defined for class Ib: 350–440 days. This value is larger for classes Iab and Ia (1400–3900 days for class Ia).

A simple explanation can be given of the existence of this critical period. Figure 2 shows the variation of the radius of a 9 $M_\odot$ star with time (Maeder, 1981), during the supergiant stage (blue loop phase). At the first maximum of the radius, following the first, quick, crossing of the HR diagram, corresponds a limiting orbital period P_{lim}: in the binary systems which have a period smaller than P_{lim} on the ZAMS, a mass exchange between the components will occur before the supergiant stage. By applying the method described in Kopal (1978) to the models of Maeder

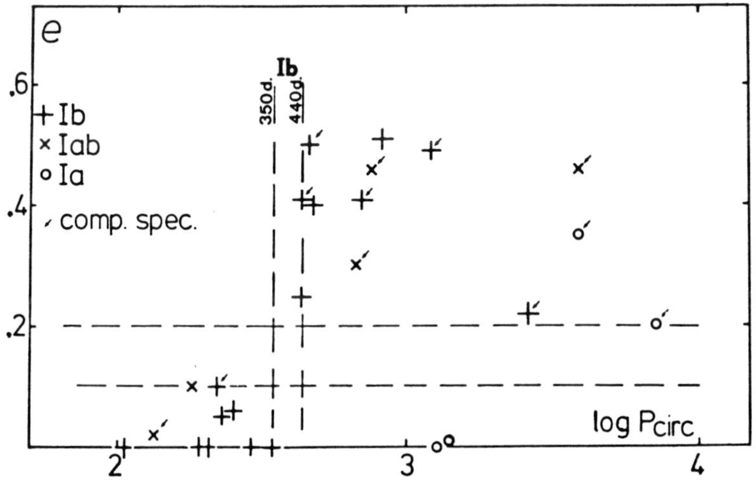

Figure 1

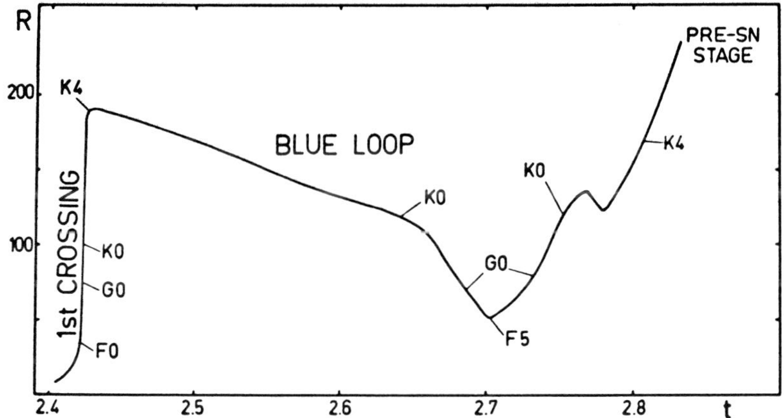

Figure 2

(1981), we found that P_{lim} is in the range of 300–500 days for the systems
containing a progenitor of Ib-type supergiant. This range for P_{lim} is in
good agreement with the critical value of the period deduced from
Figure 1. Consequently, the orbit of the systems with shorter periods
have probably been circularized by mass exchange. Note that the tidal
effects certainly operate in addition to the circularization by mass
exchange.

In Figure 1, the systems with a composite spectrum have been referred to by arrows. Almost all the systems with $e > 0.2$ have a composite
spectrum whereas only two such cases are known in the systems with
$e \leq 0.1$. In addition, the mass function $f_1(M)$ for the systems with
$e \geq 0.2$ is on an average larger than the mass function for systems with
$e \leq 0.1$ (see Burki and Mayor, 1983). These two facts indicate that: on
an average, the secondary components are less massive in systems having
been influenced by mass transfer than in the others. A probable explanation of this surprising fact is that, at least in several systems with
$e \leq 0.1$, the star actually observed as a supergiant was the less massive
component at the origin of the system.

REFERENCES

Abt, H.A., Levy, S.G.: 1978, Astrophys. J. Suppl. 36, 241.
Batten, A.H., Fletcher, J.M., Mann, P.J.: 1978, Publ. Dominion Astrophys.
     Obs. 15, 121.
Burki, G.: 1983, Astron. Astrophys. in press.
Burki, G., Mayor, M.: 1983, Astron. Astrophys. 124, 256.
Kopal, Z.: 1978, Dynamics of Close Binary Systems, D. Reidel, Dordrecht,
     Holland, p. 414.
Maeder, A.: 1981, Astron. Astrophys. 102, 401.

DISCUSSION

<u>Lodén</u>: In your statistics you got the sum of singular stars and binaries
to be exactly 100 %. What has then happened to the triple stars and
higher multiples?

<u>Burki</u>: They are included in the binary group.

OBSERVATIONAL CONSTRAINTS FROM MODELS OF CLOSE BINARY EVOLUTION

J.P. De Grève, W. Packet
Vrije Universiteit Brussel
Pleinlaan 2, B-1050 Brussels, Belgium.

The evolution of a system of 9 $M_\odot$ + 5.4 $M_\odot$ is computed from
Zero Age Main Sequence through an early case B of mass exchange, up to
the second phase of mass transfer after core helium burning. Both components are calculated simultaneously. The evolution is divided into
several physically different phases. The characteristics of the models
in each of these phases are transformed into corresponding 'observable'
quantities. The outlook of the system for photometric observations is
discussed, for an idealized case. The influence of the mass of the loser, m_{li}, and the initial mass ratio q_i is considered.

I. INTRODUCTION.

This talk deals with the following question: how does a computed model of a close binary system in a certain evolutionary phase, look
to the observer? As the present results are only the onset of a large
project, on various aspects of observing theoretical models, they are
subject to severe limitations in the treatment:
- essential one value of the initial mass of the loser is considered,
- only early case B of mass exchange,
- zero eccentricity and spherically symmetric stars,
- uniform brightness distribution for both components,
- inclination i = 90°.
Summarizing we will discuss aspects of a non-existing system, just as
most theoreticians do.

II. EVOLUTIONARY PHASES AND TIMESCALES.

The model has initial masses of $9M_\odot$ and $5.4M_\odot$, with a period of
2.98dys. Its evolution has been discussed to some extend at IAU Colloquium
No. 80 (De Grève and Packet, 1983). The following phases are defined:
A: Main sequence and hydrogen shell burning (29.27 E6 yrs or 82% of the
   total timescale computed t_t)
B: The mass exchange (2.3E5 yrs or 0.7% of t_t), subdivided in
   B1: A sd-phase at the onset of mass loss (6% of t_B)
   B2: A contact phase, lasting till the loser reaches minimum lumino-
       sity (3% of t_B)
   B3: A sd-phase, with large mass transfer rate (10% of t_B)

A. Maeder and A. Renzini (eds.), Observational Tests of the Stellar Evolution Theory, 407–409.

B4: A sd-phase, with small mass transfer rate (81% of t_B)
C: Core helium burning (5.54E6 yrs or 16% of t_t)
D: He-shell burning (3.6E5 yrs or 1% of t_t)
E: Case BB of mass transfer, sd with slow mass transfer

III. TRANSFORMATIONS TOWARDS AN 'OBSERVABLE' SYSTEM.

The theoretical results were transformed into color-magnitude diagrams for loser, gainer and for the total system, using the calibration of Flower (1977). Also the magnitude difference $\Delta M_V = M_V$ (gainer) $-M_V$ (loser) was calculated, together with the eclipse depth in magnitude, δ_l adn δ_g, during eclipse of the loser, resp. the gainer. The influence of initial mass ratio was investigated for $q_i = 0.38$, $q_i = 0.6$ and $q_i = 0.9$ (with $m_{li} = 3\ M_\odot$); comparison of results for the masses 9 and $3\ M_\odot$ gave some idea about the influence of m_{li}. The behaviour of several characteristics of both components, during the different phases, is shown in the Domino Table of Evolution (table1), for the 9 + 5.4 system. The columns represent the different evolutionary phases, the rows component parameters (black dot: loser, open circle:gainer). Upper position of a dot simply means largest value of the specific parameters. Changes of role of the components during a phase is indicated by dashed lines. It follows from the table that the system behaves different in   bolometric magnitude than in visual magnitude. Moreover there is no direct correlation between changes in radii or temperature and changes in visual magnitude.

Table 1 : The Domino Table of Close Binary evolution (see text).

Phase	H-core H-shell	RLOF	RLOF	RLOF	RLOF	He-Core	He-shell	RLOF 2
State	d pre-RLOF	sd $\dot{M}\uparrow$	c $\dot{M}\uparrow$	sd $\dot{M}\uparrow$	sd $\dot{M}\downarrow$	d	d	sd $\dot{M}\downarrow$
Mass	● ○	● ○	● ○ ○ ●	○ ●	○ ●	○ ●	○ ●	○ ●
bol. lum.	● ○	● ○ ○ ●	○ ●	○ ●	○ ●	○ ●	○ ●	○ ●
radius	● ○	● ○	● ○ ○ ●	○ ● ● ○	● ○	● ○ ● ○ ● ○	● ○	● ○
color (Te)	● ○	● ○ ○ ●	○ ●	● ○	○ ●	● ○ ○ ●	○ ● ● ○	○ ●
$-M_V$	● ○	● ○ ○ ●	○ ●	○ ●	○ ● ● ○	● ○ ○ ●	○ ● ● ○	● ○

IV. RESULTS.

a) In the colour-magnitude diagram the behaviour of the system as a whole is determined by the gainer's rise in M_V, during the first part of the mass exchange. Later it is completely dominated by the M_V evolution of the loser.

b) the colour of the system remains inaffected by the sharp increase
   in B-V of the loser during the rapid phase of mass transfer:
   $$(B-V)_t : -0.2 \text{ to } -0.1$$
   $$(B-V)_l : -0.2 \text{ to } 0.7$$
c) during the slow phase, for $m_{li} = 9\ M_\odot$ and $q_i < 1$, $\Delta M_V > o$ where as
   $M_{bol,l} > M_{bol,g}$
d) at the beginning of the slow phase the eclipse depth δ_l is small for
   large m_{li} ($0\overset{m}{.}0$ to $0\overset{m}{.}8$) large for small m_{li} ($2\overset{m}{.}1$ to $3\overset{m}{.}2$)
The eclipse depth δ_g evolves towards smaller values during the slow phase.
e) during the semidetached phase the primary minimum may be due to loser
   or gainer, to cool or hot component, to the more or the less massive
star, to the largest or the smallest one. Hence when analysing the observations and dicussing the evolutionary state of the system, we should
ask ourselves: who is who on the close binary scene?

References.
De Grève, J.P., Packet, W. : 1983, Astrophys. Space Sci ($\underline{96}$, in press)
Flower, P.F.: 1977, Astron. Astrophys. $\underline{54}$, 31.

DISCUSSION

R. Cayrel: Speaking of non existing cases, do you expect circumstances
in which the evolution produces the coalescence of the two components of
a close binary system?

de Grève: One expects coalescence in a close binary system if much
angular momentum is removed from the system. This occurs most probably
for extreme initial conditions, especially for small initial mass ratios.
Computations for these extreme cases have to be performed in the future.

ORBIT CIRCULARIZATION TIME IN BINARY STELLAR SYSTEMS

M. Mayor[1], J.-C. Mermilliod[2]
[1]Observatoire de Genève
[2]Institut d'Astronomie de l'Université de Lausanne

Summary: We have analyzed the orbital parameters of 33 red dwarf and 17 red giant spectroscopic binaries belonging to open clusters to deduce the time-scale for orbital circularization. The dynamical evolution of BD +23°635, a short period binary member of the Hyades, should result in the formation either of a cataclysmic binary or a WUMa system.

Zahn (1966, 1977) and Lecar et al. (1976) have studied the process of circularization of spectroscopic binaries orbits induced by tidal effects. Such a mechanism is especially effective for late-type stars because of the turbulent viscosity associated to the convective envelope. But owing to difficulties inherent to the treatment of convection, the time-scale for circularization is rather ill-defined. Although it has been known for a long time that the orbits of short period binaries are circular, this fact has not yet been used, mainly because of difficulties in assigning a proper age to late-type stars.

Late-type (Sp. Type $\geq$F5) binaries in open clusters offer an opportunity of determining the time-scale of orbit circularization. The measurement of radial velocities by photoelectric techniques (Griffin, 1967; Baranne et al., 1979) has recently improved our knowledge of stellar duplicity in nearby open clusters. The 29 orbits obtained so far with the radial velocity scanners (Griffin and Gunn, 1978, 1981; Griffin et al., 1982; Mermilliod and Mayor, Coravel unpublished measurements) display the capabilities of these instruments as compared to the classical photographic techniques which yielded up to now only four orbits for late-type dwarfs in open clusters. As concerns the red giants, the present state is similar, being one classical orbit against 16 photoelectric orbits.

Figure 1 shows the eccentricity vs period diagram on the one hand for red dwarf stars in the Hyades, Pleiades, Praesepe and Coma Ber open clusters (left part), and on the other hand for red giants belonging to several open clusters (right part).

A. Maeder and A. Renzini (eds.), Observational Tests of the Stellar Evolution Theory, 411–414.

<u>Late-type dwarfs</u>. The diagram for the late-type dwarfs shows a strong discontinuity near a value of the period equal to 5.7 days. The orbits with periods shorter than this critical value are circular, while all orbits with longer periods have non-zero eccentricities. Lecar et al. (1976) give the following expression for the characteristic time of circularization:

$$t_{circ,1} = 0.08 \ (M_1/M_2)(a/R)^5 (g/g_\odot) \ P^2/\lambda\eta v$$

The circularization time of the system is taken as the harmonic mean of the times associated to each component. The circularization time needed for the eccentricity to become as low as 0.01 to 0.02 is assumed to be of the order of one-third of the cluster age. Thus, for systems composed of two stars of one solar mass with ages equal to the Hyades or Praesepe ages, the limiting period should be about 5.1 days. The circularization time depends on spectral types of both components through the value of radii, masses, relative thickness λ and mass η of the convective envelope. Thus only a detailed study of systems with period close to the limiting period will provide a significant test of the circularization time-scale. Using the three systems[1] nearest to the discontinuity we can restrain the uncertainty to a factor of two around the circularization time-scale proposed by Lecar et al. (1976).

<u>Red giants</u>. By considering the two systems[2] whose periods are on both sides of the discontinuity in the (e,P) diagram, we can estimate the value of the viscous energy dissipation necessary to account for the circularization of the red giant orbits. Expression (2) may be written in the form:

$$t_{circ} = R_{LWK} \ (M_1/M_2)(a/R)^5 (g/g_\odot) \ P^2 \qquad (P \text{ in days})$$

where $R_{LWK}=0.08/\lambda R\theta$. For solar type dwarf stars, the value of R_{LWK} is about 16 y. For giant stars we estimate a value of R_{LWK} between 0.015 and 0.025 y, in good agreement with a direct estimate of $R_{LWK}=0.08/\lambda\eta v$ from K0 III models (R_{LWK}(K0 III) ~ 0.09 y).

It should be noticed that we have tried to explain the discontinuity in the (e,P) plane for dwarf stars in terms of orbital circularization during the main sequence lifetime of the stars. However, since the radii of the proto-stars are larger than those of the main sequence stars, the convective zone more developed and the gravity lower, the circularization

---

[1] (VB 22, P=5.60d, e=0.0 in the Hyades, McClure 1982; VB 121, P=5.75d, e=0.35 in the Hyades, Griffin and Gunn 1978; KW 181, P=5.87d, e=0.35 in Praesepe, CORAVEL unpublished orbit.)

[2] (NGC 752-110, P=127d, e=0.0 and NGC 7092-95, P=178d, e=0.23, both CORAVEL unpublished orbits.)

time during the pre-main sequence evolution is considerably shorter for a given period. One cannot reject the possibility that the limiting period at 5.7 d results entirely from the star evolution prior to arrival on the main sequence.

The system BD +23°635

The Hyades member system BD +23°635 presents an interesting problem: although it has a short period, P = 2.4 d, the eccentricity is significantly different from zero: e = 0.057 ± 0.005 (Griffin and Gunn, 1981). These authors also discovered that a third star belongs to this system. This fact enables us to explain the observed orbital parameters by invoking the mechanism studied by Mazeh and Shaham (1982): the non-zero eccentricity most probably results from the perturbation created by the third companion. The simultaneous action of the energy dissipation (by tidal interaction) and of the perturbation (which maintains a non-zero eccentricity) induces a secular evolution of the close binary. The orbital period shrinks, with a characteristic time ($\tau_a = \tau_{circ}/2<e_{1,2}>\simeq 2\cdot 10^9 y$) shorter than the nuclear lifetime of the stars. Therefore, such a system appears as a probable progenitor of WUMa or small mass cataclysmic systems.

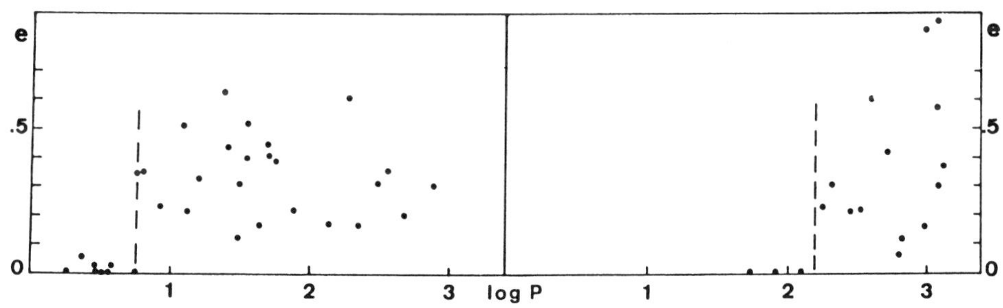

Fig. 1. (e,P) diagram for dwarfs (left) and giants (right)

REFERENCES

Baranne, A., Mayor M., Poncet J.-L.: Vistas in Astron. 23, 279
Griffin R.F.: 1967, Astrophys. J. 148, 465
Griffin R.F., Gunn J.E.: 1978, Astron. J. 83, 1114
Griffin R.F., Gunn J.E.: 1981, Astron. J. 86, 588
Griffin R.F., Mayor M., Gunn J.E.: 1982, Astron. Astrophys. 106, 221
Lecar M., Wheeler J.C., McKee C.F.: 1976, Astrophys. J. 205, 556
Mazeh T.,Shaham, J.: 1982, Astron. Astrophys. 77, 145
McClure R.D.: 1982, Astrophys. J. 254, 606
Zahn J.P.: 1966, Ann. Astrophys. 29, 565
Zahn J.P.: 1977, Astron. Astrophys. 38, 3

DISCUSSION

<u>Spruit</u>: Is the formation of a cataclysmic system from a triple likely to be a rare event?

<u>Mayor</u>: With T. Mazeh we are conducting a search of triple systems through the precession of the nodes effect. Preliminary results show the existence of a non-negligible percentage of triple systems exhibiting this effect. Evidently not all of them will have a contraction time τ_a shorter than the evolutionary time τ_e.

APSIDAL MOTION IN MAIN-SEQUENCE BINARY STARS

C. Simon Jeffery
University Observatory, St Andrews, Scotland

1. INTRODUCTION

Early studies of apsidal motion in binary stars showed a marked discrepancy between observational results and theoretical expectation (Schwarzschild, 1958; Kopal, 1965). Improvements in theory (Petty, 1973; Odell, 1974; Stothers, 1974) have led to much closer agreement. However, recent observational results (Giménez & García-Pelayo, 1982; Monet, 1980) have provided much more data with which to test theoretical results.

A number of evolutionary sequences for main-sequence stars have been computed, in which the apsidal motion constant k_2 has been obtained for each stellar model. These evolutionary sequences show good agreement with observed apsidal motion in intermediate mass main-sequence binary pairs and confirmation of some of the observations of k_2 by Giménez et al (1982). If the success of the theory is accepted, Monet's (1980) method for testing stellar models with observations of apsidal motion in non-eclipsing systems may be modified to obtain an estimate for the mass and age of a binary system from the observed apsidal motion. Details of these results are reported by Jeffery (1984).

2. THEORETICAL MODELS

The evolution of the internal density distribution of a star depends upon the mass of the star. A comparison with the evolution of the surface gravity provides a comparison with observable quantities (Giménez & García-Pelayo, 1982). The rate of evolution of the apsidal motion constant can be measured by the quantity $\partial\log k/\partial\log g$, which is found to increase with mass. Giménez et al suggested a similar conclusion after comparing their observations with appropriate ZAMS models.

A. Maeder and A. Renzini (eds.), Observational Tests of the Stellar Evolution Theory, 415–417.

For the evolution of a 10.9 $M_{\odot}$ model, we obtained good detailed agreement with Stothers (1974). This agreement persisted in explaining the observed apsidal motion of Y Cyg, AG Per and CW Cep. However, if models for a star of the appropriate mass ($\sim 4\ M_{\odot}$) are used, the discrepancy between the observed apsidal motion (Semeniuk, 1967) and the observed stellar radii (Smak, 1967) of CO Lac cannot be resolved.

3. SPECTROSCOPIC SYSTEMS

By rearranging Kopal's (1965) equations, Monet (1980) obtained an approximate equation relating orbital parameters of an apsidal motion system (P, U, e), on one side, to stellar parameters (q, R_i, k_i, M) on the other. By estimating stellar parameters from the stellar spectral type (using Adelman's (1978) effective temperature calibration and Stothers' (1974) stellar models), Monet showed that it is possible to compare observed apsidal motion with theory for a number of non-eclipsing systems. Using a complete set of evolutionary sequences, we find that apsidal motion in main-sequence pairs may usually be accounted for by the evolution of the components.

4. STELLAR MASSES AND AGES

An adaption of this approach requires that theoretical models produce relations of the form

$$S_c = S_c\ (M, q, t, X, Z)$$
$$T_c = T_c\ (M, t, X, Z) \tag{1}$$

where S is a 'structure' parameter and T is the effective temperature, t is age and X, Z represent composition. Meanwhile, observations yield

$$S_o = S_o\ (P, U, e)$$
$$T_o = T_o\ (Sp) \tag{2}$$

where Sp is spectral type. Further details may be found in Jeffery (1984). Assuming a composition and mass ratio, q, (1) may be inverted so that the observed quantities yield a mass and age

$$M = M\ (S_o, T_o)$$
$$t = t\ (S_o, T_o). \tag{3}$$

Applying this method to eclipsing systems with known masses (Popper, 1980), the 'apsidal motion' masses are in good agreement with the orbital masses (Table 1). This gives confidence in this method as an independent way of estimating masses and ages for binary systems showing apsidal motion.

TABLE 1

System	Sp	P (days)	U (years)	M (Popper)		M (apsidal motion)	
Y Cyg	O9.8IV	2.996	54	16.7	(0.5)	17.4	(2.8)
C W Cep	B0.5V	2.729	69	11.8	(0.2)	13.0	(2.1)
α Vir	B1V	4.015	130	10.8	(1.3)	12.9	(2.1)
u Her	B2.5V	2.051	107	7.3	(1.0)	6.90	(1.1)
A G Per	B5pV	2.029	76	4.53	(0.07)	4.44	(0.7)

REFERENCES

Adelman, S.J.: 1978, Astrophys. J. 222, p.547
Gimenez, A., & García-Pelayo, J.M.: 1982, IAU Coll. 69, p.37,
     eds. Kopal, Z., & Rahe, J., Reidel, Dordrecht, Holland
Jeffery, C.S.: 1984, Mon. Not. R. Astr. Soc. (in press)
Kopal, Z.: 1965, Adv. Astr. Astrophysics 3, p.89
Monet, D.G.: 1980, Astrophys. J. 237, p.513
Odell, A.P.:1974, Astrophys. J. 192, p.417
Petty, A.F.: 1973, Astr. Space Sci. 21, p.189
Popper, D.M.: 1980, A. Rev. Astr. Astrophys. 18, p.115
Schwarzschild, M.: 1958, 'Structure and Evolution of the Stars',
     Princeton University Press
Semeniuk, I.: 1967, Acta Astr. 17, p.223
Smak, J.: 1967, Acta Astr. 17, p.245
Stothers, R.: 1974, Astrophys. J. 194, p.651

APSIDAL MOTION IN EVOLVED BINARY SYSTEMS

Alvaro Giménez
Departamento de Astrofísica. Facultad de Físicas.
Universidad Complutense. Madrid-3. Spain.

The study of apsidal motions in eclipsing binaries has proven to be one
of the best methods to check the internal density concentrations of the
stars predicted by theoretical models. During the main sequence phase,
we have found a good agreement between the observed apsidal motion rates and computer-constructed stellar models provided that a realistic
consideration is made of the evolution between the lower and upper bor-
ders of the main sequence (Giménez and García-Pelayo, 1982). An obvious
extension of this work is a throughout study of the more evolved evolved systems beyond the TAMS where theoretical models are less accurate
and empirical data from different sources are largely needed (see review paper by Zahn in this volume). A preliminary report on such a study is presented.

In order to compare the available observations with the theoretical models, we have adopted those computed by Hejlesen (1980) which
also include the values of the internal structure constants k_j (j=2,3,
4) for a wide range of stellar masses and chemical compositions (Hejlesen, 1982) and make use of the opacity tables by Cox and Stewart. In
this evolutionary sequences we have found that: a) During the main sequence, $\log k_2$ decreases smoothly for decreasing log g with a slope
slightly dependent on mass (as it was empirically predicted by Giménez
and García-Pelayo, 1982), b) After the consumption of the hydrogen fuel
in the core, the inner density concentration continues to increase in
a less pronounced way until a rapid grow of the dimensions of the stars
is accompanied by a sudden increase in the $\log k_2$ value, and c) For the
range of masses of our interest, it is clear that a minimum value is
theoretically predicted for $\log k_2$ around -2.5 below which observational values are not expected to be found. Furthermore, variations in log
k_2 are found to be small for different chemical compositions, within
reasonable ranges, particularly with respect to the minimum value after
the TAMS. Comparing the evolutionary sequences given by Petty (1973)
with those by Hejlesen, we did not find significant differences.

The candidate systems to be compared with the theoretical models were selected according to the usual criteria (Giménez, 1981) but
the following difficulties were found: a) Highly eccentric systems are
not common among evolved binaries, b) To keep the system well detached

A. Maeder and A. Renzini (eds.), Observational Tests of the Stellar Evolution Theory, 419–420.

even after the large expansion of the external envelope the orbital periods should be relatively long, thus making difficult an accurate determination of the apsidal motion period, c) The obtention of absolute parameters in these binaries is generally complicated due to the photometric variations found among most of them and d) As far as the mass ratio is not very close to unity, the rapid evolution after the main se quence implies a small luminosity ratio which makes the system to be a single-lined spectroscopic binary.

Anyhow, 10 candidates could be finally selected for which it was possible to make a reasonable estimation of the involved parameters (AR Cas, V346 Cen, V380 Cyg, GN Nor, δ Ori A, o Per, EO Vel, α Vir and HR 7551). The apsidal motion periods were taken from the literature except for V346 Cen which was deduced from our own observations. The absolute parameters were estimated mainly from the light and radial velocity curves but also spectral types, colour indices, etc. It should be noticed that, when the relativistic contribution is not important, the total mass of the system is not needed to compute log k_2 and that the main involved parameters are the relative radii far more accurate than their absolute values when derived from the light curve (i.e. changing the masses will move the points left or right in the log k_2-log g plane but not upwards unless the relative radii are changed). In case the radii are poorly known, it is possible to adopt minimum values from the models (e.g. on the TAMS) which will provide upper limits for log k_2.

From the comparison we could immediately deduce the following points: a) The less evolved binaries (close to the TAMS border like AR Cas, V453 Cyg and EO Vel) there is a good agreement within their mean errors with the models, b) For the remaining more evolved systems there is a clear tendency towards more centrally condensed configurations than theoretically predicted, c) This difference is too large to be easily explained in terms of observational errors or small modifications in the procedure to build up the models, d) The deviations are clearly systematic suggesting that either the actual structure of the stars is more highly centrally condensed or the radii of the components are too large (!). Moreover, the differences become larger for the more evolved systems with smaller surface gravities.

New observations and analyses are being carried out in order to increase the accuracy of the above sketched comparison and thus provide further background to our knowledge of stellar evolution. It seems that changes in the adopted opacities may modify the picture but also the existence of some kind of mixing in the stellar interior (Maeder & Mermilliod, 1981) leading to a significant widening of the main sequence or any physical mechanism producing changes in the orbital parameters which may bias the determined apsidal motion rates.

REFERENCES

Giménez, A.: 1981. In "Photometric and Spectroscopic Binary Systems", NATO Adv. Study Inst. p 511.
Giménez, A., García-Pelayo, J.: 1982. IAU Colloq. No. 69, p 37.
Hejlesen, P.M.: 1980. Astron. Astrophys., 84, p 135.
Hejlesen, P.M.: 1982. Private Communication.
Maeder, A., Mermilliod, J.C.: 1981. Astron. Astrophys., 93, p 136.
Petty, A.F.: 1973. Astrophys. Space Sci., 21, p 189.

CONSTRAINTS ON STELLAR EVOLUTION FROM PULSATIONS

Arthur N. Cox
Theoretical Division, Los Alamos National Laboratory
University of California
Los Alamos, New Mexico 87545 USA

Consideration of the many types of intrinsic variable stars, that is, those that pulsate, reveals that perhaps a dozen classes can indicate some constraints that affect the results of stellar evolution calculations, or some interpretations of observations. Many of these constraints are not very strong or may not even be well defined yet. In this review we discuss only the case for six classes: classical Cepheids with their measured Wesselink radii, the observed surface effective temperatures of the known eleven double-mode Cepheids, the pulsation periods and measured surface effective temperatures of three R CrB variables, the δ Scuti variable VZ Cnc with a very large ratio of its two observed periods, the nonradial oscillations of our sun, and the period ratios of the newly discovered double-mode RR Lyrae variables. Unfortunately, the present state of knowledge about the exact compositions; mass loss and its dependence on the mass, radius, luminosity, and composition; and internal mixing processes, as well as sometimes the more basic parameters such as luminosities and surface effective temperatures prevent us from applying strong constraints for every case where currently the possibility exists.

Use of the pulsation existence as evidenced by observed periods is often of limited value because the actual mode of the pulsation is uncertain. For example, there are the questions whether some classical Cepheids are radial fundamental or first overtone pulsators, or whether the β Cephei variables are radial or nonradial pulsators. This problem is sometimes very embarrassing because the dispute over whether the Mira variables are radial fundamental or radial first overtone pulsators means that we cannot identify periods that are a factor of three different. For most cases, however, the principal mode of pulsation is well known, although any other modes present can be identified only with less certainty.

Perhaps the greatest successes for the constraints of pulsation theory are for the classical Cepheids, the RR Lyrae variables, and the population II Cepheids (BL Her and W Vir variables). Masses derived by some pulsation-based methods agree well with evolution theory, but there are

421

A. Maeder and A. Renzini (eds.), Observational Tests of the Stellar Evolution Theory, 421–440.
© *1984 by the IAU.*

problems that still exist. These problems might eventually be bona-fide constraints to be settled by changes in our current ideas of the internal composition structure. It seems that, except for much needed accurate surface effective temperatures for Mira variables and for the extremely hot dwarf stars and more periods for double-mode RR Lyrae variables in the field and in globular clusters, the next advances in these pulsation-based constraints will come from theoretical calculations of internal mixing and element separation and from nonlinear pulsation calculations.

Cepheid masses have been shown by Cox (1979) to be consistent with evolution calculations using the new larger distance scale of Hanson (1977) and the cooler effective temperatures of Pel (1978), Flower (1977) and Bell and Parsons (1972). We ignore for this discussion the problems for the double-mode and bump Cepheids that can indicate internal helium enhancements or magnetic fields. For periods longer than 10 days, there has been a remaining problem that the Wesselink masses used with the period-mean density relation produce masses too small by typically a factor of 0.6-0.7. We here propose as Burki (1983) recently has done, that these Wesselink radii indicate mass loss rates that can be used as a constraint on evolution calculations which include mass loss.

The measured Wesselink radius has been discussed by Burki, who shows that the presence of a companion star can greatly affect the derived radius. Therefore, it is better that only truly single Cepheids be considered. Solution of three equations, the period-mean density relation, the definition of the surface effective temperature, and a fit for the pulsation constant Q as a function of mass, radius, luminosity, and effective temperature, are solved for three unknowns, Q, luminosity and mass when the period, effective temperature and Wesselink radius are given. The Cepheids RS Pup, 1 Car, U Car, and SV Vul (possibly a binary) are shown to have a mass as small as 0.3 the so-called theoretical mass. Burki shows that the Maeder (1981) evolution tracks with moderate to large mass loss (cases B and C) bracket the current masses for these four Cepheids. Figure 1 gives these Burki masses versus luminosity for these Cepheids together with the case A (no mass loss), B and C relations for the initial masses of 9 and 15 $M_\odot$. It appears that the mass loss rate almost as high as given for case C is indicated for these four variables.

Actually both RS Pup and SV Vul have measured luminosities and effective temperatures, allowing the calculation of their pulsation masses. RS Pup shows little mass loss in Figure 1, and indeed the Wesselink and pulsation masses agree at 10 $M_\odot$ within about 2 $M_\odot$. For SV Vul, however, it appears that evolution, theoretical and pulsation masses yield about 12 $M_\odot$, but the Wesselink value is less than 4. It may be that indeed SV Vul is a binary star with a blue companion which wrongly reduces its Wesselink radius by an appreciable amount and therefore reduces the Wesselink mass to a very low, incorrect value. We do not have luminosity data for the two other Carina stars to

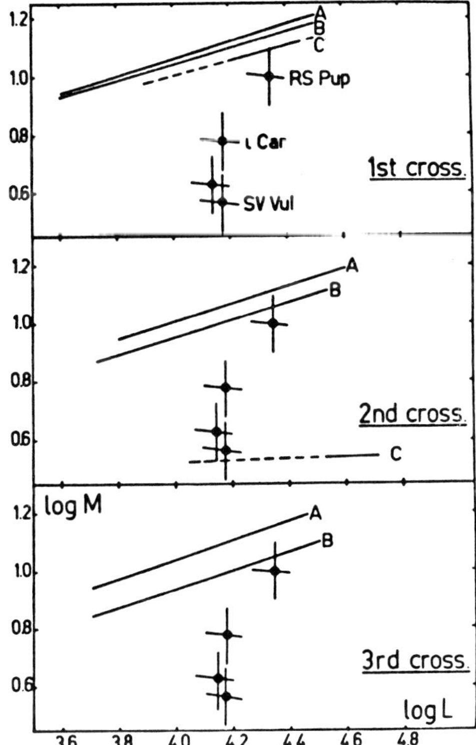

Figure 1. The luminosity-mass relation for four Cepheids with Wesselink
    masses and luminosities from the work of Burki. Also plotted are
    the Maeder theoretical relations during the first, second, and
    third crossings of the Cepheid instability strip for no mass loss
    (case A), moderate mass loss (case B) and high mass loss (case
    C).

dispute their Wesselink masses, and so factors of two mass loss for
long period Cepheids may be correct. This result constitutes a
constraint on the mass loss rates for yellow giants of original mass
about 12 M_θ.

Actually the observed SV Vul period decrease reported by Fernie (1979)
also indicates blueward evolution that occurs slowly compared to the
first and possibly the third crossings of the instability strip when
the original mass is about 15 M_θ. Since evolution tracks indicate very
rapid blueward evolution when the mass loss rate is as high as for case
C, it is very unlikely that such a short-lived Cepheid would be observable with a rapidly decreasing period. The large mass loss of case C
seems excluded by observations of the moderate long period Cepheid
period decreases.

Since the large oxygen depletion reported by Luck and Lambert (1981) no
longer seems to be correct, the large overshooting at 9 M_θ considered

by Becker and Cox (1982) in the main sequence stages is not necessary. The result that the blue loops are at a much higher luminosity than thought before is unaffected, however, and the 15 or more day Cepheid evolution masses are reduced as much as 17% as reported by Becker and Cox and by Matraka, Wasserman and Weigert (1982).

For shorter period Cepheids, there is another problem. The two periods of the double-mode Cepheids and the bumps for the 5 - 15 day bump Cepheids suggest masses considerably less than the evolutionary, theoretical, or pulsation masses. The factor is 0.6 for the bump Cepheids, while it is about 0.3 for those double-mode Cepheids that directly display two periods. The question for the last 10 years has been what is the cause of such large theoretical P_1/P_0 period ratios and such low derived masses? Are they really correct or is something else wrong? The current proposed answers are that the surface layers have an unconventional structure which produces a few percent increase of the fundamental radial mode period relative to a similar increase of the first overtone mode period. This structure can be established by an actual mass loss or an enhanced helium content, a magnetic field, or an opacity increase in the outer 10^{-4} of the mass. The first and last of these suggestions now seems to be ruled out by other considerations that we will not discuss here.

Other discussion of the cause of this anomalously small period ratio, as indeed the cause of the double-mode pulsation, we leave to another talk. Here we merely look at the position of these variables on evolutionary tracks to see what constraints exist on evolutionary theory. Balona and Stobie (1979) made careful measurements of the intrinsic colors of eight of the 11 double-mode Cepheids. Conversion of these colors by a formula given by Bell and Parsons (1972), shows that all these variables have the same effective temperature of 5940 K within a small scatter of about 130 K. More recently, Barrell (1981) using the shape of the hydrogen α line, has verified both the narrow range of the temperature and its mean value for all 11 variables. Our present interest is to see if we can learn what this means for evolution tracks.

The cause of blue loops that produce Cepheids is the existence of the μ-gradient region left behind by a shrinking convective core in main sequence evolution. Too much mass loss can disrupt this structural feature and produce only a return to near the main sequence. Also for only rather low or rather high ratios of the mixing length to pressure scale height in the surface convection zone can these blue loops bring the tracks into the Cepheid variable instability strip.

Recently Huang and Weigert (1983) have investigated this last point as well as how much overshooting is allowed from the central hydrogen burning convective core to homogenize the bottom part of the μ-gradient shell. They find that, at 5 M_θ, overshooting of no more than half of a pressure scale height above the convective core is allowed or no blue loops and no Cepheids will be predicted. Figure 2 shows the blue loops for the two cases with the core convection overshooting 0.5 and 1.0

pressure scale heights. The effects of the outer ratio of convective
mixing length to pressure scale height are also indicated for both core
overshooting cases. This limit of only about 0.5 scale height core
overshooting is reasonable because there is really a very strong
μ-barrier against the outward motion of the heavy helium-rich core
material into the lighter hydrogen-rich layers. For normal compositions then, there are these reasonable constraints: some, but not
much, core overshooting, Maeder and Mermilloid (1981), and a convective
envelope ratio of mixing length to pressure scale height not much
smaller than 1.5, just as appropriate for our sun.

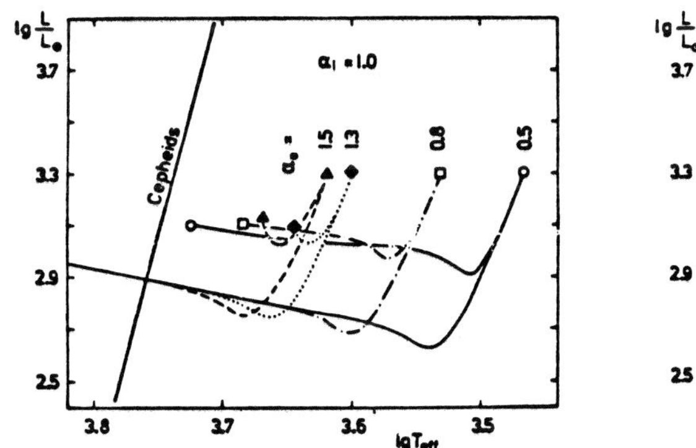
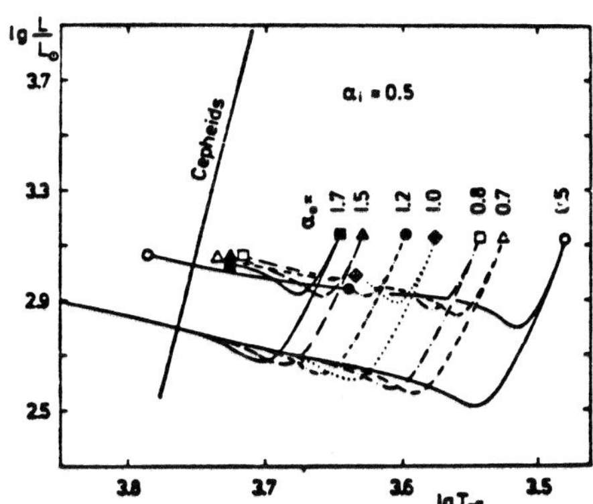

Figure 2. Theoretical Hertzsprung-Russell diagrams are given for the
     core overshooting of 1.0 and 0.5 pressure scale heights. For
     each case the ratio of the mixing length to pressure scale height
     ranges from 0.5 to 1.7 for the outer convective shell.

Unfortunately, the length of the blue loops depends also on the
composition. For low Z, the tracks can go much bluer, but such low Z
values do not seem reasonable for the recently born 5 to 15 $M_\odot$
Cepheids. <u>Figure 3</u> shows the track of a 6.0 $M_\odot$ short period Cepheid
with a period of about five days at the blue loop tip. The upper blue
loop is traversed very rapidly after the end of core helium burning.
Notice that the mass is very reasonable and that the Z=0.02 value of
the composition is normal. For lower luminosities and periods at the
blue loop tips, the luminosity needs to be as low as $\log(L/L_\odot)$ = 3.0
which occurs with a 5 $M_\odot$ model with a Z of 0.017. The very large
number of the double-mode Cepheids (maybe one in four between periods
of 2 and 5 days and almost one in two between 2 and 3 days) points to
evolution tracks with a slight mass dependent Z, older stars having a
lower Z, to give the long-lived blue loop tips (double-mode regime?) at
the observed effective temperature for all these double-mode stars.
To get the even hotter Cepheid SU Cas at an effective temperature of
over 6300 K, one needs an even lower Z value, or perhaps more reasonably an assignment of this 1.95 day period Cepheid as a first crossing
star. These tracks come from the library of tracks of Becker and
Mathews, and the assumption here is for no core overshooting. If there

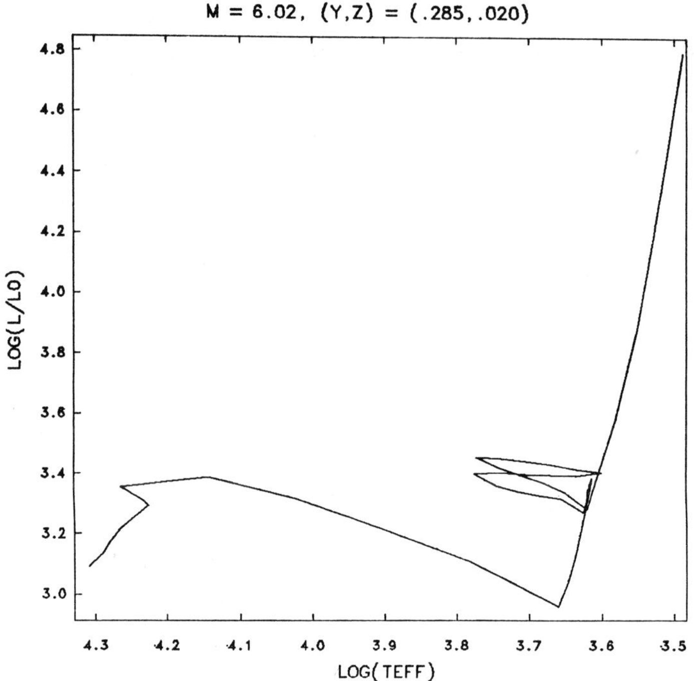

Figure 3. The evolution track for a 6 M_θ model with the Y and Z compositions, respectively, 0.285 and 0.020. The blue loops have their tips just at the effective temperature observed for the double-mode Cepheids.

is a little overshooting at 5 M_θ, then the Z value needs to be even lower than 0.017, but not excessively so.

There is an unsolved problem here, though. Paczynski (1970), Robertson (1971), and Robertson and Faulkner (1972) have considered the possible overshooting of the core convection on the blue loops in the core helium burning stage. This overshooting extends the blue loops and may upset our argument about the Z value for the composition. One nice feature of their work is that with this overshooting during the blue loops, an even larger fraction of time is spent at the blue tips. With all the proper effects included, we may find that double-mode Cepheids naturally occur at the blue loop tips for a conventional composition. The final results of Robertson (1973) need further discussion, especially since the large opacity increases over the Cox and Stewart (1965) values no longer seem justified.

Turning now to another class of variable stars, Saio and Wheeler (1983) have considered the constraint of pulsation theory on the mass of three well known R CrB stars. These hydrogen-deficient stars presumably have lost their hydrogen envelope by a high mass loss rate either from a rather massive star, say 10 - 15 M_θ, or from an older, less massive, star evolving through the asymptotic giant branch region.

<u>Figure 4</u> shows the blue edge for the fundamental radial pulsation mode on the Hertzsprung-Russell diagram for 1, 2, and 3 $M_\odot$. The cause of the sharp turn of the blue edge towards higher effective temperatures at a critical luminosity is that the envelope density there becomes very low. At this mass-dependent critical luminosity the radiatively damping region of the star loses its effectiveness, while the cyclical ionization of helium even nearer to the surface still causes pulsation driving.

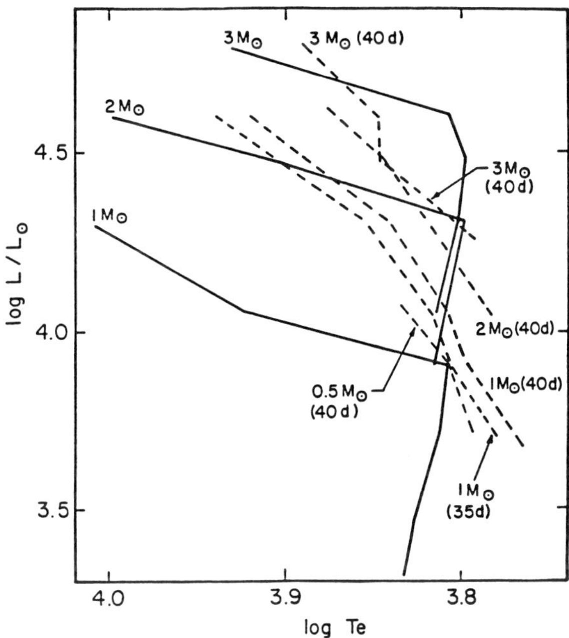

Figure 4. The Hertzsprung-Russell diagram in the region of the R CrB variables. Blue edges are marked as solid lines for 1, 2, and 3 $M_\odot$. Also given are lines of constant period for masses and periods as marked.

The figure also gives dashed lines for the constant period of 40 days and one case for 35 days. For the 3 $M_\odot$ case, we can see the sudden increase in luminosity near $\log(L/L_\odot)$ of 4.5 due to mode bumping. This bumping occurs when stars become very centrally concentrated at high luminosity and periods decrease compared to those for less concentrated models. This decrease of a so-called longer period strange mode period makes it close to that for the fundamental which is less decreased in its period. Therefore, the potential collision of the two periods is solved by the star when the fundamental period is bumped down in period. This means that to retain the 40-day period, the radius and luminosity need to be increased suddenly.

The observed periods and surface effective temperatures for R CrB and RY Sgr are 44 and 39 days and 7000 K and 7100 K. This first effective temperature is uncertain by only 250 K, whereas the second has its uncertainty as 600 K. XX Cam has similar properties, but it does not pulsate, probably because its luminosity is just a bit lower than for the other two stars and it never enters the pulsation instability strip.

Mass and luminosity limits can be obtained for the two variable stars by noting where the line of the observed period intersects the nearly horizontal blue edge at possible effective temperatures of log T_e between 3.829 and 3.860 for R CrB and 3.813 and 3.886 for RY Sgr. Above about 3 M_θ the variable stars would not be predicted to pulsate because the line of constant period of 40 days is always in the stable, less luminous, region for all possible effective temperatures. Since RY Sgr can be a little bluer than R CrB, its mass and luminosity may possibly be a bit higher. For lower masses, the lines of constant period always are in the unstable region. In these cases, the luminosity limits always come from the observed effective temperature limits with, of course, a larger range for RY Sgr.

Figure 5 presents the relation between the mass and luminosity limits. At low mass, if it all is in a helium core, there is an upper limit to the luminosity that can be produced according to Paczynski (1971), and that is the dash-dot line. The Eddington limit for the luminosity at any given mass is the upper solid line. Recently Cottrell and Lambert (1982) have given for R CrB a value for the gravity that can be combined with the measured effective temperature to constrain the mass-luminosity relation, and that is the heavy dotted line. We see that the two R CrB variables that pulsate have masses between about 0.8 and 3.1 M_θ and luminosities between $\log(L/L_\theta)$ of 4.1 to 4.8. Further work on nonlinear pulsation calculations is now being done by Saio.

We now turn to some constraints from pulsation theory for δ Scuti variables. Here we consider the slow settling of helium below the bottom of the hydrogen-helium ionization convection zones. A 2 M_θ star evolves slowly enough that this draining of helium can operate even in the presence of mixing processes such as those from rotation and turbulent diffusion. During this stage the convection zones remain homogeneous due to the strong motions of convection, but the helium content steadily decreases. When the helium mass fraction decreases to about 0.08, then the deeper helium convection zone disappears. At this low helium content the opacity is now low enough so that all the internally generated luminosity can be carried by radiation alone. However, the outer hydrogen convection zone is, if anything, slightly enlarged because the high opacity of the hydrogen-helium mixture is even now increased with less dilution by helium which does not contribute so much to the opacity at 10,000 K.

New evolution tracks have been computed by Andreasen, Hejlesen, and Petersen (1983) allowing the helium to settle to depths of about 300,000 K and 1,000,000 K. As one might expect, the tracks are not greatly changed on the Hertzsprung-Russell diagram. However, these

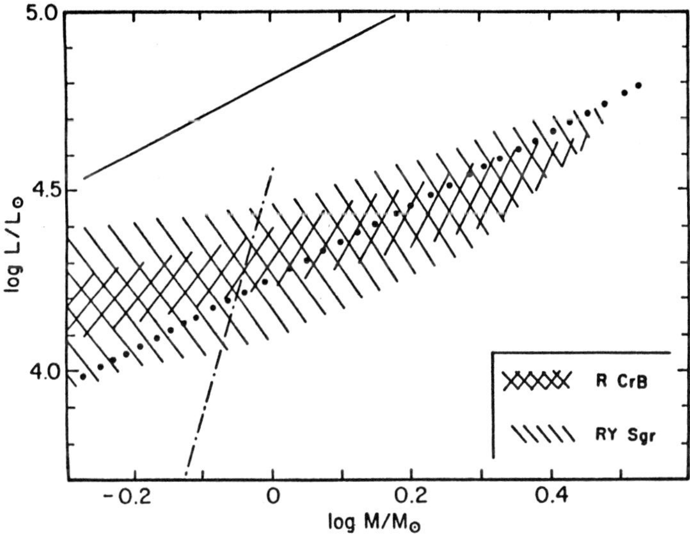

Figure 5. The allowed regions of mass and luminosities for R CrB and RY
    Sgr from pulsation theory are given as hatched areas. The maximum
    luminosity as a function of mass is also drawn for low masses.
    The solid line is the Eddington limit as a linear function of
    mass. The dotted line is the mass-luminosity relation derived
    from the observation of the effective temperature and gravity of
    R CrB.

authors have found that as the stars expand in their normal evolution
and their pulsation periods increase, the period ratio of the first
radial overtone to the fundamental increases to much larger values than
ever considered before. Their period ratio reaches 0.79 compared to
just over 0.76 for 1.7 to 2.0 $M_\odot$ when the helium is depleted down to
about 300,000 K. Deeper helium depletion returns the period ratio back
to or even slightly lower values.

This behavior of the overtone to fundamental period ratio is just as
expected from the studies by Cox, King, and Hodson (1979) for decreasing the period ratios for double-mode Cepheids. In that case, surface
enrichment was needed to reduce the ratio by about 0.01 to 0.05 for
masses, respectively of 6 and 4 $M_\odot$. As discussed above, the physical
effect to change the period ratio is a change in the apparent concentration of the star as seen differently by the two modes. For the
double-mode Cepheids the enrichment seems best for depths of 250,000 K,
not too much deeper than we will see is appropriate for VZ Cnc our δ
Scuti variable.

With McNamara from New Mexico State University, I began wondering if
the large-period ratio observed for the 0.178 day δ Scuti variable VZ
Cnc might indicate this helium settling. This star has been reobserved

recently by McNamara to compare with data obtained by Fitch (1955). Both observers get the ratio to be 0.8006, but the modern data give clearly a change in the amplitude ratio of the two modes with the overtone now 0.326/0.276 larger relative to the then existing fundamental amplitude. The accuracy of the periods themselves is not great enough to detect the few parts in a million period increase expected in the 30-year time interval, and therefore, a check that the expected redward evolution is really occurring is not yet possible. Anyway, it appears that the overtone is getting appreciably stronger rather than weaker as the star evolves to the red and to a more pure fundamental mode domain.

Figure 6 gives the period ratio being discussed for five calculated depths of the helium depletion in a model for a 2 M_θ star with the observed surface effective temperature and a luminosity such that the observed period is matched to the radial fundamental. To reproduce the observed period ratio for VZ Cnc, a depletion profile is needed that has a surface helium mass fraction, Y, of either 0.00 or 0.08 down to almost 200,000 K. Deeper the Y increases to 0.08 until a depth is reached where the temperature is almost 300,000 K. Then we set Y=0.18. Finally Y=0.28, the normal value, is used for depths deeper than about 400,000 K.

The apparently correct interpretation for VZ Cnc is that its helium has settled to below 200,000 K. This makes the blue edge of the instability strip for such a star just at 7000 K, and the red edge is also probably at about 7000 K. Thus this star just barely has an instability strip, and it is driven by hydrogen γ and κ effects alone. Note that as the helium settles, the fundamental mode blue edge evolves to the red due to the depletion of helium. When a star is at or near this

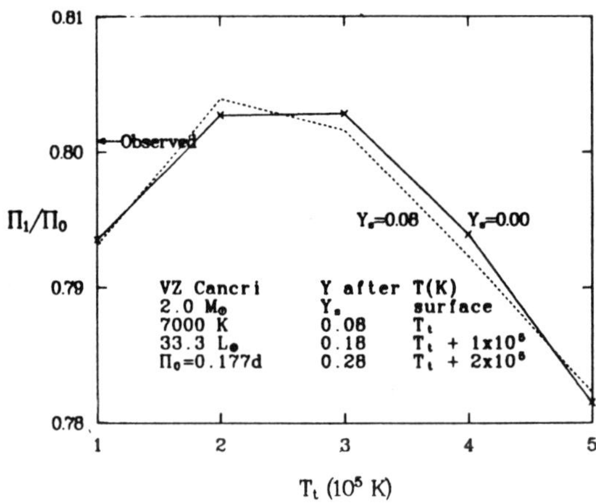

Figure 6. The period ratio versus transition temperature is plotted for five depths of settling of helium in a 2 M_θ model at 7000 K with the period of 0.177 day. Two surface helium content mass fractions have been considered. The observed period ratio is also indicated.

blue edge there is a strong tendency to move over to an overtone pulsation as we see for double-mode Cepheids and double-mode RR Lyrae variables. It appears that this star has a redward evolution of its fundamental mode blue edge that is faster than the evolution of the star itself to the red, because the overtone mode is getting stronger. Helium depletion to depths such as we have found seem reasonable from the current results of diffusion theory.

One of the current interests in the constraints of pulsation theory on evolution theory is in the solar 5 minute oscillations. The low degree modes with $\ell=0$ to 5 give sensitivity to the central temperature and density structure, and that in turn bears on the persistent problem of the higher than observed predictions for the neutrino flux from the sun. The reason for the sensitivity of the periods of low degree oscillations is that the spacing between nodes near the center of the sun for, say the 10th overtone, is comparable to the horizontal distance between node lines when $\ell=1$ to 5. Thus a change in the central conditions results in a change in the nodal structure which reflects itself in changes of periods.

<u>Figure 7</u> gives the internal structure of a solar model calculated by us at Los Alamos. The logarithms of the temperature, opacity, and density are plotted versus the external mass fraction. The central temperature for this model is about 15 million kelvin, and the central density is about 122 g/cm^3, both somewhat different from other recent models. These differences may be due to the equation of state of the solar material that we used. Some coulomb corrections for the pressure have been included, but maybe not in as elaborate way as Ulrich and his collaborators or Shibahashi and his collaborators have done. The very high opacity across most of the figure produces a convective shell down to a temperature of just over 2.5 million kelvin.

<u>Figure 8</u> shows the hydrogen composition mass fraction profile for four different models. The basic profile is due to Christensen-Dalsgaard (1982) who has made an actual evolution calculation using standard methods. The solid stepped line shows our Los Alamos model which attempts to track the Christensen-Dalsgaard model as well as possible. The differences are needed due to slight differences between the equation of state and opacities. Note that we need a smaller amount of helium everywhere and it is not clear whether the sun produces that 10% smaller amount during its lifetime. The figure also shows two partially mixed models, one with a homogeneous shell between q=0.3 and 0.5 and another with the central 15% of the mass homogeneous as shown. The pulsation periods for low degree, high overtone modes will later be calculated, and the effects of these two mixed layers on the periods will be given.

Observations of these modes with frequencies between 900 and 4500 μHz are given in <u>Figure 9</u>. The plus signs, circles and triangles give data for many of the modes, and the dots are the latest Shibahashi, Noels, and Gabriel (preprint) theoretical results. The figure is really only

Figure 7. The logarithms of
the temperature, opacity, and
density are plotted for our
solar model versus surface
mass fraction. The units are:
temperature (K), opacity
(cm^2/g), density (g/cm^3).

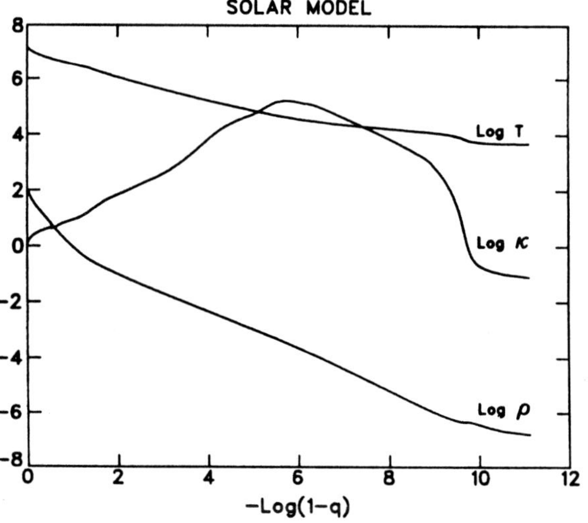

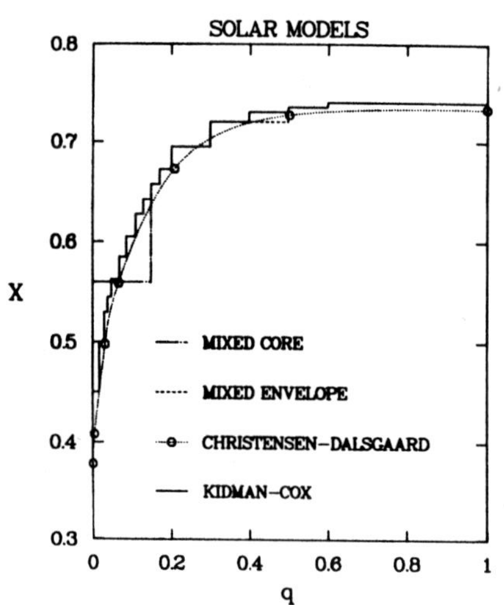

Figure 8. The hydrogen mass
fraction, X, is plotted
versus internal mass fraction
for four models including
one with a mixed region in
the envelope between q=0.3
and 0.5 and one homogenized
in the central 0.15 of the mass

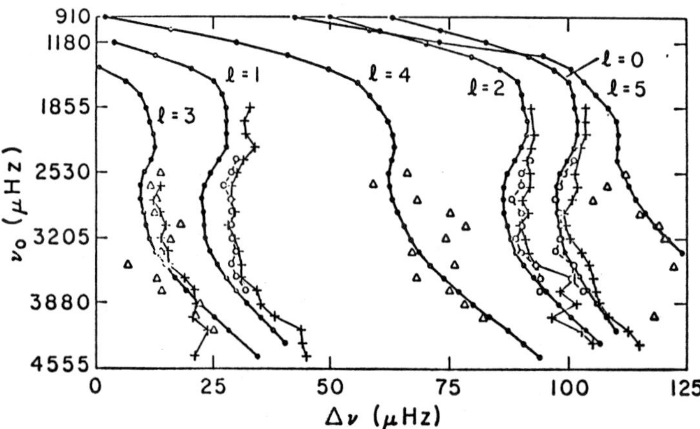

Figure 9. Frequencies of solar oscillations calculated by Shibahashi,
   Noels and Gabriel are plotted in 135 µH sections and placed above
   one another. The observed nonradial modes are also given as
   presented by Claverie et al. (1981) (open circles), Grec et al.
   (1980) (crosses), and Sherrer et al. (1982) (triangles).

one-dimensional, but the line of increasing frequency is cut into seg-
ments of 135 µH and these lines are displayed one above the other. The
match between theory and observation is good for these adiabatic
periods. Our nonadiabatic periods are typically 0 to 7 µHz smaller in
frequency. A thing to note about this figure is that the even ℓ values
are close together and so are the odd ones if the l=5 curve had been
plotted one raster line below. As shown by Vandakurov (1967) and
recently by Tassoul (1980), the difference in frequency between two
modes separated by +2 in ℓ and -1 in n (overtone) is almost zero. The
splitting of these modes is a good probe of the internal solar structure and its possible mixing either now or in the past.

The table of frequencies give our values for our standard, mixed
envelope, and mixed core models. We have been limited at the present
time to only 330 mass shells, but we believe that maybe 100 more zones
are required to get accurate frequencies even for these longer than
average periods, and lower than average order. Thus we get frequencies
too large by 1.0 to 1.5%. To approximately correct for this defect in
order to assess the effects of the model structure on frequency
splitting, we have corrected all of our frequencies by the ratio of the
measured to standard frequencies. This should reduce our splitting
prediction errors to much less than 1%.

The second table, with the separation of successive modes at a fixed ℓ
value, averaged over all the four overtones, and with the above
mentioned ℓ splitting averaged over four pairs shows the effect of the
two mixed models. Our models still seem to have the separations
slightly too large, though there is apparently no sensitivity to the
possible internal mixing as expected. This separation is determined by
the model structure in the outer 10% of the mass. We do see that
homogenizing the envelope shell seems to slightly increase the ℓ
splitting. An even larger increase, as also discussed recently by
Ulrich and Rhodes (1983), is seen for the core mixed model. It seems
that the ℓ splitting is sensitive to the core structure. Mixing
increases the splitting, but our core spacial resolution is too coarse
to match observed splittings. No mixing at all seems to be a more
correct model for our sun.

As the final subject, we discuss the masses of the RR Lyrae variables.
The discovery of the double-mode RR Lyrae variables in M15 and M3, as
well as in the field (AQ Leo), in M68, and in the Draco galaxy, has
allowed us to measure their masses. The value of 0.65 M_θ seems just
right for the Oosterhoff group II cluster M15. Only two stars are
available for the Oosterhoff group I cluster M3, and they seem to indicate 0.55 M_θ. This mass is too low for current ideas of the 0.8 M_θ
stars evolving with some unknown mass loss rate on the red giant branch
and their helium-rich core mass when they experience the core flash.
If the low mass for M3 is correct, there is a strong constraint for red
giant evolution theory. At 0.55 M_θ, stars are very blue and not in the
pulsational instability strip unless their helium core mass is also
quite low, 0.425 rather than the 0.475 M_θ, as currently accepted as
appropriate.

NONADIABATIC FREQUENCIES (μHz)

DEGREE	ORDER	STANDARD MODEL	MIXED ENVELOPE	MIXED CORE	MEASURED
0	10	1595	1595	1597	
	11	1732	1732	1734	
	12	1867	1868	1869	1824
	13	2002	2003	2004	1954
	14	2138	2139	2140	2094
1	10	1645	1645	1645	
	11	1782	1782	1783	1753
	12	1919	1919	1920	1888
	13	2056	2056	2057	2022
	14	2193	2193	2194	2157
2	10	1698	1699	1698	
	11	1835	1835	1835	1812
	12	1971	1972	1971	1947
	13	2110	2110	2110	2082
	14	2247	2248	2247	
3	10	1749	1749	1749	
	11	1887	1887	1887	
	12	2025	2025	2024	
	13	2163	2163	2162	
	14	2301	2301	2301	
4	10	1795	1794	1794	
	11	1934	1934	1934	1918
	12	2074	2074	2074	2052
	13	2213	2212	2212	
	14	2352	2352	2351	2323

Table 1. Frequencies of our solar model are given for the longer
observed periods in the 5 minute region.

AVERAGE CORRECTED SEPARATIONS AND SPLITTINGS (μHz) (N=10−14)

L	STANDARD MODEL	MIXED ENVELOPE	MIXED CORE	MEASURED
$\overline{\Delta\nu}(0)$	134	134	134	135
$\overline{\Delta\nu}(1)$	135	135	135	135
$\overline{\Delta\nu}(2)$	135	135	135	135
$\overline{\Delta\nu}(3)$	136	136	136	
$\overline{\Delta\nu}(4)$	137	138	137	134
$\overline{\delta\nu}(0)$	31	32	33	10
$\overline{\delta\nu}(1)$	31	32	32	18
$\overline{\delta\nu}(2)$	36	38	37	30

Table 2. Separations and ℓ-splitting in units of μHz are given
for ℓ values of 0 - 4.

Figure 10 shows how photometric data can reveal double-mode behavior.
From a Sky and Telescope report we see on top the Sandage, Katem, and
Sandage (1981) blue magnitude data for M15 variable V31. The period
used to phase the observations made over three summers is 0.408231 day.
The scatter in the data indicates that maybe the derived period is not
accurate enough. Cox, Hodson, and Clancy (1981,1983) noticed that the
scatter in this light curve and many others with periods near 0.4 days
revealed the presence of two periods of the star. The separation of
the two components is also in the figure with the periods of the overtone and fundamental modes indicated. The lower two panels show bolometric magnitudes rather the photographic blue magnitudes above. There
are 17 RR Lyrae variables with periods between 0.389 and 0.429 day in

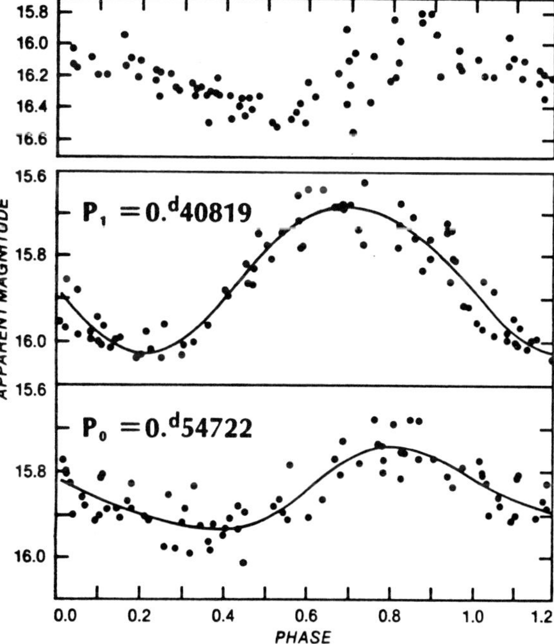

Figure 10. In the upper panel are given blue magnitudes for V31 as published by Sandage, Katem and Sandage. Below are phased bolometric magnitude data for the two separate modes with their periods indicated.

M15, and there are indications that all pulsate with two modes simultaneously. For the 12 cases where a good period ratio can be determined, the overtone is always the stronger mode.

From ideas first presented by Jorgensen and Petersen (1967), it is possible to use the period ratio to derive a mass for each star. This is physically possible because the two modes sample the outer structure differently due to their different oscillation amplitude variations. The best way to portray this mass dependence is to show in Figure 11 the period ratio versus period. This diagram depends only slightly on the model luminosity and effective temperature, and so it is usable because it has been calculated for these two parameters being close to those for real RR Lyrae variables. In addition to 12 M15 variables, we show also in the figure the period ratio for the field RR Lyrae variable AQ Leo, a representative point for nine Draco galaxy RR Lyrae variables, a point representing the only two known M3 variables, and a point for the single known M68 double-mode variable. It appears that the M15 stars have a mass of 0.65 M_θ as well as those nine (only one plotted) in Draco and for AQ Leo. The intermediate Oosterhoff group cluster M68 seems to have a mass for its single known double-mode variable near 0.60 M_θ. According to Nemec at this meeting V75 in Draco also has a mass of 0.60 M_θ. Finally the two M3 variables and V165 in Draco point to 0.55 M_θ. Since M3 is representative of an Oosterhoff group I cluster, there is the possible interpretation that group I clusters have horizontal branch stars of 0.55 M_θ and the group II clusters, with certainly higher luminosity and lower metal composition stars, have masses of their stars of 0.65 M_θ.

Figure 11. For all the listed
double-mode RR Lyrae variables
are plotted their period ratios
versus their fundamental and
overtone periods. The lines of
constant mass are computed from
linear nonadiabatic radial
pulsation theory.

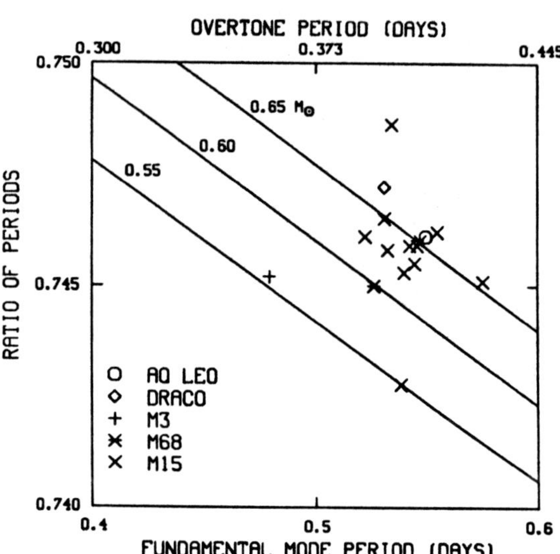

With a mass of 0.65 M_θ, it is possible to make a theoretical
Hertzsprung-Russell diagram with the RR Lyrae variable instability
strip blue and red edges shown. Figure 12 also gives the lines of
constant fundamental mode period and lines of constant period ratio. A
standard composition called King Ia has been used to derive the periods
and blue edges from linear pulsation theory. The red edge comes from
the calculations of Deupree (1977) which show how dependent the red
edge is upon the helium content. On this diagram, we have plotted the
mean luminosities of the M15 RR Lyrae variables with F for the fundamental mode variables and O for the overtones. The double-mode vari
ables are plotted as + signs. Here we use only the relative luminosities and periods as observed. The observed effective temperatures are
not used. A distance modulus of 15.28 for M15 results in the luminosities that are plotted. Possible errors in the magnitudes and that from
a change in mass by 0.05 M_θ from the assumed 0.65 M_θ are indicated. The
interesting thing is that the the eight plotted double-mode variables
all lie between the fundamental and overtone pulsators suggesting that
they are the result of some kind of mode switching.

From the mass of 0.55 M_θ, we can construct another Hertzsprung-Russell
diagram that can be used for plotting the M3 RR Lyrae variables. For
this globular cluster we need a distance modulus of 15.00 to fit the
variables into the theoretical instability strip. This is Figure 13
that again shows the separation of the fundamental and overtone pulsators. The two double-mode stars have not been observed on the photometric scale of Sandage (1981), and therefore, are not plotted. The
appearance of overtone pulsators in the theoretical fundamental mode
region is a problem for both M15 and M3. We would expect that overtone
pulsation could occur slightly redward of the fundamental blue edge as
the overtone there may be able to grow in a fundamental mode pulsator.

But the overtones and the double-mode pulsators seem to be much too red to match the best currently available blue edges and the fundamental-overtone transition line.

Figure 12. The RR Lyrae pulsation instability strip is plotted on the Hertzsprung-Russell diagram for a mass of 0.65 $M_\odot$ and the King Ia mixture composition. Mean luminosities over the pulsation cycle are plotted for the M15 RR Lyrae variables with the fundamental pulsators marked by F, the overtones by O, and the double-mode variables by +.

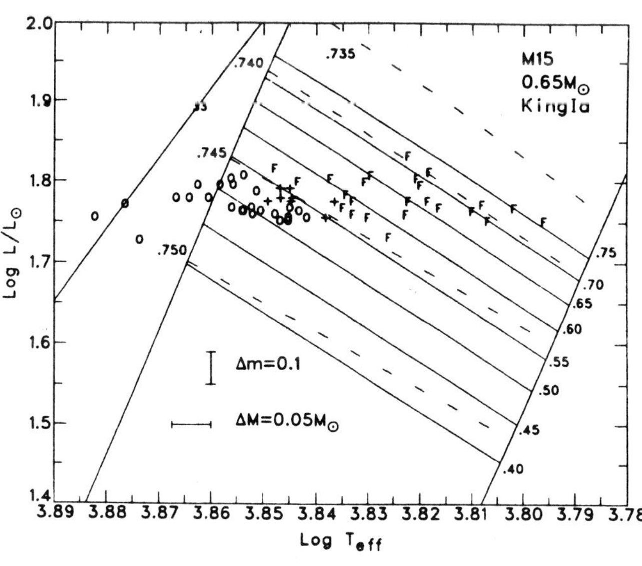

Figure 13. The RR Lyrae pulsation instability strip is plotted on the Hertzsprung-Russell diagram for a mass of 0.55 $M_\odot$ and the mixture composition with Y=0.279. Mean luminosities over the pulsation cycle are plotted for the M3 RR Lyrae variables with the fundamental pulsators marked by F, the overtones by O, and the double-mode variables by +.

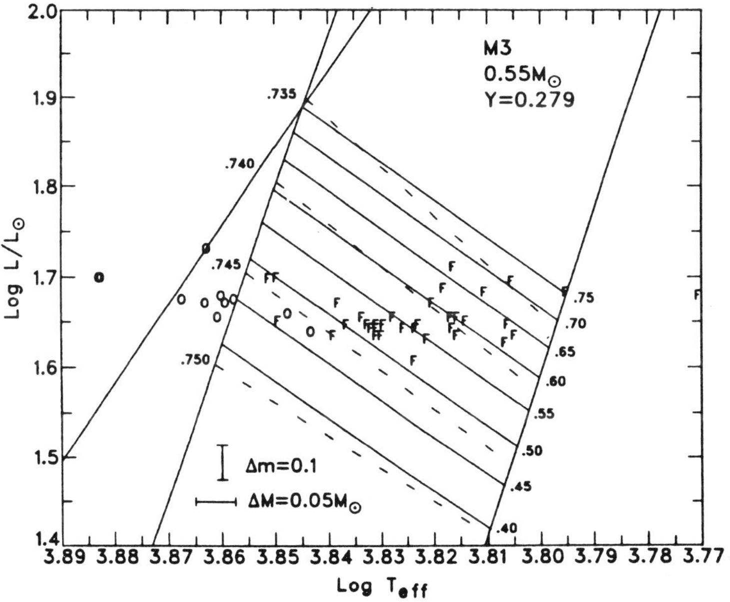

The constraint on evolution theory is that maybe the mass of the RR
Lyrae variables differs with Oosterhoff groups, being lower in the
higher metal content Oosterhoff group I clusters. Actually, with a
core mass of about 0.475 $M_\odot$, as currently preferred, a mass of 0.55 $M_\odot$
would be born on the horizontal branch and evolve at much bluer positions than the instability strip. The only way to get RR Lyrae variables then is to reduce the helium-rich core to values as low as 0.425
$M_\odot$ according to the calculations of Sweigart and Gross (1976). Does
this mean that the higher Z clusters have more mass loss and the core
helium flash occurs at an earlier time in its evolution when less
helium has been produced in the core?

We conclude from this discussion that there are some constraints on
evolution theory from the various results of pulsation theory. The
solar problem is interesting in that it reinforces our standard ideas
of evolution, leaving the neutrino problem unsolved. The two other
potentially strong constraints are from the double-mode Cepheids and
the double-mode RR Lyrae variables. Future work for the first of these
will be in theoretical studies of blue loop parameters. For the latter
case, more accurate observations of RR Lyrae period ratios will eventually define the horizontal branch star masses to produce very strong
constraints for red giant and horizontal branch evolution calculations.

References

Andreasen, G.K., Hejlesen, P.M., and Petersen, J.O. 1983, Astron. Ap. in press.

Balona, L.A. and Stobie R.S. 1979, M.N.R.A.S. 189, 659.

Barrell, S.L. 1981, M.N.R.A.S. 196, 357.

Becker, S.A. and Cox, A.N. 1982, Ap.J. 260, 707.

Bell, R.A. and Parsons, S.B. 1972, Ap Letters, 12, 5.

Burki, G. 1983, Astron. Ap. 1983, in press.

Christensen-Dalsgaard, J. 1982, M.N.R.A.S. 199, 735.

Claverie, A., Issak, G.R., McLeod, C.P., Van der Raay, B., and Roca Cortes 1981, Nature 293, 443.

Cottrell, P.L. and Lambert, D.L.,1982, Ap.J. 261, 595.

Cox, A.N. 1979, Ap.J. 229, 212.

Cox, A.N., Hodson, S.W., and Clancy, S.P. 1981, in IAU Colloquium 68, "Astrophysical Parameters for Globular Clusters," eds. A.G. Davis Philip and D.S. Hayes L. Davis Press, Inc, Schenectady, New York p337.

Cox, A.N., Hodson, S.W., and Clancy, S.P. 1983, Ap.J. 266, 94.

Cox, A.N., King, D.S., and Hodson, S.W. 1979, Ap.J. 228, 870.

Cox, A.N. and Stewart, J.N. 1965, Ap.J. Suppl., 11, 22.

Deupree, R.G. 1977, Ap.J. 214, 502.

Fernie, J.D. 1979, Ap.J. 231, 841.

Fitch, W.S. 1955, Ap.J. 121, 690.

Flower, P.J. 1977, Astron. Ap. 54, 31.

Grec, G., Fossat, E., and Pomerantz, M. 1980, Nature 288, 541.

Hanson, R.B. 1977, in IAU Symposium 80, "The H-R Diagram" eds. A.G. Davis Philip and D.S. Hayes, Reidel, Dordrecht, p154.

Huang, R.Q. and Weigert, A. 1983, Astron. Ap. in press.

Jorgensen, H.E. and Petersen, J.O. 1967, Zs.Ap. 67, 377.

Luck, R.E. and Lambert, D.L. 1981, Ap.J. 245, 1018.

Maeder, A. 1981, Astron. Ap. 102, 401.

Maeder, A. and Mermilloid, J.C. 1981, Astron. Ap. 93, 136.

Matraka, B., Wassermann, C., and Weigert, A. 1982, Astron. Ap. 107, 283.

Paczynski, B. 1970, Acta Astr. 20, 195.

Paczynski, B. 1971, Acta Astr. 21, 1.

Pel, J. 1978, Astron. Ap. Suppl. 31, 489.

Robertson, J.W. 1971, Ap.J. 170, 353.

Robertson, J.W. 1973, Ap.J. 185, 817.

Robertson, J.W. and Faulkner, D.J. 1972, Ap.J. 171, 309.

Saio, H. and Wheeler, J.C. 1983, Ap.J. Lett. 272, L25.

Sandage, A. 1981, Ap.J. 248, 161.

Sandage, A., Katem, B., and Sandage, M. 1981, Ap.J. Suppl. 46, 41.

Sherrer, P. H., Wilcox, J. H., Christensen-Dalsgaard, J. and Gough, D. O. 1982, preprint.

Sweigart, A.V. and Gross, P.G. 1976, Ap.J. Suppl. 32, 367.

Tassoul, M. 1980, Ap.J. Suppl. 43, 469.

Ulrich, R.K. and Rhodes, E.J. 1983, Ap.J. 265, 551.

Vandakurov, Y.V. 1967, Ast.Zh. 44, 786.

DISCUSSION

Schatzman: The model with core mixing, or with mixing of any layer does
not mimic at all the $\nabla\mu$ distribution which comes out from turbulent
diffusion mixing.

Cox: This is true. Our very preliminary study only looked at two arbitrary mixing scenarios - an envelope shell and a core mixing.

Maeder: Although mass loss and overshooting are two different physical
effects, they have some similar consequences on the structure of stellar
models as both increase the core mass fraction. Thus, I would be careful
in interpreting all the deviations between models and observations of
massive stars in terms of mass loss. What is your own appreciation of
this kind of problem?

Cox: If we are talking about the Burki data, it must be mass loss because the mass is directly measured by the Wesselink radius. In the
R CrB case again the current mass is measured. If a track on the HRD is
matched to derive period changes, the mass is still needed, but it has
less effect.

Roxburgh: It is interesting that people can arrive at different conclusions from the same data! My conclusion from your results is that
the so-called "standard model" has difficulties in explaining the
oscillation data. It may be that the other proposals you studied also
have difficulties but this does not support the standard model! The
standard model also has problems in that it cannot explain the solar
neutrino flux.

Lynas-Gray: I would like to comment on the RY Sgr period decrease reported by Kilkenny (1982, Mon. Not. R. Astr. Soc. 200, 1019) and Marraco
and Milesi (1982, Astron. J. 87, 1775). This places a constraint on the
future evolution, and agrees with what is expected from Schönberner's
(1977, Astron. Astrophys. 57, 437) evolutionary tracks.

CEPHEID PERIOD CHANGES AND STELLAR EVOLUTION

J.D. Fernie
David Dunlap Observatory
University of Toronto

The question of whether or not the period changes observed in
some classical cepheids are due to evolution across the HR diagram
remains controversial. On the one hand, it is known that the stars are
indeed evolving across the diagram and are therefore changing their radii
and must therefore be changing their periods. On the other hand, only a
fraction of cepheids appear to have period changes, and some of these
seem to be in the form of sudden jumps. Is it then that the evolutionary
changes are too small to be detected, and the observed changes due to
something else? This note explores the question.

First we consider the period changes predicted by stellar
evolution theory. Second, we shall compare these predictions to the
observed changes. Finally, we discuss briefly the problems of sudden
jumps and the small number of cepheids showing period changes.
For theory I have used the models of Becker, Iben, and Tuggle
(1977: hereafter BIT). They show detailed evolutionary tracks on the
theoretical HR diagram near the cepheid instability strip for 3, 5, 7,
and 9 solar masses and a variety of chemical compositions. For each
track I have read off the diagram the values of log L and log Te at each
point where the track enters and leaves the instability strip. The
corresponding fundamental period of pulsation was then calculated from
equation (4) of Iben and Tuggle (1975), which gives P as a function of
luminosity, temperature, and mass. Tables in BIT give the ages of the
models located at the edges of the instability strip. Thus one obtains
the change in period and age across the strip, and hence an average
value of dP/dt at an average period.

An important point must be made here. The above data also
give the time taken by a cepheid to cross the instability strip. These
crossing times are very dependent on chemical composition and it is
difficult to generalize from the rather coarse grid of models, but it is
clear that no one particular crossing of the strip is completely dominant.
It was long believed that virtually all cepheids are in their third
crossing of the strip, based on the rapidity of other crossings. The
BIT models, however, do not bear this out. Space precludes a full
discussion, but as an example a 5 M⊙ model of high helium content
(Y=.36, Z=.02) spends 940 thousand years on the second crossing and 560

A. Maeder and A. Renzini (eds.), Observational Tests of the Stellar Evolution Theory, 441–444.
© *1984 by the IAU.*

thousand on the third crossing. The same mass with low helium (Y=.20,
Z=.02) has second and third crossing times of 620 and 680 thousand years
respectively. Similar trends are to be found at other masses. This may
have serious implications for studies involving the 'effective' mass-
luminosity law for cepheids. Its importance for this study is that we
need not be deterred by finding a thorough mix of positive and negative
period changes, because the probabilities of a cepheid being on its
second or third crossing are similar and the second crossing will produce
a negative period change while the third will show a positive change.

The calculated period changes (expressed in sec/yr) are shown
in Figure 1 as a function of mean period. The numbers refer to the
crossing, and one may note that even-numbers imply $\dot{P}<0$ while odd-numbers
imply $\dot{P}>0$. They lie in a band shown roughly by the two slanted lines,
the slope of which indicates that $\dot{P}$ increases as P^3. No segregation of
the data by chemical composition is evident.

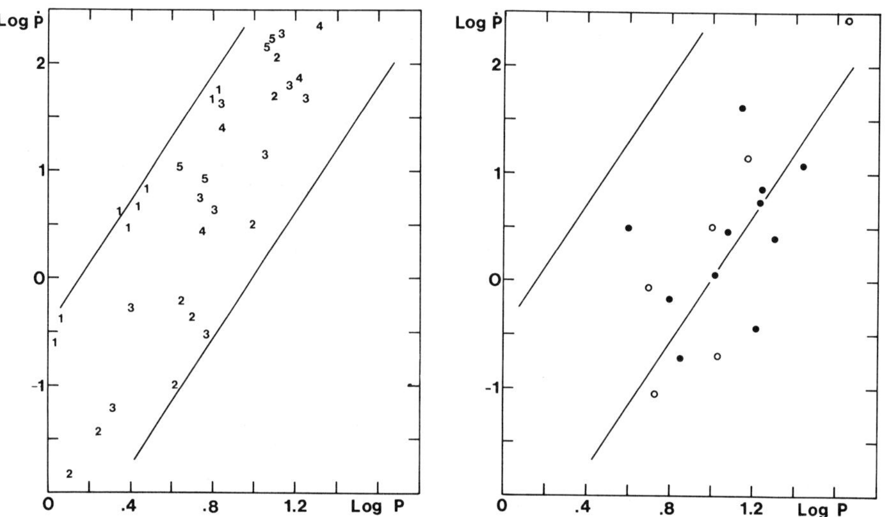

Fig. 1.   Theoretically predicted   Fig. 2.   Observed rates of
          rates of period change.   period change.

I have made use of the extensive material of Szabados (1977,
1980, 1981) to compile the observed period changes shown in Table 1.
This, of course, is seriously incomplete, including only those northern
cepheids he happened to observe himself.

Table 1

Star	P	$\dot{P}$ (sec/yr)	Star	P	$\dot{P}$ (sec/yr)
α UMi	3.97	+3.2	SZ Cas	13.64	+46
V Lac	4.98	−0.85	RW Cas	14.79	−14
δ Cep	5.37	−0.089	X Cyg	16.39	+0.38
RS Cas	6.30	+0.67	CD Cyg	17.07	+5.5
η Aql	7.18	+0.19	SZ Aql	17.14	+7.4
SY Aur	10.14	+1.1	VX Cyg	20.13	+2.5
ζ Gem	10.15	−3.2	T Mon	27.03	+12
Z Lac	10.89	−0.20	SV Vul	45.0	−290
RY Cas	12.14	+2.8			

The observed values are shown plotted in Figure 2, with open circles representing negative changes, filled circles positive changes. The slanted lines come from Figure 1. As expected, no observed values lie near the upper line, which is mainly set by the very fast first crossings. Even so, the theoretical lower boundary is high − by about a factor of 5, in fact. Nevertheless, considering the many orders of magnitude over which these quantities could range, agreement is not unreasonable. The important conclusion, I think, is that the observed period changes are almost certainly due to evolution; otherwise where are the evolutionary changes, they being at least as large as any change observed?

Finally, the problems of sudden period jumps and the low percentage of cepheids showing period changes. The case of Polaris is illuminating; it has been quoted in the literature for many years as having shown a sudden period jump around 1927, yet a new, careful study (Arellano Ferro 1983) now shows that to have been simply a miscount of cycles. Polaris, it turns out, has historically always shown a smooth period increase. This is not to say that sudden jumps cannot occur, but I suspect their frequency is less than appears and perhaps inadequately substantiated.

Similarly, I believe we currently have only a poor idea of how many cepheids really show period changes. One is struck by the number of very well-known cepheids in Table 1, the implication being that if a star is studied long enough and often enough and carefully enough, a period change is revealed. Clearly, a detailed objective study of these problems would be worthwhile.

REFERENCES

Arellano Ferro, A. 1983, Ap.J., Nov. 15 issue.
Becker, S.A., Iben, I., and Tuggle, R.S. 1977, Ap.J., 218, 633.
Iben, I., and Tuggle, R.S. 1975, Ap.J., 197, 39.
Szabados, L. 1977-81, Mitt.Sternwte Ungar.Ak.Wiss., no. 70, 76, 77.

DISCUSSION

<u>Tayler</u>: Am I correct in thinking that you plotted the average theoretical $\dot{P}$ for each crossing? If so, since the actual $\dot{P}$ will vary between a small multiple of the average and zero, you must expect a larger spread below the theoretical value in log $\dot{P}$ than above it.

<u>Fernie</u>: Yes, it is the average P that is plotted. I agree the scatter would be larger below than above the lower theoretical envelope.

OBSERVATIONS OF EVOLUTIONARY CHANGES IN THE PULSATION PERIOD OF CEPHEIDS

L. Szabados
Konkoly Observatory of the Hungarian Academy of Sciences
Budapest, P.O. Box 67
H - 1525 Hungary

In spite of the fact that Cepheid variables pulsate quite reg-
ularly their pulsation period remains constant only in the first approximation. The pulsation period is subject to variations because of stellar evolution. The calculations made by Hofmeister (1965) predicted that
the evolutionary period changes of classical Cepheids should be observed
on a time scale of several decades or longer. No detailed analysis of the
observed period changes has been made since Hofmeister's fundamental theoretical work.

The available data concerning the period variations observed in
classical Cepheids did, however, result in several useful pieces of information. Petit (1978) attempted to study the period changes of Cepheids
on the basis of the remarks in the General Catalogue of Variable Stars
(Kukarkin *et al.*, 1969-1970). One of the problems that was not solved,
however, was that the use of inhomogeneous data resulted in less reliable
statistics. Erleksova and Irkaev (1980) suspected that the observed progressive period changes were of evolutionary origin but they approximated
the O-C diagrams by sections of straight lines instead of parabolas. Wayman (1981) estimated the period variations on the basis of independent
determinations of period. Since the differences in the period much ex-
ceeded the theoretical period changes in several cases, he suggested that
some causes of period change other than stellar evolution across the HR-
diagram ought to be sought. The observed period variations of individual
Cepheids do indicate, however, that the changes are mainly due to stellar
evolution, at least in the case of the rapidly evolving long period Cepheids (see e.g. Fernie, 1979).

Extensive analysis of the O-C diagrams of Cepheids (Szabados,
1977, 1980, 1981 and 1983) made it possible to search for evolutionary
period changes. The large frequency of parabolic O-C graphs (i.e. continuous period changes) serves as qualitative proof of its observability
(see Figure 1). Parabolic O-C graphs occur more frequently with longer
period classical Cepheids since the rate of evolution is faster for these
stars. On the other hand, the numerical values of period changes as a
function of the pulsation period also supports the assumption that stellar evolution can be caught by observations.

A brief summary of the theoretical and observed period changes

445

A. Maeder and A. Renzini (eds.), Observational Tests of the Stellar Evolution Theory, 445–448.
© *1984 by the IAU.*

                                                           L. SZABADOS

Table I

Theoretical period variations due to stellar evolution
(after Hofmeister, 1965)

Mass	Crossing	Average period	$\left[\dfrac{\Delta P}{P}\right]_{100}$
$5\,\mathfrak{M}_{\odot}$	1	$1\overset{d}{.}8$	+0.0014
	2	3.5	±0.00001
$9\,\mathfrak{M}_{\odot}$	1	12.4	+0.027
	2	22.4	−0.0054
	3	23.9	+0.0030
	4	27.0	−0.0079
	5	16.5	+0.024

Table II

Observed period changes

Average period	$\left[\dfrac{\lvert\Delta P\rvert}{P}\right]_{100}$
$4\overset{d}{.}2$	0.000035
6.0	0.000052
9.1	0.000068
16.0	0.00023
36.0	0.0049

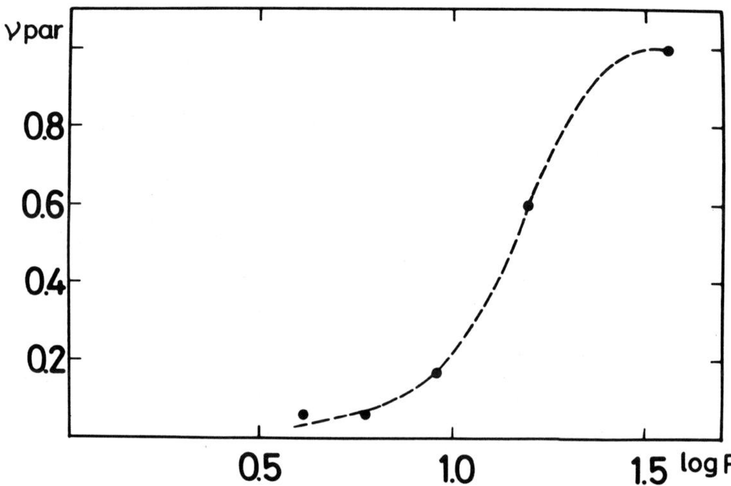

Figure 1. Relative frequency of parabolic O-C diagram vs. the
logarithm of the period.

is to be found in Tables I and II, respectively. In Table I the mass and crossing of different models as well as the average period and the predicted relative period variation within 100 years are indicated in the successive columns. Its observational counterpart in Table II gives the average period and the absolute value of the observed relative period variation within 100 years.

The agreement between the theoretical and the observed values is noteworthy especially in the case of longer period Cepheids. According to the O-C diagrams at an average period P = 36^{d}0 the relative period change within 100 years is 0.0049 , while the theoretical value (averaged from the second, third and fourth crossings of a 9 $\mathcal{M}_\odot$ Cepheid) is 0.0054 at an average period of 24^{d}4 . (The first and fifth crossings have not been considered here since Cepheids have a much shorter pulsation period in these particular cases.)

The short period Population I Cepheids show a somewhat larger period variation than predicted by the stellar evolution. This may be caused by other kinds of period change for which the term 'period noise' is used. Since the shortest period Cepheids have the smallest evolutionary period changes this kind of variation may be commensurable with the period noise. In the case of the long period Cepheids the rate of period change caused by stellar evolution is far larger than the period noise, the latter kind of period change does not contribute significantly to the relative period variation within 100 years. The period noise, however, can be seen in the O-C diagrams of long period Cepheids, as well. It occurs as an irregular variation superimposed on the parabola.

Cepheid variables in the Magellanic Clouds seem to behave in a different way, viz. they are strongly affected by period noise, although parabolic O-C graphs also exist among the MC Cepheids (van Genderen, 1983). It is hoped that further O-C diagrams covering a longer time base will mitigate this difference.

REFERENCES

Erleksova, G.E., and Irkaev, B.N.: 1980, 5th European Regional Meeting on
          Astronomy, Abstract of Papers, B.1.8, Liège.
Fernie, J.D.: 1979, *Astrophys. J.* 231, p. 841.
Genderen van, A.M.: 1983, *Astron. Astrophys. Suppl. Ser.* 52, p. 423.
Hofmeister, E.: 1965, *Kl. Veröff. Bamberg*, Vol. IV, No. 40, p. 244.
Kukarkin, B.V., Kholopov, P.N., Efremov, Yu.N., Kukarkina, N.P., Kuroch-
          kin, N.E., Medvedeva, G.I., Perova, N.B., Fedorovich, V.P., and
          Frolov, M.S.: 1969-1970, General Catalogue of Variable Stars,
          3rd Ed., Sternberg Astronomical Institute, Moscow.
Petit, M.: 1978, *Inf. Bull. Var. Stars*, No. 1402.
Szabados, L.: 1977, *Mitt. Sternw. ung. Akad. Wiss.*, Budapest, No. 70.
Szabados, L.: 1980, *Mitt. Sternw. ung. Akad. Wiss.*, Budapest, No. 76.
Szabados, L.: 1981, *Commun. Konkoly Obs. Hung. Acad. Sci.*, Budapest, No.77.
Szabados, L.: 1983, *Astrophys. Space Sci.* 96, 185.
Wayman, P.A.: 1981, *Irish Astron. J.*, 15, No.2, p. 121.

DISCUSSION

<u>Burki</u>: The O-C diagram is also very useful to detect cepheid binaries.
How many new binaries did you find with our diagrams?

<u>Szabados</u>: The apparent period changes due to light-time effect were de-
tected in five cases. Moreover, a stepwise O-C diagram ("rejumping"
period) characterizes some of the binary cepheids.

CEPHEID COMPANIONS AND THE MASSES OF CEPHEIDS

Erika Böhm-Vitense, Susan Borutzki and Hugh Harris
Astronomy Department, University of Washington, Seattle, WA

ABSTRACT

We have observed in the ultraviolet the hot companions of the
Cepheids SV Per, RW Cam, SY Nor and KN Cen. The study of the absolute
and relative intensities reveals that all, except the companion for
KN Cen are evolved stars which should fit on almost the same mass track
as the Cepheid. We find however that with generally accepted reddening
values the companions of at least SV Per and RW Cam are too faint.
Either the Cepheid loops are more luminous than presently calculated or
the reddening is larger than presently accepted.

We have obtained low resolution short wavelengths IUE spectra for
the hot companions of the population I cepheids SV Per, RW Cam, SY Nor
and KN Cen. The ultraviolet fluxes of the companions are lower than
expected.

The true energy distribution of the companions can only be obtained
after correction for interstellar extinction. The correction depends
on the color excess and on the ultraviolet extinction law. We have used
Seaton's average galactic extinction law.

Generally we can say the lower the assumed value for E(B-V) the lower
will be the true stellar flux and the lower will be the luminosity of
the star which matches the observed flux. For small E(B-V) we find that
the companion has to be a main sequence star, for larger E(B-V) the
companions is found to be a giant or even supergiant. The overall fit
of the energy distribution and the limits for the visual magnitude must
help us to decide which of the possibilities is the correct one.

For a given E(B-V) the absolute flux at 1350 $\overset{\circ}{A}$ determines a relation between T_{eff} and R. The observed relation between R and T_{eff} for
main sequence stars or for average giants gives another relation between
R and T_{eff}. For a given E(B-V) the R and T_{eff} can thus be determined
uniquely either for main sequence stars or for average giants.

449

A. Maeder and A. Renzini (eds.), Observational Tests of the Stellar Evolution Theory, 449–452.
© *1984 by the IAU.*

The final value of E(B-V) and thereby T and R is determined from the best fit of the overall observed energy-distribution with model energy distributions as seen in Figure 1.

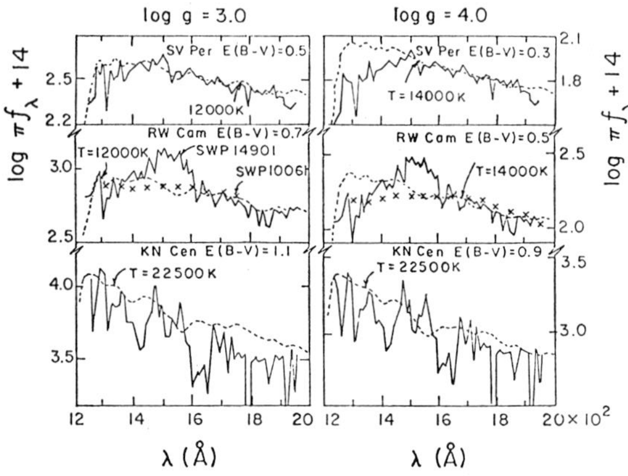

Figure 1 shows a comparison between observed and theoretical energy distributions for giant companions. Also shown is the similar comparison for main sequence companions.

The fit is much better for giant stars. For main sequence stars we also derive too small a flux for the companion in the optical region. We conclude that the companions must be giants.

In the T_{eff}, L diagram the cepheids and the companions should fit on one isochrone. Figure 2 shows that this is not the case for SV Per, RY Nor and especially RW Cam.

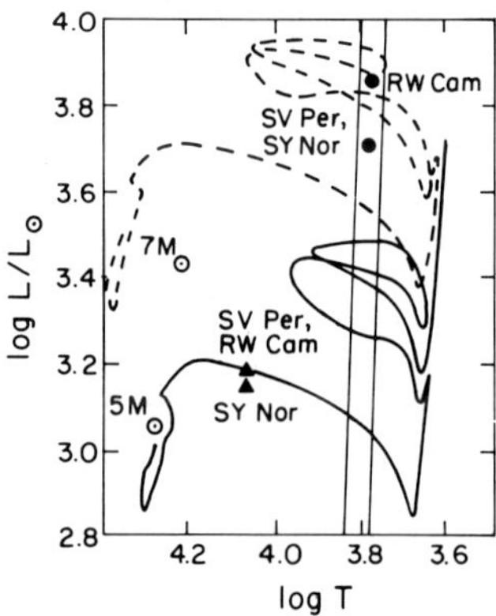

Figure 2. The positions of the Cepheids and their giant companions in the T_{eff}, luminosity diagram. Also shown are evolutionary tracks for different masses according to Becker, Iben and Tuggle 1977.

It appears that the cepheids are too luminous for this. We see basically two possibilities to resolve this discrepancy:

a) The cepheids have indeed lower masses and the loops occur at higher luminosities than calculated so far, or
b) we have used the wrong relation R(T) and therefore perhaps too small a value for E(B-V).

The reddenings required for the best overall fit temperatures are given in Table 1, where we also give the measured values for E(B-V). The required values appear rather high, though not completely out of the range of previously determined values.

TABLE 1. Measured and required values for the color excess E(B-V)

Star	$E(\langle B\rangle-\langle V\rangle)$[1] measured best value	$E(\langle B\rangle-\langle V\rangle)$ range	E(B-V) giants	E(B-V) required T=12000K	E(B-V) required T=13000K
SV Per	0.37	0.32-0.65	0.45	0.53	0.55
RW Cam	0.53	0.53-1.03	0.66	0.79	0.82

[1]According to Dean, Warren and Cousins 1978.

Whether possibility a) or b) is the correct answer to our problem must be decided by accurate determinations of E(B-V).

ACKNOWLEDGEMENTS

We are grateful to the staff of the IUE Observatory for their continued help and support with the IUE observations and reduction. This study was supported by a NASA grant NSG 5398 which is gratefully acknowledged.

REFERENCES

Allen, C. W.: 1973, Astrophysical Quantities, 3rd ed. The Athlone Press, London.
Becker, S. A., Iben, I. and Tuggle, R. A.: 1977, Astrophys. J. 218, 633.
Dean, C. A., Warren, W. H. and Cousins, A. W.: 1978, Monthly Notices Roy. Astr. Soc. 183, 569.
Hofmeister, E.: 1967, z.f. Astrophys. 65, 194.
Kurucz, R.: 1979, Astrophys. J. Suppl. 40, 1.
Seaton, M. J.: 1979, Monthly Notices Roy. Astr. Soc. 187, 73p.

DISCUSSION

<u>Frogel</u>: Are there any Cepheids with <u>red</u> companions?

<u>Böhm-Vitense</u>: Yes, but we only considered Cepheids with blue companions which can be observed in the UV where the intensity of the Cepheid proper is very small. Red companions are hard to observe unless they are red giants of large luminosity. Red main sequence stars could not be seen.

<u>Lloyd Evans</u>: A study using infra-red photometry of suspected cases of Cepheids with red companions gave no convincing detections. However, the observational test is harder than with hot companions, because the colour difference is smaller.

<u>Conti</u>: How sensitive are the conclusions to the adoption of a standard UV extinction law? Savage and associates have just published studies showing substantial variations from place to place in the Galaxy. Even within a cluster, the UV extinction seems to vary, as compared to E(B-V).

<u>Böhm-Vitense</u>: In order to put the companions on the evolutionary track required by the Cepheid with accepted E(B-V) values we would need to increase the UV extinction to values larger than observed anywhere in the galaxy. The largest observed extinction (ξ Oph) will not do it. I do not know whether it would be possible that the Cepheids have some unusual circumstellar extinction.

<u>Lynas-Gray</u>: Did you use the Kurucz (1979, Astrophys. J. Suppl. <u>40</u>, 1) model atmospheres?

<u>Böhm-Vitense</u>: Yes.

<u>Waelkens</u>: Did all the Cepheids that showed peculiar loops in the colour-colour diagrams and that you studied turn out to have blue companions on your ultraviolet spectra? If not, how can the loops be explained?

<u>Böhm-Vitense</u>: Not all the Cepheids with open loops in the two colour diagram do have observable blue companions. For none of the population II Cepheids with open loops did we see a blue companion. The open loops appear to be intrinsic. Several population I Cepheids with small open loops also do not have observable blue companions.

HR 7308 (V473 Lyr): THE STRANGEST CEPHEID

G. Burki
Observatoire de Genève

The variability of HR 7308 was discovered by Breger (1969), and
Percy et al (1979) found this star to be a new cepheid. The correct
period and the variation of amplitude were first recognized by Burki and
Mayor (1980 a,b). Further studies of this star were published by Percy
and Evans (1980), van Genderen (1981), Breger (1981), Fernie (1981),
Burki et al (1982) and Henriksson (1983). Here the up-to-date radial
velocity observations made with the spectrophotometer CORAVEL are
presented.

Figures 1 and 2 clearly demonstrate the unique characteristic of
HR 7308: the amplitude varies by over a factor of 15 in a time of about
1200 d. Note that the minimum of amplitude observed during July-September
1982 was about twice as small as the previous one. The essential results
obtained for this star in Geneva are the following:

- The period is the shortest known for a classical cepheid:
  1.49101 d.
- The amplitude is strongly variable.
- The mean radius is $34 \pm 5 \; R_\odot$ (Wesselink analysis). The constancy of
  the period implies the constancy of the mean radius.
- The pulsation occurs in a radial mode, probably in the 2nd overtone. The variation of amplitude is not due to the beating of two
  modes.
- Apparently, HR 7308 is not a member of a binary system.

Concerning the physical interpretation of the variation of amplitude, it was suggested (Burki and Mayor, 1980 b) that HR 7308 could be
a star which presently becomes a cepheid or leaves the cepheid instability
strip. Indeed, what happens to a supergiant star when arriving into or
getting out of the cepheid instability strip? Are there transition periods
during which pulsation is successively initiated and stopped?

A. Maeder and A. Renzini (eds.), Observational Tests of the Stellar Evolution Theory, 453–455.

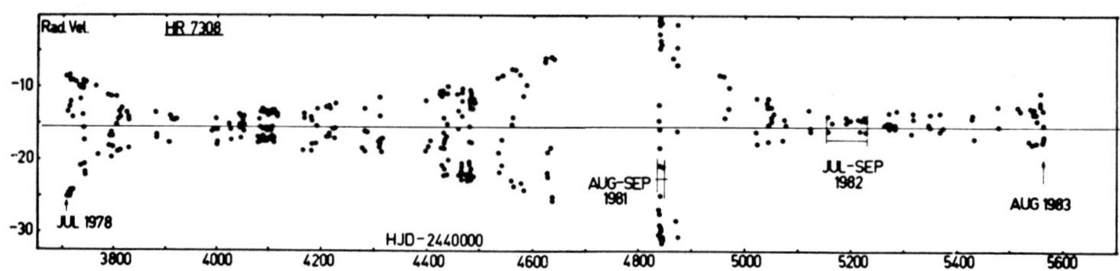

Figure 1

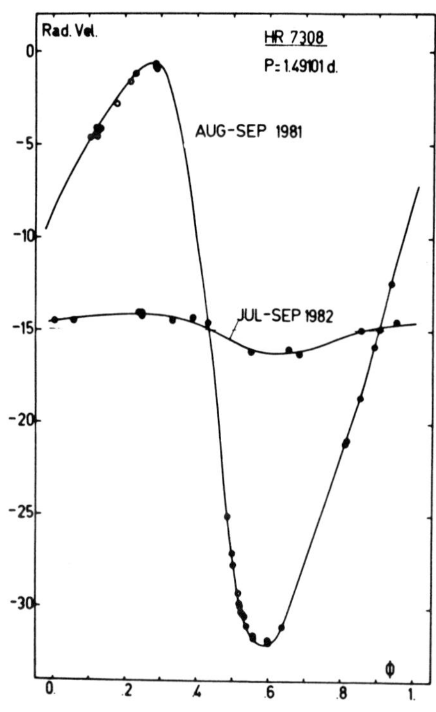

Figure 2

REFERENCES

Breger, M.: 1969, Astrophys. J. Suppl. 169, 79.
Breger, M.: 1981, Astrophys. J. 249, 666.
Burki, G., Mayor, M.: 1980a, Inf. Bull. Variable Stars, no 1728.
Burki, G., Mayor, M.: 1980b, Astron. Astrophys. 91, 115.
Burki, G., Mayor, M., Benz, W.: 1982, Astron. Astrophys. 109, 258.
Fernie, J.D.: 1982, Publ. Astron. Soc. Pacific 94, 537.
van Genderen, A.M.: 1981, Astron. Astrophys. 99, 386.
Henriksson, G.: 1983, Uppsala Astron. Rep., no 26.

Percy, J.R., Baskerville, I., Tremorrow, D.W.: 1979, Publ. Astron. Soc.
     Pacific 91, 368.
Percy, J.R., Evans, N.R.: 1980, Astron. J. 85, 1509.

DISCUSSION

Cox: When one gets strange behaviour such as a second overtone, as in
δ Sct stars, one suspects non-radial modes. Are you sure the pulsation
of HR 7308 is radial?

Burki: The differences in phase between light, colour and velocity curves
strongly indicate a radial pulsation. Moreover, no indication of a second
mode of pulsation has been found.

K. Barlai
Konkoly Observatory, Budapest

The O-C curves - widely used for investigating the period behaviour
of variable stars - are generally constructed under the assumption that
the period changes are linear in time. If it is indeed so, the main
trend of the diagram should outline a positive or negative parabola and
β values are determined to characterize the curvature of the parabolae,
i.e. the rate of change of the period. This rate, however, even in the
case of ω Cen does not indicate exact parabola (Belserene, 1973).

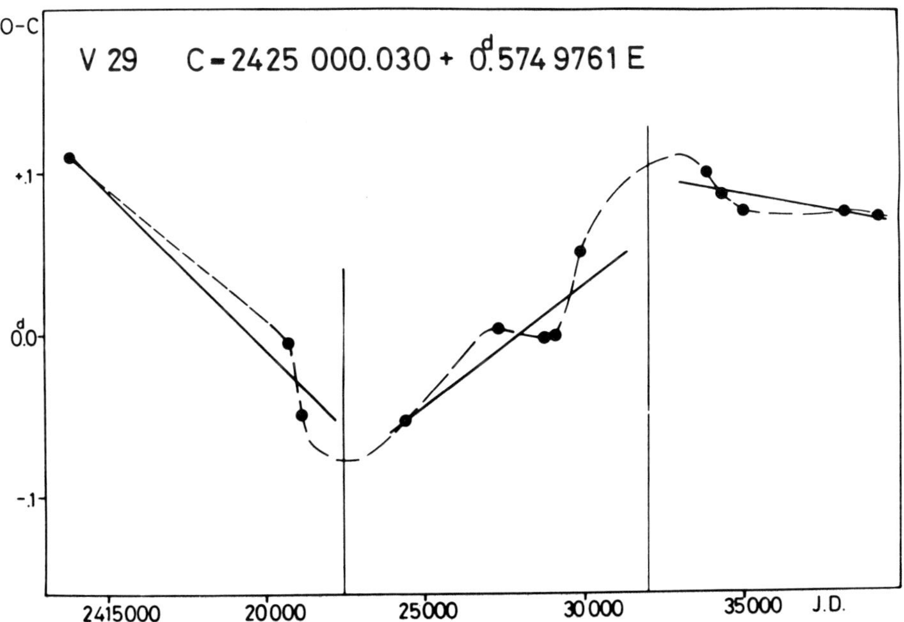

Fig. 1. An example for determination of the average periods.

A. Maeder and A. Renzini (eds.), Observational Tests of the Stellar Evolution Theory, 457–460.

The various shapes of the O-C diagrams obtained for 49 RR Lyrae variables in the globular cluster M 15 convincingly suggest (Barlai, 1984) that the concept of a linear period change does not work in our case. Only a minor part of the O-C residuals could be approximated by parabolae or straight lines (Szeidl, 1975). In view of this, when investigating the trend of the period changes in M 15 a method was worked out which would need no specific preliminary assumptions concerning the character of the period changes.

The time interval covered by the observations (J.D.2413700 - J.D. 2440000) has been divided into three parts. Within these time intervals three average periods have been determined for each RR Lyrae variable from the points of their O-C residuals. By subtracting the average periods belonging to the subsequent intervals two series of ΔP values have been obtained. For each star the single ΔP values have been related to the time elapsing between the corresponding average epochs, i.e. the average abscissa values belonging to the straight line representing the average periods valid within the subsequent sections of time. (See Fig.1 as an example.)

Having performed this calculation for all the RR Lyrae variables measurable on our plates in M 15, two series of $\Delta P/\Delta t$ values have been obtained. The members of one single series practically all belong to the same section of time. We have concerned ourselves here with the distribution of these $\Delta P/\Delta t$ values.

Both of the resulting histograms (Fig.2) have a prominent peak at zero, are roughly symmetrically situated around the zero point, and this symmetrical part contains the majority of the data. These features and the fact that the observed period changes can be explained by different random processes going on inside an RR Lyrae star (Sweigart and Renzini, 1979) do not contradict a normal distribution. So a Gaussian fitting has been attempted although the data available are not very numerous from a statistical point of view.

Having fitted a Gaussian curve upon the central part of the distribution it became obvious that the broad wings cannot be included in the curve. Single stars are to be found even at a distance as far as $\sim 15\sigma$ (V 15). A considerable part of the variables studied (11) belongs to the wings and these stars underwent strong and rapid period changes during the seven decades of the observations. In several cases a jump from the left wing to the right can be observed and vice versa. For example variables 2 and 13 appear in the left wing of distribution I and are to be found on the right wing of distribution II.

The stars belonging to the Gaussian distribution exhibit "noisy" period behaviour. The evolutionary trend does not emerge above this noise level.

The method outlined here has enabled us to take into account all kinds of period changes without being prejudiced against the shape of

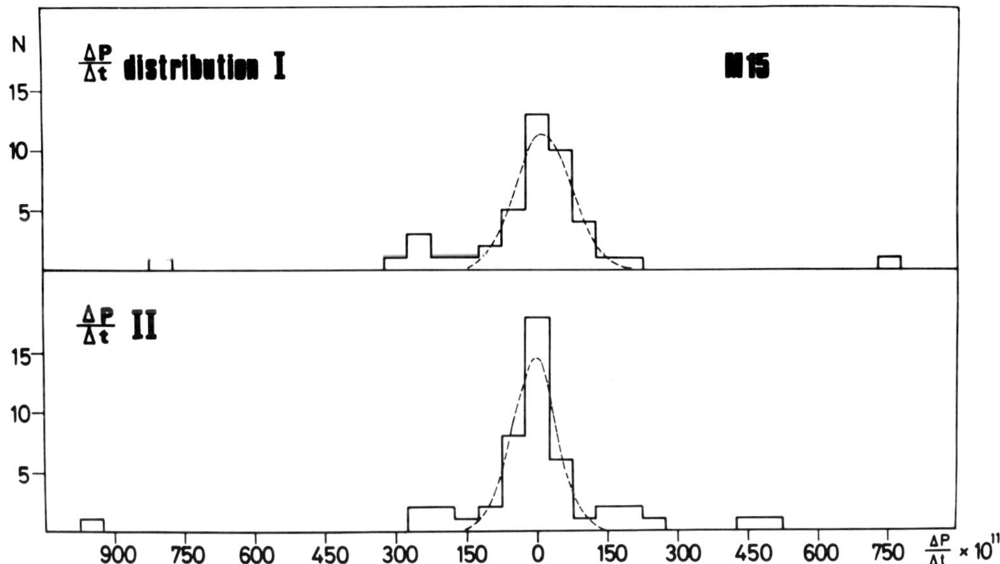

Fig. 2. For distribution I σ=60, x_0=+15;
        for distribution II σ=46, x_0=-7.

the resulting O-C diagrams, even if these shapes are irregular or unusual. If the O-C residuals are only considered which do not contradict
a parabola-like curve we might be deprived of evaluating the entire
material available (Smith and Sandage, 1981). In the course of time
further distributions can be made and the inclusion into the investigation of more clusters richer in variables will give more significant
results.

        Thanks are due to Drs. K. Olah, L. Pocs and B. Szeidl for helpful
discussions and advice.

REFERENCES

Barlai, K.: 1984, Comm. Konkoly Obs., Budapest, No. 85 (in press)
Belserene, E.P.: 1973, in Variable Stars in Globular Clusters and in
        Related Systems, IAU Coll. 21, ed. J.D. Fernie, D. Reidel,
        Dordrecht, pp. 105-111
Smith, H.A. and Sandage, A.: 1981, Astron. J. 86, pp. 1870-1881
Sweigart, A.V. and Renzini, A.: 1979, Astron. Astrophys. 71, pp. 66-78
Szeidl, B.: 1975, in Variable Stars and Stellar Evolution, IAU Symp. 67,
        eds. V.E. Sherwood, L. Plaut, D. Reidel, Dordrecht, pp. 545-552

DISCUSSION

<u>Cox</u>: Did you look at any of the double mode variables?

<u>Barlai</u>: Yes, I did. From the 10 double mode RR Lyrae studied in your paper six have been measurable on our plates: V39, 26, 30, 31, 54 and 17. One of them, V26, makes a big jump from the left wing of distribution I to the right wing of distribution II. V39 with a distance of ~4σ between its positions gets from the Gaussian part of distribution I to the right wing of distribution II. The remaining variables are included in the "noisy" part of the distributions.

CONVECTIVE EFFECTS AND THE RR LYRAE INSTABILITY STRIP

R. F. Stellingwerf
Mission Research Corporation
Albuquerque, New Mexico

ABSTRACT: The effects of nonlinear, nonlocal, diffusive convection have
been included in the structure equations of pulsational hydrodynamic
analyses of a series of RR Lyrae models. We find a well-defined red
edge of the unstable region that depends on mode of pulsation. It is
shown that this result has strong implications for the nonlinear
behavior of RR Lyrae stars.

1. CONVECTIVE STRUCTURE

Nonlinear RR Lyrae models containing convection as derived in
Stellingwerf, 1982a,b have been computed for a range of effective
temperatures, fundamental and first overtone modes, and helium abundances of $Y = 0.2, 0.3$. Most models have been studied at small amplitude (1 km/s) to determine the static convective structure and estimate
the stability properties of the small amplitude pulsation. The structure of a typical convective envelope is shown in Figure 1. The
unstable regions near the hydrogen and helium ionization zones are seen
to be very thin (dashed line = L(c)/L, dotted line = T'/T), but the convective velocity (solid line) shows a much broader structure due to
overshooting. Roughly half of the flux is carried by convection in the
H zone, with only 10% being carried in the He zone.

The work curve for this model is shown in Figure 2. The total work
(solid line) is decreased somewhat relative to the purely radiative
model, but the turbulent pressure work (dashed line) contributes positive driving, and the eddy viscosity work (dotted line) is very small.
This type of structure is typical of models in the instability strip.
Only well to the red of the unstable temperature range do we find thick,
efficient convective zones.

A. Maeder and A. Renzini (eds.), Observational Tests of the Stellar Evolution Theory, 461–463.
© *1984 by the IAU.*

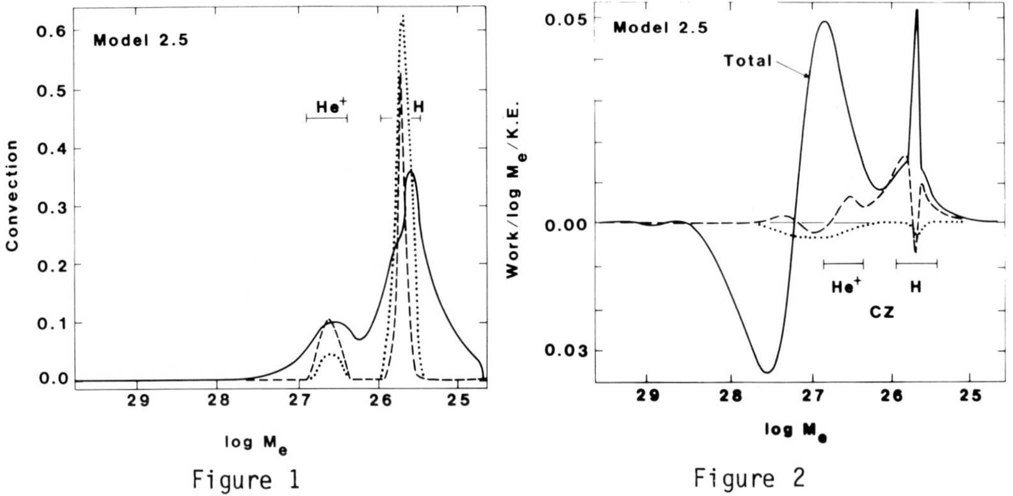

Figure 1   Figure 2

The growth rates for a sequence of models with Y = 0.3 spanning the
instability strip are shown in Figure 3, where the dotted lines depict
the growth rates of the same models without convection. The curves
labeled "0" refer to the fundamental mode and the curves labeled "1"
refer to the first overtone. We see that the effect of convection is to
strongly stabilize the pulsation toward the red, with the overtone edge
occurring well to the blue of the fundamental edge. At the blue extreme
of the instability strip the effect is the opposite: convective models
are slightly more unstable than radiative models.

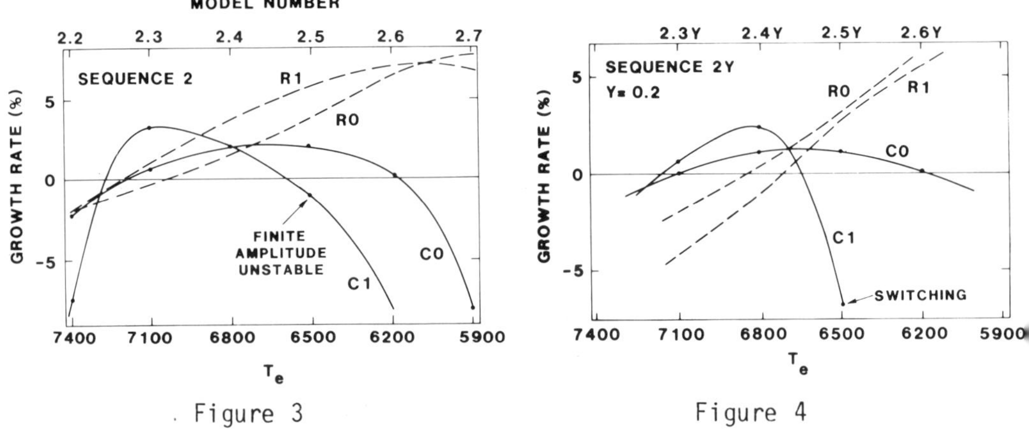

Figure 3   Figure 4

Models were also computed with lower helium abundance: Y = 0.2.
The growth rates for these convective and radiative models are shown in
Figure 4. It is seen that the convective models in this case show
rather strongly enhanced pulsation toward the blue, so much so that the
blue edges move only very slightly as Y is varied. If these preliminary
results are correct, then determination of helium abundance from the
temperature of the blue edge in globular clusters will be very difficult, if not impossible.

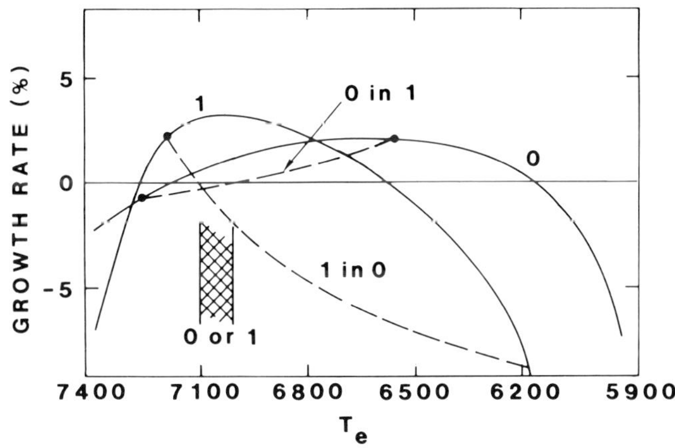

Figure 5

The structure of the growth rate diagrams shown in Figures 3 and 4
have implications for the nonlinear stability of the full amplitude
pulsation. The reason for this is that the full amplitude switching
rates from one mode to the other must coincide with the small amplitude
growth rates at the edges of the instability strip. This phenomenon is
well verified for radiative models. In the convective case, the
switching rates are constrained at both edges of the strip, as shown in
Figure 5. Here the solid lines are duplicated from Figure 3, and the
dashed lines represent hypothetical switching rates: the growth rate of
mode 0 in the full amplitude mode 1 (0 in 1), and vice versa. These
curves must originate and end at the large dots on the solid curves.
Nonlinear stability analyses must be calculated to find their behavior
between these points. Figure 5 shows one possible configuration. It is
clear that the dashed curves must cross the neutral growth rate line
somewhere near the vicinity of the center of the instability strip. The
points at which they cross this line are transition edges - in Figure 5
we have overtone pulsation toward the blue, fundamental pulsation toward
the red, and a region in which both modes are stable (an hysteresis
region) in the center (shaded). If the two curves cross exactly on the
neutral line, then a sharp transition between modes is indicated. If
the dashed curves cross above the neutral growth rate line, then a
region of mixed mode behavior is indicated, as found observationally in
the clusters M15 and M3. These are the only possibilities. Note that
the transition lines must fall in the central region of the strip, and
therefore must approximately parallel the edges of the strip. Clusters
may have different types of mode switching behavior depending on luminosity or other parameters. Presently, it is believed that the presence
of mixed mode stars indicates a stable mixed mode region, rather that a
simple mode switch.

REFERENCES

Stellingwerf, R.F., 1982a, Astrophys.J., 262, 330.

Stellingwerf, R.F., 1982b, Astrophys.J., 262, 339.

DOUBLE-MODE RR LYRAE STARS IN THE DRACO DWARF GALAXY
AND IN OTHER STELLAR SYSTEMS

James M. Nemec
Dominion Astrophysical Observatory
Herzberg Institute of Astrophysics
and
Department of Astronomy, University of Washington

Eleven double-mode (dm) RR Lyrae stars, with periods midway between the periods for the c-type and ab-type RR Lyrae stars, have been identified in the Draco dwarf galaxy by reanalyzing the photometry of Baade and Swope (1961) for 35 stars. The stars are V11, 72, 75, 83, 112, 138, 143, 156, 165, 169 and 190, three of which previously were noted as dm RR Lyraes by Goranskij (1982). The methods of Stellingwerf (1978) and Stobie (1970) were used to find the periods. The period ratios, periods and amplitudes suggest that the stars radially pulsate simultaneously in the fundamental and first overtone modes. The beat masses, estimated from the P_1/P_0 vs. P_0 diagram (Petersen 1973), using as calibration the King Ia (Y=0.299, Z=0.001) models of Cox, Hodson and Clancy (1983, hereafter CHC), are M_{beat}/M_0=0.65 for nine stars, =0.60 for V75 and =0.55 for V165. If the mass loss rate prior to arriving on the horizontal branch is proportional to the metal abundance (Stobie 1971, CHC), V75 and V165 should be more metal rich than the other nine dm RR Lyraes. In the period-amplitude (P-A) diagram, at a given amplitude, V165 has a smaller period shift relative to the standard M3 line than do the higher mass stars. Subsequently, the correlations of Preston (1959) and Sandage (1982) suggest that it is more metal-rich than the other nine dm RR Lyrae stars in Draco. Furthermore, reanalysis of the P-A and period-mean magnitude relations of all the RR Lyrae stars in Draco shows evidence for a widespread range in the metal abundances. Stars with large period shifts relative to the M3 relation are found to be more luminous than stars with smaller period shifts. The frequency of variable amplitude (ie. Blazhko effect) RR Lyrae stars is greatest for the ab-type stars with short periods. The brightnesses of the highest maxima appear to fit the P-A relation, and the amplitudes of the lowest maxima are variable, with V123 being an extreme example. Figure 3 of Szeidl (1975) shows the analogous situation for M3 RR Lyrae stars with the Blazhko effect. These findings suggest that the amplitude variability is related to the mode-switching activity of the dm RR Lyrae stars.

In the Ursa Minor dwarf galaxy, five probable dm RR Lyrae stars

A. Maeder and A. Renzini (eds.), Observational Tests of the Stellar Evolution Theory, 465–466.

have been identified from among the stars measured by van Agt (1967, 1968). The stars are: V44,V49,V57,V58 and V83. Until more accurate photometry is available for these stars it will not be possible to estimate their masses. A dm RR Lyrae star with M_{beat}/M_o=0.65 also has been discovered in the LMC globular cluster NGC 2257 (Nemec, 1983; Nemec, Liller and Hesser 1984). Reanalysis of the Sandage, Katem and Sandage (1981) photometry for seven M15 dm RR Lyraes confirms the CHC period ratio for V17. Other P_1/P_0 ratios found include: 0.7461 for V26, 0.7469 for V39, 0.7439 for V51, 0.7432 for V54, 0.7450 for V58, and 0.7454 for V61. Preliminary analysis of the Martin (1938) photometry for 54 stars in the Galactic globular cluster ω Centauri (Nemec and Norris 1983) shows no evidence for double-mode RR Lyrae stars.

REFERENCES

Baade, W. and Swope, H.H.: 1961, A.J. **66**, p. 300.
Cox, A.N., Hodson, S.W. and Clancy, S.P.: 1983, Ap.J. **266**, p. 94.
Goranskij, V.P.: 1981, I.B.V.S. No. 2007.
_____. 1982, Astron. Circ., Kazan, No. 1216, p. 5.
Martin, W.C.: 1938, Ann. Sterrew. Leiden 17, Part 2, p.1.
Nemec, J.M.: 1983, Ph.D.Dissertation, University of Washington.
Nemec, J.M., Liller, M.H. and Hesser, J.E.: 1984, in I.A.U. Symp. 108, page 39.
Nemec, J.M. and Norris, J. 1983: in preparation.
Petersen, J.O.: 1973, Astron. and Astrophys. **27**, p. 89.
Preston, G.: 1959, Ap.J. **130**, p. 507.
Sandage, A.: 1982, Ap.J. **252**, p. 553.
Sandage, A, Katem, B. and Sandage, M.: 1981, Ap.J.Suppl. **46**, p. 41.
Stellingwerf, R.F.: 1978, Ap.J. **224**, p. 953.
Stobie, R. S.: 1970, Observatory, p. 20.
_____. 1971, Ap.J. **168**, p. 381.
Szeidl, B.: 1975, in I.A.U.Colloq. No. 29, Budapest, p. 133.
van Agt, S.L.Th.J.: 1967, Bull. Astr. Inst. Neth. 19, p. 275.
_____. 1968, Bull. Astr. Inst. Neth. Suppl. 2, p. 237.

DISCUSSION

<u>Hesser</u>: John Norris became so interested in these results when he visited Victoria this summer that he punched up all the old ω Cen RR Lyrae data. Nemec analysed them and found no double-mode variables. This was quite disappointing to those of us who think ω Cen shares many properties in common with the dwarf spheroidals.

B V PHOTOMETRY OF RR LYRAE VARIABLES IN M15

E.A. Bingham[1], C. Cacciari[2,4] R.J. Dickens[1,3] F. Fusi Pecci[4]

[1] Royal Greenwich Observatory,Herstmonceux Castle, UK
[2] ESTEC Astronomy Division, Villafranca, Spain
[3] Rutherford Appleton Laboratory, Chilton, UK
[4] Osservatorio Astronomico Universitario, Bologna, Italy

New BV photographic photometry for 56 RR Lyrae variables in M15, combined with original measures by Sandage, Katem and Sandage (1981), yield data of improved precision which enable a rediscussion of the RR Lyrae characteristics and Oosterhoff problem.

Pulsation theory is used to derive a mass-luminosity ratio:
Log L = 1.92 ($\pm$0.03) + 0.81 + Log M, which gives M_V(RR) = 0.40 assuming a mean mass of 0.65 $M_\odot$ obtained by Cox, Hodson and Clancy (1983) for the multimode variables. This luminosity implies an age a few billion less than the current estimates, but such a derivation is still affected by many uncertainties.

The observed distribution of stars in the HR diagram (Fig. 1) is in good agreement with the morphology of the appropriate evolutionary tracks of Sweigart and Gross (1976). However, the mass-luminosity ratios derived from evolutionary models are consistent with those derived by pulsation theory only if the helium abundance is Y $\gtrsim$ 0.30. If Y $\lesssim$ 0.25 as might be

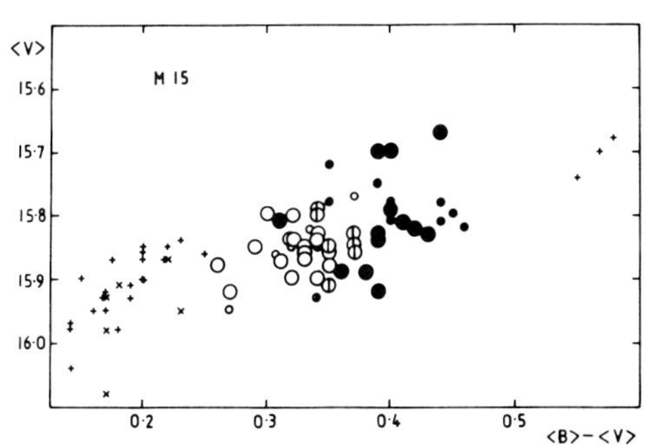

Fig. 1   RR Lyrae variables in the c-m diagram. Non variable stars are also plotted (+,×). c-type variables are plotted as open circles or open circles with bar (multimode behaviour), ab-types as filled circles. The blue edge of the instability strip is clearly demarcated.

467

A. Maeder and A. Renzini (eds.), Observational Tests of the Stellar Evolution Theory, 467–469.
© *1984 by the IAU.*

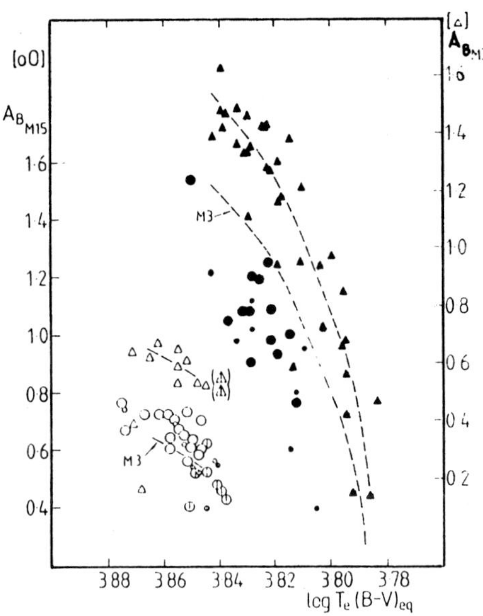

Fig. 2 Blue amplitude as a function of Log Teff for RR Lyraes in M15 (circles) and M3 (triangles).
The thick dashed line shows the position of the mean M3 relations referred to the left hand ordinate.
Effective temperatures are derived using temperature scales from BDE and reddenings of E(B-V) = 0.00 (M3) and 0.10 (M15).

expected from independent evidence, the evolutionary models predict a
lower luminosity ($\Delta \mathrm{Log}\ L \simeq 0.1$) than obtained from pulsation theory.
This discrepancy can be removed if the effects of stellar rotation and
CNO enhancement are taken into account.

Comparison with the RR Lyraes in M3 shows that the Osterhoff effect
between these two clusters cannot be simply described in terms of a period
shift, since the distribution of amplitude and number with period and temperature differ in detail, in particular a unique amplitude-temperature
relation for both c- and ab-type variables is not possible (Fig. 2). The
present analysis confirms Sandage's result that the correlation of period
shift with metal abundance over many clusters requires an anticorrelation
of metal abundance with helium abundance within the current frame of evolutionary models.

Comparison with the morphology of the appropriate theoretical tracks
reveals that many of the RR Lyraes in M15 begin their HB evolution within
the instability strip, spending much longer in the center of the strip than
variables in M3-like clusters which evolve more rapidly bluewards across
the strip. The distribution in M15 peaks just bluewards of the transition
temperature giving rise to the many c-type variables found there. The multimode variables occur precisely at the transition temperature (the same
in M15 and M3) and appear to be a "stable" mode of pulsation.

REFERENCES

Bingham et al., Mon.Not.R.astr.Soc. in press (full paper).
Cox et al.,1983, Astrophys.J. in press.
Sandage et al.,1981, Astrophys.J. Suppl. 46,41.
Sweigart, A.V. and Gross, P.G.,1976, Astrophys.J. Suppl. 32,367.

DISCUSSION

Cox: The brightness of the horizontal branch in M15 gives a rather high
luminosity. That is M_{bol} = 0.3. This corresponds to a small age of
only 10^{10} years. Can you comment?

Cacciari: This value of luminosity for the horizontal branch comes from
pulsation theory, assuming a mass 0.65 $M_\odot$. It certainly leads to an
estimated age smaller than 15 billion years. On the other hand, this
larger value of age is based on the assumption that the luminosity of
RR Lyrae stars in M3 is M_V = 0.80 mag, which is probably too faint by
approximately 0.1-0.2 mag, as suggested by a number of independent
estimates.

FG SAGITTAE, A TEST OF THE THEORY OF EVOLUTION OF DOUBLE
SHELL SOURCE STARS

Yu. A. Fadeyev
Astronomical Council
USSR Academy of Sciences

FG Sge is a unique stellar object possessing a number
of unusual properties which are as follows: (1) FG Sge is a
core of a planetary nebula, (2) at least since 1894 this
star has crossed the HR diagram from the left to the right,
(3) the atmosphere of FG Sge was enriched in s-elements,
(4) FG Sge is a pulsating variable the period of which
progressively increases from 15 days in 1962 to 110 days in
1980.

As it was primarily supposed by Paczyński (1971) FG Sge
moves in the HR diagram along the horisontal loop due to a
helium shell flash in its interior. This hypothesis provides
not only the explanation of enriched surface abundances of
s-elements but allows also to estimate both the distance and
the age of the planetary nebula surrounding FG Sge. According
to Flannery and Herbig (1973) the expansion velocity and the
angular radius of the planetary nebula are 34 km/s and 18
arcsec, respectively. Thus, the age of the planetary nebula
τ is related to the distance d by

$$\log \tau \ (\text{yr}) = 0.400 + \log d \ (\text{pc}) \tag{1}$$

Supposing that the planetary nebula was formed at the pre-
ceding helium flash, the interflash period - core mass
relation (Paczyński, 1975) and the core mass - luminosity
relation (Paczyński, 1970) can be used for obtaining
another relation between the age of the planetary nebula
and its distance:

$$\log \tau = 5.201 - 4.79 \cdot 10^{-8} \ d^2. \tag{2}$$

The relations (1) and (2) are shown in Fig. 1 by the curves
(1) and (2), respectively. From equations (1) and (2) it
follows that the distance is d = 4.83 kpc and the age is
τ = 12100 yr. Thus, FG Sge's luminosity is L = 14 700 $L_\odot$.

A. Maeder and A. Renzini (eds.), Observational Tests of the Stellar Evolution Theory, 471–473.

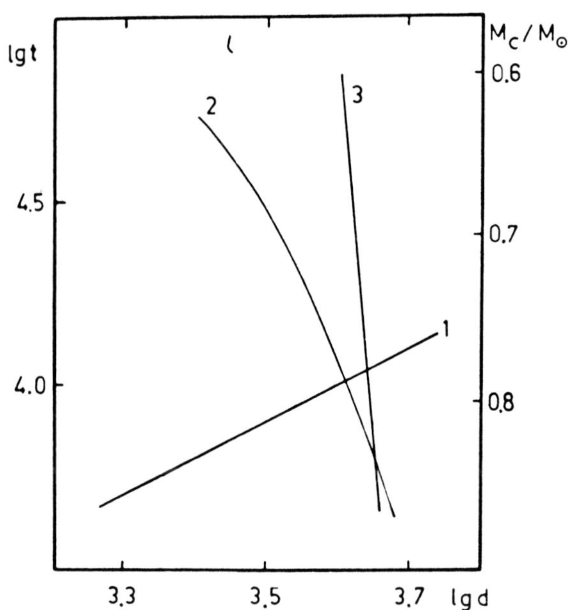

Fig 1   The distance - age diagram.

It should be noted that this estimate of the distance by
2 times exceeds the one observed.

Pulsation of FG Sge provides another way for estimating
its distance by using the period - mean density relation.
Expressing the stellar radius in terms of the effective
temperature T_e and the distance and fixing time of observation we find that d depends on the total stellar mass M,
only. The hydrodynamic computations for the models of
pulsating yellow supergiants showed that the appropriate
value of the pulsation constant is Q = 0.045 days (Fadeyev,
1982). Neglecting the envelope mass and using the inter-
flash period - core mass relation we show this dependence
in Fig. 1 by the curve (3) corresponding to t=1977 when
P=90 days. The solution of this dependence and the relation
(1) coincides with that of (1) and (2) within the interval
0.15 mag.

To decrease influence of uncertainties in observable
parameters on the luminosity estimation averaging over the
time interval is more preferable. Combining the period -
mean density relation with the core mass - luminosity
relation we get the period - luminosity relation for
double shell source stars:

$$P = 4.707 \cdot 10^{13} Q(L/L_\odot)^{3/4} T_e^{-3}(L/L_\odot+30930)^{-1/2} \qquad (3)$$

where the approximation $M_c = M$ is used. The decrease of T_e

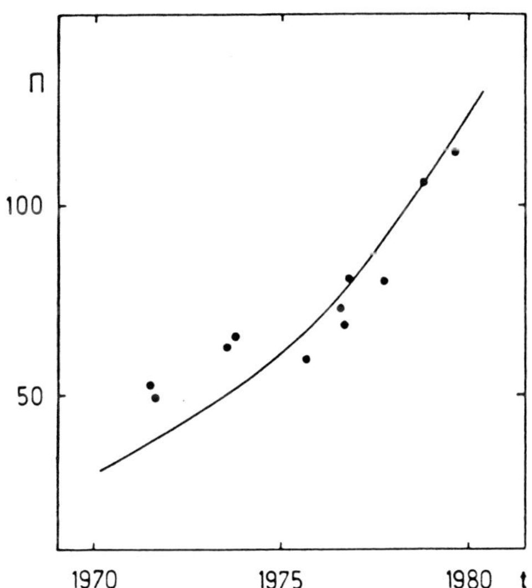

Fig. 2 Observational (filled circles) and theoretical
       (solid line) temporal dependence of FG Sge's period.

as a function of time is known from the scanning
observations (Stone, 1979), so the relation (3) with
Q=0.045 days gives the temporal dependence of FG Sge's
period as $P=P(t,L)$, where L is an unknown parameter. The
observational estimates of FG Sge's period obtained in the
interval from 1971 to 1979 are best fitted by the relation
(3) for L=19 500 $L_\odot$(see Fig. 2).

    Thus, the estimate of FG Sge's luminosity infered from
the relations for double shell source stars is in a close
agreement with that obtained from the period - mean density
relation for pulsating stars. The difference between these
estimates is about 0.3 mag and may be caused by uncertainties in the interstellar extinction and the bolometric
correction as well as period variability of semiregular
pulsation.

REFERENCES

Fadeyev, Yu.A.: 1982, Astrophys.Space Sci. 86, pp. 143.
Flannery, B. and Herbig, G.H.: 1973, Astrophys.J. 183,
     pp. 489.
Paczyński, B.: 1970, Acta Astron. 20, pp. 47.
Paczyński, B.: 1971, Acta Astron. 21, pp. 417.
Paczyński, B.: 1975, Astrophys.J. 202, pp. 558.
Stone, R.P.S.: 1979, Publ.Astron.Soc.Pacific 91, pp. 389.

THE ROLE OF ROTATION IN STELLAR EVOLUTION

Jean-Louis Tassoul,
Département de Physique, Université de Montréal

1. Introduction

During the seventeenth century, in the wake of the discovery of the
solar differential rotation, some scientists argued that stellar variability was a direct consequence of axial rotation, the spinning body
showing alternately its bright (unspotted) and dark (spotted) hemispheres
to the observer (Brunet 1931). Although this idea did not withstand
the passage of time, it is nevertheless an interesting one because it
is clearly indicative of the kind of fascination stellar rotation has
aroused since its inception. And yet, at this writing there is no
longer any doubt that spherically symmetric models do explain the major
observed properties of stars. Moreover, if one excepts the very early
and very late moments of a star's lifetime, the effects of rotation on
stellar structure are apparently dynamically unimportant (e.g., Tassoul
1978, hereafter T.R.S.; Moss and Smith 1981, and references therein).
What is the purpose, then, to discuss the role of rotation on the main-
sequence and post-main-sequence phases of stellar evolution?

As we shall see below, perhaps the most interesting effect of rotation is to generate small-scale, eddy-like motions as well as large-
scale meridional currents wherever radiative equilibrium prevails.
The importance of these motions lies in the fact that, under certain
conditions, they may lead to some degree of mixing of the stellar material, or that they may prevent the gravitational sorting of the elements
in the surface layers of most (but not all!) early-type stars. However,
since I disagree with the current habit of placing the cart in front
of the horse, I shall not discuss these practical implications without
making first a thorough discussion of the state of motion in a stellar
radiative zone. This is definitely not an exhaustive review of the
recent literature. Rather, I shall try to summarize in nontechnical
terms the theoretical work I'm pursuing in collaboration with my wife
(Tassoul and Tassoul 1982a, b, c, 1983a, b, c,1984a, b; hereafter
Papers I, II, ..., VIII, respectively). Our personal approach, which
we borrowed from geophysics, resolves in a very simple manner the many
contradictions and inconsistencies that have beset the theory of rotating stars.

475

A. Maeder and A. Renzini (eds.), Observational Tests of the Stellar Evolution Theory, 475–489.
© *1984 by the IAU.*

2. Stability Considerations

In principle, by making use of the basic equations of astrophysical fluid dynamics, one should be able to obtain at every instant the angular momentum distribution within a star. For evident reasons, this is an impossible task, even were the initial conditions known. The standard procedure is to calculate in an approximate manner an equilibrium structure that corresponds to some prescribed rotation law, ruling out (in principle) those configurations that are dynamically or thermally unstable with respect to axisymmetric disturbances. Although such an ad hoc approach may be used to estimate the gross effects of rotation on stellar evolution, it is totally inadequate for the following two reasons: (i) the instantaneous angular momentum distribution does not follow from the equations of motion, and (ii) no dynamically stable model can possibly exist when non-axisymmetric disturbances are taken into account. Because the role of these ever-present dynamical instabilities has been overlooked in astrophysics, a brief review of the relevant concepts is thus in order.

If a rotating star was a barotrope, then its angular velocity Ω would be constant on cylinders centered on the rotation axis, i.e., $\Omega = \Omega\,(\omega, t)$, where ω is the distance from the rotation axis. In this case, the dynamical stability of the configuration depends on the specific angular momentum $j = \Omega\omega^2$ and the Richardson number $Ri = N^2/S^2$, where N is the buoyancy frequency and S is the shear in the linear velocity $\Omega\omega$. (Since we consider radiative zones only, we have $N^2>0$ everywhere.) Instability with respect to axisymmetric disturbances occurs wherever the j-distribution decreases outward, that is, $dj^2/d\omega<0$. In geophysics, it is called the condition for symmetric instability, and it merely generalizes the Rayleigh criterion for an incompressible fluid (e.g., T.R.S., p. 167). As was shown by Lorimer and Monaghan (1980), the symmetric instability is a violent one in the sense that, given an initially unstable j-distribution, a slowly rotating barotrope will at once generate meridional currents and non-axisymmetric motions in the nonlinear regime, the resulting flow becoming chaotic with a very slow trend to equilibrium. Equally well known is the fact that a rotating barotrope may also become dynamically unstable with respect to non-axisymmetric disturbances. It is the so-called shear-flow instability that occurs wherever $Ri \lesssim 1/4$. This instability is a mild one in the sense that it merely generates turbulence when the shear is large enough to overcome the stabilizing influence of the density stratification. No dynamical instability occurs in a rotating barotrope when $Ri \gtrsim 1/4$ and $dj^2/d\omega>0$ (e.g., Turner 1973; see also Chimonas 1979).

As we shall see below, a real rotating star is not a barotrope but a barocline; that is, the isothermal surfaces are in general inclined over the isobaric surfaces so that, in cylindrical coordinates (ω, ϕ, z), one has $\Omega = \Omega\,(\omega, z, t)$ or $\partial\Omega/\partial z \neq 0$. In this case, then, the violent symmetric instability occurs wherever the j-distribution decreases outward on the surfaces of constant specific entropy (e.g., T.R.S., p. 168), and the mild shear-flow instability sets in wherever

Ri $\lesssim$ 0(1). In sharp contrast to the ideal case of a barotrope, however,
a barocline is also dynamically unstable with respect to non-axisymmetric disturbances wherever Ri $\gtrsim$ 0(1). It is the so-called <u>baroclinic
instability</u>, depending essentially upon the difference between the isothermal and isobaric surfaces. This instability, which draws its kinetic energy mainly from the potential energy of the unperturbed rotational motion, has received the most attention in geophysics because of
its importance to the dynamics of the Earth's atmosphere and oceans
(e.g., Charney 1973; see also Paper VI, n. 3, and references therein).
It is a <u>mild</u> instability in the sense that it continuously generates
small-scale, time-dependent motions that propagate in the azimuthal direction. As it is done in geophysics, we shall merely assume that the
ever-present baroclinic instability produces anisotropic turbulence
in the radiative zone of a rotating star. Lacking any better description of these irregular motions, we shall further assume that the eddy
flux of momentum can be represented parametrically by means of suitable
coefficients of eddy viscosity. Evidently, in this representation the
ever-present <u>thermal instabilities</u> (such as the <u>feeble</u> GSF instability)
play no important role because they are overshadowed by the ever-present
shear-flow and baroclinic instabilities, which have time scales of the
order of the rotation period (Paper I, pp. 337-338).

3. <u>Rotation and Circulation</u>

     All theoretical speculations about the angular momentum distribu-
tion within a rotating star have their roots in von Zeipel's paradox,
which states that the conditions of hydrostatic and radiative equilibrium are in general incompatible in a rotating barotrope. This paradox can be solved in two different ways: <u>either</u> one allows for a slight
departure from barotropy and choose the angular velocity $\Omega = \Omega$ (ω, z, t)
so that strict radiative equilibrium prevails at every point, <u>or</u> one
allows for large-scale motions in meridian planes. The first alternative (originally developed by M. Schwarzschild) is mainly of academic
interest because, as we pointed out in Paper I (p. 341), these circulation-free models are dynamically unstable with respect to non-axisymmetric motions; hence, the slightest disturbance will generate three-
dimensional motions and, as a result, a large-scale meridional circulation will commence. The second alternative was independently suggested
by Vogt and by Eddington, who pointed out that the breakdown of strict
radiative equilibrium tends to set up slight rises in temperature and
pressure over some areas of any given level surface and slight falls
over other areas, the ensuing pressure gradient between the poles and
the equator causing a flow of matter along the level surfaces. In other
words, it is the small departures from spherical symmetry in a rotating
star that lead to unequal heating along the polar and equatorial radii,
which in turn causes large-scale circulatory currents in meridian planes.
Eddington's (1925) farsightedness is particularly apparent from the following remark: "The coefficient of viscosity in a star is rather high,
and it seems likely that when the currents attain a moderate speed a
steady rate will result; of course, the fundamental equations of equi-

librium are then modified by the addition of viscous stresses to the
pressure-system, and the star is relieved from von Zeipel's condition."
It is my purpose to show that this sentence contains the essence of our
global solution, although obviously the nature of this "rather high
viscosity" has to be specified more clearly.

I have already referred to the many inconsistencies that can be
found in the literature on meridional streaming in a stellar radiative
zone (T.R.S., pp. 198-207; Paper IV, p. 301). As is well known, in
most papers the main consensus is that viscous friction can be neglected
in a radiative zone because: (i) the microscopic (molecular and radiative) viscosity is negligibly small in a star, and (ii) laminar motions
always prevail outside a (turbulent) convective zone. The first statement is correct, as we showed in Paper I. The second one, which most
probably stems from man's propensity to idealize what he cannot see,
is an unacceptable oversimplification in the present context. Our global solution rests essentially on a dynamical linkage between the ever-
present eddy-like motions (which we called "anisotropic turbulence"
because they are predominantly two-dimensional) and the mean flow (that
is, the differential rotation and concomitant meridional currents). To
be more specific, strict radiative equilibrium prevents a rotating star
from being a barotrope; hence, a radiative zone is necessarily filled
with small-scale transient motions that are caused by the non-axisymmetric instabilities. This anisotropic turbulence, in turn, generates
viscous boundary layers so that the circulation velocities do not become infinite near the boundaries of the radiative zone. (The presence
of unwanted singularities was the main defect of Sweet's [1950] laminar,
inviscid solution.) Moreover, the turbulent friction acting on the
differential rotation can be made to balance (in part or _in toto_) the
transport of angular momentum by the large-scale meridional currents
(Papers IV and VI).

By making use of the _eddy-mean flow interaction_ which takes place
continuously in a radiative zone, one can obtain a simple but adequate
description of the mean state of motion in a rotating star. For example, in Papers I, IV, and VI we have considered the case of a chemically homogeneous, nonmagnetic, early-type star, assuming that departures
from spherical symmetry are not too large. As we have shown, if one excludes the most rapid rotators on the verge of equatorial break-up, one
can rightfully expand about hydrostatic equilibrium in powers of the
small parameter $\varepsilon = \Omega_0^2 R^3/GM$, neglecting all terms of $O(\varepsilon^2)$ or smaller.
Here Ω_0 is the (constant) overall angular velocity, R the radius, G the
constant of gravitation, and M the mass. (Note that $\varepsilon \approx 0.4$ at the
centrifugal limit.) To make a long story short, let us say that the
overall rotation, of $O(\varepsilon^{\frac{1}{2}})$, generates a large-scale meridional circulation, which is of $O(\varepsilon)$ because the centrifugal force is of that order; these currents, in turn, react back on the overall rotation so that,
correct to $O(\varepsilon^{3/2})$, the actual angular velocity takes the form

$$\Omega = \Omega_0 [1 + \varepsilon (A + B \sin^2 \theta)], \tag{1}$$

in spherical coordinates (r, θ, ϕ). The problem of evaluating the mean circulation velocities is thus neatly separated from that of evaluating the mean departure from solid-body rotation, i.e., the functions $A(r)$ and $B(r)$ in equation (1). This fact explains why some of the early results remain partially valid, even though their meanings are changed because of the essential eddy-mean flow interaction.

In Paper I we have illustrated the meridional currents in the turbulent radiative envelope of a $3M_\odot$ early-type star. Clearly, to 0 (ϵ) the circulation pattern does not depend on the overall rotation rate; it consists of a single cell extending from the core-envelope interface to the surface, with rising motions at the poles and sinking motions at the equator. Because of the presence of viscous boundary layers, there are no singularities in the mean flow; hence, the circulation velocities remain uniformly small everywhere in the radiative zone. The typical speed of these currents is of the order of $\epsilon\, LR^2/GM^2$, where L is the total luminosity. This result confirms the belief that the time scale of the large-scale circulatory currents established by rotation in a chemically homogeneous radiative zone is everywhere of the order or the Eddington-Sweet time, $t_{ES} = t_{KH}/\epsilon$, where t_{KH} ($= GM^2/RL$) is the Kelvin-Helmholtz time.

The nicest feature of our mean circulation pattern is that the meridional velocities do not depend on the (largely unknown) eddy viscosities in the bulk of a radiative zone. Of course, these velocities depend on the radial (i.e., along gravity) coefficient of eddy viscosity (μ_t, say) in the core and surface boundary layers. Fortunately, they depend respectively on $\mu_t^{1/7}$ and $\mu_t^{1/10}$ in these layers. Hence, because of the presence of the small exponents 1/7 and 1/10, the dependence on the poorly understood coefficient μ_t is appreciably reduced near the boundaries. To be specific, an uncertainty of 0 (10^3), say, on μ_t near the surface leads to an error of 0 ($10^{0.3}$) $\approx$ 2 on the meridional velocities. The same accuracy was achieved in our discussions of meridional streaming in tidally distorted stars (Papers II and III) and cooling white dwarfs (Paper V). In all cases, there are no singularities in the meridional flow, the circulation velocities remaining uniformly small everywhere in the star. Owing to the extreme smallness of these currents, however, they are without any doubt dynamically unimportant. Yet, because they continuously transport angular momentum, they play a crucial role in establishing the actual amount of differential rotation in a star.

The derivation of the rotation law is a much more intricate problem (see eq. [1]). In principle, the function Ω may be obtained by merely stating that the transport of angular momentum by the meridional flow is balanced by the turbulent friction acting on the differential rotation and a change in time of the mean azimuthal motion. (This is the meaning of the ϕ-component of the equations for the mean motion, see eq. [4] of our Paper VI.) Even assuming that a mean steady state has been reached ($\partial\Omega/\partial t \equiv 0$), one is still faced with the inescapable fact that the functions A and B are roughly proportional to $1/\mu_t$. Since it is impossible at this time to perform a meaningful evaluation of the eddy coefficients in a radiative zone, there is thus no hope to calculate the departure from solid-body rotation with any accuracy. (A similar difficulty occurs in the theory of the solar rotation in the outer convective envelope; but, then, it is at least possible to adjust the theoretical rotation law to the observed surface rotation; see, e.g.,Monin and Simuni 1982, and references therein.) The problem is even more severe when one considers a radiative zone in which the mean rotation is not steady in time (because the star is very young or rapidly evolving, or because it loses angular momentum). In this case, as we showed in Papers VI and VIII, the (non-uniform) overall angular velocity, of $O\ (\varepsilon^{\frac{1}{2}})$, must be derived also from the ϕ-component of the Reynolds equations. Hence, until such time as the anisotropic turbulence in a radiative zone has been thoroughly investigated, it is <u>impossible</u> to derive from first principles only the instantaneous rotation law in a star. Perhaps the only sure thing is that, although the departure from uniform rotation may be quite small (if μ_t is quite large), there is no longer any reason to believe that a radiative zone is spinning exactly like a solid body, no more than there is to claim that strict uniform rotation is forced upon a star by the simple presence of a magnetic field. Uniform rotation may be a convenient (and reassuring) assumption in some cases; in no case is it compatible with the conservation principles of astrophysical fluid dynamics.

4. <u>The Effects of a μ-gradient</u>

As is well known, Mestel (e.g., 1965) has convincingly argued that the nonspherical distribution of mean molecular weight μ set up by meridional streaming in a rotating star tends to choke back the motion in meridian planes. According to Mestel, then, there should be no tendency for spontaneous mixing of matter between a chemically inhomogeneous region and the rest of a star (except perhaps for rapidly rotating stars on the verge of equatorial break-up). Since Mestel's analysis is mainly a qualitative one, in Paper VII we have discussed the effects of a μ-gradient on the meridional currents that pervade the turbulent radiative zone of a single, nonmagnetic, main-sequence star. To the best of my knowledge, this is the first attempt to actually <u>solve</u> the time-dependent differential equations that govern the meridional flow in an evolving star.

To illustrate the main features of meridional streaming in a chemically inhomogeneous star, we have considered the simple problem of a hydrogen-burning core for which the μ-gradient develops quasi-statically from the center outwards. In (almost) agreement with Mestel's finding, we found that the meridional flow is the superposition of "Ω-currents" (i.e., rotationally-driven currents that are nevertheless modified by the radial component of the μ-gradient) and "μ-currents" (i.e., currents that try to restore spherical symmetry to the μ-distribution). When the departure from spherical symmetry is not too large, the corresponding speeds-v (Ω) and v (μ), say – are of the order of $\varepsilon LR^2/GM^2$

and $(\varepsilon LR^2/GM^2)$ $(\Delta\mu/\mu)$, respectively, where $\varepsilon\Delta\mu/\mu$ is a measure of the
latitudinal μ-variations over the isothermal surfaces. As was expected,
both speeds are proportional to the ratio ε because both currents are
caused by the nonsphericity of the rotating star. Hence, to 0 (ε) the
streamlines of the meridional flow do not depend on the overall rotation rate.

The numerical work indicates further that, almost from the start,
the μ-currents oppose the Ω-currents, the large-scale currents dying
out as the μ-gradient spreads through the core of a solar-type star.
Within the numerical accuracy of our calculations, then, a μ-gradient
virtually kills off the meridional flow. Yet, although no substantial
mixing of matter may take place between the inner (inhomogeneous) and
outer (homogeneous) regions, it is found that the μ-gradient does not
perfectly insulate these regions. Evidently, to 0 (ε) we were unable
to substantiate Mestel's claim that, in spite of a μ-gradient, Ω-cur-
rents do exist in stars on the verge of equatorial break-up. This conjecture is plausible but, short of a second-order analysis in ε, it
remains unproven.

How do we explain, then, that Kippenhahn (1974) was able to derive
a simple, first-order criterion for rotational mixing in chemically inhomogeneous zones? His argument is as follows. To 0 (ε), the speed of
the Ω-currents is v $(\Omega) = \varepsilon LR^2/GM^2$, which is Sweet's original result.
As for the μ-currents, Kippenhahn evaluates v (μ) by means of a local
analysis, claiming that v (μ) is about equal to (LR^2/GM^2) $(D\mu/\mu)$, where
$D\mu$ is the difference in molecular weight between a blob of matter and
its surroundings. In this crude picture, a μ-gradient kills off the
meridional currents if v $(\Omega) \lesssim$ v (μ); hence, it follows at once that
no mixing will occur if $\varepsilon \lesssim D\mu/\mu$. The fallacy of this criterion lies
in the fact that, as I said, to 0 (ε) one has v $(\mu) = (\varepsilon LR^2/GM^2)$ $(\Delta\mu/\mu)$.
Accordingly, because both speeds are proportional to ε in a first-order
calculation, no critical value of ε can possibly be found to this order.
In other words, Kippenhahn's derivation is incorrect because his local
analysis does not include the fact that departure from spherical geometry is the ultimate cause of the global μ-currents. This is also the
reason why Huppert and Spiegel (1977)'s analysis did not confirm
Kippenhahn's.

To sum up, in Paper VII we have found that the large-scale meridional currents virtually die out from the center outwards as the μ-
gradient steadily grows in a late-type star. Correct to order ε, our
quantitative discussion thus amply confirms Mestel's picture of μ-bar-
riers that prevent substantial mixing rates in slowly rotating stars,
although these barriers may be penetrated to some extent. We have not
yet calculated the circulatory currents in the radiative envelope of an
evolving early-type star. Technically speaking, however, the two problems are quite similar. Accordingly, one readily sees from the appropriate equations that, to 0 (ε), the pattern of meridional streaming
in an early-type star does not depend on the overall rotation rate;
that is, to this order there is no critical rotation rate above which

unimpeded mixing may take place in its turbulent radiative envelope.
(This obvious fact was overlooked by Mestel and Kippenhahn because
they did not make the proper scaling and ordering in powers of ε.)
Moreover, the μ-gradient should inhibit the large-scale currents in
the bulk of the radiative shell interior to the initial core boundary.
However, in view of our numerical results, there is no reason to believe that this region cannot be penetrated to some degree. Hence, one
must allow for some rotational mixing between the homogeneous and inhomogeneous parts of the radiative envelope. Of course, the ability of
the circulatory currents to bring CNO-processed material to the outer
surface will depend primarily on their speed, which is proportional to
the squared overall angular velocity Ω_0^2. Hence, _if_ the radiative envelope of an early-type star is rotating almost uniformly, abnormal abundances could be observed at some stage, when the mixing time t_{ES} (=
$G^2M^3/LR^2v_e^2$) is shorter than the star's main-sequence lifetime, the
mixing efficiency of the meridional circulation increasing as the _squar-
ed_ equatorial velocity v_e^2 (= $\Omega_0^2R^2$). As I pointed out in Section 3, however
there is absolutely no guarantee that the observed equatorial velocity
v_e (or, for that matter, v_e sin i) is a correct measure of the star's
inner rotation, which cannot be predicted on pure theoretical grounds
only.

5. Applications

Quite understandingly, an observer would like to receive clear-cut
answers to the following two questions: (i) how does a star rotate? and
(ii) what are the effects of this rotation on the observable parameters?

As far as the first question is concerned, it has long been accepted that the detailed rotation law in a convective core is much uncertain, because our description of turbulent convection in a rotating
fluid necessarily involves some free parameters that cannot be calculated from first principles alone. Our knowledge of the state of motion
in an outer convective envelope is also incomplete; but then, as it is
done in geophysics, the free parameters may be selected so that the
theoretical rotation rate agrees with the observed surface motions.

In the case of a radiative core or envelope, it was implicitly assumed that the rotation law would not involve free parameters because,
as it was thought, motions are always laminar outside a convective zone.
In this lecture I have advocated the idea that the prevalent baroclinic
instability in a stellar radiative zone continuously generates anisotropic turbulence, which interacts with both the rotation law and the
concomitant meridional currents. (A similar eddy-mean flow interaction
has been known to the geophysicists since the late 1940s!) Because the
study of transient, eddy-like motions in a radiative zone is a new development, it is not yet possible to obtain the eddy viscosities and,
thence, the mean rotation law. (As we pointed out in Paper VI, the geophysicists themselves cannot evaluate their eddy viscosities from first
principles alone; unless one is willing to rely upon doubtful hand-
waving arguments, there is thus no hope to evaluate ours at this time!)

Fortunately, thanks to a caprice of nature, the slow but inexorable meridional currents in a turbulent radiative zone are almost independent of the smaller-scale motions. Hence, the circulation velocities presented in Papers I and II are amply adequate to discuss the problem of circulation versus diffusion in (single and double) early-type stars. This was done with great clarity by Michaud and his collaborators; I shall not repeat their arguments that point in favor of both the diffusion model and large-scale meridional streaming in turbulent radiative envelopes (Michaud 1982, Michaud et al. 1983). Of equal importance is the problem of rotational mixing in evolving stars, which I will now briefly consider.

Although meridional circulation has the potentially important effect of mixing the composition of the stellar material, it is known that most stars do not mix because the assumption of homogeneous stellar evolution does not explain the existence of the giant branch in the HR diagram. Mestel's idea that μ-barriers prevent the mixing of matter (except perhaps for stars on the verge of equatorial break-up) was thus all the more plausible, and the ill-formulated problem of rotationally-driven meridional currents could be swept under the rug. By necessity, however, the concept of slow rotational mixing was evoked by Paczyński (1973) and again by Sweigart and Mengel (1979) to explain CNO anomalies in some early-type stars and red giants (see also, e.g., Norris 1981, and references therein). What can we add to these phenomenological discussions in the light of the quantitative results presented in Papers I-VIII?

First of all, there is some confusion about the fact that, supposedly, there should exist a simple recipe which indicates when rotational mixing will or will not occur. As I have explained in Section 4, for all stars rotating well below the break-up velocities, the streamlines of meridional circulation do not depend on the overall rotation rate. Furthermore, our detailed numerical calculations clearly show that, although a μ-gradient virtually kills off large-scale currents, it does not perfectly insulate a chemically inhomogeneous region from its surroundings. In other words, some mass exchange may take place, after some lapse of time, whenever the Eddington-Sweet time in the homogeneous region is smaller than the evolutionary time scale.

The extreme slowness of the circulatory currents is the main reason why rotational mixing is utterly inefficient when it comes to bringing a sizable amount of hydrogen-rich material into the burning core of a solar-type star (unless, of course, its inner rotation period is no more than a fraction of a day!) Accordingly, rotational mixing cannot possibly explain the solar neutrino problem. But then, as I have shown in Section 2, one must keep in mind that rotation always generates a whole spectrum of eddy-like motions in a radiative zone. Hence, turbulent diffusion mixing (as originally suggested by Schatzman) remains a plausible explanation to this problem, even though the numerical modelling of Schatzman et al. (1981) might appear inadequate, as was shown by Ulrich and Rhodes (1983). In my opinion, because the (non-uniform) turbulent transport coefficients in the Sun are still too much uncertain, no unequivocal solar model can be made at this time; and there is not a single good reason why the corresponding Reynolds number should take a universal value in all main-sequence stars.

In early-type stars one has in general $\varepsilon \approx 0.01-0.1$, whereas in solar-type stars ε may be two or three orders of magnitude smaller. This is the main reason why even a moderate rotation rate can bring CNO-processed material to the outer surface in an evolving early-type star, in spite of the presence of a μ-gradient near the core-envelope interface. The idea that slow rotationally-driven currents may be the ultimate cause of abnormal atmospheric abundances in some stars is a most plausible one, therefore. Not unexpectedly, we found it impossible to derive a precise criterion stating under what condition(s) and to what extent meridional streaming may indeed change atmospheric CNO abundances, either on the main sequence or on the giant branch. It is my firm belief that <u>no</u> simple, black-or-white criterion exists, because this problem involves too many (unknown) parameters. (For example, the inner rotation rate of a star may be larger than that implied by its observed $v_e \sin i$ value, so that the mixing currents may be much faster than those expected on the basis of the results presented in Paper I.) Yet, I'm looking forward to the time when CNO surface abundances (and perhaps other seemingly unrelated observations) will be used to probe the <u>inner</u> state of motion in a stellar radiative zone, which always consists of highly anisotropic turbulence, rotation, and much slower circulatory currents.

Since mixing has also been mentioned in connection with the blue stragglers, I shall conclude this lecture by commenting briefly on these puzzling objects. As we know, Saio and Wheeler (1980) and Maeder and Mermilliod (1981) have shown empirically that partial <u>ad hoc</u> mixing is a viable assumption to account for their existence. On the basis of our self-consistent results about meridional streaming, it would appear most unlikely that slow rotational mixing can indeed play a role in this problem, being overshadowed by convective overshooting, which is a much more efficient stirring mechanism. However, because there is no compelling reason why overshooting should occur in some stars but not in others, one should not exclude the possibility that blue stragglers have rapidly rotating cores. If so, I do not think that rotational mixing would be of much importance either, extended main-sequence lifetime being then caused (in part or <u>in toto</u>) by a much reduced temperature in the central region (as compared to their slowly rotating counterparts). Of course, observation indicates that the blue stragglers do not appear to be abnormally rapid rotators (e.g., Smith and Hesser 1983). But then, as I have already said, these measurements refer to the surface rotation rates, which may not be indicative in all cases of the inner rotation rates. Since it is most likely that there exists a spectrum of core rotational velocities in stars, the assumption of rapidly rotating cores in the blue stragglers is therefore not unreasonable. And, again, it could be used as an indirect way to evaluate unobservable inner motions.

To sum up, the main leitmotiv of this lecture was that neither an observer nor a theoretician can give a firm, direct information about the state of motion deep inside an evolving star. The main difficulty for the theoretician is that the motions within a rotating star are always turbulent; hence, given the present state of knowledge about turbulence in a compressible fluid, the mean rotation rate necessarily depends on poorly known parameters (such as the eddy viscosities). Accordingly, instead of studying the gross effects of ad hoc rotation laws on stellar evolution, one should perhaps accept rotation as a plausible explanation of some observations and, thence, use the data to evaluate the mean and fluctuating motions within a star. In other words, one should not overlook the fact that some unrelated phenomena (such as the solar 5-minute oscillations, the CNO abundances, or perhaps the blue stragglers) might provide a clue as to what is the specific angular momentum within an evolving star. Moreover, because direct measurements of turbulence and large-scale motions can now be made in the Earth's atmosphere and oceans, the theoretician who has a personal interest in stellar rotation should avail himself of the much more rapid related advances that are being made in geophysics. A genuine background in fluid mechanics may be of some use too.

REFERENCES

Brunet, P. 1931, L'introduction des théories de Newton en France au XVIIIe siècle, pp. 223-228 (Genève: Slatkine Reprints, 1970).
Charney, J.G. 1973, Dynamic Meteorology, ed. P. Morel (Dordrecht: Reidel), p. 251.
Chimonas, G. 1979, J. Fluid Mech., 90, 1.
Eddington, A.S. 1925, The Observatory, 48, 73.
Huppert, H.E., Spiegel, E.A. 1977, Ap.J., 213, 157.
Kippenhahn, R. 1974, IAU Symposium 66, Late Stages of Stellar Evolution, eds. R.J. Tayler, J.E. Hesser (Dordrecht: Reidel), p. 20.
Lorimer, G.S., Monaghan, J.J. 1980, Proc. Astr. Soc. Australia, 4, 45.
Maeder, A., Mermilliod, J.C. 1981, Astr. Ap., 93, 136.
Mestel, L. 1965, Stellar Structure, eds. L. H. Aller, D.B. McLaughlin (Chicago: Univ. of Chicago Press), p. 465.
Michaud, G. 1982, Ap.J., 258, 349.
Michaud, G., Tarasick, D., Charland, Y., Pelletier, C. 1983, Ap.J., 269, 239.
Monin, A.S., Simuni, L.M. 1982, Proc. Natl. Acad. Sci. USA, 79, 3903.
Moss, D., Smith, R.C. 1981, Rep. Prog. Phys., 44, 831.
Norris, J. 1981, Ap.J., 248, 177.
Paczyński, B. 1973, Acta Astr., 23, 191.
Saio, H., Wheeler, J.C. 1980, Ap.J., 242, 1176.
Schatzman, E., Maeder, A., Angrand, F., Glowinski, R. 1981, Astr.Ap., 96, 1.
Smith, H.A., Hesser, J.E. 1983, P.A.S.P., 95, 277.
Sweet, P.A. 1950, M.N.R.A.S., 110, 548.
Sweigart, A.V., Mengel, J.G. 1979, Ap.J., 229, 624.
Tassoul, J.L. 1978, Theory of Rotating Stars (Princeton: Princeton Univ. Press) (T.R.S.).

Tassoul, J.L., Tassoul, M. 1982a, Ap.J. Suppl., 49, 317 (Paper I).
---. 1982b, Ap.J., 261, 265 (Paper II).
---. 1982c, Ap.J., 261, 273 (Paper III).
---. 1983a, Ap.J., 264, 298 (Paper IV).
Tassoul, M., Tassoul, J.L. 1983b, Ap.J., 267, 334 (Paper V).
---. 1983c, Ap.J., 271, 315 (Paper VI).
---. 1984a, Ap.J., 279, in press (Paper VII).
---. 1984b, in preparation (Paper VIII).
Turner, J.S. 1973, Buoyancy Effects in Fluids (Cambridge: Cambridge
      Univ. Press).
Ulrich, R.K., Rhodes, E.J., Jr. 1983, Ap.J., 265, 551.

DISCUSSION

Tayler: I agree that in principle you need to make the careful time-dependent treatment of μ-currents. However, is it not true that it is only really essential when $t_{ES} \approx t_{nuclear}$; if $t_{ES} \ll t_{nuclear}$ the star is kept homogeneous, if $t_{ES} \gg t_{nuclear}$ the μ distribution will evolve independent of the ES circulation and you can then ask what effect the prescribed μ has on the circulation?

J.-L. Tassoul: Let me rephrase our results in another manner. There are three time scales: $t_{nuclear}$, $t_{ES} = t_{KH}/\varepsilon$, and $t_1 = t_{KH}/\varepsilon_1$, where $\varepsilon_1 (\approx 0.1$, say) is the value of ε above which our first-order expansions break down. We have made a careful discussion of the <u>simultaneous</u> interaction between the effects of a μ-distribution on the <u>meridional</u> currents and the effects of this circulation on the μ-distribution in an evolving 1 $M_\odot$ star. We have found that, when $t_{ES} \gtrsim t_1$, the <u>unsteady</u> circulation pattern is the same in all cases ($t_{ES} \ll t_{nuclear}$ or $t_{ES} \approx t_{nuclear}$). We found almost no penetration across the outwardly moving μ-barrier because, when $t_{ES} \gtrsim t_1$, the radial part μ_o of the μ-distribution produces a large effect on the circulation pattern whereas the back reaction $\varepsilon\mu_{1,2}$ of the currents on the μ-distribution is a small effect only. (Incidentally, this is why our first-order expansions are self-consistent) In other words, to order ε it takes a quite small departure from spherical symmetry in the μ-distribution to kill off the meridional currents in the central inhomogeneous region, where nuclear burning takes place. (McDonald reached exactly the same conclusions in his <u>correct</u>, first-order discussion of the μ-distribution that is needed to obtain steady, circulation-free solutions; see Ap. Sp. Sci., <u>19</u>, 309, 1972). Now, if $t_{ES} < t_1$, second-order series in ε are needed. In this more complex initial-value problem, one does not expect to find almost perfect insulation in all cases, because the streamlines of meridional circulation then depend on ε. This is another time-dependent problem that has not yet been solved. On physical grounds, new behaviours are previsible when $t_{ES} \ll t_1$. However, unless one is willing to make use of <u>unsound</u> first-order series in ε, these more complex circulation patterns can be described by second-order expansions only. Our calculation is strictly first-order in ε. I do not want to claim more than one can achieve on the basis of a <u>consistent</u> first-order analysis (see also our Paper VII).

Spruit: In the beginning you said that for ordinary microscopic viscosity "there is no solution". Does that mean that the mathematical problem stated has no solution, or does it mean that the solution just looks unappealing?

J.L. Tassoul: I should have said: "there is no acceptable steady solution when ordinary microscopic viscosity is taken into account." The reason why we did not want to retain our laminar solutions is that we wanted to restrict ourselves to mean steady motions in our Paper I. Later on, we realized that one can also obtain consistent solutions with a much smaller viscosity, provided one allows for unsteadiness in the mean azimuthal motion. This is explained at length in our Papers IV and VI.

Spruit: From the fact that the solution for low ν has large differential rotation one cannot conclude that there is turbulence. This has to come from a different argument.

J.L. Tassoul: Our argument that the motions are turbulent rather than laminar stems from the fact that, no matter what kind of viscosity one assumes, the rotation law necessarily depends on ϖ and z. Accordingly, one must consider baroclinic models, which may easily sustain a wide range of small-scale, eddy-like motions. This is shown in our Papers I (App. B) and VII (App.).

Spruit: Your statement that baroclinic instability will always be present in a barocline is not justified. Also, when it is present it need not produce a significant turbulence in the bulk of the star. See Spruit and Knobloch (A & A, 1983, "Baroclinic instability in stars"). This paper also describes why baroclinic instability in stars is so much less important than in the Earth's atmosphere and oceans. Only when Ω is of the order N can baroclinic instability become important in stars.

J.L. Tassoul: In the Earth's atmosphere, baroclinicity is caused by the pole-equator temperature difference due to solar heating. As we explained in our paper I (App. B), the thermal-wind relation and the geostrophic approximation do not apply in a stellar radiative zone. Indeed, in such a system, baroclinicity is a consequence of strict radiative equilibrium at every point, which forces large-scale meridional currents and a mean rotation law of the form $\Omega = \Omega(\varpi, z, t)$. By virtue of the Poincaré-Wavre theorem, this condition implies that the isothermal surfaces are in general inclined over the isobaric surfaces, so that there always exist some unstable baroclinic modes in a stellar radiative zone. In other words, because the basic mechanisms that lead to baroclinicity (and, hence, to baroclinic instability) are not the same in geophysics and in astrophysics, one should not make use of approximations that are valid in the former case to discuss baroclinic instability in a rotating star. The fact that this instability may be less effective in a star is of no concern to us either, because we do not want to describe stellar weather waves. In fact, all what we need are some small-scale disturbances superimposed on the large-scale flow, no matter how large or small these eddy-like motions may be (see our Papers IV and VI).

<u>Zahn</u>: What I highly appreciate in your model of a rotating star
(Tassoul and Tassoul, 1982) is its mathematical consistency, achieved
for the first time. But one has also to worry about its physical consistency, as mentioned by I. Roxburgh. I came to the conclusion that the
conservation of angular momentum, in the stationary state you have described, requires the turbulent Prandtl number to be of order one
(13th Advanced Course of Saas-Fee, 1983). This would have dramatic consequences on the structure of the star, since the transport of heat by
turbulent motions would be of the same order as the radiation transfer.
How can you overcome this difficulty?

<u>J.L. Tassoul</u>: During the 13th Saas-Fee Course, by making use of an order-
of-magnitude argument, Zahn has explained why he did not believe that
our solutions were physically consistent. The big flaw in his argument
can readily be seen as follows. By virtue of equations (121) and (122)
of our Paper I, one has $\beta_i = 0(ur/\nu_t)$, where $\nu_t = \mu_t/\rho$ and $j = 1,3$.
According to Zahn, then, one can write $ur = 0(R^2/t_{KH}) = 0(\chi_r/\rho c_v)$, so
that $\beta_i = 0(P_t^{-1})$, where $P_t = c_v \mu_t/\chi_r$. His conclusion follows at once
from this formula because, for consistency, one must have $\beta_i = 0(1)$. Now,
by making use of the numerical results of our Paper I, one readily sees
that ur/ν_t is a bounded function that vanishes at $r = R_c$ and $r = R$; on
the contrary, in our models one has $P_t \propto T$, so that P_t^{-1} varies as the
inverse of the temperature in the bulk of a stellar radiative zone!
Since T drops by <u>many</u> orders of magnitude from the core-envelope interface to the surface, it is now pretty much obvious that Zahn's relation
$\beta_i = 0(P_t^{-1})$ is largely in error in a realistic stellar radiative zone.
In other words, Zahn has routinely made an order-of-magnitude discussion
that applies to a Boussinesq fluid, without noticing that such an argument does <u>not</u> apply to our solutions because we have considered highly
<u>non</u>-Boussinesq fluids. The following conclusions can thus be made:
(i) the solutions reported in our Papers I-VIII are physically and
mathematically consistent, (ii) because we have shown that the turbulent
transport of energy is everywhere much smaller than the transport of
energy by radiation, Sweet's solution adequately describes the mean
meridional flow in the bulk of a nonmagnetic, chemically homogeneous
radiative zone (see our Paper VI), (iii) viscous boundary layers that
depend weakly on the magnitude of the prevailing eddy viscosity prevent
the mean circulation velocities from having unwanted singularities at
the boundaries (see our Paper I), and (iv) the back reaction of meridional streaming on the overall rotation rate can be obtained in a consistent manner in a nonmagnetic star, no matter how large or how small
the prevailing eddy viscosity is (see our Paper IV).

<u>Frogel</u>: CNO abundance anomalies have been observed in giants that evolve
from solar-type stars. You said core rotation is too slow in the sun to
cause mixing. How then can it cause mixing in stars which have solar-type
stars as their predecessors?

J.L. Tassoul: My comment applies only to rotational mixing in the radiative core of a main-sequence, solar-type star. Once such a star leaves the main sequence, however, the core contraction modifies its (unobservable) inner rotation rate, so that rotational mixing may then no longer be negligible during some phases of post-main-sequence evolution. (Remember that the circulation speeds are proportional to the squared overall inner rotation rate).

R. Cayrel: 1) I was confused about the value of the meridional circulation for the sun: was it 10^{-5} cm s^{-1} or 10^{-9} cm s^{-1}?
2) Can you tell us the order of magnitude of the turbulent velocity you expect to be associated with your solution?

J.L. Tassoul: 1) The circulation speed is less than 10^{-9} cm s^{-1} because, in our solar models, $v(\Omega) = \varepsilon|u|$, where $\varepsilon \approx 10^{-4}$ and $|u| \lesssim 10^{-5}$ cm s^{-1} ;
2) At this time, no firm statement can be made about the speed of the turbulent eddies (see also our Paper IV, p. 300).

PHYSICAL MECHANISMS OF MIXING IN STELLAR INTERIORS

Evry SCHATZMAN

Observatoire de Nice , B.P. 139 Nice-Cedex 06003 France

Abstract

The different mechanisms by which mixing can take place in
stellar interiors are considered : the classical Rayleigh-Benard
instability with penetrative convection and over-shooting, semi-
convection , gravitationnal and radiative settling,turbulent
mixing . The latter mechanism is thoroughly described, from the
driving force of turbulent mixing to its influence on   stellar
structure , stellar evolution and the analysis of the corresponding observationnal data .
Turbulent mixing has to be considered each time the building up
of a concentration gradient takes place, either by gravitationnal
or radiative settling or by nuclear reactions . Turbulent mixing,
as a first approximation, can be described by an isotropic diffusion coefficient . The process is then governed by a diffusion
equation . The behaviour of the solution of the diffusion equation needssome explanation in order to be well understood .
A number of examples concerning surface abundances of chemical
elements are given (3He, 7Li, Be, 12C, 13C, 14N), as well as a
discussion of the solar neutrinos problem .
The building up of a μ-barrier , which stops the turbulence
allows stellar evolution towards the giant branch and   explains
nitrogen abundance at the surface of giants of the first ascending branch .
Turbulent mixing is also of some importance for the transfer of
angular momentum and has to be taken into account for explaining
the abundance of the elements in Wolf-Rayet stars .

A. Maeder and A. Renzini (eds.), Observational Tests of the Stellar Evolution Theory, 491–512.

1. Introduction

Mixing inside the stars is not a new idea.Already   L . Biermann
(1937) suggested the existence of some sort of turbulent diffusive mixing in stellar interiors. However, the question of the
observationnal proof of chemical elements transport was not considered seriously until the discovery of the presence of the
radio-active element Technecium in the atmosphere of S stars by
Merrill (1952). The gravitationnal sorting of heavy elements in
white dwarfs (Schatzman , 1945) , due to their large gravitationnal field was never questionned , but the extension of this physical mechanism to other stars did not come up until the work of
Chapman and Aller (1960).

We are interested here in the various mixing processes which take
place inside a star and the observationnal test of their presence.
We shall certainly not ignore the physics which is underlying
these mixing processes, as it will certainly provide us with some
indication of the evolutionnary phase at which they can show up ,
either through surface abundance of the chemical elements, or
through global properties otherwise unexplained . Let us add that
the evidence of new physical processes acting inside the stars
give the possibility of building up new consistent models of
stellar evolution .

Let us stress this point . The hydrodynamical processes we are
concerned with are not only interesting by themselves, and in
many cases are worth devoting a large amount of work and efforts
to understand them; but they have far reaching consequences for
stellar evolution, redistribution of the elements in the galaxy ,
and consequently for the so-called cosmic abundances of the
elements and therefore for nucleo-synthesis and our understanding
of the origin of the Universe .

2.The convective zone .

The abundance of the elements at the surface of the stars has not
a unique cosmic value . Enhancement of some elements, depletion
of others, let us suspect the presence of a variety of physical
processes leading to either one or the other of these differences.
There is little doubt , nowadays that gravitationnal or radiation
settling on one hand, nuclear processing on the other hand are at
the origin of the observed variety of the abundance of chemical
elements, when it cannot be explained by the time and space
dependance of the chemical composition of the Galaxy .

However, after these two statements a number of questions are
raised, which have to be answered in a consistent way. Let us
consider first the question of gravitationnal or radiative
settling. If we describe the outer layers of a star (spectral
type later than AO or therabout) , we have a radiative atmosphere, a convective zone, deep or shallow , eventually double ,
according to the spectral type, and a radiative zone below, with

some amount of penetration of the convection and overshooting .
The penetration of the convection can be of the order of one
pressure scale height , according to Toomre et al (1976) for the
shallow convective zone of main sequence A stars, and , according
to Latour et al (1983) of the order of 1/10 of a pressure scale
height for a deep convective zone , like in the Sun. The convective zone is considered as being of a uniform chemical composition . However, before any further consideration, let us study
this question . To that effect, and in order to introduce orders
of magnitude, take a two-dimensions plane parallel fluid , with a
diffusion coefficient of diffusion D , and, at t=0 , introduce
along the axis y a certain amount of a contaminent of concentration c . The perturbation can be described as

$$c(x,z) = c .\delta(x,z)$$

where δ is a delta function, and at t=0, c=1 . The well known
solution, with the boundary conditions $\partial c/\partial z = 0$ for z = h
and z = -h is :

$$c = (4\,D\,t)^{-\frac{1}{2}}\ e^{-(x^2/4Dt)} \sum_j \cos(j\pi z/h)\ e^{-Dj^2\pi^2 t/h^2}$$

The horizontal distance x is reached by the contaminent over a
time t of the order of $(x^2/4Dt)$ and the contaminent is spread
vertically over a time $(h^2/\pi^2 D)$. The Rayleigh number in any convective zone is so large that we can assume that it is highly
turbulent . In the case of the solar convective zone, we can
estimate D to be of the order of 10^{13} cm^{2}s^{-1}, which gives for
the vertical spreading a characteristic time of the order of 10
days and for the horizontal spreading over the whole solar surface
a characteristic time of the order of 30 years . These durations
are very short compared to the stellar life time, and the assumption of a uniform chemical composition of a convective zone is
physically sound .

It should be noticed that the assumption of an isotropic D can be
a fair approximation for the diffusion of contaminents and a very
poor one for the transport of mechanical quantities like angular
momentum.

3.Elements sorting.
Let us come back to the problem of gravitationnal or radiation
settling . The convective zone can be considered as a reservoir
where the contaminent under consideration is stored . Gravity or
radiation pressure can the contaminent downwards, out of the convective zone or push the contaminent upwards inside the convective
zone (Michaud et al, 1976).

The effect of the penetration of the motions from the convectively
stable zone into the radiatively stable zone is to generate a

a fully mixed region which is larger than the convective zone stricto sensu. Let us assume a sharp boundary between the radiative zone and the fully mixed region. Consider the case of a test element : no retroaction of the test element on the background. Then, the equation of continuity in the radiatively stable region can be written

$$\frac{\partial}{\partial t}\rho c + \text{div } \rho v\, c = 0$$

where v includes the diffusion and the drag , and can be written:

$$cv = -D \left\{ \frac{\partial c}{\partial z} + \frac{c}{h} \right\}$$

the flow $-D\,(c/h)$ represents the drag due to the gravity, the pressure gradient, the electric field and the radiation pressure. The sign oh (D/h) depends whether the gravity exceeds the radiation pressure $((D/h) > 0)$, or the reverse $((D/h) < 0)$; and the quantity (D/h), which is independant of the concentration for a test element, is concentration dependent through saturation effects in the radiative transfer, for a finite concentration (and nevertheless small with respect to the background).

The boundary condition

$$-\rho\, c\, v = m\, \frac{\partial c}{\partial t}$$

where m is the mass, per unit surface, of the fully mixed region, expresses the fact that the depletion of the convective zone is due to the gravitationnal drag . Conversely, it can express the fact that the filling of the convective zone is due to radiation pressure .

When the diffusion is only due to the microscopic diffusivity,the major effect is the drag, $-(D/h)$. The time scale turns out to be (Schatzman, 1969) of the order of $t_{mic} = (H\, h\,/\,D_{mic})$. H is the equivalent height of the fully mixed region .

In the case of pure gravitationnal drag,

$$h = H_p/(1.692\ i^2 + 3.870\ A - 2.203\ i - 2.666)$$

where the sorting is supposed to take place in pure hydrogen , and i is the degree of ionization and A the atomic weight .

Considering the value of D_{mic} we find that as soon as the convective zone is sufficiently shallow, the time scale of depletion, t_{mic} , becomes very short, compared to the stellar life time. We are therefore facing the situation that gravitationnal or radiative settling can become a major phenomenon for all main sequence stars earlier than F2. The question is not so much to include or not to include the sorting of the elements but to understand the reason for which the phenomenon is observed in such a small number of stars, and why it turns out that in "normal" stars the effect is negligible .

Let us consider again a simplified problem in which the essential of the physics is present, which is mathematically simple and can be solved exactly . We have a fully mixed region with a uniform chemical composition; at the lower boundary we find the presence of graviationnal sorting and turbulent diffusion, with a coefficient D_t . Assuming that the turbulent diffusion coefficient and the gravitationnal drag are independant of z, the evolution of the concentration c is governed by the equation

$$D_t \frac{\partial^2 c}{\partial x^2} + f \frac{\partial c}{\partial x} - \frac{\partial c}{\partial t} = 0$$

The boundary condition at the bottom of the convective zone says that the rate at which the element flows downwards is equal to the rate of depletion of the convective zone :

$$- D_t \frac{\partial c}{\partial x} - f c = H \frac{\partial c}{\partial t}$$

f is the (D/h) used above , and we can write $f = (D_{mic}/h)$. When $D_{turb} \gg D_{mic}$ the change of the concentration in the convective zone as a function of time is given for $(t/t_{turb}) \ll 1$ by

$$c = c_0 \left(1 - \frac{2}{\sqrt{\pi}} (t/t_{turb})^{1/2} \right)$$

and for $(t/t_{turb}) \gg 1$ by

$$c = c_0 \frac{2}{\sqrt{\pi}} (t/t_{turb})^{-1/2} \exp(-t/t_{turb})$$

It should be noticed that for t/t_{turb} small, the turbulent diffusion mixing slows down the sorting, wheras for t/t_{turb} large, the gravitationnal sorting is accelerated ! This is due to the disappearance of a concentration gradient .

The characteristic time t_{turb} is given by

$$t_{turb} = (4 h H D_{turb} / D_{mic}^2)$$

Let us consider the behaviour, as a function of the mass, of turbulent diffusion mixing. To that effect, we obtain values of $(1/2) t_{star}/t_{turb}$, taking for D_{turb}, $D_{turb} = Re^* \nu$, $Re^* = 100$ and ν being the microscopic viscosity (molecular plus radiative) as a function of the mass M. For Pop.I stars the behaviour is given in fig 1.

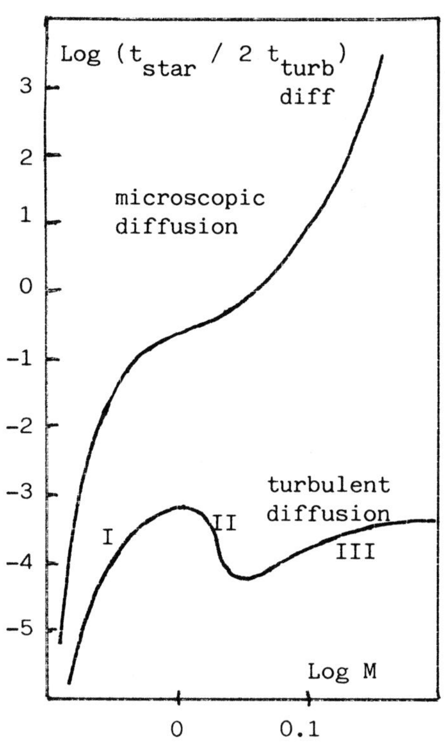

Fig.1. The time scale of element (Li) sorting in the presence of microscopic diffusion only and with the presence of turbulent diffusion mixing. In the section I of the curve, the size of the convective zone decreases and the viscosity is the molecular viscosity. In section II the viscosity changes to the radiative viscosity, the turbulent diffusion coefficient increases , in section III, the radiative viscosity dominates, the convective zone is shallow. For population I stars with turbulent diffusion mixing the element separation does not take place

We shall come later to the question of the origin of this turbulence. For the time being, we have still to consider the effect of penetrative convection and overshooting, with a slight difference in the two terms. Penetrative convection is the penetration of large scale lines of flow in the radiatively stable zone; overshooting is the penetration of a turbulent element which, reaching the boundary with a finite velocity, is there slowed down in the radiatively stable region by negative boyancy force and friction . A phenomenological description of overshooting has been given by Shaviv and Salpeter (1973), Roxburgh (1978),Maeder (1975). Overshooting launches turbulent elements over a finite distance, and it seems likely that beyond that point the convective zone does not induce any mixing. Similarly, it has been shown by Latour et al (1983) that the lines of flow do not penetrate beyond a finite distance from the boundary of the convective zone. Due to the very large value of the turbulent diffusion coefficient in the convective zone, the stochastic motions produced in the radiatively stable zone generate a large turbulent diffusion coefficient. Even if the energy density of the motion is small, and such that the radiative transfer still dominates, the effect of the motion is to keep the region of penetration chemically homogeneous. This is the reason for introducing the concept of fully mixed region, different from the convective zone it self.

4.Lithium

Let us consider now another astrophysical puzzle, the case of the
Lithium abundance. The problem of the surface abundance of Lithium
has been considered now for a long time, starting with Herbig
(1965). The abundance of Lithium in clusters (Duncan,1981), in
field stars (Boesgard, 1976), show that the abundance is mass
dependant and time dependant. For one solar mass stars, except
perhaps a $\sqrt{t}$ dependance at small t (Skumanich,1972), the obvious
law, as given by Duncan (1981) is $dLog(n(Li))/dLog\ T = -0.63$.
The range of spectral type where the Li λ 6707 doublet can be
observed is relatively small, from about 6000°K to 5300°K (Cayrel
et al, 1983). The lines are observed in dwarfs and in giants. The
abundance of Lithium in giants, according to Scalo and Miller
(1980) can be interpreted as the result of the dredge up taking
place in stars which have destroyed their Lithium on the main
sequence, confirming Schatzman picture (1977). Further questions
are raised by the measure by Spite and Spite (1982) of the abundance of Lithium in old population II stars , in the range 6200°K
-5200°K. They find a quasi uniform abundance, Log Li = 1 (for
Log NH = 12) wheras for other stars, the abundance varies from
Log Li = 3 to very small values.

Is it possible to put some order in this disorder ? The idea of
pre-main sequence destruction of Li has been considered by
Bodenheimer (1965) and more recently by Mazitelli(1983) . It is
assumed that the convective zone before receding to its main
sequence extension , reaches the temperature of Lithium burning
with the right time scale . The interesting result of Mazitelli
is that , depending on the extension of the overshooting, and
with an extreme sensitivity to that extension, either pre-main
sequence burning does not take place, or the Lithium is entirely
burnt .

This result can easily be understood if one considers the rate of
Lithium burning . At $T = 2.5 . 10^6$ °K, the reaction rate t_R is
highly temperature dependant, $t_R \sim T^{-20}$. The range of temperature
at the bottom of the convective zone of pre-main sequence stars,
the time spent at the maximum teperature would require such a
minute adjustment to obtain the final abundance that we can definitely consider that Li burning at the bottom of the convective
zone is ruled out . Similarly, the Li deficiency on the main
sequence cannot be explained by the rate of nuclear reactions at
the bottom of the fully mixed region . The result of Cayrel et al
(1983) for the main sequence stars of the Hyades shows beautifully
that another explanation has to be found .

We are then bound to consider again turbulent diffusion mixing.
The equations governing the problem are (1) the diffusion equation
with chemical reactions; (2) the boundary conditions . The search
of a solution of the form exp(- s t) leads to an eigen value

problem for s .
It turns out , as shown by numerical study of the problem by
Baglin and Morel (1983) that the z dependance of ρ and D is relati-
vely unimportant . We can therfore consider the system

$$D \frac{\partial^2 c}{\partial z^2} + (s - K \rho T^n) = 0$$

$$D \frac{\partial c}{\partial z} = s H c , \text{ boundary of the fully mixed region}$$

c = 0 deep inside
An approximate solution of the WKB type has been obtained
(Schatzman, 1983). In the solar case, we have the following values
H (equivalent height of the fully mixed region) = 5.6 10^9 cm ,
T(bottom fully mixed region) = 2 . 10^6 °K,
T(Li burning, $t_{nucl} = t_*$) = 2.5 . 10^6 °K . The distance from the
bottom of the fully mixed region to $T(t_{nucl} = t_*)$ is 1.5 10^9 cm .

As a function of the depletion factor, we have the following
results :

Depletion factor	s t_*	D	Re*
1/250	5.521	1100	55
1/100	4.6	650	35

In the case of the Hyades, Schatzman (1983) has obtained a solution which is model free, as it is expressed as a function of the
temperature at the bottom of the fully mixed region (fig.2).

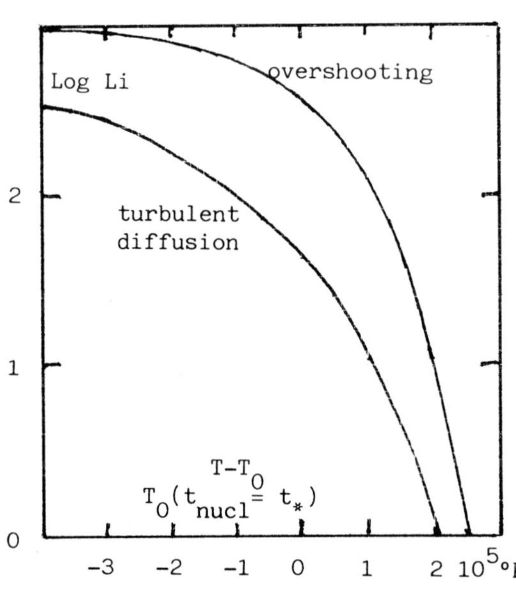

For the same temperature
at the bottom of the
fully mixed region,tur-
bulent diffusion gives
more depletion than
burning ; as a function of
the temperature,diffusion
gives a depletion which
varies more slowly than
burning . This is in
agreement with the results
of Baglin and Morel and
fits with the results of
Cayrel et al.The exact
solution obtained by
Baglin and Morel (1983)
shows that the exact
fitting of the theoretical
curve with the observed
points is possible .

Fig.2. Li depletion

We are not through the mystery story of the Lithium destruction
and hiding . Coming back to the old population II stars observed
by Spite and Spite (1982), it is very likely that the presence of
Lithium is due to the fact that, as a consequence of the low metal
content, the convective zone is not as deep as the convective zone
of Population I stars. Turbulent diffusion, if it is there, cannot,
even in 10^{10} years carry the Lithium to the burning region .However
this not the whole story, as gravitationnal drift of Lithium takes
place slowly, but takes place, as well as metal deposition, even
in the presence of turbulent diffusion .

This reasoning leads to the conclusion that the Li abundances
found by Spite and Spite (1982), with respect to the primordial
Li abundances, may be underestimates. The constraint on the big-
bang nucleosynthesis can therefore be considered as less rigid.
No final conclusion can be reached before quantitative calculations
have been carried with well established models of the outer layers
of these stars .

5.The solar Helium ^{3}He .
The problem of the surface abundance of He is closely related to
the problem of the primordial abundance of the elements . Geiss
et al (1972) have shown that there is a significant increase in
the ^{3}He trapped in the lunar ashes, beginning something like
3 . 10^9 y ago . The surface abundance of ^{3}He in the Sun,
(^{3}He/^{4}He) = 4 to 5.10^{-4} includes the product of deuterium
burning in the convective zone. The low value of the present ^{2}D
abundance (Laurent, 1983), (^{2}D/^{1}H) $\simeq 5.10^{-6}$ suggest that the
surface abundance ratio (^{3}He/^{4}He) which has to be explained is
about 3.5 to 4.5 . 10^{-4} .

In the presence of turbulent diffusion mixing, the distribution of
the main isotopes in the Sun is greatly modified . Fig.3 shows,
according to Schatzman and Maeder (1981) the internal distribution
of ^{1}H and ^{3}He at the age of 4.57 . 10^9 y, assuming a turbulent
diffusion coefficient D = Re*ν , where ν is the microscopic viscosity and Re* is respectively 0,100 and 200.

The internal distribution of ^{3}He presents in the standard model a
peaked distribution which is illustrated in fig 3. During the
evolution, the peak moves outwards and increases in size. At
t=4.57 . 10^9 y the maximum reaches 2.8 . 10^{-3} at M_r/M = 0.58.
With turbulent diffusion mixing, the distribution of ^{3}He below
M_r/M = 0.4 is mainly determined by the rapid equilibrium between
creation and destruction, as in the standard case. Above M_r/M = 0.4
(T 8.3 . 10^6 °K), the transport by turbulent diffusion has a
tendency to uniformise the ^{3}He distribution : the peak is reduced
and ^{3}He is spread through the star and its abundance considerably
increased in the outer layers. At t = 4.57 . 10^9 y the surface
abundance in ^{3}He is 8.5 . 10^{-4} for Re* = 200 and 3.8 . 10^{-4} for

for Re* = 100. An interpolation formula, applicable for the case
of diffusion from a source,

$$\text{He} \quad = \quad (A/\text{Re*}^{\frac{1}{2}}) \ \exp \ (-B/\text{Re*})$$

gives the following results :

Re*	^{3}He/^{4}He
30	2.80. 10^{-4}
31	3.12. 10^{-4}
32	3.44. 10^{-4}

The measured surface abundance of ^{3}He fits with a turbulent diffusion coefficient given by Re*$\simeq$ 30

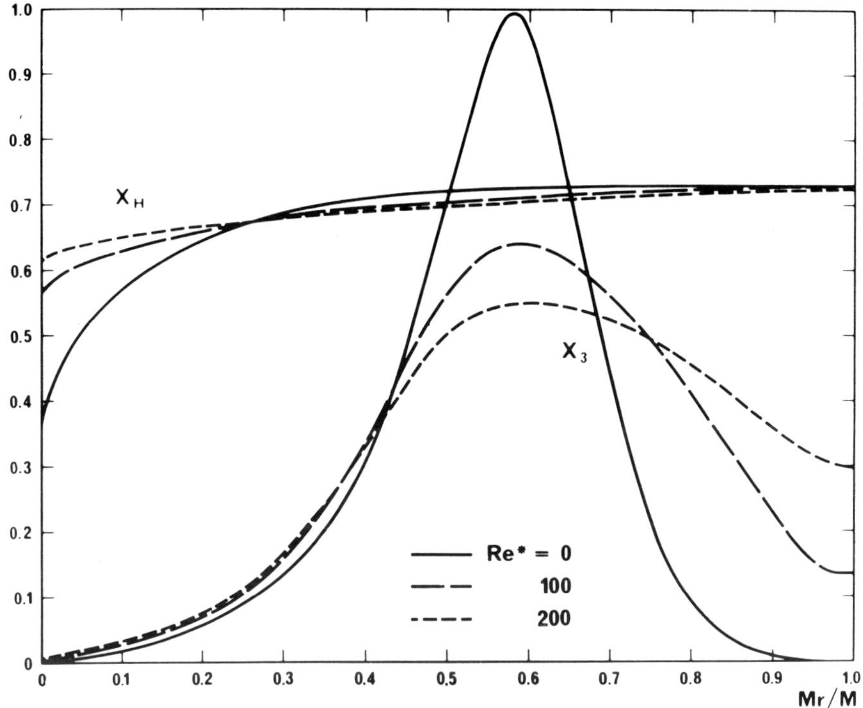

Fig.3 The hydrogen ^{1}H and ^{3}He concentrations in a 1 solar mass
star at the age of 4.6 . 10^9 y (according to Schatzman and Maeder,
1981). The curves are plotted for various values of the parameter
Re* which describes the efficiency of the turbulent diffusion .

6. <u>The Carbon isotopes</u> .
The explanation of the abundance ratio (^{12}C/^{13}C) meets in the
classical models a number of difficulties which can be solved by
introducing the turbulent diffusion mixing .

Let us consider a one solar mass star at the end of its evolution
on the main sequence (Dearborne et al , 1976). Outside the region
$q = (M_r/M*) = 0.4$, ^{13}C has been buildt at the expense of ^{12}C.
Inside the region $q = (M_r/M*) = 0.3$, ^{13}C is destroyed and the
first two reactions of the Carbon cycle are follown by the production of ^{14}N.

It has been noticed for a long time (see, for example Iben, 1977)
that in the first ascending branch, when the deep convective zone
is generated, the dredge up of ^{13}C decreases the $(^{12}C/^{13}C)$ ratio;
however, starting with an initial abundance ratio of the order of
80 to 100, the mixing in the deep convective zone cannot make the
abundance ratio lower than 40 or therabout , and in any case
cannot bring it to the observed value of 10 to 25 (Lambert et al
1980).

The difficulty concerning the interpretation of the $(^{12}C/^{13}C)$ratio
is due to the fact that one finds,in the same area of the HR
diagram stars which have different masses and are in different
evolutionnary stages:(i) stars on the first ascending branch,(ii)
stars of the horizontal branch : according to Faulkner (1966) for
population I stars, the stars of the ZAHB are gathered in a clump
which sits very close to the first ascending branch, (iii) stars
on the giant asymptotic branch . For stars belonging to the first
ascending branch, the observations lead to an abundance ratio
$(^{12}C/^{13}C)$ of the order of 15 to 25, definitely smaller than the
abundance ratio predicted by the standard models.

The idea that turbulent diffusion mixing is important is obvious
if on looks at the ^{13}C concentration as a function of the radius
instead of the lagrangian variable, the mass. The concentration
gradient is larger and the expectancy is that the flow of ^{13}C ,

$$ - 4 \pi r^2 \rho D_{turb} \frac{\partial(^{13}C)}{\partial r} $$

outside the region of ^{13}C production will be appreciable.During
the life of the star on the main sequence, ^{13}C is carried slowly,
driven by the concentration gradient outside of the region of ^{13}C
burning, enriching the outer radiative zone in ^{13}C. When the
dredge up takes place, the ^{13}C which has been stored in the outer
radiative zone shows up and the abundance ratio $(^{12}C/^{13}C)$ is
smaller than in the standard models.After the preliminary results
of Genova and Schatzman (1979), Bienaymé et al (1983) have
obtained the exact solution of the diffusion equation for station-
nary stellar models with 1;1.5; 2 and 3 solar masses.It appears
from the exact solution that turbulent diffusion mixing cannot
bring the $(^{12}C/^{13}C)$ ratio to values smaller than 15 to 20. The
need of stopping some time the turbulent diffusion mixing in the
stellar core is obvious if one considers the physics of the evolution towards the giant branch. This is confirmed by the fact that

continuous turbulent diffusion mixing brings ^{14}N in the outer
radiatively stable regions, wheras a proper inhibition of turbulent diffusion mixing prevents the appearance, at the giant
stage, of anomalous abundances of Nitrogen .

The inhibition of turbulence is likely to be due to the generation
of a gradient of molecular weight. As soon as the gradient $\nabla\mu$
(d log μ/d log P) exceeds some critical value, the turbulence
stops, and turbulent diffusion mixing meets an unpenatrable
barrier. Bienaymé <u>et al</u> have explored for a 1.6 solar mass star
the effect on abundance ratios (C/N) and ($^{12}C/^{13}C$) of the critical
value of the gradient of molecular weight (fig.4). The best fit

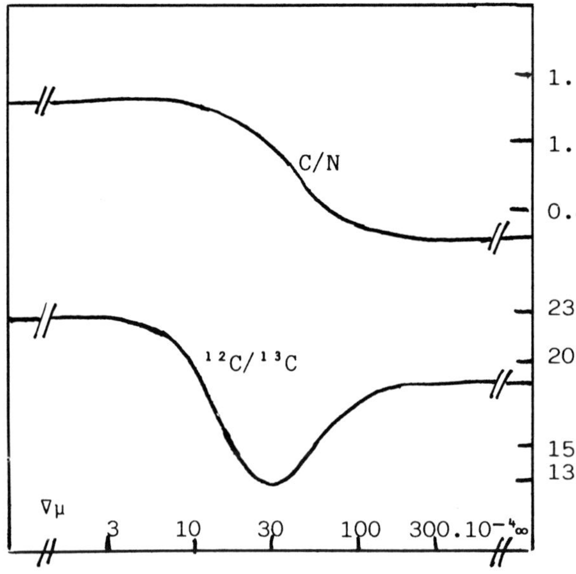

Fig.4. Surface
abundance ratios for
1.6 solar mass as a
function of $\nabla\mu$ crit.
$\nabla\mu=\infty$: no inhibition
of diffusion
$\nabla\mu=0$: identical to
Re* = 0
(from Bienaymé <u>et al</u>
1983)

has been obtained for Re* = 100.

The case of the more evolved stars is different. Sweigart and
Mengel 1979) noticed that the Carbon anomalies become more
prominent with increasing luminosity. They suggest that, as far as
the Carbon anomalies are concerned, there is no compelling observationnal evidence for an additionnal mechanism operating beyond
the Red Giant Branch phase .

In order to explain the anomalies, they suggest that the Eddington
Sweet flow carries matter from the convective envelope to the
level of Carbon burning, where the ^{13}C is generated at the expense
of ^{12}C. This obtained by assuming a very fast meridional circulation and consequently requires a swift rotation.

However, it is possible to explain the mixing from the convective
envelope to the Carbon burning region by turbulent diffusion
mixing. The gradient of μ, $\nabla\mu$ is very small in the region going
from the bottom of the convective envelope to the Carbon burning
region . The time scale of the turbulent diffusion mixing is of
the order of

$$t_{diff} = [\int D_{turb}^{-\frac{1}{2}} \, dz]^{-2}$$

Using the tables of Sweigart and Gross (1978) it is possible to
obtain estimates of t_{diff} . The table gives t_{diff} as a function of
$(t_{fl} - t)$ for one of the population I models of Sweigart and Gross.

t_{diff} as a function of $(t_{fl}-t)$
(t_{fl} : He flash epoch)
$(M;Y;Z) = (0.90; 0.30; 0.01)$; $Re^* = 100$

$t_{fl}-t$	t_{diff}
$23 . 10^6$ y	$3.4 . 10^6$ y
3	0.67
0	0.55

It appears clearly that almost until the Helium flash the
diffusion time can be shorter than the evolution time scale .
The result would be then a reduction of the Carbon abundance and an
increase of the $(^{12}C/^{13}C)$ ratio, like in the model of Sweigart and
Mengel (1979), but with a different physical processwhich has the
advantage of being more consistent with the general picture of
stellar evolution with turbulent diffusion mixing .

7. The solar neutrinos problem.
Compared to standard models, turbulent diffusion has two consequences :
(i) it increases the central concentration of hydrogen by turbulent
diffusion mixing along the gradient of concentration of hydrogen;
(ii) it increases the central concentration of the isotope ^{3}He by
turbulent diffusion mixing along the gradient of concentration of
the isotope (fig.3). This last effect is not included in partially
mixed models of Shaviv and Salpeter (1968), Ezer and Cameron (1968)
Shaviv and Beaudet(1968), Bahcall et al (1968) .

The effect of turbulent diffusion is then to make available more
energy per proton than in the standard models. The small amount of
^{3}He which is brought by turbulent diffusion mixing to the central
regions, due to the large amount of energy involved in the ^{3}He
reactions, contributes appreciably to the energy generation in the
solar core; the final result is a decrease of the central temperature and consequently a considerable decrease in the neutrino flux.

Three models have been calculated by Schatzman and Maeder (1981),
for a turbulent diffusion coefficient $D = Re^*\nu$, with Re^* resp.
equal to 0,100 and 200.

More models are needed . However, it is possible to obtain, by an interpolation formula the neutrino flux as a function of Re*.The solution of the diffusion equation with chemical reactions show that for D close to zero the partial derivative $\partial/\partial D$ is very large. We then obtain the following table :

Re*	Φ_ν (SNU)		
0	11.6	6	4.6
20	5.76	2.97	2.28
30	4.95	2.56	1.96
50	3.83	1.98	1.33
100	2.39	1.23	0.95
200	1.43	0.74	0.57

whe the interval 11.6 to 4.6 for the predicted flux is about the error interval given by Bahcall <u>et al</u> (1982) . The observed neutrino flux (Davis, quoted by Bahcall <u>et al</u>, 1980),

$$\Phi_\nu = 2.2 \pm 0.4$$

gives as a central value Re* = $40 \begin{smallmatrix} +100 \\ -20 \end{smallmatrix}$.

8.<u>The origin of turbulent diffusion mixing</u> .
Zahn(1983) has given a simple and natural explanation of the slow turbulence which is at the origin of the mixing process which we have considered here. With a turbulent diffusion coefficient D = Re* ν , Re* = 25 we obtain for the Sun characterestic velocities of the order of 20 μ s^{-1}and characteristic lengths of the order of 200 meters.

The basic idea is to consider the baroclinic instability. When a star does not rotate cylindrically,$(\partial\Omega/\partial z) \neq 0$, the surfaces of constant entropy do not coincide with the level surfaces, and there are always motions, those inside the wedge between the isentropic surface and the horizontal which are unstable. The instability generatesa turbulence which is first dominated by the almost horizontal nature of the motion, and is a two-D turbulence. This holds for all sacles for which the Coriolis force dominates over the inertial terms,

$$\Omega > 2.5 \; (u/1) \qquad\qquad\qquad\qquad (I)$$

the numerical coefficient being issued from an experiment carried by Hopfinger <u>et al</u> (1982).The cascade towards smaller scales is accompanied by a transition to 3-D turbulence, when the condition (I) is violated.

Assuming that the kinetic energy of the differential rotation,fed itself by advection of the angular momentum,is converted into turbulent energy. Zahn obtains the following value for the vertical turbulent diffusion coefficient, due only to the 3-D turbulence ,

$$D = Re^* \; \nu$$

with

$$Re^* = \frac{K}{\nu} \; \frac{\Omega^2 r}{g} (\nabla_{ad} - \nabla_{rad})^{-1}$$

where K is the thermal diffusivity, ν the microscopic viscosity, (ν/K) is the Prandtl number, which is small . In the solar case, this gives near the region of Li-burning , $Re^* = 60$, assuming Ω to be the same than at the surface of the Sun.

This can be considered as a reasonable estimate, the exact value depending, on the one hand on the precise choice of numerical coefficients of the order of 1, and on the other hand, on the complete solution of the flow diagram, including the feed back mechanisme rotation $\rightarrow$ turbulence $\rightarrow$ rotation and a proper choice of the rate of loss of angular momentum by the star.

The question of the stabilizing μ-gradient has been extensively discussed recently by Knobloch and Spruit (1982,1983). When the μ-gradient is small, the Goldreich-Schubert-Fricke instability is the most important. This is suppressed for a very small μ-gradient, $\nabla\mu \cong (H_p \; \Omega^2 / \; g) << 1$. However, if the turbulent motion is present, we do not know under which conditions it will stop, either by a change in the distribution of Ω, or by a change in the μ-gradient. Knobloch and Spruit (1982) suggest that the turbulent motions, which have been generated at an earlier phase of stellar evolution, could remain present for a long time after the linear condition of stability has been reached during stellar evolution .

In this respect, the phenomenological theory of turbulent diffusion mixing, by the quantitative astrophysical estimates which are obtained, can be considered as a guide to a full understanding of the generation and disappearance of turbulence in a rotating star .

9. <u>He deposition</u> .
The problem of the discrimination between Am stars and δ Scuti stars, which sit in the same area of the H-R diagram, has been discussed several times, for example by Baglin <u>et al</u> (1973) . Breger (1979). Let us recall the suggestion of <u>Baglin</u> (1972) that the ^{4}He settling in a non turbulent star leads to the disappearance of the He-convective zone (Vauclair <u>et al</u> , 1974) and allows then the dripping in of the heavy elements . Mixed stars, being in the instability strip, can pulsate, wheras demixed stars, will show the Am characteristics.

This discrimination is perhaps the most striking indication of the presence or absence of turbulent diffusion mixing .

The baroclinic instability which appears to be at the origin of the turbulent mixing in stars, in a star of uniform chemical composition, for a very slow rotation. As an order of magnitude, for a star where the viscosity is the radiative viscosity,

$$\Omega^2 \geq 4 \frac{g^2}{c^2} - (\nabla_{ad} - \nabla_{rad}).$$

For Log g = 4.5, $(\nabla_{ad} - \nabla_{rad}) = 0.1$,

$$P \leq 60 \text{ days.}$$

However, this does not mean that the 2-D turbulence is generated. We need to have a shear sufficiently strong to generate the 2-D turbulence; when the viscosity is the radiative viscosity,

$$\Omega^2 \geq \frac{g^2}{2 c^2} R_c (\nabla_{ad} - \nabla_{rad})$$

where R_c is the critical Reynolds number. In other terms, we have

$$\Omega(\text{2-D turbulence}) \cong (R_c/8)^{1/2} \Omega(\text{baroclinic instability})$$

For $R_c = 2000$,

$$P(\text{2-D turbulence}) \leq 4 \text{ days}$$

$$V_{eq} \geq 20 \text{ km s}^{-1}$$

The order of magnitude is comparable to the estimate of Baglin (1972) : 50 km s^{-1}. The difference rests in fact on a different value of the critical Reynolds number .

This result suggests that all Am stars are slow rotators. There are, however, a few fast rotating stars (Abt and Moyd, 1973). This can be explained in the following way. According to Abt (1961, 1967) all Am stars belong to a binary system. For close binaries, as shown by Zahn (1977), "solid body roatation" can be achieved, the axial period of rotation becoming equal to the period of the orbital motion . In this cas, the velocity of the circulation is, with respect to the ratio $\chi = (\Omega^2 R^3 / G M)$, one order higher . We then have for the instability of the shear (for radiative viscosity):

$$\Omega > \left[\frac{g}{R}\right]^{\frac{1}{2}} \left[\frac{g R}{2 c^2} R_c (\nabla_{ad} - \nabla_{rad})\right]^{1/4}$$

$$P \leq 1 \text{ day}$$

$$V_{eq} \geqq 70 \text{ km s}^{-1}$$

We then can conclude to the possibility of two classes of Am stars, depending on the period of the binary. In wide binaries, the star behaves as if it were isolated; and only slow rotators can present the Am characterestics . In close binaries, on the contrary, almost solid rotation is present, the circulation is much more slow and relatively fast rotation can be present without preventing the gravitationnal settling to take place.

10.<u>Wolf Rayet stars</u> .
The time scale for turbulent diffusion mixing of large mass stars has been considered by Maeder (1982) taking the turbulent diffusion coefficient $D = Re^* \nu$, with $Re^* = 100$. If we take the expression of Re^* given by Zahn (1983) and the privileged distribution of angular velocity given by Spruit and Knobloch (1982) we obtain

$$Re^* \cong \frac{4 \, r \, g \, \mu}{(k_B/m_H)T}$$

which gives for M = 30 solar masses , $Re^* \cong 2.10^2$ in the middle of the star . We can the compare , with Maeder , the time scale of mixing, mass loss and nuclear main sequence time scale (fig.5). It is quite clear that the time of mixing is likely to be smaller than any other characteristic time for M > 50 M(sun) . Evolutionnary models for initial masses of 30 and 60 M(sun) have been computed by Maeder (1982) including mass loss, overshooting and turbulent diffusion mixing, leading to the evolutionnary tracks shown on fig.6.As a consequence of mixing, the 60 M(sun) star remains almost homogeneous and definitely turns to
the left of the HR diagram before the end of he core burning phase.

The frequancy of the stars generated by mixing is of the order of five times larger than when only mass loss is present . Maeder notices that mixing appears necessary in view of explaining the observed frequency of transition WN 7-9 stars.

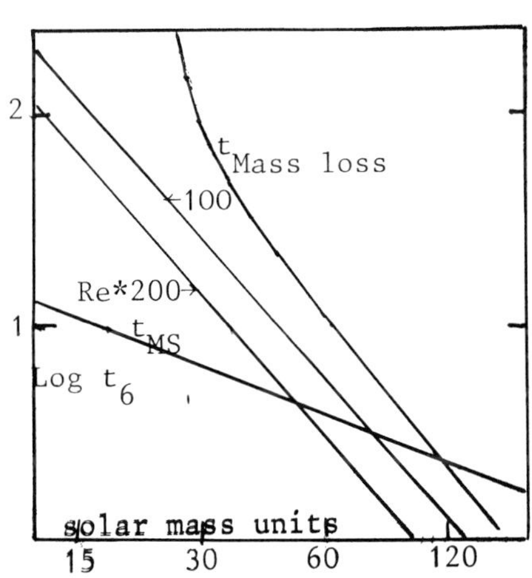

Fig. 5. The different time scales for large mass stars : mass loss, mixing with Re* =100 and 200, main sequence life time. Mixing is faster than other scales for M>50 solar masses.

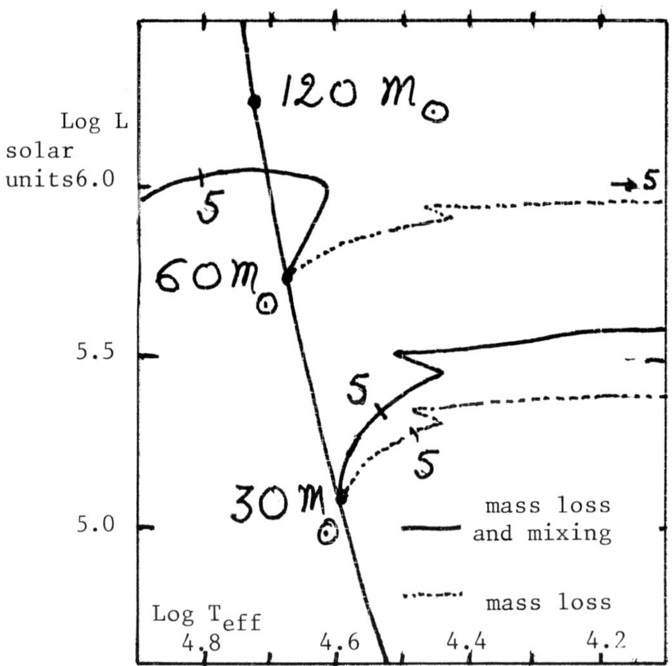

Fig.6. Evolutionnary tracks of massive stars with mass loss
(dotted line) and with mass loss and TDM (continuous line).

11.Conclusion

Turbulent diffusion mixing seem to be able to explain a variety
of stellar properties in a consistent way . It is not an ad hoc
hypothesis, but is based on a consistent physical model. We shall
conclude from the above given examples that turbulent diffusion
mixing (TDM) is there and that it is necessary to deal with it .

More work has to be done, to determine exactly the conditions
under which TDM appears and disappears, its possible connection
with the magnetic field, the dynamo mechanism, the transfer of
angular momentum, its anisotropy and its relation with meridional
circulation and differential rotation ...

For the time being, we shall consider that the existence of a
turbulent diffusion coefficient ,

$$D_{turb} = Re*$$

is established, with values of Re* in the range 30 to 200 .

BIBLIOGRAPHY

Abt H.A. 1961, Ap.J. Suppl.6,37.
          1967, Magnetic and related stars, Cameron Ed.,Monobook
Corporation, Baltimore, p.173.
Baglin A. 197Z, Astr. Astrophys.19,45.
Baglin A., Breger M.,Chevalier C., Hauck B., Lecontel J.M.,
Sareyan J.P., Caltier C., 1973, Astr. Astrophys.,23,221.
Baglin A., Morel P., This symposium
Bahcall J.N., Bahcall N., Ulrich R.K., 1968, Astrophys. L. 2.91
Bahcall J.N., Heubner W.F., Lubow S.H.,Magee N.H.,Jr, Merts A.L.,
Parker P.D., Roeshyai B., Ulrich R.K., Argo, M.F., 1980,
Phys.Rev Lett., 45, 945.
Bahcall J.N., Heubner W.F., Lubow S.H., Parker P.D., Ulrich R.K.,
1982, Rev.Mod.Phys., 54, 767.
Bienaylé O., Maeder A., Schatzman E., 1983, Astr. Astrophys., to
be published, preprint.
Biermann L., 1937, Astr. Nachr. 263, 185.
Boesgard A., 1976, P.A.S.P., 88, 353.
Breger M., 1979, P.A.S.P., 91, 5.
Cayrel R., Cayrel de Strobel G., Campbell B., Däppen W., 1983,
preprint .
Chapman S., Aller L.H., 1960, Astrophys.J., 132, 461.
Dearborne D.S.P., Eggleton P.P., Schramm D.N., 1976, Astrophys.J.
203,455.
Duncan D.K., 1981, Astrophys.J., 248, 651.
Ezer D., Cameron A.G.W., 1968 , Astrophys.L., 1, 177.
Faulkner D., 1966, Astrophys.J., 144, 978.
Geiss J., Buhler F., Cerutti H., Eberhardt P., Filleux C.H.,
1972, Apollo 16 Prel.Sci.Rep. NASA (SP 315).
Genova F., Schatzman E., 1979, Astr. Astrophys. 78, 323.
Herbig G., 1965, Astrophys.J.,141, 188.
Hopfinger E.G.,Browand F.K., Gagne Y., 1982, J. Fluid Mech.,
125,505.
Iben J., 1977, Lectures at Saas Fe, Société Suisse d'Astronomie.
Knobloch E., Spruit H.C., 1982, Astr.Astrophys., 113, 261
                         1983, preprint.
Lambert D.L., Dominy J.F., Sivertson S., 1980, Astrophys.J.,
235,114.
Latour J., Massaguer J.M., Toomre J., Zahn J.P.,1983, preprint.
Laurent C., 1983, ESO workshop on primordial Helium, P.A.Shaver,
D.Kunth, K.Kjär Ed. ESO Observatory , p.335.
Maeder A., 1975, Astr. Astrophys. 40, 303.
           1982, Astr. Astrophys., 105, 149.
Mazzitelli I., 1983, Private communication to R.Cayrel.
Merrill   , 1952,Astrophys.J. 116
Michaud G., Charland Y., Vauclair S., Vauclair G., 1976,
Astrophys. J. 210, 447.
Roxburgh I.W., 1978, Astr.Astrophys., 65,281.
Scalo M.S., Miller G.E., Astrophys.J., 239,953.
Schatzman E., 1945, Ann . d'Ap., 8, 143.

Schatzman E., 1969, Astr. Astrophys. 3, 331.
          1977, Astr. Astrophys., 56, 211
          1983, to be published.
Schatzman E., Maeder A., Astr. Astrophys., 96, 1.
Shaviv G., Beaudet G., 1968, Astrophys. L.,2, 17.
Shaviv G., Salpeter E.E., 1968, Phys. Rev. Lett., 21, 1602.
Shaviv G., Salpeter E.E., 1973, Astrophys. J., 184, 191.
Skumanich A., 1972, Astrophys. J., 171,565.
Spite F., Spite M., 1982, Astr. Astrophys., 115, 357.
Sweigart A.V.,Gross P.G., 1978, Astrophys. J. Supplt., 36, 405.
Sweigart A.V., Mengel J.G., 1979; Astrophys. J., 229, 624.
Toomre J., Zahn J.P., Latour J., Spiegel E.A., 1976, Astrophys.J.
207, 545.
Vauclair G., Vauclair S., Pamjatnick A., 1974, Astr. Astrophys.
31, 63.
Zahn J.P., 1977, Astr. Astrophys. 57, 383.
          1983, Cours à Saas Fe, Société Suisse d'Astronomie Ed.

DISCUSSION

Cox: Does the He lifting problem limit the value of Re^*? Can that energy come from the ^3He burning?

Schatzman: The problem of the transport of ^4He upwards from the place where it has been formed is certainly a difficulty. The amount of energy available in the unburnt ^3He is at least one order of magnitude larger than the energy which is necessary to lift ^4He.

Vanbeveren: I thought that already with standard evolution one is able to produce WR stars. I therefore do not understand your fourth conclusion where you state that "T.D. is a must" in order to form WR stars!

Schatzman: It is what I have understood from Maeder's paper: it is much easier to obtain Conti's scenario with mixing than without, and it is much easier to compel with the number of transition WN 7-9 stars.

Renzini: Since we don't have a theory of stellar mixing in non-fully convective regions, it is legitimate to parametrize the mixing and ask whether such parametrized models give a better account for the pertinent observations. However, when doing so all the isotopes which can potentially be affected by mixing should be explicitly included in model calculations. For instance, in the case of the Sun, besides ^7Li and ^3He, also ^6Li, Be, B and the CNO isotopes should be considered, and the results compared with the observational data. To my knowledge this has never been performed, and fitting one observable with one free parameter does not say much about what is going on.

<u>Tayler</u>: The internal conditions in the different types of stars which you have discussed differ greatly and whatever process gives rise to the turbulence must also be very different. Presumably you would be unhappy if your parameter Re^* was too similar from star to star.

<u>Schatzman</u>: There is no doubt that the turbulent diffusion coefficient varies from star to star and is certainly not a constant inside a star. However, if turbulence results from a feed-back mechanism, in which we are always very close to the instability limit, we can expect little variation from place to place and from star to star.

<u>Gurm</u>: For any such formulation in the continuum, the presence of the magnetic field, particularly in the case of Am and Sun, is going to affect the turbulence and mixing.

<u>Schatzman</u>: The presence of a magnetic field can certainly affect mixing in various ways. In the presence of a weak magnetic field, the viscosity of the medium will be increased. In fact, it can easily be recognized that turbulence by itself, being able to carry chemical contaminents, is unable to carry enough angular momentum. A factor of the order of 10 is actually missing, and the way of solving this discrepancy is to introduce a turbulent magnetohydrodynamic viscosity.
On the other hand, the effect which you mention, the rising of magnetic loops exists certainly in a convective zone: see the Sun! But this is related to the strong dynamo effect which is present in the solar convective zone. Do we have the same kind of instability taking place in radiatively stable regions? Is the differential rotation strong enough to generate unstable loops of magnetic fields? A weak mixing due to upward motion of flux tubes may eventually explain some of the difficulties of Am stars: in the absence of any mixing, some elements would be overabundant or very underabundant by a large factor, which is not the case. On the other hand, it does not seem to be possible to explain these difficulties with a unique turbulent diffusion coefficient, that the mechanism you suggest would provide. Weak mixing in the outer layers of Am stars must be there, but it must be a very delicate and complex mechanism.

<u>R. Cayrel</u>: Si je peux intervenir sur la question du rapport $^{6}Li/^{7}Li$ dans les Hyades je peux préciser que le ^{6}Li n'est certainement pas dominant et que l'on a une limite supérieure de ce rapport de l'ordre de 1/4.

<u>Andersen</u>: In your model of Li depletion, would you expect the $^{6}Li/^{7}Li$ isotope ratio to go down with the Li abundance, and is that observed in the Hyades?

<u>Schatzman</u>: The isotope ratio would go down with the lithium abundance, as the surface of ^{6}Li burning is closer to the boundary of the fully mixed region.

SOME REMARKS ON THE THEORY OF ROTATING STARS

L. Mestel
Astronomy Centre, University of Sussex

1. MIXING BY THE EDDINGTON-SWEET CIRCULATION

The principal question studied in Mestel (1953) was: under what
conditions will the E-S circulation through the radiative envelope
ensure that a Cowling-type main sequence star stays effectively
homogeneous? The E-S velocities v_Ω are of order $(L/Mg)(\Omega^2 r/g)$, so that
the circulation time t_{circ} is $\simeq t(\text{Kelvin-Helmholtz})/(\Omega^2 r/g)$, where
the bar indicates a mean value. The circulation advects nuclear-
processed material from the convective core, and so sets up horizontal
variations $\Delta\mu$ in mean molecular weight μ. The condition of
hydrostatic support requires corresponding horizontal temperature
variations; the consequent local breakdown in radiative equilibrium
yields a "μ-current" (analogous to the E-S "Ω-current") with
velocities v_μ estimated roughly as $(L/Mg)\,(\Delta\mu/\mu)$. The total velocity
v is the linear superposition $\underset{\sim}{v}_\Omega + \underset{\sim}{v}_\mu$, each constructed using first-
order perturbation theory. The μ-distribution that fixes $\underset{\sim}{v}_\mu$ is itself
determined by the circulation:

$$\partial\mu/\partial t = -\underset{\sim}{v}.\nabla\mu. \tag{1}$$

If the whole star is to be kept steadily mixed, then $\partial\mu/\partial t$
$\sim L/M(c^2/125) \sim 1/t_{nucl}$. It is known that near the convective
core the μ-current opposes the Ω-current, so that for steady mixing
to be possible we require $v_\mu < v_\Omega$, whence from (1)

$$\Delta\mu/\mu \simeq r/(v_\Omega\, t_{nucl}) \simeq (t_{KH}/t_{nucl})/(\overline{\Omega^2 r/g}) \simeq 10^{-2}/(\overline{\Omega^2 r/g}). \tag{2}$$

Note that in this problem the correct asymptotic form for $\Delta\mu/\mu$ <u>varies
inversely</u> with $\overline{\Omega^2 r/g}$. "Rapid rotation" in this context means: $\overline{\Omega^2 r/g}$
large enough to keep $\Delta\mu/\mu < \overline{\Omega^2 r/g}$, so that the assumption $v_\mu < v_\Omega$ is
vindicated. From (2), this yields

$$\overline{\Omega^2 r/g} > (t_{KH}/t_{nucl})^{1/2} \simeq 1/10. \tag{3}$$

(This should be contrasted with the estimate $\overline{\Omega^2 r/g} > t_{KH}/t_{nucl} \simeq 10^{-2}$
that follows from ignoring the μ-currents, and just comparing

513

A. Maeder and A. Renzini (eds.), Observational Tests of the Stellar Evolution Theory, 513–515.
© *1984 by the IAU.*

t_{circ} with t_{nucl}). The value (3) is high, but not so as to invalidate
the use of linear perturbation theory over the bulk of the star; and
in any case, the above meaning of rapid rotation is logically quite
distinct from "requiring second-order terms".

The accurate treatment requires solution of the angle-dependent
part of (1), subject to the appropriate boundary conditions $\Delta\mu = 0$
at the core surface and at the star's surface, and with $\partial\mu/\partial t$ again
given by assuming steady mixing. The critical angular velocity is
found to be given by

$$(\Omega^2 r/g)_{core\ surface} \simeq 1/30. \tag{4}$$

(An extension of the analysis to take account of Legendre coefficients
beyond the second will almost certainly increase this value somewhat).
If a star could rotate more rapidly than this, then it would evolve
up and to the left of the ZAMS. The Tassouls (1983) fail to find
a critical angular velocity because they impose from the beginning
$\Delta\mu/\mu \propto \Omega^2 r/g$. They state that the horizontal μ-gradient is just a
consequence of the rotational distortion of a zero-order radial
μ-gradient, whereas in the present problem both gradients are determined
by the circulation, and both decrease with increasing $\Omega^2 r/g$. However,
if the star rotates nearly uniformly, as is probable, then condition
(4) implies that $\Omega^2 r/g$ at the stellar surface is near unity, simply
because in a Cowling model the mean density within radius r decreases
from core to surface by a factor $\simeq 30$. Thus it seems very unlikely
that a star can in fact rotate at the necessary rate, but the
limitation is set by the conditions at the stellar surface; in the
region near the core where the μ-currents oppose the Ω-currents, linear
perturbation theory would suffice. Similar results hold for shell-
source stars (Mestel 1957).

If continuous steady mixing by the E-S circulation is ruled out,
how does a rapidly rotating star evolve? The original suggestion
was that the circulation would continue in the radiative envelope,
but would be deflected horizontally in a viscous boundary layer just
outside a "μ-barrier" surrounding the central nuclear-processing region.
However, it has proved difficult to construct such a model satisfying
all the boundary conditions, and so it was later suggested alternatively
that the circulation would be subject to "creeping paralysis"; the
Ω-currents would steadily build up a μ-distribution that leads
ultimately to $v_{\tilde{\Omega}} + v_{\tilde{\mu}} = 0$. The Tassouls in fact arrive at this picture
from their analysis. In a rapidly rotating star, the associated
radial μ-variations could have a significant effect on structure and
so on position in the H-R diagram.

2. GENERAL COMMENT ON THE THEORY OF ROTATING RADIATIVE ZONES.

Studies of non-magnetic rotating radiative zones implicitly impose
severe upper limits on the strength of any poloidal magnetic field B_p.
It was early recognized that to keep the E-S circulation flowing in a
chemically nearly homogeneous region, one needs something to off-set

the departures from uniform rotation caused by advection of angular
momentum by the circulation itself. Provided the Alfvén speed
$B_p/(4\pi\rho)^{1/2} \gg v_\Omega$, the non-uniform rotation will be kept small;
also, over the bulk of the star one can expect $B_p^2/4\pi\rho \ll \Omega^2 r^2$,
so that the implicit neglect of the magnetic disturbance to hydrostatic
and thermal equilibrium is justified. It is because $v_\Omega \ll \Omega r$ that a
field of less than 1 gauss can be simultaneously negligible for gross
hydrostatic equilibrium and overwhelmingly dominant in determining
the rotation field. Further, magnetic forces may be locally important
for hydrostatic and thermal equilibrium. I agree that the standard
E-S treatment is likely to yield large horizontal velocities in boundary
layers, which in a non-magnetic star must be subject to locally large
viscous forces. However, in a star with even a weak large-scale
magnetic flux distribution, continuity plus flux-freezing can build up
fields that locally exert forces comparable with the centrifugal force.
For example in their approximate self-consistent steady-state models of
rotating magnetic stars, Mestel and Moss (1977) and Moss (1983) find
that in the low-density surface regions, the magnetic field satisfying
the kinematic hydromagnetic equation (with finite resistivity necessarily
retained) must be included in the hydrostatic and thermal equations.

In this Symposium we are especially interested in anything that
can cause mixing of material and so affect the direction of stellar
evolution. Over the long time-scales available, even a weak large-scale
magnetic flux F in a rotating star may be significant. As an example,
I refer to studies of the "Internal dynamics of the oblique rotator"
(Mestel et al 1981; Nittmann and Wood (1981, 1983)). The motions
enforced on the star can be analysed into the Eulerian nutation, of
frequency $\omega \simeq \Omega(F^2/\pi^2 GM^2)$, and consequent divergence-free "ξ-motions"
that preserve hydrostatic equilibrium. The radial amplitude
$\tilde{\xi}_r \sim (\Omega^2 r/g)\lambda$, where λ is the local scale-height. Provided the
$\tilde{\xi}$-motions persist, an oblique, rapidly rotating star may experience
significant mixing, either directly through the oscillatory ξ-motions,
or indirectly through the destruction of the μ-field that would kill
off the E-S circulation.

REFERENCES

Mestel, L. 1953, M.N.R.A.S., 113, 716; 1957, Ap.J., 126, 550.
Mestel, L. and Moss, D.L. 1977, M.N.R.A.S., 204, 575.
Mestel, L., Nittmann, J., Wood, W.P. and Wright, G.A.E. 1981, M.N.R.A.S.
     195, 979.
Moss, D.L. 1983, preprint.
Nittmann, J. and Wood, W.P. 1981, M.N.R.A.S., 296, 491; 1983, Nature,
     301, 46.
Tassoul, J. -L and Tassoul, M. 1983, preprint.

ON THE TASSOUL APPROXIMATION SCHEME FOR DETERMINING THE STRUCTURE OF
ROTATING STARS

Ian W. Roxburgh
Department of Applied Mathematics, Queen Mary College, London

The method used by Tassoul and Tassoul assumes a "turbulent viscosity"
which is important in determining the dynamics, but is unimportant in
heat transport. This approximation is inconsistent, the argument is as
follows:
The rotation is expanded about uniform rotation in a series.

$$\Omega = \Omega_0 + \Omega_1 + \ldots, \text{ where } \Omega_1/\Omega_0 \ll 1 \tag{1}$$

The leading term Ω_0 drives a circulation, the Eddington Sweet circulation
with velocity

$$V_{circ} \simeq \varepsilon V_{thermal} = \varepsilon \frac{\kappa}{R} \tag{2}$$

where κ is the thermometric conductivity, R the radius of the star, and
ε the ratio of the kinetic energy of rotation to the gravitational energy
(or equivalently the ratio of centrifugal force to gravity).
The toroidal component of the equation of motion requires that the advection of angular momentum be balanced by the viscosity acting on the
differential rotation Ω_1

$$v_{circ} \, \Omega_0 R \simeq \nu \Omega_1 \tag{3}$$

hence

$$\frac{\Omega_1}{\Omega_0} \simeq \frac{V_{circ} R}{\nu} = \frac{\varepsilon \kappa}{\nu} = R_e \tag{4}$$

where R_e is the Reynolds number of the circulation. Thus the approximation
scheme is actually an expansion in the Reynolds number which must therefore be small.

$$R_e = \frac{\varepsilon \kappa}{\nu} \ll 1 \tag{5}$$

A. Maeder and A. Renzini (eds.), Observational Tests of the Stellar Evolution Theory, 517–518.

If now it is assumed - as is done by Tassoul and Tassoul that $\Omega_1/\Omega_0 = 0(\varepsilon)$, then it follows that $\kappa/\nu \sim 1$.

But $\kappa\nabla T$ is the energy flux in the form of radiation, $\nu\nabla T$ is the energy flux carried by the turbulence, hence with $\kappa \sim \nu$ it follows that the energy transport by the turbulence has a major effect on the structure of the star. Neglecting this turbulent flux is incorrect.

However it does not follow that $\Omega_1/\Omega_0 = 0(\varepsilon)$. It is in fact $0(\varepsilon/\sigma)$ where $\sigma = \nu/\kappa$ is the Prandtl number. Condition (5) must be satisfied for an expansion procedure of this form to be self consistent, hence

$$\frac{\nu}{\kappa} >> \varepsilon \tag{6}$$

Consequently the turbulent energy flux $\nu\nabla T$ is always greater than $\varepsilon\kappa\nabla T$, that is the variation of the radiative flux that drives the circulation. Now the turbulent viscosity ν is assumed to be produced by the instabilities of the differential rotation. The differential rotation Ω_1 varies from one part of the star to another, and in particular with latitutde. We must therefore expect that ν varies by a factor of order 2 over a spherical surface. In this case the variation of the turbulent energy flux is larger than the variation of the radiative flux that drives the circulation. To neglect the turbulent energy flux is therefore inconsistent.

ON TURBULENT MIXING

Ian W. Roxburgh
Department of Applied Mathematics, Queen Mary College, London

Several authors (myself included!) have suggested that turbulent
mixing takes place in some, if not all, stars, and in particular
that such mixing can explain the low solar neutrino flux. This
turbulence is thought to be caused by differential rotation produced by braking due to angular momentum loss in a stellar wind,
and/or to the effect of meridional circulation currents in re-
distributing angular momentum. Whilst such instabilities may exist
even in the presence of a stabilizing distribution of chemical
composition, they do not necessarily cause mixing. To be effective
in mixing, the energy available to the instability be it differential rotation or any other mechanism, has to be sufficient to
lift the helium rich matter in the interior of the star to the
outer regions. This requires

$$\left(\frac{E_{rot}}{E_g}\right) > \left(\frac{\tau_{th}}{\tau_{nuc}}\right)^{\frac{1}{2}} \sim \frac{1}{50} \tag{1}$$

where E_{rot} is the kinetic energy in rotation, E_g the gravitational
energy, τ_{th} the thermal time scale and τ_{nuc} the nuclear evolution time
scale of the star.
The rate of change of the gravitational energy E_g due to the conversion
of hydrogen to helium is of order

$$\frac{dE_g}{dt} = \frac{E_g}{\tau_{nuc}} \tag{2}$$

where τ_{nuc} is the nuclear time scale. For mixing we require

$$\frac{dE_{rot}}{dt} \geqslant \frac{dE_g}{dt} \tag{3}$$

A. Maeder and A. Renzini (eds.), Observational Tests of the Stellar Evolution Theory, 519–521.
© *1984 by the IAU.*

Now the energy in rotation is either decreasing due to a stellar wind,
or replenished due to meridional circulation. In this latter case we have

$$\frac{dE_{rot}}{dt} = \frac{E_{rot}}{\tau_{rot}} \; , \quad \tau_{rot} = \tau_{th} \frac{E_{rot}}{E_g} \tag{4}$$

Inequality (3) then becomes

$$\left(\frac{E_{rot}}{E_g}\right)^2 > \frac{\tau_{th}}{\tau_{nuc}} \; .$$

The ^{3}He instability

There is however one source of energy that is able in principle to overcome the stabilization due to chemical composition gradients. This is
the reservoir of ^{3}He that is built up away from the centre of the star.
Gough and his coworkers have shown that this could drive an overstable
g mode oscillation after some 3.10^8 years.
To demonstrate this, it is sufficient to note that in solar models some
$0.03 \, L_\odot$ is due to H $\rightarrow \, ^3$He reactions that do not complete burning to ^{4}He.
Thus the rate of energy release from burning this ^{3}He by bringing it to
higher temperature is of the same order. Thus mixing is energetically
possible provided

$$0.03 L_\odot > \frac{dV_g}{dt} = \frac{V_g}{\tau_{nuc}} \; .$$

But $V_g/L_\odot = \tau_{th}$, hence this condition becomes

$$0.03 > \frac{\tau_{th}}{\tau_{nuc}} \sim 2 \cdot 10^{-4}$$

Thus mixing is indeed energetically possible.

But will it take place, and if so, how efficient will it be? One possibility could be that the oscillation sets down at finite amplitude, the
^{3}He produced at one layer being carried down to regions of higher temperature where it is burnt. Since ^{3}He burning varies with temperature
like T^{20}, the oscillation will settle down with an amplitude of about
$0.05 \, H_T$ where H_T is the temperature scale height. This gives $\Delta r/R \sim 0.01$
and no effective mixing, just a mechanism for burning ^{3}He.
Alternatively, this oscillation could break down to subscale turbulence.
The Reynolds number of the oscillation is very large. In this case the
turbulence would diffuse ^{3}He to higher temperature regions at such a
rate to ensure that the ^{3}He created at lower temperatures is destroyed
at higher temperatures. Since the time scale of creation is about 3.10^8
years, and the distance scale for burning ^{3}He is 0.01 R, then we would
expect a turbulent diffusion coefficient of

$$\nu_t \sim \frac{(0.01\ R_\odot)^2}{3.10^8 \text{ys}} = 60.$$

Such a diffusion (corresponding to $R_e \sim 15$) will be effective in mixing the ^{3}He layer, but not effective enough to cause substantial mixing in the central regions.
But perhaps there is another way of converting the ^{3}He energy source into a more effective mixing mechanism!

EVOLUTION OF THE SUN WITH MIXING BY HYDRODYNAMIC INSTABILITIES

Wai-Yuen Law, E. Knobloch, H.C. Spruit
Max-Planck-Institut für Physik und Astrophysik
8046 Garching bei München, FRG

Following Schatzman and Maeder (1981) we compute the evolution of the sun with partial mixing by hydrodynamic instabilities. Instead of simply assuming a turbulent diffusion coefficient which is a constant multiple of the viscosity, we incorporate some of the properties of hydrodynamic instabilities. This puts limits on the amount of diffusion that can be obtained, and makes it dependent on time and position in the star.

The hydrodynamic instabilities considered are due to differential rotation, which in turn is due to the assumed spin-down of the star's convection zone by a magnetic stellar wind.

In the present study, we consider mixing by the ABCD (Knobloch and Spruit, 1983) instability, since it is the most generally occuring instability. Two different recipes were considered for its mixing efficiency, both of which imply a significant mixing effect. Since nothing is known about the actual nonlinear development of this instability, it is still possible that both recipes overestimate the mixing effect. In both recipes the rotation rate $\Omega_s(t)$ of the convection zone is specified as a function of time, in such a way that it agrees roughly with the observed rotation rates of G stars of various ages (Skumanich, 1972). Ω is assumed to be a function of r (the radial coordinate) and t only. The condition for marginal stability is then of the form

$$\partial\Omega/\partial r = f(r,\Omega,\nu), \tag{1}$$

where ν is the microscopic viscosity. In recipe A (the most optimistic recipe) it is assumed that strong turbulence results if $\partial\Omega/\partial r$ exceeds (1) even by a small amount, so that (1) is approximately satisfied all the time. $\Omega(r,t)$ can then be calculated directly from (1) and $\Omega_s(t)$. The angular momentum transport that is taking place then follows from $\Omega(r,t)$, and it is assumed that the effective diffusion coefficient for chemical constituents is the same as the effective angular momentum transport coefficient. In the more "realistic" recipe B it is assumed that the instability reaches such an amplitude that it is in a condition

A. Maeder and A. Renzini (eds.), Observational Tests of the Stellar Evolution Theory, 523–524.
© *1984 by the IAU.*

of marginal stability under the influence of the turbulence which it has created (ν replaced by ν_t in (1)).

For the models we calculated the neutrino flux N_ν, the quadrupole moment coefficient J_2, the lithium-depletion factor f_{Li} and the period spacing P_0 of g-modes. The results are:
i) A substantial reduction of N_ν occurs with recipes A and B, though not as large as the observed reduction.
ii) The lithium depeletion is much too high. This is due to the rapid initial spin-down rate which causes a high diffusion rate.
iii) J_2 is below its observed upper limit only for the most optimistic recipe A.
iv) The period spacing of g-modes is too high in models with a low N_ν.

From these results we conclude that:
1. In order to satisfy both the constraints on Li-depletion and J_2, a mechanism is needed which transports angular momentum much more effectively than Li. This is presumably a magnetic field.
2. The low neutrino flux can probably not be explained by turbulent mixing (see also Berthomieu et al., 1983).

REFERENCES

Berthomieu, G., Provost, J., Schatzman, E., preprint.
Knobloch, E., Spruit, H.C.:1983, Astron. Astrophys. 125, p. 59.
Schatzman, E., Maeder, A.: 1981, Astron. Astrophys. 96, p. 1.
Skumanich, A.: 1972, Astrophys. J. 171, p. 565.

DISCUSSION

Cox: What are the suggestions?

Spruit: The suggestions we make are:
1) There is something in the star which transports angular momentum without mixing, presumably a magnetic field. 2) The low ν-flux is probably not due to mixing. The reason for the first of these is that with a diffusion coefficient that is required for the lithium-depletion, not enough angular momentum can be transported out of the stars so J_2 is much too high. The reason for the second is the limited success of the recipes used in mixing in the center. Stronger constraints however follow from observations of the period spacing of g-modes, which allows only a small amount of mixing.

STELLAR EVOLUTION WITH TURBULENT DIFFUSION MIXING IN LOW MASS STARS AND
$^{12}C/^{13}C$ RATIO IN GIANTS OF THE FIRST ASCENDING BRANCH

O. Bienaymé[1], A. Maeder[2], E. Schatzman[3]
[1]Leiden Observatory, Postbus 9513, 2300 RA Leiden,
 The Netherlands
[2]Observatoire de Genève, Ch-1290 Sauverny, Suisse
[3]Observatoire de Nice, BP No. 252, F-06007, Nice, France

We consider stellar evolution in low mass stars (1-3 M_o) near the main
sequence with the hypothesis that mild turbulence is present within the
all star. Turbulent transport of the elements is modeled by diffusion
equations where the diffusion coefficient is chosen to be $D = R_e^* \nu$
where ν is the kinematical viscosity and R_e^* is a Reynolds number. We
consider the effects of the growth of the gradient of the mean molecular
weight on turbulence. The main consequences of diffusion on stellar
evolution are (1) an increase of the life time near the main sequence
and (2) a change of the radial distributions of chemical species (^{12}C,
^{13}C, ^{14}N, ^{16}O) (figure 1). The inhibition of the turbulence, when the
gradient of mean molecular weight reaches a certain critical value,
allows the evolution towards the red giant branch. When stars evolve
towards the giant branch, chemical species are dredged up to the surface.
At this stage models with and without diffusion, predict substantially
different surface abundances (in particular the $^{12}C/^{13}C$ and C/N ratios).
Comparison between models and the available data on giants during the
first dredge-up show that abundance anomalies can be explained if
turbulent mixing is present during the main sequence phase (figure 2).

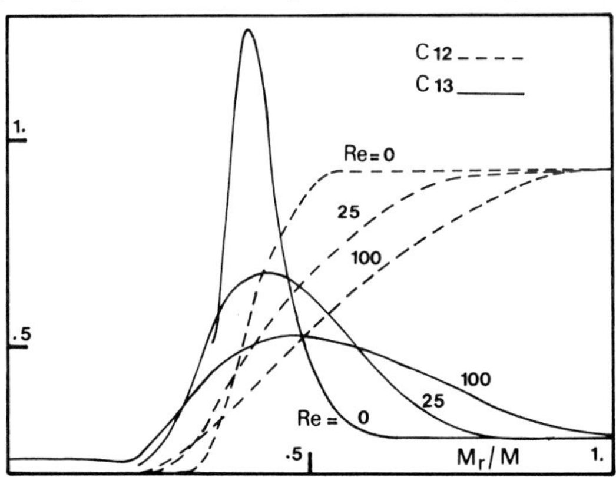

Figure 1. ^{12}C and ^{13}C abundances (in mass fraction) in 1.5 M models at
T = 1.69 10^9 years for different mixing strength R_e^* = 0, 25, 100.

A. Maeder and A. Renzini (eds.), Observational Tests of the Stellar Evolution Theory, 525–527.
© *1984 by the IAU.*

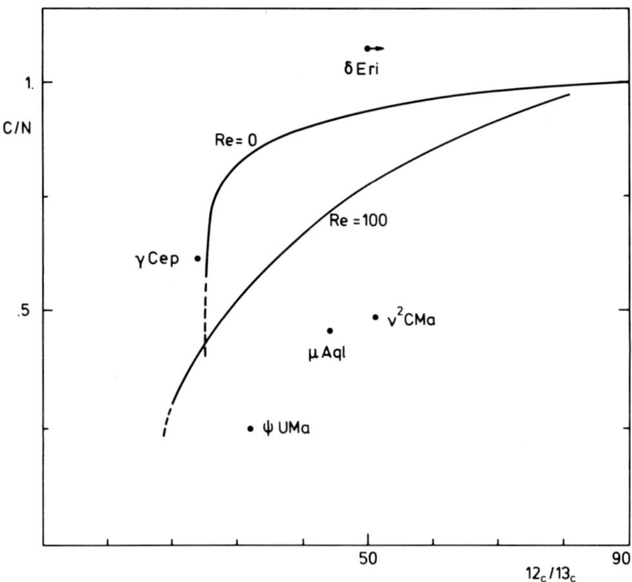

Figure 2. Evolution tracks from the end of the main sequence to the top of the AGB for our 1.5 M_O models. The tracks give the C/N ratios normalized to the solar value versus $^{12}C/^{13}C$ ratios that would be observed at the surface of 1.5 M_O stars for two hypothesis of mixing ($R_e^* = 0$ and 100). Stars plotted are subgiants observed by Lambert et al. (1981).

Bienaymé, O., Maeder, A., Schatzman, E.: 1983, Astron. Astrophys. preprint.
Lambert, D.L., and Ries, L.M.: 1981, Astrophys. J. <u>248</u>, p. 228.

DISCUSSION

R. Cayrel: If you put the right amount of turbulent diffusion to obtain the observed neutrino emission in the sun, by how much do you increase the life-time of the star on the main-sequence? (in relation with the age of globular clusters compared to the age of the Universe).

Maeder: Among the cases we have computed the one with $R_e^* = 100$ has an MS lifetime of 15 billion years. However, this value of R_e^* is probably too high; for a value of $R_e^* \leqslant 50$, more compatible with observations the increase of the MS lifetime is probably smaller than 30 %.

Cannon: I too would like to know how much the turbulence increases the lifetimes of near main-sequence stars, and whether this makes a discrepancy with the observed numbers of stars in old open clusters?

Maeder: (See answer to Cayrel for the first part of the question).
2) On the contrary, turbulent diffusion may improve the agreement with observations. In particular, stars where turbulent diffusion is always strongly active (due for example to their high differential rotation) may become blue stragglers, objects which have not received any satis- factory physical explanation so far. An important problem remains the determination of when in the course of evolution the μ-gradient is sufficient (for a given degree of differential rotation) to avoid turbulent diffusion and thus allow small mass stars to become red giants. For stars more massive than 1.5 $M_\odot$, there is no problem because the MS lifetime becomes smaller than the characteristic time of turbulent diffusion. But below 1.5 $M_\odot$ it seems that we could have two kinds of evolutionary sequences: one in which turbulent diffusion remains active throughout and leads to blue stragglers, the other in which diffusion is very mild and stopped some time by the μ-gradient which would lead to standard red giant evolution. All the comparisons with observations will of course depend on the exact criterion for the existence or non- existence of turbulent diffusion.

Tayler: Why did you take your turbulent viscosity to be some multiple of the radiative viscosity rather than some multiple of the product of speed of sound and pressure scale height which would depend differently on ρ and T?

Schatzman: The answer can be found in my review paper as well as in the references quoted.

LITHIUM ABUNDANCES IN LATE MAIN SEQUENCE STARS. THE CASE OF THE HYADES

A. Baglin and P.J. Morel
Observatoire de Nice
B.P. 139
06003   NICE CEDEX, France

It is well known that light elements like lithium are good indicators of the hydrodynamical behaviour of the outerlayers of the stars. As they are nuclearly destroyed at low temperature, i.e. close to the surface, their surface abundances reflect the nature of the transport process at work between the photosphere and the nuclear destruction region.

For intermediate temperature stars ($T_e \simeq 6\ 000$ °K) neither microscopic diffusion nor laminar circulation can account for the observed depletion (Fig. 1). A mild turbulence has been suggested by Vauclair, Vauclair, Schatzman, Michaud (1978). In low mass stars Strauss, Blake and Schramm (1976) concluded that a variable convective overshoot from 0 to 3 pressure scale height could explain the observed abundances.

Recent observations by Cayrel et al. (1983, this Symposium) on the Hyades cluster allow to reexamine the problem. They give abundances for a homogeneous set of main sequence stars of the same age in the interval $5\ 200 < T_e < 6\ 000$ °K.

I THE MODEL.

- Envelope models.
    Main sequence model envelopes have been used, computed with a code originally constructed by Latour (1970). The chemical composition is chosen as X = .73, Y = .25, Z = .02. Opacities are from the Los Alamos opacity library. Luminosities and effective temperatures are derived from the observations (Cayrel et al., 1983). The masses are deduced from a fit of the observed and theoretical Hyades main sequence (Perrin et al., 1977; Cayrel de Strobel, 1980). The age of the Hyades stars is taken as $8\ 10^8$ years (Patenaude, 1978).

- Transport processes.
    Two different transport processes are considered:
. A mild turbulence described by a turbulence diffusion coefficient

A. Maeder and A. Renzini (eds.), Observational Tests of the Stellar Evolution Theory, 529–531.

$D = R_e^* \nu$, as suggested by Schatzman (1977).

. An extension of the fully mixed region below the classical boundary of the convective zone : h = Ov.Hp (Hp pressure scale height). This extension is probably due to overshooting but we will not discuss this point here. R_e and Ov are the two parameters of the problem to be adjusted. It is assumed that they have the same values for all the models. Determinations of R_e^* have been made in several cases (initial region of the Sun, outer layers of A stars) and lead to $R_e^* \approx 30$, 100. The penetrative convection has been estimated using more or less refined description of the convective motions. The most recent computations (Latour et al., 1981 ; Hulburt et al., 1983) seem to indicate that Ov could not be larger than several tenth.

The concentration of Li is supposed to be constant from the surface fown to the lower boundary of the fully mixed layers. Below that it satisfies a diffusion equation in which the nuclear destruction is taken into account :

$$\rho \frac{\partial c}{\partial t} = \frac{\partial}{\partial z} (D\rho \frac{\partial c}{\partial z}) - K(\rho, T) \rho c$$

z geometrical depth, c Li concentration, $\rho = \rho(z)$, $T = T(z)$ density and temperature (given by the model),
D diffusion coefficient, $D = R_e^* \nu$, ν viscosity, $\nu = \nu_{mol} + \nu_{rad}$,
$\nu_{mol} = 2.21 \ 10^{-15} \ T^{5/2} \ \rho^{-1}$ (1+7X)/ (8 × lnΛ), from Spitzer (1962) for a mixture of ionized hydrogen and helium ln $\Lambda \approx 5$.
$\nu_{rad} = 6.728 \ 10^{-26} \ T^4/x^2$, $x(\rho,T)$ Rosseland mean opacity.
$K(\rho, T)$ is the nuclear destruction rate taken from Fowler et al. (1975).

It is assumed that no nuclear depletion occurs in the pre-main sequence stage and that the zero age main sequence lithium abundance is log N_o(Li) = 3.0.

II   RESULTS.

-1- In the hottest models the turbulent diffusion region is quite extended and a reasonable amount of overshooting Ov $\lesssim$ .5 does not affect significantly the surface abundance.

A first order determination of R_e^* is then possible from these models leading to 50 $\lesssim R_e^* \lesssim$ 70.

-2- In the coolest models the convective zone extends to hotter and denser regions and the distance between its lower boundary and the nuclear destruction region is much smaller. The Li surface abundance is then very sensitive to the amount of overshooting. Using the first determination of R_e^*, Ov can then be determined .3 $\lesssim$ Ov $\lesssim$ .4.

-3- This determination of Ov is then reintroduced in the hottest models to fix R_e^* and a value of $R_e^* \approx 70$ is obtained.

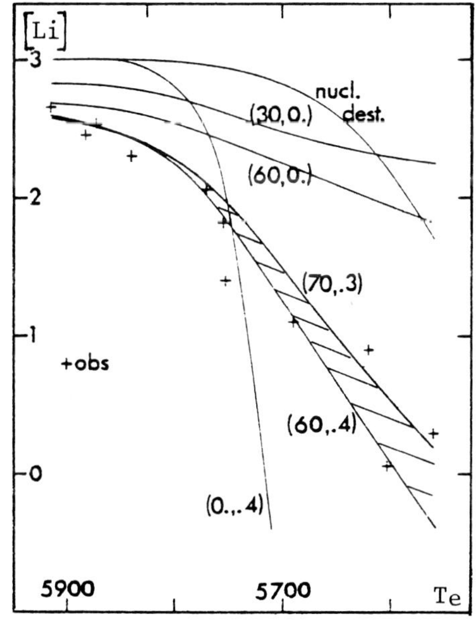

-4- The set of parameters $R_e^* \simeq 70$, $0v \simeq .3$ is then used all along the sequence. Figure 1 shows the fit with the observed values.

Figure 1.- Comparison of observed and predicted abundances. Models with turbulent diffusion only. Models with overshooting only. Models with $R_e^* = 60$, $0v = .4$. Models with $R_e^* = 70$, $0v = .3$.

III   CONCLUSIONS.

The slight depletion in the hottest models cannot be reproduced by any kind of reasonable overshooting; a mild turbulence is necessary. Note that this model is also consistent with the lithium abundances observed in the Sun. The fast depletion at low temperature is explained by the fast deepening of the convective zone with effective temperature. But a model containing only overshooting leads to a complete absence of lithium around $T_e \simeq 5\ 300$ °K due to the very sharp temperature dependance of the nuclear burning rate of lithium. The turbulent diffusion extends the region of nuclear burning and allows for the smooth decrease of Li abundance a low temperature. This preliminary study is being extended to other test observations as lithium abundances in Pop II stars, Be and B in Pop I stars.

BIBLIOGRAPHY.

Cayrel de Strobel G. (1980) IAU Symposium "Stars Clusters", p. 91.
Hulburt N., Massaguer J., Toomre J. (1983) Astron. Astrophys., in press.
Fowler W.A., Caughlan G.R., Zimmerman B.A. (1975) Ann. Rev. Astron. Astrophys. 13, 69.
Latour J. (1970) Astron. Astrophys. 9, 277.
Latour J., Toomre J., Zahn J.P. (1981) Astrophys. J. 248, 1081.
Patenaude M. (1978) Astron. Astrophys. 66, 225.
Perrin M.N., Heglesen P.M., Cayrel de Strobel G., Cayrel R. (1977) Astron. Astrophys. 54, 779.
Schatzman E. (1977) Astron. Astrophys. 56, 211.
Spitzer L. (1962) Physics of Fully Ionized Gases.
Strauss J.M., Blake J.B., Schramm D.N. (1976) Astrophys. J. 204, 481.
Vauclair S., Vauclair G., Schatzman E., Michaud G. (1976) Astrophys. J. 223, 567.

A RANDOM WALK APPROACH TO THE PROBLEM OF TURBULENT DIFFUSION
AND LITHIUM DESTRUCTION IN MAIN SEQUENCE STARS

Roger Cayrel

Observatoire de Paris

1. INTRODUCTION

It has been suggested by Schatzman (1977) that the empirical
lithium content/age relationship found by Herbig (1965) could be
explained by exchange of matter between the convective zone and
deeper layers, where the rate of lithium burning is higher, by
turbulent diffusion. The claim by Tassoul and Tassoul (1983) that
such a turbulence is necessary to obtain a self-consistent description
of stellar rotation and meridional circulation has given some momentum
to Schatzman's proposal, as well as other successes of the turbulent
diffusion hypothesis (Schatzman et al. 1981).

We describe here a way of computing the evolution of the lithium
content in the convective zone of a main sequence star, using a random
walk model for the mixing instead of the differential equation for
diffusion. The advantage of this approach is that the finite scale of
turbulent eddies can be taken into account. The burning rate of
lithium does increase so fast with temperature that it cannot be taken
for granted that the free path of turbulent motions is small compared
to the scale-height of the burning rate (this last being of the order
of $R_\odot$ /30 or a sixth of the pressure scale-height Hp$)$ H_e .

2. THE MODEL

We assume that the convective zone exchange matters with the
underlying radiative zone (RZ) at a rate of r = $\delta m / \delta t$.
We assume that the path of matter in the RZ is a random walk of step L
and time flight τ. The resulting velocity of turbulent eddies is
therefore v = L/τ and this sets also the rate r by:

$$r = 2\pi \rho_c R_c^2 v$$

ρ_c and R_c being respectively the density and the radius of the
bottom of the convective zone (CZ). After a time t = nτ three kinds
of paths have occured: A) some matter has left the CZ without returning to it; B) some matter has left the CZ and has returned to it

A. Maeder and A. Renzini (eds.), Observational Tests of the Stellar Evolution Theory, 533–535.
© *1984 by the IAU.*

after being exposed to a high destruction rate in the RZ; C) some
matter initially in the RZ has merged into the CZ. It is interesting
to note that systemic motions are a particular case of this approach
(when $t < \tau$) as well as microscopic diffusion (when $L \ll H_\omega$).

3. RESULTS

We give in Table 1 the results obtained for $t = 0.66 \times 10^9$ years
(age of the Hyades), $L = Hp/2$ and two values of the velocity v. The
computation was done using models supplied by W. Däppen ($1/Hp = 1.5$,
$Z = 0.016$, $X = 0.73$) and the rates of Fowler et al. (1975).

<u>Table 1</u>

Lithium depletion after 0.66×10^9 years for various
cases (in dex). L is taken equal to Hp/2.

Stellar mass (in $m_\odot$)	mere burning in the CZ	burning plus over shoot (to $0.30 \times Hp$)	random walk $v=1.6 \times 10^{-6}$ cm sec^{-1}	random walk $v=3.0 \times 10^{-6}$ cm sec^{-1}	Observed (Hyades)
1.0	-10^{-4}	-0.005	-0.24	-0.49	-0.24
0.9	-0.005	-0.362	-0.67	-1.13	-1.08
0.8	-0.042	-2.70	-1.08	-1.91	$\leqslant -2.7$

CONCLUSION

It can be seen from table 1 that the lithium depletion produced
by turbulence and computed using the random walk algorithm has a considerably smoother variation with mass that burning with or without
overshooting. The producted slope is in much better agreement with
observations. However, a moderate amount of overshooting would improve
the fit with the observations.

REFERENCES

Schatzman E., 1977 Astron. Astrophys. <u>56</u>, 211.
Schatzman E., Maeder A., Angrand F., Glowinski R., 1981 Astron.
Astrophys. <u>96</u>, 1.
Fowler W.A., Caughlan G.R., and Zimmerman B.A. 1975 Ann. Rev.
Astron. Astrophys. <u>13</u>, 69.
Tassoul M. and Tassoul J.L., 1983 Astrophys.J. <u>271</u>, 315.

DISCUSSION

Nissen: What is the relation between your velocity of random walk (10^{-6} cm sec^{-1}) and other hydrodynamical velocities in stellar interiors?

R. Cayrel: I have not yet attempted to connect my velocity with other hydrodynamical velocities in stellar interiors. But I know that in the tomorrow session somebody else will speak about velocities induced by differential rotation and we may then connect the two things. For the time being the velocity I gave has the meaning that with $\ell \simeq H_p$ such a velocity gives rough agreement between the predicted and observed lithium depletion in the Hyades.

AN OBSERVATIONAL TEST ON STELLAR INTERIOR MIXING: THE LITHIUM
DEPLETION IN TWELVE HYADES DWARFS

R. and G.Cayrel[1], B.Campbell[2] and W.Däppen[3].

1. Observatoire de Paris, France
2. Dominion Astrophysical Observatory, Victoria, B.C.
   Canada.
3. Institute of Astronomy, Cambridge, U.K.

1, 2, Visiting astronomer, Canada-France-Hawaii Telescope,
operated by the National Research Council of Canada, the Centre
National de la Recherche Scientifique of France, and the University
of Hawaii.

High signal to noise reticon spectra have been taken at the
CFHT coudé spectrograph for 11 Hyades main sequence solar type stars
and one Eggen's Hyades group star. A differential detailed analysis
is being carried out using as comparison star the integrated solar
spectrum obtained through the asteroid Ceres. The depletion of the
Li-abundance along the main sequence has been determined as the
depth of the convective zone deepens towards the right of the
H-R diagram. The equivalent width of the lithium doublet decreases
from 83 mÅ (VB 73) to below 3 mÅ (VB 46) at KOV - K1V. This represents an abundance reduction by a factor over 100 between G1V to
K1V. Theoretical work has been started to account for this slope
(see the posters of R.Cayrel and A.Baglin) in this conference and
the paper: 'The lithium abundance of Hyades main sequence stars'
by Cayrel, Cayrel de Strobel, Campbell and Däppen, 1984, ApJ
(in press).

A. Maeder and A. Renzini (eds.), Observational Tests of the Stellar Evolution Theory, 537.
© *1984 by the IAU.*

IV. RELATION WITH CHEMICAL EVOLUTION
OF GALAXIES AND COSMOLOGY

RELATIONS BETWEEN THE GALACTIC EVOLUTION AND THE STELLAR EVOLUTION

Jean AUDOUZE
Institut d'Astrophysique du CNRS, Paris-France
and Laboratoire René Bernas, Orsay-France.

1. INTRODUCTION

The question which has been raised in many chapters of this book is about the existence of constraints on stellar evolution coming from related topics like cosmology or in the case of the present chapter the chemical evolution of the galaxies. As it will be seen in this contribution it seems wiser to consider that chemical evolution of galaxies is indeed related to the problem of stellar evolution discussed here but is not going to provide as many constraints on it as one would expect. The purpose of this presentation is therefore to outline the principal relations between these two fields and to discuss the impact of some recent works on them.
After a quick definition of the galactic evolution and a summary of the basic ingredients (namely the abundances of the chemical elements observed in different astrophysical sites), the parameters directly related to the stellar evolution which govern the galactic evolution are outlined. They are the rates of star formation, the initial mass functions and the various nucleosynthetic yields. The "classical" models of chemical evolution of galaxies are then briefly recalled. Finally, the emphasis is made in three recent contributions interesting both the galactic evolution and the stellar evolution. They are (i) some prediction of the rate of star formation for low mass stars made from the planetary nebula abundance distribution (ii) the chemical evolution of C, O and Fe and (iii) some very recent work dealing with the chemical evolution of the galactic interstellar medium performed by Gusten and Mezger, 1983.

2. GALACTIC EVOLUTION : DEFINITION AND INGREDIENTS

In models of galactic evolution one tries to understand the variations and the evolution with the location and the time t of the functions $N_A(r,t)$ where N_A designates the observed abundances of the element A. In all the subsequent discussion it is assumed that the Universe has been formed about 15×10^9 years ago through primordial hot and dense phases (the Big Bang model).

A. Maeder and A. Renzini (eds.), Observational Tests of the Stellar Evolution Theory, 541–548.
© *1984 by the IAU.*

From the simple definition of the models of galactic evolution one
can realize already that the relation with stellar evolution is
very important since all elements with atomic mass $\geq$ 12 come from
stellar nucleosynthesis.
The ingredients of any model of galactic evolution are therefore
the abundances of the different chemical species which can be observed
in old objects (halo stars, globular clusters), in the population
I stars of various ages, in the interstellar medium which characterizes
the composition of the most recently formed material. It is well
known that the solar system composition of age 4.55 10^9 years is
of invaluable value in building up such models.
Further ingredients directly related to these abundances like the
photometry in different colors of many stars (see eg. Twarog 1980)
the age metallicity relations, the stellar metallicity distributions,
the determination of isotopic ratios in many locations or the discovery
of abundance gradients along the disks of spiral galaxies play also
an important role in the selection of galactic models.

3. THE THREE MOST IMPORTANT PARAMETERS OF THE GALACTIC EVOLUTION

The three major parameters governing the chemical evolution
of galaxies are
(i) the initial mass function of stars (IMF) generally described
by a relation similar to that designed by Salpeter

$$\Phi(m) = \varphi(m) \, dm \ \alpha \ m^{-(1+x)} \, dm$$

where m is the mass of stars in unit of solar mass and the exponent
x taken as equal to 1.35 is deduced from stellar population analysis
(see eg. Tinsley 1980 for a recent review of this topic).
(ii) the rate of stellar formation (SFR) given by a relation similar
to that sketched by Schmidt :

$$\psi(t) X \sim \mu^n$$

where $\psi(t)$ is the rate of stellar formation as a function of time; μ is
the gas density of the considered galactic region ($\mu = m_{gas}/m_{tot}$ and
n is an exponent the value of which ranges from 1 to 2.
(iii) the nucleosynthetic yields y_i which represent the amounts
of any given nuclear species i which are released by stars and evaluated
per unit of mass locked into stars

$$y_i = \frac{\underset{i \neq j}{\Sigma} \ X_j \int_{m_L}^{m_U} Q_{ij}(m) \, \varphi(m) \, dm}{I - \underset{j}{\Sigma} \ X_j \int_{m_L}^{m_U} Q_{ij}(m) \, \varphi(m) \, dm}$$

In this expression, the X_j terms are the mass fractions of the elements
j and the $Q_{ij}(m)$ are the mass fractions of the stars m which are
transformed from j to i.

One should recall now the important distinction between primary and secondary elements. Primary elements are those which can be produced in principle in a pure hydrogen stars (examples: C, O, Fe...). For those elements the yields are simply proportional to their observed abundances.By contrast, the secondary elements can only be synthetized from primary elements (examples: ^{13}C, part of ^{14}N, the s process elements) their yield is proportional to the metallicity deduced from the abundances of the primary elements. These yields are also related to the constraints coming from the nucleosynthesis. Some of them are for instance :

(i) the depletion of D during the galactic evolution by factors at least equal to 2,

(ii) the significant enrichment into ^{7}Li consequence of the recent determinations of the Li abundance concerning halo stars performed by Spite and Spite (1982),

(iii) the differences in the evolution of C, N, O and Fe which will be discussed later

(iv) an interesting difference between the behaviour of the abundance of the s process elements (like Ba) and that of the r process elements (like Eu) as noticed by Spite and Spite (1978): the [Ba/Fe] ratio is proportional to [Fe/H] which shows that the s process elements are of secondary origin or that they should be formed mainly in low mass stars. By contrast [Eu/Fe] is in first approximation independant of [Fe/H] which would mean that the r process elements are primary (which is an astonishing conclusion since the seed for their formation is Fe itself) or that they are only formed in very massive stars.

4. THE MODELS OF GALACTIC EVOLUTION.

The classical models dealing with the chemical evolution of galaxies have been reviewed in many articles (Pagel and Patchett 1975, Audouze and Tinsley 1976, Tinsley 1980). It might be sufficient here to remind the reader with the main features of the so called "simple" model and show how the model builders circumvent the drawbacks of this approach.

4.1. The features of the "simple" model and its drawbacks.

In the so called "simple" models which attempt to sketch the chemical evolution of one closed galactic zone, one adopts the following hypothesis

(i) the considered zone is well mixed and closed which means that there is no further addition (or ablation) of gas by infall, inflow or sweeping mechanisms,

(ii) at time t=0 there are no metals and no stars, the gas density is $\mu=1$ (only gas),

(iii) the rate of star formation follows the Schmidt low i.e. $dS/dt= \mu^{n}$ with $1 \leqslant n \leqslant 2$

(iv) the initial function is assumed to be constant with respect to time and follow the Salpeter law $\varphi(m)dm = \zeta(x-1) m^{-x} dm$

(v) the stars are assumed to evolve more rapidly than the galactic

zone itself. This hypothesis is called the instantaneous recycling approxim-
tion. Another way to express it is to say that the individual stellar
lifetimes are assumed to be negligible which is not the case for
low mass stars which last also very long.
In these conditions, the metallicity Z evolves with time according
to the relation $Z = p \ln(\frac{1}{\mu})$ where p is the yield and μ is related
to time t either in $e^{-t/\tau}$ for n=1 or in $1/(1+t/\tau_0)$ for n=2.
The simple model presents two major difficulties
(i) the metallicity increases too much with time while the observations
show a plateau for the metallicity of disk stars,
(ii) it predicts too many stars of low metallicity

4.2. In order to alleviate these two difficulties, many proposals
have been advanced,
(i) the assumption of infall and/or inflow of external material
which leads to a plateau in the metallicity
(ii) a varying initial mass function either by assuming that the
initial mass function is more devoided in high mass stars now than
in the past. Another way to express the same assumption is to speculate
about the existence of a first generation of massive stars which
leads to a prompt initial enrichment of metals. (No star with a
zero metallicity has ever been observed so far). With J.L.Puget
and G. Malinie ,   we are currently investigating models in
which the upper and lower limit of the initial mass function may
vary with time. It is assumed that the upper limit of the IMF may
decrease (by being inversely proportional to the metallicity) while
the lower limit could have been as high as $2M_\odot$ when the metallicity
was very low and decrease quickly after the release of some metals
by the first generation stars.
All these approaches solve effectively the two difficulties outlined
before which are suffered by the "simple" model.
At this point   one should be easily convinced that many relations
exist obviously between the galactic evolution and the stellar evolution. Stars are responsible for the metal enrichment of the interstellar
gas; their mass govern not only their evolution but also the evolution
of the galactic zone to which they belong.

5. A FEW SPECIFIC PROOFS OF THE CLOSE RELATION BETWEEN GALACTIC
   AND STELLAR EVOLUTION.

   Many contributions would deserve to be mentioned to show the
strong relation between the galactic and the stellar evolution.
I have selected here three recent works dealing respectively with
an attempt to deduce the stellar formation rate of low mass stars
from planetary nebulae, with the relative production and chemical
evolution of C, O and Fe and with a very recent model concerning
the chemical evolution of the interstellar medium.

5.1. Star formation rates deduced from planetary nebulae
This is an attempt proposed by G. Malinie, myself and M. Dennefeld
to deduce this parameter from the abundance distribution of a sample
of several planetary nebulae. Planetary nebulae possess two interesting

features : they are numerous,they are bright and they have low mass
progenitors. The procedure that we propose is the following:
(i) given a good sample of planetary nebulae construct the histogram
N(Z) here Z is taken as the oxygen abundance,
(ii) use a Z(t) distribution like the one determined by Twarog (1980)
(iii) deduce N(t) and from that the fraction of planetary nebulae
with progenitors of age larger than t

$$F(t) = \frac{\int_T^{t_G} \varphi(m_t) \; \psi(t) \; dm_t/dt \; \; dt}{\int_t^{t_G} \varphi(m_t) \; \psi(t) \; dm_t/dt \; \; dt}$$

where t_G is the age of the galaxy, $\psi(t)$ is the stellar formation
rate $= \varphi_\odot \exp(\psi t/\tau)$, $\varphi(m) \sim m^{(1+x)}$ and the lifetime of stars is
$t = m_t^b$ with b=3

$$F(t) = \frac{\int_T^{t_G} t^{x/b - 1} \; e^{t/\tau} \; \; dt}{\int_{t_{min}}^{t_G} t^{x/b - 1} \; e^{t/\tau} \; \; dt}$$

From the PN samples studied so far $\tau = 4^{+2}_{-1} \; 10^9$ years, which means
a fairly significant decrease of the stellar formation rate with
time if this method is found to be applicable.

5.2. The relative production and chemical evolution of C, O and
     Fe.

Clegg, Lambert and Tomkin (1981) have observed a sample of about
20 F and G main sequence stars with a high resolution reticon and
for which 0.9 < [Fe/H] +0.4. They found :

[C/H] = (0.84 ± 0.08) [Fe/H] - (0.02 ± 0.03)
[N/H] = (1.31 ± 0.25) [Fe/H] + 0.07 (±0.07)
[O/H] = (0.52 ± 0.07) [Fe/H] + (0.03 ± 0.03)
[S/H] = [Fe/H]

Clegg et al (1981) interpret their results by arguing that O is
less deficient than Fe in metal poor stars because it might be produced
by heavier stars.
Two other analyses of this observational set of data have been proposed:
 Twarog and Wheeler (1982) have proposed a model of chemical evolution
(the simple model with infall) adopting the nucleosynthetic prescriptions of Arnett (1978) with constant IMF and yields. With a production
rate of 2.4 Fe$_\odot$ pc^{-2}Gyr^{-1}, they overproduce O, C, Ne and Mg relative
to Fe. In order to solve this overproduction, they propose a change
in the IMF slope and a upper limit in the IMF fairly low at 25-40M$_\odot$.
 Chiosi and Matteucci (1984) have very recently proposed an alternative
and argue that the nucleosynthetic yields may vary during the galactic
evolution.
Their nucleosynthetic prescriptions are the following: for stars with
1 < m < 9 they adopt those of Renzini and Voli (1981): they assume

that these stars are important sources of He, ^{12}C, and primary N. The stars with masses from 9 to 120 M$_\odot$ produce C, O and Fe while the heavier stars would produce only O <u>but not Fe</u>.
In their galactic model, they therefore make use of varying nucleosynthetic yields and possible significant stellar mass loss rates during the stellar evolution. With those assumptions they manage to explain the difference of behaviour of the O and Fe abundances.

5.3. A very recent model of the chemical evolution of the interstellar medium

Güsten and Mezger (1983) have made a very interesting proposal to solve the difficulties encountered by the simple model. They propose
(i) a continuous infall of unprocessed material,
(ii) a bimodal star formation. They assume that in the galactic arms and the galactic interarms the star formation is different. They quantify this effect by assuming that the lower limit of the IMF in the arms is 2 M$_\odot$ while it is 0.1 Mo in the interarms. This assumption is fairly similar to that of Malinie, Puget and I quoted above.The net result of this last assumption is to increase the yield in the arms.
As a result their model provides a good account of the age metallicity relation like that of Twarog (1980); it also reproduces well the gradients of abundances observed along the galactic disk from the external up to the central regions. Finally since the yields increase for a given star formation rate, these rates can be lower and reproduce well the Lyman continuum photon production rate.

6. FINAL REMARKS.

From the above developments , the reader can easily be convinced that models of chemical evolution of galaxies are constrained by stellar evolution models. <u>The reverse is highly debatable</u> because there are too many free parameters for the present amount of relevant observations. In any case progress made in one of these fields benefits eventually to the other.

REFERENCES

Arnett,W.D.,1978,ApJ.,<u>219</u>,pp.1008.
Audouze,J. and Tinsley,B.M.,1976,   Ann. Rev. Astron. Astrophy.,<u>14</u>,43.
Chiosi,C. and Matteucci,F., 1984 , in Formation and Evolution of
     galaxies and large structures of the Universe, eds J. Audouze
     and J. Tran Thanh Van, Reidel, Dordrecht , p. 401.
Clegg,R.E.S., Lambert,D.L. and Tomkin,J., 1981, ApJ. <u>250</u>,pp.262.
Güsten,R. and Mezger,P.G., 1983, preprint.
Malinie,G., Audouze,J. and Dennefeld,M., 1984, Astron. Astroph.,
     submitted.
Pagel,B.E.J. and Patchett,B.E., 1975, M.N.R.A.S., <u>172</u>,pp.13.
Renzini,A. and Voli,M., 1981, Astron. Astroph, <u>94</u>,pp.175.
Spite,M. and Spite,F., 1978, Astron.Ap.,<u>67</u>,pp.23.

Spite,F. and Spite,M., 1982, Astron. Astroph., 115, pp.357.
Tinsley,B.M., 1980, Fund. Cosm. Phys., 5, pp.287.
Twarog,B.A., 1980, ApJ. 242, pp.242.
Twarog,B.A. and Wheeler, J.C., 1982, ApJ., 261.pp.638.

ACKNOWLEDGEMENTS

I thank A. Maeder for having given me the opportunity to present
this contribution.

DISCUSSION

G. Cayrel: Could the introduction of initial mass function in the models
of evolutionary scenarios avoid the formation of too many stars of low
metallicity?

Audouze: There are many scenarios which have been proposed so far to
avoid the formation of too many stars of low metallicity: they all belong
to the family of models assuming that the initial mass function varies
with time: In all these scenarios one assumes that when the metallicity
was lower more massive stars were formed. They all succeed in solving the
so-called F-G dwarf problem.

Schatzman: Source of information for the observed abundance of Europium?
It should be noticed that Europium abundance is very sensitive to radiation pressure and mixing (or non-mixing). Did you include these effects
in your discussion?

Audouze: (i) The observed abundances of Europium discussed here have been
determined by Spite M. and Spite F. (Astron. Astrophys. 67, 23, 1978).
(ii) In this discussion the only effects which have been taken into
account are the nucleosynthetic ones, i.e. the fact that Europium is an
r-process element.

Renzini: Concerning the use of PNe as tracers of the past SFR, one should
be aware that the PN lifetime is likely to be a rather sensitive function
of the mass of the precursor. So this should produce a systematic error
in the estimated SFR.

Audouze: I would agree with you that this is a possible problem in the
type of analysis we try to propose although it is not proved at all that
such an effect (the sensitivity of the PN lifetime with the mass of precursor) does really exist. Moreover, one could, as an exercice, solve
this problem if it exists by making the most reasonable assumption that
the PN lifetime is a decreasing function of the mass of the precursor.

<u>Vanbeveren</u>: Is the last model you proposed, where it is suggested that more massive stars are formed when the metallicity is low, not somewhere conflicting with the observations of R. Humphreys presented Wednesday? I have the impression that the lack of massive stars in the SMC could suggest that the lower the metallicity, the lower the probability that more massive stars are formed!

<u>Audouze</u>: The apparent difficulty which worries you can be in fact easily alleviated by noticing that the stars observed by Roberta Humphreys and other participants like Peter Conti are much more massive ($\sim 30-40\ M_\odot$) than those which are needed to solve the problem of the lack of low metallicity stars (a few $M_\odot$). For instance J. Silk argues that when the metallicity is very low the lower limit of the IMF might be 2 $M_\odot$ instead of .01 $M_\odot$ (or so) when the metallicity is normal.

<u>Weidemann</u>: If the new initial-final mass relation which I presented (Weidemann and Koester, Astron. Astrophys. <u>121</u>, 77, 1983) is correct, the amount of mass locked up and the yield becomes very different in the important mass range 3 to 8 $M_\odot$, with supernova production only beyond 8 $M_\odot$, and much more unprocessed material returned to the ISM.

<u>Maeder</u>: The initial mass limit for black hole formation is also a critical one in this context as above this limit the stars contribute to the galactic enrichment only by their winds, while most of their remaining mass is removed from our visible universe. We do not know where this limit lies and this may affect considerably the models of galactic chemical evolution.

THE RELATION BETWEEN STELLAR EVOLUTION AND COSMOLOGY

R. J. Tayler
Astronomy Centre, University of Sussex, UK

ABSTRACT

Observations of star clusters combined with the theory of stellar evolution enable us to estimate the ages of stars while cosmological observations and theories give us a value for the age of the Universe. This is the most important interaction between cosmology and stellar evolution because it is clearly necessary that stars are younger than the Universe. Stellar evolution also plays an important rôle in relating the present chemical composition of the Universe to its original composition.

1. INTRODUCTION

Although the title of my review is quite general, I shall restrict myself mainly to discussing the relation between stellar evolution and the big bang cosmological theory because there is such a good qualitative agreement between the hot big bang theory and observations. There are two principal ways in which stellar evolution and cosmology interact within the framework of the big bang theory. The first is concerned with the initial chemical composition of galaxies and the second with the ages of objects in the Galaxy. In the first case we need to understand the evolution of stars of very high mass as well as the processes which lead to their formation. In the second case we are concerned with the evolution of stars of solar mass and a little less.

The initial chemical composition of galaxies may differ from the primaeval composition of the Universe if some stars were formed before galaxies. In the hot version of the big bang theory, the primaeval composition is about three parts hydrogen to one part helium by mass with no heavy elements. In the cold version the composition is pure hydrogen. No stars have been discovered without heavy elements. Were the heavy elements in the oldest observed stars produced in Population III stars which preceded galaxy formation or in a first generation of galactic stars, which are no longer observed? In the cold big bang with

A. Maeder and A. Renzini (eds.), Observational Tests of the Stellar Evolution Theory, 549–562.
© *1984 by the IAU.*

no primaeval helium, there must be pregalactic Population III stars which
produce not only the heavy elements but also the helium which we deduce
to be present in the oldest stars. In either case we are interested in
the process of star formation in the early Universe and we are
particularly concerned with those stars which were sufficiently massive
to evolve and to release processed matter in the time available before
or during galaxy formation. It is possible that Population III contained
a significant number of stars which were much more massive than any
stars observed today.

The second main point of interaction concerns stars which were
formed essentially the time of galaxy formation and which are at an
interesting stage of their evolution today. Stellar evolution studies
can give us an estimate of the ages of such stars, specifically the stars
in globular clusters, and, if the standard cosmological theory is to be
correct, the ages of these stars must be less than the age of the
Universe as determined from Hubble's constant and the deceleration
parameter. In particular the ages of all objects must certainly be less
than the reciprocal of Hubble's constant. If that is not the case, a
non-standard cosmology must be invoked, possibly involving a non-zero
cosmological constant. An earlier apparent discrepancy between the age
of the Universe and the age of the Sun was one factor in the development
of the Steady State theory. If it were possible to make a direct
observation of the helium abundance in old unevolved stars, one would be
able to check whether it is comparable with that supposed to have been
produced in the hot big bang. As there are no such observations,
stellar evolution theory must also be used to obtain an estimate of the
helium abundance. In fact the determination of age and helium abundance
is part of a single process.

2. COSMOLOGICAL BACKGROUND

I shall discuss observations in terms of the big bang theory, in
particular the hot big bang theory in which a primaeval composition of
about three parts hydrogen to one part helium by mass is attributed to
nuclear reactions in the first few minutes and in which the cosmic
microwave radiation is another indicator of a hot origin. However most
of my remarks will be equally applicable to the cold big bang model in
which both the helium and the microwave radiation must be attributed to
stars formed before galaxies.

In any version of the big bang theory in which the cosmological
constant, Λ, is zero, there is a simple relation between Hubble's
constant, H_0, the deceleration parameter, q_0, and the age of the
Universe, t_0. Instead of q_0 one can use the present mean density of the
Universe, ρ_0, which for $\Lambda = 0$ is related to H_0 and q_0 by

$$\rho_0 = 3H_0^2 q_0 / 4\pi G \ . \tag{2.1}$$

At present there are not really reliable values of any of H_0, q_0, ρ_0.

q_0 is difficult to measure because of the lack of reliable *standard candles* at large distances, while ρ_0 must include not only visible matter and matter which is detected by its gravitational influence but also hidden matter such as low mass elementary particles. H_0 is generally believed to lie in the range

$$50 \leqslant H_0/\text{kms}^{-1} \text{ Mpc}^{-1} \leqslant 100 \ , \qquad (2.2)$$

although more extreme values cannot be totally excluded. Whatever is the value of q_0, t_0 is less than $t_H(\equiv 1/H_0)$ and, using (2.2),

$$10^{10} \leqslant t_H/\text{yr} \leqslant 2 \times 10^{10} \ . \qquad (2.3)$$

The higher the value of q_0, the lower is the value of t_0/t_H. If the Universe is just closed, $q_0 = 1/2$ and $t_0 = 2t_H/3$.

Because this is a symposium on stellar evolution rather than on cosmology, it is not my present concern to discuss the accuracy of estimated values of H_0, q_0 and t_0. Instead I shall consider the age of the Galaxy and in particular of the globular star clusters and ask whether it is compatible with suggested values of these quantities. The most obvious problem arises if H_0 is so large that the ages of the globular clusters are found to be greater than t_H. There is then a clear contradiction between the predictions of stellar evolution theory and the standard cosmology. However, it is also desirable that the age of the Universe should not be too much greater than the ages of the oldest objects in the Galaxy, because it is not easy to see how galaxy formation and star formation can be delayed for an arbitrarily long time. If the age of the Universe does appear to be less than that of the globular clusters, it may be necessary to consider theories with positive Λ in which it is possible for t_0 to exceed t_H. There are possible objections to such theories which we shall discuss briefly below.

It should be noted that for t_0 to exceed t_H, Λ must necessarily exceed $4\pi G\rho_0$. For given value of H_0 the age is increased by increasing Λ or by decreasing ρ_0 or by doing both simultaneously. A positive value of Λ can only give an age in excess of t_H if q_0 is negative; this is a necessary but not sufficient condition. For non-zero Λ equation (2.1) is replaced by

$$q_0 = (4\pi G\rho_0 - \Lambda)/3H_0^2 \ , \qquad (2.4)$$

so that $q_0 < 0$ for $\Lambda > 4\pi G\rho_0$. Table 1 shows the present age of the Universe in terms of t_H for various values of the two parameters $\Lambda/8\pi G\rho_0$ and $8\pi G\rho_0/3H_0^2$. These specific results were obtained by Bukhari (1983) although similar results have been obtained by many previous authors. It should be noted that the second column is the ratio of ρ_0 to the density required to close the Universe in the $\Lambda = 0$ case.

Table 1

$\Lambda/8\pi G\rho_0$	$8\pi G\rho_0/3H_0^2$	t_0/t_H	q_0
3.75	0.4	1.28	-1.3
2.86	0.7	1.36	-1.65
10.0	0.1	1.38	-0.95
20.0	0.05	1.57	-0.975
6.0	0.25	1.70	-1.375
3.64	0.55	1.89	-1.725

It can be seen from Table 1 that ages reasonably in excess of t_H can be obtained if q_0 is of order -1 or -2. de Vaucouleurs (1983) has for example suggested that $\Lambda/8\pi G\rho_0$ = 16.4 and $8\pi G\rho_0/3H_0^2$ = 0.07 is consistent with H_0 = 100 kms^{-1} Mpc^{-1} and the accepted ages of globular clusters. It is necessary to ask whether the values of ρ_0 and q_0 listed are reasonable. The minimum acceptable value of $8\pi G\rho_0/3H_0^2$ is probably between 0.05 and 0.1 but the actual value could be considerably higher. In particular, if the inflationary model of the very early Universe (see e.g. Guth 1981) is correct, the value should probably be unity. In such a case even higher values of $-q_0$ would be required to give ages much in excess of t_H. Is it in fact reasonable to suppose that q_0 is negative? Hoessel et al.(1980) by assuming that first ranked galaxies in clusters are standard candles obtained the formal result $q_0 = -0.55 \pm 0.45$. They pointed out that corrections are needed both because isolated galaxies would have been more luminous in the past and because the galaxies which they studied might have increased their luminosity by merger with smaller galaxies. Two corrections in the opposite direction, which they estimate to be $1 \leqslant \Delta q_0 \leqslant 1.7$, are required to the observed q_0. It does not therefore seem possible to rule out values of q_0 comparable to those in Table 1. In contrast, a similar investigation by Kristian et al.(1978) gave q_0 = 1.6 $\pm$ 0.4 so that the position is far from clear.

It is next necessary to ask whether there are any other possible objections to a finite value of Λ. There are first some observational constraints. When Λ is zero, the cosmological scale factor R(t) increases smoothly from zero to infinity if the Universe is open or increases smoothly from zero to some R_{max} and then decreases to zero again if the Universe is closed. If $\Lambda > 4\pi G\rho_0$, there is a point of inflexion in the R(t) curve at some past time and there may have been a considerable period in which R changed only slightly. This means that there is a plateau in the relation between redshift and time since, for radiation received now which was emitted at time t,

$$R_0/R(t) = 1 + z, \tag{2.5}$$

where z is the redshift. There might therefore be expected to be a bunching of objects observed at the plateau redshift (see e.g. Tytler 1981). The observations might be capable of ruling out the model or making it **implausible** if the plateau redshift has a value for which observations are easy. In addition, if the Universe is closed and if Λ

is so large that the age of the Universe is significantly greater than t_H, it might be possible for light to travel more than once round the Universe during its present lifetime so that long-lived objects could be observed at different stages of their evolution and a bunching at one redshift could be repeated at another. This could place additional constraints on $\Lambda > 0$ models.

There are also theoretical reasons for being concerned about non-zero values of Λ. For example, the inflationary model of the very early Universe which has been mentioned earlier does not appear to provide an explanation for a value of Λ of the size required. In addition the interesting values of $\Lambda/8\pi G\rho_0$ being close to unity provides us with another cosmological coincidence between small numbers which must be explained since Λ is a constant while ρ_0 varies as the Universe evolves.

The above considerations apply equally to the hot and cold versions of the theory. These theories differ in how the initial chemical composition of galaxies is determined. The hot big bang theory predicts that there was a primaeval composition of about three parts hydrogen to one part helium by mass after the first few minutes of the expansion of the Universe. It is well-known (see e.g. Tayler 1982) that the precise value of the helium abundance, Y, depends on the present mean baryon density of the Universe and on the total number of species of low mass neutrinos or other weakly interacting particles. The theoretical predictions are compared with the primaeval helium abundance deduced from observations. This abundance is usually obtained from a study of gas clouds in our own and other galaxies. There appears to be a correlation between helium and heavy element abundances and extrapolation to zero heavy element abundance then gives the original value of Y. A value of 0.23 ± 0.01 has been obtained (Pagel 1982) and this is in reasonable agreement with theory provided that the baryon density is significantly less than closure density and that there are no more than three species of light neutrino.

One problem with this discussion is that we cannot definitely assert that we are determining the primaeval abundance. At best we are determining the original disk composition of our own and nearby galaxies. Unfortunately it is not possible to obtain an observational value for the helium content of unevolved halo stars although, as will be mentioned below, indirect discussions do suggest $Y \sim 0.25$ and no obvious difference between the original helium abundance of the disk and halo. Some evolved halo stars have been reported with very much lower values of Y, which looks like a serious contradiction with cosmology, but it seems that these surface abundances can be explained in terms of a settling of helium similar to that discussed by Cox for δ Scuti stars in this Symposium. Halo stars contain some heavy elements whereas the predicted primaeval composition contains no heavy elements. Their origin must be placed before that of the observed globular cluster stars and a possible site is pregalactic Population III stars, which might also produce some helium. Kunth and Sargent (1983) have recently deduced $Y = 0.245$ for dwarf metal-poor galaxies and this is perhaps comparable with a halo

abundance but even this is not obviously a primordial abundance; it may
be noted that Kunth and Sargent do not find the correlation between
helium and heavy element abundance claimed by previous authors.

There are reasons for believing that the formation of galaxies might
have been preceded by a generation of pregalactic stars. Such stars
could then evolve and produce the heavy elements in the oldest stars
presently observed. In the hot big bang theory, that is all that is
required. There have, however, also been suggestions of a cold origin
for the Universe with not only the heavy elements but also the helium
and the microwave radiation being produced by pregalactic stars and we
shall have more to say about this possibility below.

3. GLOBULAR CLUSTERS

Globular clusters play a key role in the comparison between stellar
evolution and cosmology because they are the oldest known objects in the
Galaxy. Of importance are estimates of their ages and of their chemical
compositions. Their metal abundance can be observed directly and,
although there have been recent significant changes in estimated
metallicities as a result of high-resolution spectroscopy reducing the
spread in metallicity from cluster-to-cluster, two things are clear;
the globular clusters are very significantly metal-deficient when compared
with the Sun and other younger galactic stars and the metal abundance is
clearly non-zero with a minimum value of at least 10^{-5} by mass. In
addition there is a variation of metal content with position in the
Galaxy with those clusters nearer to the centre tending to have more
metals. The existence and size of the metal abundance lies at the
boundary between cosmology and galactic evolution. At what stage in the
history of the Universe were these metals produced?

Of even greater importance than the metal abundance is the helium
abundance but this cannot be observed directly in old unevolved stars
and it must therefore be deduced from theory. To a first approximation
we can say that we must use a comparison between theory and observation
to deduce values of cluster ages and cluster helium abundances and then
compare these both with the age of the Universe and the primaeval helium
abundance from the hot big bang theory and with the deduced original
helium abundance of the galactic disk. Relative cluster ages are almost
as important as absolute ages. If the globular cluster system was formed
in the initial collapse phase of the Galaxy, it is difficult to see how
there can be a spread in cluster ages greater than a few x 10^{8} years.
This leads to a related point. Most of the recent concern about the ages
of globular clusters has concentrated on the possibility that they might
be greater than the age of the Universe as deduced from Hubble's constant
and the conventional big bang. It would, however, be almost equally
worrying if the ages of the globular clusters were substantially less
than the age of the Universe because it is difficult to see how galaxy
formation can have been delayed long enough; only almost as bad because
this might be a failure of understanding rather than a sharp disagreement.

The comparison between theory and observation for globular cluster stars involves the use of theoretical and observational HR diagrams. The theoretical diagram is an isochrone constructed by assuming that all stars in the cluster have the same initial chemical composition and age, or rather that the spread in age is negligible compared with the present age of the cluster. There are three factors in the comparison of theoretical and observed diagrams. These are:

(a)   absolute position of diagram;
(b)   shape of diagram;
(c)   relative populations of different parts of diagram
        including the observed widths.

There is no way of determining the distances to globular clusters which does not involve assuming something about the absolute magnitudes of stars in the cluster, for example either the RR Lyrae variables or main sequence stars. This means that (a) is mainly used to determine cluster distance. Because there is no independent method of obtaining distances to clusters there is some trade-off between Y and distance in main sequence fitting. The age and helium content are mainly determined from (b) with the age usually being deduced from the position of the turnoff point from the main sequence. (c) is a much stricter test of the comparison between theory and observation because, unless the initial mass function of the cluster has very peculiar properties, the population of different parts of the diagram above the main sequence should be directly proportional to the time spent in any particular evolutionary phase. Comparisons of population with theory tend to lead to problems such as the at present unexplained gaps in the observational diagram near the foot of the giant branch in some clusters, which have been mentioned by Cannon earlier in this Symposium.

In any very precise comparison between theory and observation, the theoretical models should ideally be transformed to the observational diagram rather than *vice versa*; i.e. observed colours and bolometric corrections should be calculated from the theoretical spectrum of the stars rather than use being made of standard transformations, although these will certainly be good enough for rough estimates of ages. There are of course dangers in this approach because it implies that the entire theory including its weakest points is being compared with observation. There are certainly many remaining uncertainties in theoretical models which include:

(a)   the continuing solar neutrino problem;
(b)   the theory of convection;
(c)   opacity at low temperatures;
(d)   mixing processes;
(e)   treatment of stellar atmospheres.

Most of these effects have been mentioned earlier in this Symposium. An ultimate resolution of the solar neutrino problem may affect Population II stars as well as Population I stars unless it is concerned

with the properties of the neutrinos themselves. There is no *a priori*
reason to believe that the mixing lengths used in a crude convection
theory should be the same for stars of different masses and compositions,
for the same star at different stages of its evolution or at different
depths in a star, although Vandenberg (at this Symposium) has made some
reassuring comments about the use of a constant ratio of mixing length
to scale height on the giant branch. Changes in gross stellar properties
produced by a change in the value of the mixing length may also be
produced by changes in, for example, chemical composition which is not
completely known. Several speakers at the Symposium have stressed the
need to determine the correct value of the opacity at $T \sim 10^6$K. We have
also been introduced by Schatzman to possible mixing processes inside
stars. Stringfellow et al. (1983) have suggested that gravitational
and thermal diffusion of helium can reduce the estimated ages of clusters
by 25 per cent. This would be a very important effect if true, but it is
not clear what effect such settling would have on other stars such as
the Sun. Although the treatment of stellar atmospheres is not going to
have an important effect on the bolometric luminosity of a star, it could
be important in specific wavelength ranges affecting the conversion to
observed magnitudes and colours.

Recent studies of the ages of globular star clusters were
summarised by Cannon (1983) at the IAU General Assembly at Patras.
Table 2 is a slightly modified version of the results which he presented.

Table 2

Reference	Age/10^9 year	Clusters
Demarque and McClure (1977)	14-16	Metal-poor clusters
Saio et al. (1977)	18	M92 (Metal-poor)
Carney (1980)	18	Metal-poor clusters
	10	47 Tuc, M71
van Albada et al. (1981)	11-14	Various
Sandage (1982)	17±2	All clusters
Vandenberg (1983)	15-18	All clusters

It can be seen that there are considerable differences between the
results obtained by different authors, although there is a general
agreement that ages exceed 10^{10} yr with most results being substantially
greater. One particular source of disagreement is that most early work
suggested that the metal-rich globular clusters were substantially
younger than the metal-poor clusters, whereas recently several authors
have concluded that there is no good evidence of an age difference
between clusters. Whilst it seems reasonable that the metal-rich clusters
concentrated towards the centre of the Galaxy should be slightly younger,
the age spreads reported seem difficult to fit within any model of
galaxy formation and what seems like a reasonable age spread (few $\times 10^8$yr)
from that point of view would be equivalent to no age difference to the
accuracy to which the ages can be determined. It should be noted that

all of the ages discussed by Cannon are incompatible with the standard
model of the big bang if $H_0 \simeq 100$ kms^{-1} Mpc^{-1} and that many of them are
even incompatible with $H_0 \simeq 50$ kms^{-1} Mpc^{-1} if the Universe is closed.
It is therefore clearly of great importance to know whether the ages are
as large as those in Table 2 and whether H_0 has a large or small value.
We have already mentioned the reconciliation of a large H_0 and the
stated ages by introduction of non-zero Λ by de Vaucouleurs (1983).

The theoretical information about the globular cluster helium
abundance is still somewhat unclear. In main sequence fitting there is
a trade-off between helium abundance and distance and the remainder of
the HR diagram is not critically dependent on the value of Y. Most
results give a value of Y between 0.2 and 0.3 suggesting that there was
a substantial helium abundance at the time that the globular clusters
were formed and that there is no reason to worry about incompatibility
with the predictions of the hot big bang theory or with the original helium
abundance of the galactic disk. It has been suggested that the best
way of determining Y would be to compare observed and predicted widths
of the RR Lyrae instability strip but results presented by Stellingwerf
(this Symposium) suggest that is unlikely to be true. Sandage (1982) has
reported an anti-correlation between helium and metal abundances in his
best-fit models for different clusters. This appears difficult if not
impossible to understand but it certainly needs to be cleared up.

There have been several further discussions of globular cluster
ages since the Patras meeting. Janes and Demarque (1983) find an age of
$16.6 \pm 0.5 \times 10^9$yr for all clusters and $Y \simeq 0.2$, although they say that
the actual uncertainty in the age is larger than the formal error.
Sandage (1983) finds an age of $18 \pm 2 \times 10^9$yr for M92 and M15. Renzini
(this Symposium) has argued that use of the main sequence turn-off alone
to estimate cluster ages is unreliable and that the best method is one
used by Sandage involving the difference between the bolometric
luminosity of the RR Lyraes and at turn-off. This appears to be
essentially independent of metallicity and is a measure of age. The
helium abundance is not determined by this technique but,if
$Y = 0.23 \pm 0.02$, the age of all clusters is $16 \pm 3.5 \times 10^9$yr.

All of these investigations lead to a high age for the globular
cluster system, with serious discrepancy with standard cosmology with
$H_0 = 100$ kms^{-1} Mpc^{-1} and with a possible discrepancy if $H_0 = 50$ kms^{-1} Mpc^{-1}.
In one further recent paper Flannery and Johnson (1982) have also found
a high value for the age but they have also argued that the theoretical
uncertainties, such as I have listed earlier, are such that it is not
possible to rule out an age below 10^{10}yr. However a very small age
might imply a helium content which is uncomfortably small. It is clearly
vital to have agreement about the uncertainties in the estimated ages
as well as better agreement concerning the cosmological parameters
H_0, q_0 and ρ_0.

4. POPULATION III STARS

The oldest observed metal-poor stars in our Galaxy have a heavy element abundance (Z) of order or greater than 10^{-5} by mass whereas the big bang theory suggests that the primaeval composition of the Universe included no heavy elements. The simplest solution is that there was star formation earlier than the observed Population II stars and that these Population III stars produced the heavy elements. The Population III stars have to be largely or all of high mass so that no low mass zero-metal stars are observed today. It is unclear whether the Population III stars are to be associated with the time of galaxy formation or whether their formation and evolution preceded and possibly influenced galaxy formation.

There are both theoretical and observational reasons why most interest in Population III stars has concentrated on the possibility that they formed slightly after recombination, or the equivalent phase in a cold Universe. The theoretical reason is that discussions of the formation of structure in the Universe have indicated that condensations probably formed which were more massive than ordinary stars but less massive than galaxies. It is suggested that these condensations either directly or after some fragmentation formed a first generation of very massive objects. The various observational reasons are summarised in a forthcoming paper by Carr et al. (1984), which contains references to all of the earlier work. Population III stars might produce the metals in the oldest observed stars, a small distortion in microwave background radiation or even the whole of the radiation, and some or all of the original galactic helium. Their radiation might reionize the intergalactic medium or stimulate the formation of galaxies and their remnants could account for missing mass in clusters of galaxies and galactic halos.

In the simplest discussion it is assumed that the hot big bang theory is correct and that only heavy elements with possibly a small distortion in the microwave radiation and a very small amount of helium are required from Population III stars. Thus it is assumed that the helium abundance of the material after Population III stars have evolved is indistinguishable from the primaeval abundance but that there is now $Z \sim 10^{-5}$. There are, however, more radical suggestions that the Universe had a cold origin and that all of the primaeval helium and the cosmic microwave radiation were produced by Population III stars. There are obvious disadvantages of losing the natural explanation of $Y \sim 0.25$ and the black body radiation but it is argued that there may be compensations in a better understanding of the origin of structure in the Universe (see e.g. Hogan 1982). Because it is essential that $Z \sim 10^{-5}$ is produced in the cold big bang just as in the hot big bang, it is necessary either that there is a very different spectrum of masses in this case or that the predicted element production by stars of different masses is very different from what has previously been assumed.

A crucial role in this discussion is played by star formation, a topic which has not been discussed at this Symposium. Cosmological theory should predict the initial spectrum of irregularities from which the Population III stars can form but we then need to know what initial mass function is formed by fragmentation. There are several reasons why the initial mass function should be different from what is observed in the solar neighbourhood today. One is the absence of heavy elements and another, in the hot big bang, is the much greater intensity of the background radiation at a time close to recombination. Both of these are thought to encourage the formation of more massive stars. Once the stars have formed it is crucial to know at what stage in their evolution they lose mass and what is the composition of that mass. Here two things must be recognised. We do not at present have a good theoretical understanding of mass loss from stars in the mass range that can be studied observationally and we do not have any observations relevant to the evolution of objects of the supposed Population III masses, particularly, if as has been suggested earlier in this Symposium, R136a is a star cluster and not a single massive object.

It is naturally to be hoped that eventually the predicted mass spectrum of Population III stars in both the hot and cold versions of the big bang will be understood and that it will be possible to discover whether their evolution produces element abundances and microwave radiation in agreement with observations in either case. It must, however, be recognised that results other than the two extreme ones could be relevant. For example, Population III stars in the hot big bang might produce a small amount of heavy elements, a small distortion in the cosmic microwave radiation and a small but significant amount of helium. Suppose the Population III stars produce about 2 per cent helium as well as the minimum Population II metal abundance; while different authors are prepared to consider that they produce either essentially no helium or 25 per cent helium, it is not obvious that this value or any other intermediate value can be ruled out except as a result of prejudice. Carr et al. (1984) mention the possibility that Population III stars could impose fluctuations of a few per cent on the primordial helium abundance. This would imply that the original value of $Y \sim 0.23$ to 0.25 deduced by a study of gaseous nebulae in nearby spiral galaxies or of dwarf compact galaxies need not be the true helium abundance produced in the big bang and it could make the constraints on the hot big bang model that much more stringent.

While there remains the possibility that Population III stars might produce enough helium to modify the primaeval abundance, it must be premature to pass final judgement on the consistency between theory and observation of the hot big bang. It is partly for this reason that a proper understanding of the formation and evolution of Population III stars is vital and that theoretical studies must be made of a type of star which cannot be observed. Because it is necessary to understand the origin of the Population II metal abundance, they cannot simply be ignored.

A further comment on this metal abundance is necessary. As is well known, all globular clusters do not have the same metallicity and

there is some correlation between metallicity and place of origin in the
Galaxy, with the higher metallicity clusters being concentrated closer to
the galactic centre. Because of this correlation of metallicity with
galactic properties, it is difficult to suppose that it has arisen because
of irregularities in enrichment due to pregalactic Population III stars.
Instead it must be supposed that the pregalactic metal abundance has in
any case been enhanced by the evolution of early Population II stars.
Assuming that these are stars which are no more massive than stars which
we observe in the disk today, there is no reason to suggest any
significant further enhancement of the helium abundance and we might
expect all globular clusters and the oldest disk stars to have essentially
the same Y. The main point is that the assumption of a significant rôle
for Population III stars does not remove the need for early Population
II stars to affect others that we now observe. The evolution of massive
stars takes a time short compared with the period of galaxy formation so
that it is possible that Population III stars have not been important
and that all of the changes from original abundances have been produced
inside galaxies.

5.   CONCLUSIONS

     In this review I have concentrated on the relation between stellar
evolution and the big bang cosmological theory because at present this
does seem to be in good qualitative agreement with the observations.
Obviously there are also important interactions between stellar evolution
and other cosmological theories one obvious case being that of theories
in which the laws of physics change with time, in particular *variable G*
theories. Calculations have been made of stellar evolution in some such
theories and theoretical isochrones have been compared with cluster HR
diagrams but I do not feel that it is really necessary to study them in
detail at present. Equally I do not personally regard the cold big bang
as a likely viable alternative to the hot big bang.

     I have indicated that the evolution of stars in two mass ranges is
of crucial importance. The first range contains stars close to and below
a solar mass which can just have evolved significantly in the present
life of the Galaxy. They can give information about the age and initial
chemical composition of the Galaxy. The second group are stars which are
sufficiently massive that they can have evolved either before galaxies
were formed or during the process of galaxy formation, because they can
relate the initial composition of the Galaxy determined from the oldest
low mass stars to the original cosmological abundances. In the case of
low mass stars we can observe the stars in whose evolution we are
interested. In contrast the early high mass stars can only be studied
theoretically and progress may be hampered by lack of a good understanding
of the processes of star formation and stellar mass loss.

     We have seen that most estimates of globular cluster ages are too
high to be compatible with the big bang theory with zero cosmological
constant if $H_o \sim 100$ kms^{-1} Mpc^{-1} and that they might even be incompatible

with $H_0 \sim 50$ kms^{-1} Mpc^{-1}. The origin of the metals in the lowest
metallicity globular clusters may lie in pregalactic Population III stars
but it is necessary to establish that the production of metals is not
accompanied by a production of helium sufficient to modify the primordial
abundance, unless of course the Universe had a cold origin.

Several topics have deliberately not been discussed in this review.
They include the origin of the observed abundances of the light elements
and isotopes such as ^{2}H, ^{3}He and ^{7}Li. In addition it should be recognised
that any clear observation of a helium abundance well below $Y = 0.25$ in
any unevolved object would cause serious problems for the cosmological
theory. Also there is another source of information about the age of the
Galaxy in the heavy radioactive elements.

ACKNOWLEDGMENTS

The final version of this paper has benefited from many other papers
given in the Symposium as well as from informal discussions with other
participants. I thank F. A. Bukhari for his calculations of the ages of
$\Lambda \neq 0$ models.

REFERENCES

Bukhari, F. A. : 1983, M Sc Dissertation, University of Sussex.
Cannon, R. D. : 1983, Highlights of Astronomy, 6, pp 109-117.
Carney, B. W. : 1980, Astrophys. J. Suppl. 42, pp 481-500.
Carr, B. J., Bond, J. R. and Arnett, W. D. : 1984, Astrophys. J. in press.
de Vaucouleurs, G. : 1983, Astrophys. J. 268, pp 468-475.
Demarque, P. R. and McClure, R. D. : 1977, In *The Evolution of Galaxies
     and Stellar Populations*, eds B. M.Tinsley and R. B. Larson,
     Yale University Observatory, pp 199-215.
Flannery, B. P. and Johnson, B. C. : 1982, Astrophys. J. 263, pp 166-186.
Guth, A. H. : 1981, Phys. Rev. D, 23, pp 347-356.
Hoessel, J. G., Gunn, J. E. and Thuan, T. X. : 1980, Astrophys. J. 241,
     pp 486-492.
Hogan, C. J. : 1982, Astrophys. J. 252, pp 418-432.
Janes, K. and Demarque, P. R. : 1983, Astrophys. J. 264, pp 206-214.
Kristian, J., Sandage, A. and Westphal, J. A. : 1978, Astrophys. J. 221,
     pp 383-394.
Kunth, D. and Sargent, W. L. W. : 1983, Astrophys. J. 273, pp 81-98.
Pagel, B. E. J. : 1982, Phil. Trans. Roy. Soc. (Lond.) A, 307, pp 19-34.
Saio, H., Shibata, Y. and Simoda, M. : 1977, Astrophys. Space Sci. 47,
     pp 151-162.
Sandage, A. : 1982, Astrophys. J. 252, pp 553-573.
Sandage, A. : 1983, Astron. J. 88, pp 1159-1165.
Stringfellow, G. S., Bodenheimer, P., Noerdlinger, P. D. and Arigo, R. J.
     : 1983, Astrophys. J. 264, pp 228-236.
Tayler, R. J. : 1982, Phil. Trans. Roy. Soc. (Lond.) A, 307, pp 3-17.
Tytler, D. : 1981, Nature, 291, pp 289-292.

van Albada, T. S., Dickens, R. J. and Wevers, B. M. H. R. : 1981.
     Monthly Notices Roy. Astron. Soc. 196, pp 823-833.
Vandenberg, D. A. : 1983, Astrophys. J. Suppl. 51, pp 29-65.

DISCUSSION

Audouze: I have two comments on population III stars: 1) The short one:
When you speak about the formation of population III stars you refer
obviously to isothermal fluctuations. 2) About the problem of cold Big
Bang let me mention the work that Jo Silk and I try to pursue about this
problem. In the frame of the cold Big Bang theory you have not only to
worry about ^{4}He and the metals but also about D and Li. We make the
hypothesis that the cold Big Bang leaves rise only to pure hydrogen. If
massive stars of pure hydrogen are formed at red shifts of Z ~ 100, they
are going to form helium and metals, some of them will be accelerated
like pregalactic cosmic rays and could account for the observed D and ^{7}Li.
We circumvent by this way the difficulty presented by the He + He reaction
which could form too much lithium (see our abstract in the book "Primordial Helium" edited by P. Shaver and D. Kunth and published by ESO,
1983). This possibility seems to be still very ad hoc and fairly unlikely
but cannot be ruled out in the present stage of our knowledge.

Demarque: You have mentioned the question of the metal-rich globular
cluster ages. The main uncertainty in this regard has been due to uncertainties in the metallicity scale for globular clusters. A related problem, also important for our understanding of the evolution of the Galaxy,
is the apparent large difference in age between the globular cluster
system and the oldest disk clusters.

THEORETICAL PREDICTIONS FOR THE RADIAL DISTRIBUTION OF OXYGEN IN SPIRAL GALAXIES

M. TOSI[1] and A.I. DIAZ[2,3]

[1] Osservatorio Astronomico Universitario, Bologna, Italy
[2] Royal Greenwich Observatory, Herstmonceux, U.K.
[3] Astronomy Centre, University of Sussex, U.K.

To the aim of reproducing the oxygen abundances and gradients observed in the disk of our Galaxy and four nearby spirals (M31, M33, M83, and M101), we have computed numerical models of galactic evolution assuming only two free parameters: the infall of gas from outside the disk and the law of star formation (e.g. proportional to some power of the gas density, decreasing with time, constant, etc.).

Input data required for our models are: a) the present total mass distribution observed in the disk; b) the present gas distribution observed in the disk; c) the stellar yields, i.e. the amount of heavy elements ejected by stars of any mass during and at the end of their lives; d) the initial mass function, IMF. The present oxygen abundances resulting from the models have been compared with those observed in the HII regions of each galaxy. Details concerning both the models and the references for the observed data are given by Diaz and Tosi (1983).

In Figure 1, the oxygen abundances observed in the disk of the Galaxy are compared with the results of different models. The dashed line in panel **a** shows that the combination of a star formation rate slowly decreasing with time $(\psi = \psi_0 \exp(-t/\tau)$, $\tau = 15$ Gyr) with a constant infall $F = 3 \cdot 10^{-3}$ $M_{\odot}$ kpc^{-2} yr^{-1} (the value derived by Oort (1970) from measurements of high velocity clouds) gives results in very good agreement with both the observed gradient and the absolute abundances. The same model also reproduces (Tosi 1982) the age-metallicity relation derived by Twarog (1980) in the solar neighbourhood and the metallicity gradients obtained by Mayor (1976) for old stars and by Panagia and Tosi (1981) for young stars and clusters. The dotted line represents a 'simple' model with constant star formation and no infall. Both models have been computed by assuming stellar yields derived from stellar evolution theory taking mass loss into account (Chiosi and Caimmi 1979, Renzini and Voli 1981). The other three panels of Fig.1 show the effect of varying the main model assumptions: in panel **b**, the same models of

A. Maeder and A. Renzini (eds.), Observational Tests of the Stellar Evolution Theory, 563–566.
© 1984 by the IAU.

panel **a** are presented, but with stellar yields computed without mass loss (Arnett 1978). Panel **c** shows the effect on the F=3, τ =15 model of reducing the upper mass limit of the IMF. Since the oxygen is mainly produced by stars in the range 10-30 $M_\odot$, this effect becomes significant when the limit is lower than 30 $M_\odot$. The curves of panel **c** refer to a uniform IMF. By adopting an IMF varying with galactocentric distance the effect would be a steepening (flattening) of the gradient slope if we assume more (less) massive stars toward the center. Panel **d** presents the abundance distributions resulting from models with star formation proportional to the nth power of the gas density and the same infall as in panel **a**.

   It is apparent that the only choice of parameters leading to a good agreement between model results and observations is that of panel **a**. Furthermore, any other reasonable variation of the parameters would not recover the discrepancies between models and observations in panels **b,c,d**. For this reason, we present for the other galaxies only the same class of models as in Fig.1a, i.e. with exponentially decreasing star formation and constant infall.

   For each galaxy of Figure 2, the upper dotted line represents a 'simple' model with constant mass yields, while in all the other cases mass loss has been taken into account. In particular, the lower dotted line is again a 'simple' model and the other curves represent the models

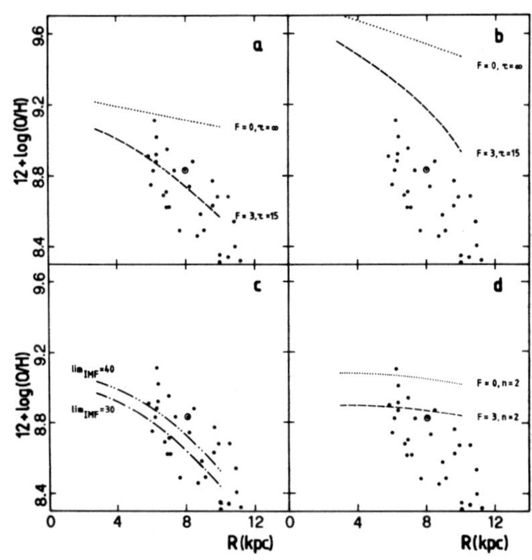

Fig.1. The Galaxy: Radial distribution of the oxygen abundances as derived from the indicated models. The infall rate F is in units of 10^{-3} $M_\odot$ kpc^{-2} yr^{-1} and the star formation e-folding time τ in Gyr. The dots represent the observed HII region abundances while the solar abundance is indicated by the usual sun symbol. See text for details on each panel.

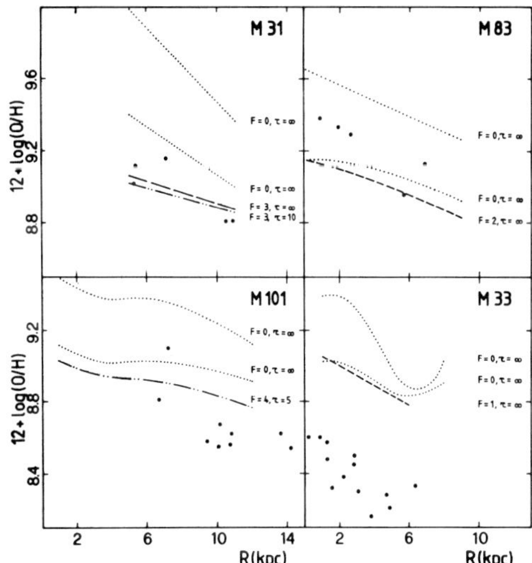

Fig.2. Radial distribution of the oxygen abundances in M31, M33, M83 and M101 as derived from the indicated models. Symbols are as in Fig.1.

in better agreement with the oxygen distribution observed in each galaxy (see Diaz and Tosi 1983 for details). The strong discrepancy apparent in M33 might be due to: a) the very high uncertainty on the gas and mass observational data, b) the fact that M33 is a sort of big dwarf galaxy which might evolve in a way completely different from that of normal spirals, c) an IMF with much less massive stars than here adopted, although there is no observational evidence for this latter case.

From Figures 1 and 2, we can derive the following main results:
i) the star formation rate is slowly decreasing with time in most of these spirals and is not simply proportional to some power of the gas density;
ii) the IMF is independent of time and galactocentric distance in all our spirals, having in both the Galaxy and M31 the same slopes as derived for the solar neighbourhood by Tinsley (1980), slightly more massive stars in M83 and slightly less in M101;
iii) the observed abundances and gradients are reproduced only by models taking mass loss into account.

REFERENCES
Arnett, W.A.: 1978, Astrophys.J. **219**, 1008
Chiosi, C., Caimmi, R.: 1979, Astron.Astrophys. **80**, 234
Diaz, A.I., Tosi, M.: 1983, M.N.R.A.S. in press

Mayor, M.: 1976, Astron.Astrophys. **48**, 301
Oort, J.M.: 1970, Astron.Astrophys. **7**, 381
Panagia, N., Tosi, M.: 1981, Astron.Astrophys. **96**, 306
Renzini, A., Voli, M.: 1981, Astron.Astrophys. **94**, 175
Tinsley, B.M.: 1980, Fund.Cosmic Phys. **5**, 287
Tosi, M.: 1982, Astrophys.J. **254**, 699
Twarog, B.A.: 1980, Astrophys.J. **242**, 242

DISCUSSION

Humphreys: Your results on M33 are very surprising. Elly Berhuijsin
has determined an IMF based on the Humphreys and Sandage (1980) catalogue
which is consistent with the galactic IMF. Also in M101 the observations
of its brightest stars show that there are many very massive stars in
that galaxy.

Tosi: In fact, I think that the present discrepancy between our model
results and the observed abundances in M33 is due much more likely either
to the high uncertainty on the mass data or to a completely different
galactic evolution than to a very steep IMF.

Audouze: By what factors the SFR decreases in your models?

Tosi: Our best model for the Galaxy (exponentially decreasing SFR with
e-folding time τ = 15 Gyr) gives a ratio ψ_o/ψ_{now} = 2.4 for the initial
over present SFR in the solar neighbourhood. This value is well within
the range derived by many authors (e.g. Mayor and Martinet, Miller and
Scalo, Twarog) from different observational constraints: $0 \leqslant \psi_o/\psi_{now} < 7$.

^{3}He$^+$ IN GALACTIC H II REGIONS: POSSIBLE EVIDENCE FOR NON-CONVECTIVE MIXING IN LOW MASS STARS

Robert T. Rood
University of Virginia

Thomas M. Bania
Boston University

Thomas L. Wilson
MPIfR, Bonn

ABSTRACT

We have detected the 8.7 GHz hyperfine line of ^{3}He$^+$ in the HII regions W43, W51, and W3 and obtained significant upper limits for M17, W49, and Orion A. Even though the abundances derived from measured line parameters can have large uncertainty due to the detailed structure of the sources, it seems almost certain that the ^{3}He$^+$ abundance in W3, 12 kpc from the galactic center is significantly higher than in sources closer to the galactic center. If this is the result of general chemical evolution of the galaxy, it is exactly the opposite of what is expected.

I. OBSERVATIONS

Our observations were mostly done in August 1982 with the 43 m telescope of the NRAO at Green Bank. The excellent signal/ noise achieved is typified by the spectrum for W3 shown in the figure.

Since the strength of the ^{3}He$^+$ feature depends on n_3/n_H x $\int n_e dl$ and all measures of the total amount of ionized material in our beam depend on $\int n_e^2\, dl$, obtaining an abundance ratio from the observations is non-trivial. We have modeled the regions we observed and find that the uniform sphere models previously used sometimes significantly underestimate the ^{3}He abundance. Not only are the corrections for W3 and W51 large; they are quite uncertain with just the 1σ error in the measured continuum temperature introducing a 40% error in the structure correction. It is difficult to assign errors to the abundances. From the structure corrections alone, the W51 value could range from 6 - 14 x 10^{-5} and the W3 value could range from 30-70 x 10^{-5}. It does seem safe to conclude that the abundance in W3 is significantly higher than in W43 and that there are source to source variations. The W43 value appears to be the best determined. Our results are summarized in the table.

A. Maeder and A. Renzini (eds.), Observational Tests of the Stellar Evolution Theory, 567–570.
© *1984 by the IAU.*

Source	Galactocentric distance (kpc)	$T_B(^3He^+)$ (mK)	Structure Corrections	$^3He/H$ (x 10^{-5})
W43	6	11	0.9	4
M17S	8	< 6	1.0	< 2
W51	8	9	2.4	10
W49	9.5	< 5	1.2	< 2
Orion	10.5	< 8	1.0	< 6
W3	12	15	2.7	50

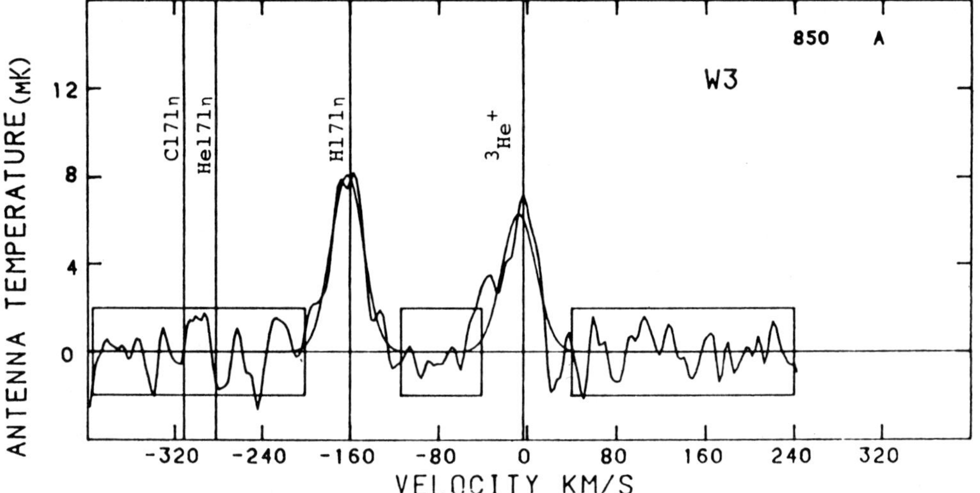

Figure 1. The region of the spectrum containing the 171η recombination lines and the He line. The expected rest frequencies are denoted by vertical lines. The spectrum has been smoothed by a gaussian smoothing function FWHM = 8 km s^{-1}.

II. CONSEQUENCES

Rood, Steigman, and Tinsley (1976) suggested that ^3He in the present ISM should be enhanced over the solar value. Schatzman (1984) has suggested some problem with the surface ^3He in the standard solar model. Both of these arguments require a knowledge of the protosolar $^3He/H$ which is generally derived by subtracting some estimate of the protosolar D/H from the current solar wind ^3He. Because of the uncertainty in D the protosolar $^3He/H$ could range from 0 to 4 x 10^{-5}. We thus see that there may or may not have been any enrichment of the ISM in the past 5 billion years. Further from the range of ^3He now, we conclude the sun could well have formed with whatever ^3He is required to match the current value. No matter what other problems may exist with the standard solar model, $^3He/H$ certainly is not a problem.

The puzzle at the moment is the high abundance in W3. Is it a pecularity of the source itself? W3 is indeed an odd H II region. W3A is a ring-like source containing a 50-100 M$_\odot$ of gas. If all of our signal comes from W3A then ^{3}He/H $\sim 10^{-3}$ in that source. This value is near the maximum which might occur in planetary nebulae with $\sim$1M$_\odot$ progenitors.

If the high abundance in W3 is due to the general chemical evolution of the galaxy, the result is exactly the oposite of what would be expected. This is shown by observations of ^{13}C which should be produced in the same stars which produce ^{3}He. Observations of H$_2$CO (Henkel, Wilson, Bieging, 1982) show ^{13}C/^{12}C enhanced by about a factor of two at galactocentric distances of 5 kpc over the approximately solar value found near W3. This seems to agree with the generally held belief that there is more stellar processing toward the galactic center. The simplest interpretation of our observations of ^{3}He which is consistent with the ^{13}C observations is that stars destroy ^{3}He. This would place an important constraint on diffusive mixing models, since, for example, the formulation for turbulent diffusion described by Schatzman (1983) does not lead to the destruction of ^{3}He. It seems possible that a diffusive mixing may occur just maintaining the ^{3}He gradient required for the onset of the overstable oscillations first found by Dilke and Gough (1972). Such a mechanism might lead to the destruction of ^{3}He and the transport of ^{13}C and perhaps Li. Further it would address the comments made by Roxburgh at this meeting pointing out that the ^{3}He peak provides an adequate energy reservoir for diffusive mixing whereas rotation, for example, does not.

We should emphasize that the role of ^{3}He as a probe of stellar or cosmological nucleosynthesis or of stellar mixing mechanisms will be determined only by further observations.

RTR was partially supported by NSF 81-08418.

REFERENCES

Dilke, F. W. W., and Gough, D. O. 1982, Nature, 240, 262.
Henkel, C., Wilson, T. L., and Bieging, J. H. 1982, Astr. Ap., 109, 344.
Rood, R. T. Steigman, G., and Tinsley, B. M. 1976, Ap. J. Letters, 207, L57.
Schatzmann, E. 1984, this volume, p. 491.

DISCUSSION

<u>Bochsler</u>: There is direct evidence for the existence of solar-system-^{3}He prior to deuterium burning by the sun: Some carbonaceous meteorites contain "planetary" helium with a uniform ^{3}He/^{4}He-ratio of $\sim 1.4 \cdot 10^{-4}$ (cf. Anders et al. 1970, Geochim. Cosmochim. Acta <u>34</u>, 127; Eberhardt, 1982).

<u>Rood</u>: Yes, this was also noted by Black (1972, Geochim. Cosmochim. Acta <u>36</u>, 347), who identified three components of ^{3}He in carbonaceous chondrites: implanted solar wind ^{3}He/^{4}He $\sim 3.9 \cdot 10^{-4}$, solar flare $\sim 4.1 \cdot 10$ and "primordial" $\sim 1.5 \cdot 10^{-4}$.

<u>Renzini</u>: Let me mention that there are stellar models which destroy ^{3}He. These are AGB stars experiencing the so-called Envelope-Burning process.

M GIANTS AT TWO LATITUDES IN THE NUCLEAR BULGE OF THE GALAXY

Jay A. Frogel and V.M. Blanco
Cerro Tololo Inter-American Observatory

A.E. Whitford
Lick Observatory, University of California, Santa Cruz, CA

I. Introduction

We are engaged in a long term study of the stellar component in
the nuclear bulge of the Galaxy. Regions of low obscuration at different
galactic latitudes above the galactic nucleus are surveyed for M giants
with a transmission grating and thin prism at the prime focus of the
CTIO 4-m reflector. Photographic and infrared photometry are obtained
for samples of the stars found. Spectroscopic observations are used
to investigate band strengths and radial velocities.

Work on Baade's Window (BW; Baade 1963) around NGC 6522 (latitude
-3°.9, 0.6 kpc below the plane) is essentially complete. Infrared photometry has been obtained for a sample 229 M giants (Blanco and McCarthy
1983; Frogel, Whitford, and Blanco 1983, hereafter FWB). Initial
results have been reported by Whitford and Blanco (1979), Frogel (1981),
and Frogel and Whitford (1982). Whitford and Rich (1983) have discussed
the metallicities of the K giants in this window and have shown that
most are super-metal-rich (SMR) with a mean $[Fe/H] = 0.3$. The spectroscopic survey of the van den Bergh and Herbst window (vHW, 1974) at
-8° (1.3 kpc below the plane) is nearly complete for the M5-9 giants
and infrared photometry has been obtained for 49 of them.

Objectives of the program include a study of the stellar distribution, content, and metallicity in the bulge, application of these results
to the construction of the population models of other galaxies (Whitford
1977), and a comparison with evolutionary models for giants.

II. The Color-Magnitude Diagram

Figure 1 shows that both windows have a significant number of
stars with luminosities greater than the top of the giant branch of
the metal rich cluster 47 Tuc. Because of their luminosity, they must
be asymptotic giant branch stars. An important question is whether they
are old (>10 Gyr), but luminous because of their high metallicity and
slow evolutionary rate as suggested by Frogel and Whitford (1982),

A. Maeder and A. Renzini (eds.), Observational Tests of the Stellar Evolution Theory, 571–575.

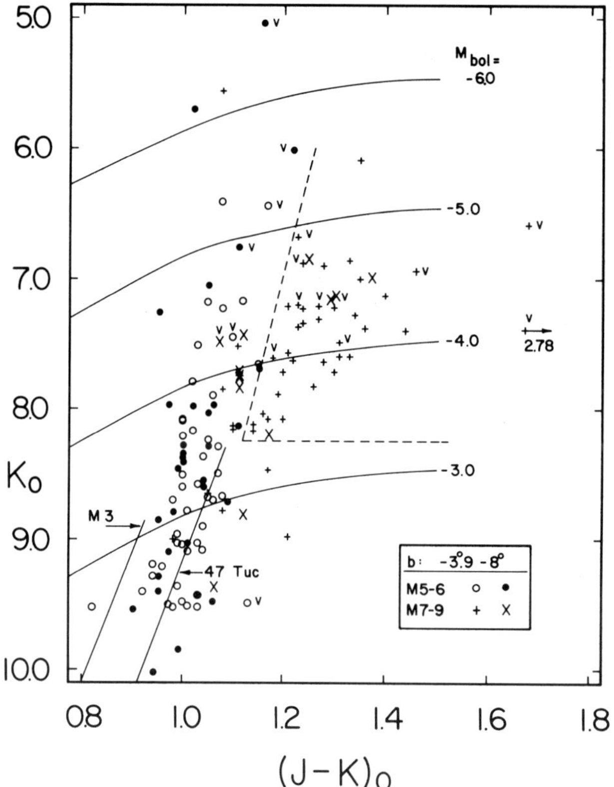

Figure 1. Photometry of an unbiased sample of M5-9 stars in two Galactic bulge windows. Variables in this sample are indicated with a v. The wedge-shaped region (dashed line) is discussed in the text.

or whether they are relatively young and massive as argued by Iben and Renzini (1983) and as Wood and Bessell (1983) suggest for the red variables in BW. Mould's (1983) kinematic data support the former interpretation.

Stars in each of the two windows are divided into two groups based on the survey spectra – M5-6 and M7-9. The relative numbers, median magnitudes, and CO and H_2O indices of stars in each of the two groups in the two windows are quite similar. Only the J-K colors of the M7-9 giants differ (cf. Fig. 1). Consider the stars more luminous than the top of the 47 Tuc giant branch. The wedge shaped region to the right contains only M7-9 stars whereas to the left stars of all spectral types are found. For the complete spectroscopic samples the ratio (vHW/BW) of stars in the wedge shaped region is 0.09 whereas it is 0.21 for the bluer giants to the left.

Lloyd Evans (1976) search BW for red variables. There are 17 in our surveyed area. Plaut's (1971) list of variables includes 6 Miras and SRas in the vHW area we surveyed. Thus the ratio of numbers of variables, vHW/BW, is 0.35.

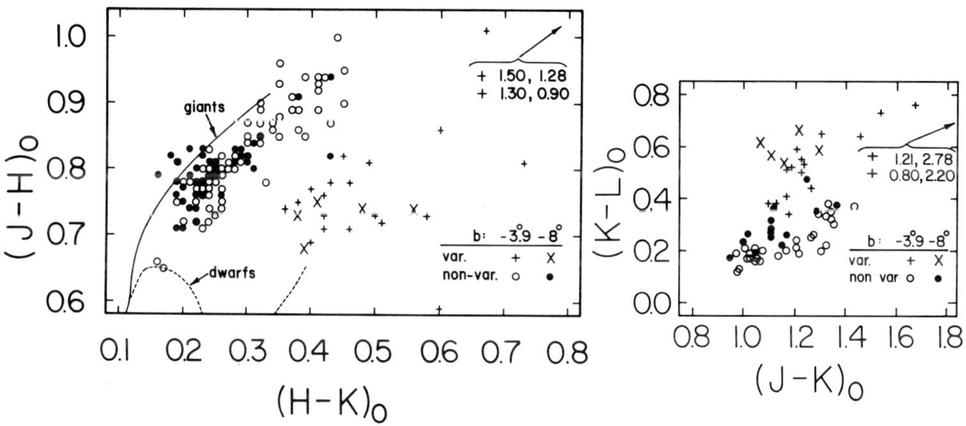

Figures 2 and 3.   Samples of M5-9 giants in two Galactic bulge windows.
All variable stars observed in both windows are shown.

    If the spheroidal stellar component of the Galaxy has a surface
brightness which falls off like $R^{-\frac{1}{4}}$ and a characteristic length of
2-3 kpc, the ratio of projected stellar densities at central distances
of 1.3 and 0.6 kpc is 1/3. Whereas the variables and possibly the
bluer bright giants have such a distribution, the giants in the wedge
shaped area of Fig. 1 are decreasing in numbers too rapidly to be in
this distribution. These red stars may represent a distinct, much
more condensed spheroidal component with a metallicity higher than that
of the remainder or they could be the tail end of a single distribution
with a steep metallicity gradient. They could also be the very central
part of an exponential disk component with a scale height of not more
than a few hundred parsecs. Van den Bergh and Herbst (1974) have in
fact concluded that their window is somewhat less metal rich than BW.

III.   JHKL Colors and the Variables

    Essentially all of the BW giants lie to the right of the mean
line for solar neighborhood giants in Figure 2. Globular cluster giants,
on the other hand, almost all lie to the left of this line (Frogel,
Persson, and Cohen 1983, Fig. 3). Frogel and Whitford (1982) suggested
that this displacement of the BW giants could arise from blanketing
effects on J-H and H-K colors due to their being SMR since they must
have evolved from the SMR K giants found by Whitford and Rich (1983).

    The variables are distinctly separated from the non-variables in
Figs. 2 and 3 in each of the two windows. In Fig. 3, the distributions
of the two windows are shifted in J-K as expected from Fig. 1. Glass
and Feast's (1982) data are consistent with the location of the variables
in the figures. The separation of the variables which lie below and
to the right of the non-variables in Fig. 2 can be accounted for by
the effects of stellar H_2O absorption in the JHK bandpasses (Frogel,

Persson and Cohen 1983). For K-L the separation at constant J-K is
due more to the presence of circumstellar emission from hot (800-1200
K) dust rather than from H_2O effects (FWB). The sequence of variables
extending towards the reddest colors in both Figures can be accounted
for by superimposing emission from hot dust on normal photospheric
radiation (FWB). This sequence is quite close to that defined by solar
neighborhood Miras with OH emission and circumstellar shells (Hyland,
et al. 1972; Wilson et al. 1972). The 10μm emission from 11 of the
BW variables (FWB) is what would be predicted from the K-L colors
based on similar observations of the local Miras. The effects of hot
dust and H_2O absorption make temperature estimates for the variables
based on near infrared colors unreliable (cf. Wood and Bessell 1983).

It is not clear at this time whether the presence in BW of very
red variables and their absence in vHM is due to the smaller numbers
of stars observed in vHM or if it is related to the presence of the
very red M7-9 stars in BW.

REFERENCES

Baade, W.: 1963, in Evolution of Stars and Galaxies , ed. C.P.
    Gaposchkin (Cambridge: Harvard University Press), p. 279.
Blanco, V.M., and McCarthy, M.F.: 1983, A. J., submitted.
Frogel, J.A.: 1981, in Physical Processes in Red Giants, ed. I. Iben,
    Jr. and A. Renzini (Dordrecht: Reidel), p. 63.
Frogel, J.A., Persson, S.E., and Cohen, J.G.: 1983, Ap. J. Suppl., in
    press.
Frogel, J.A., and Whitford, A.E.: 1982, Ap. J. (Letters) 259, p. L7.
Frogel, J.A., Whitford, A.E., and Blanco, V.M.: 1983, in preparation.
Glass, I.S., and Feast, M.W.: 1982, M.N.R.A.S. 198, p. 199.
Hyland, A.R., Becklin, E.E., Frogel, J.A., and Neugebauer, G.: 1972,
    Astr. Ap. 16, p. 204.
Iben, I., Jr., and Renzini, A.: 1983, Ann. Rev. Astr. Ap., in press.
Lloyd Evans, T.: 1976, M.N.R.A.S. 174, p. 169.
Mould, J.R.: 1983, Ap. J. 266, p. 255.
Plaut, L.: 1971, Astr. Ap. Supp. 4, p. 75.
van den Bergh, S., and Herbst, E.: 1974, A. J. 79, p. 603.
Whitford, A.E.: 1977, Ap. J. 211, 527.
Whitford, A.E., and Blanco, V.M.: 1979, Bull. AAS 11, p. 675.
Whitford, A.E., and Rich, R.M.: 1983, Ap. J., in press.
Wilson, W.J., Schwartz, P.R., Neugebauer, G., Harvey, P.M., and
    Becklin, E.E.: 1972, Ap. J. 177, p. 523.
Wood, P.R., and Bessell, M.S.: 1983, Ap. J. 265, p. 748.

DISCUSSION

Wing: Would you comment on the possibility of picking up some foreground
stars - perhaps from the 3 kpc arm - in your fields? They might help
account for the brightest stars, especially the ones that are not very
red.

Frogel: I have not checked this quantitatively yet, but Arp's analysis and Blanco's statistics indicate that the percentage of field stars will be quite low.

Richer: Is there any hope of seeing the turn-off in the high latitude field?

Frogel: It will be much easier there than in the -4° field because the star density is lower by a factor of 3.

Renzini: Do you know whether there is any heavy element abundance determination for Sgr A (or any other cloud in the galactic nucleus), and if so, how do they compare with the stellar abundances of Rich and Withford?

Frogel: Nothing has been done optically because of the obscuration. Some radio recombination line work may have been done but I don't recall any results.

McCarthy: 1) In a Grism survey similar to those mentioned by Frogel, Blanco, Meier, de Graeve and I have found in a region even closer to the direction of the galactic center a very large number of late-type M stars: more than 750 stars in an area of 0.12 square degrees. We have not detected any carbon stars. This region is one of the windows cited originally as clear by Walter Baade; it was studied for variables by Oosterhoff and is called Sagittarius I. We have discovered the spectra and determined positions for the stars; no photometry has been carried out.
2) I wish to comment the excellent photometry reported here by Frogel and his co-workers at CTIO. Especially in the dusty and crowded fields toward the galactic center photographic photometry is difficult, dangerous and should be discouraged, if possible. Frogel's work is doing so much to allow Baade's windows to shed more light on us from the galactic center.

PRIMEVAL OXYGEN OVERABUNDANCE AND TYPE I-1/2 SUPERNOVAE

F.Matteucci   and A.Tornambé
Istituto di Astrofisica Spaziale,C.N.R.,Frascati,Italy

Abstract: Models of chemical evolution of the Galaxy have been computed
by taking into account the different roles played by TypeI-1/2 (single
stars suffering degenerate C-ignition) and TypeII supernovae in the chemical enrichment. The overabundance of oxygen observed in the Halo stellar
population has been well reproduced.

Introduction: Little attention has been devoted to the role played by Ty-
peI-1/2 or I supernovae in the chemical history of our Galaxy.
Following some recent results and suggestions on the mass spectrum of single stars suffering Carbon-deflagration and on its behaviour versus the
initial metal content of the progenitor star (Tornambé,1983),we have computed models of galactic chemical enrichment starting from a pure Hydrogen
and Helium protohalo mixture.
As is well known,degenerate Carbon-ignition gives rise to a thermonuclear
runaway in a core of about 1.4 $M_\odot$;one half of this core can reach nuclear statistical equilibrium and,if the star is finally destroyed,a large
amount of Iron-peak nuclei can be ejected into the interstellar medium.
The question concerning the exact amount of matter processed is still under debate,because fully hydrodynamical codes and several physical processes are required to follow correctly the last evolutionary stages of the
quoted stars; current ideas suggest an estimate of 0.7 $M_\odot$ of Iron peak
nuclei and 0.35 $M_\odot$ each of Carbon and Oxygen (Nomoto,1984). These amounts
have been chosen as inputs for our code.
The mass spectrum of TypeI-1/2 supernovae falls between 4 and 10 $M_\odot$ for $Z \leq 10^{-8}$
becoming narrower as the initial metallicity of the progenitor star incre-
ases(i.e. between 8 and 9 $M_\odot$ for $Z=10^{-2}$),(Tornambé,1983b).

Results: We have used the chemical evolution model performed by Chiosi
and Matteucci(1982),in which the basic assumptions are: i) no instantaneous approximation ; ii) detailed description of the variation over the

577

A. Maeder and A. Renzini (eds.), Observational Tests of the Stellar Evolution Theory, 577–578.

   galactic lifetime of a set of elements (He,C,O,Si+Fe),due to stellar nucleosynthesis,stellar mass ejection and inflow of primordial gas.
The initial mass function is expressed as a power law($\varphi(m) \propto m^{-x}$),and two cases are considered: costant slope($x=1.35$) over the galactic lifetime and
a time variable slope,as suggested by Melnick and Terlevich,1983.
One of the most straightforward tests for our chemical model was to account for recent observations concerning the overabundance of Oxygen with
respect to Iron in metal deficient stars(Sneden et al.1979,Clegg et al.
1981). The other properties of the solar vicinity(i.e. the G-dwarf distribution and the age-metallicity relationship) have been also fitted likewise. The model results can be summarized as follows:

i) TypeI-1/2 supernovae are required
to explain the Oxygen overabundance
with respect to Iron in metal poor
stars.In fact,they restore to the interstellar medium a substantial fraction of Iron with a temporal delay
due to their longer lifetimes.
However,the Iron ejected by TypeII
supernovae($M \geqslant 10\ M_\odot$),although it must
be produced in a smaller amount(one
half) with respect to the one predicted by Arnett(1978),is also required.
ii) Our best model contains both Type
II and I-1/2 supernovae,no mass loss
in massive stars and constant initial
mass function over the galactic lifetime.

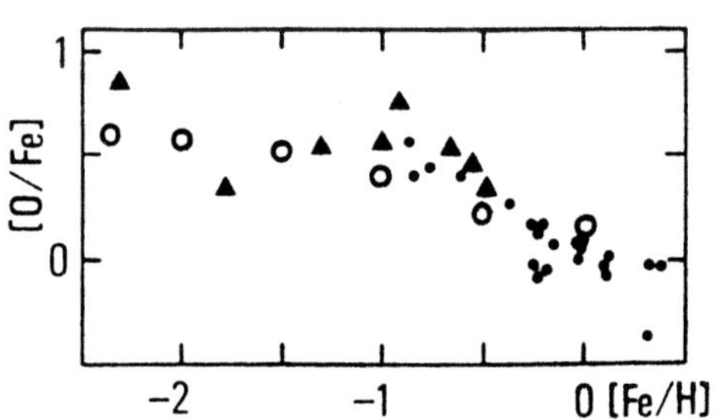

Observational data(filled points
and triangles) by Clegg et al. and
Sneden et al. Open circles represent our theoretical best model.

iii) The model computed with variable initial mass function well fits the
overabundance of Oxygen during the Halo phase but Oxygen and Iron during
the Disk phase are underestimated.

References:
Arnett,W.D.,1978,Ap.J.219,1008
Chiosi,C.,Matteucci,F.,1982,Astron.Astrophys.105,140
Clegg,R.E.S.,Lambert,D.L.,Tomkin,J.,1981,Ap.J.250,262
Melnick,J.,Terlevich,R.,1983,preprint
Nomoto,K.,1984,in "Stellar Nucleosynthesis",Reidel.Publ.Comp.
Sneden,C.,Lambert,D.L.,Whitaker,R.W.,1979,Ap.J.234,964
Tornambé,A.,1983, M.N.R.A.S.,in press
Tornambé,A.,1983b,in "Population Synthesis",Frascati workshop atVulcano

EVOLUTIONARY TRACK OF AN INTERMEDIATE MASS FIRST-GENERATION STAR

D. Eryurt - Ezer
Department of Physics, Middle East Technical University
Ankara, Turkey

Abstract: An evolutionary track for a metal-free star of intermediate mass
with parameters $(M/M_\odot, X, Z) = (3, 0.80, 0.00)$ has been constructed. The evolutionary behaviour of the star in the theoretical H-R diagram is compared
with that of a normal star with the same mass and its implication on the
chemical evolution of the interstellar medium is discussed.

On the early enrichment of the Pre-Galactic medium with the nucleosynthesis products the evolutionary lives of the first-generation stars
(called Population III or Pregalactic stars) which are supposed to have
formed essentially with no heavy elements, play very important role. Very
little study has been carried out for the evolution of first-generation
stars (e.g.Ezer and Cameron, 1971; Ezer,1972,1981;Guenther and Demarque,
1983; El Eid et.al.1983). The mass ejected from the stars, generally is
the surface layers of the stars. There are different processes which are
responsible for the changing of the abundances of the surface material
during the evolution of the stars. In the study of the evolution of normal stars, Becker and Iben(1979) distinguished three convective dredge-up
phases during which the composition of surface layers might be largely
affected.

In this paper, the evolutionary history of a first-generation
star of $3M_\odot$ has been presented from the main sequence up to the appearance
of the first thermal pulse. The evolutionary behaviour of the star during
the occurance of the dredge-up phases is examined and its implication is
discussed. The method of computation and input physics were given in Ezer
(1981).

The result of the evolutionary study is presented in the theoretical H-R diagram in Figure 1 and compared with that of a normal star of
the same mass, taken from the works of Becker et al (1977) and Becker and
Iben (1979). The positions of the stars on the evolutionary tracks, during
which the derdge-up phases occur, are indicated by Romen numbers.

In the evolution of a first generation star of $3M_\odot$, hydrogen
burning near the main sequence occurs at a higher temperature via the

A. Maeder and A. Renzini (eds.), Observational Tests of the Stellar Evolution Theory, 579–582.

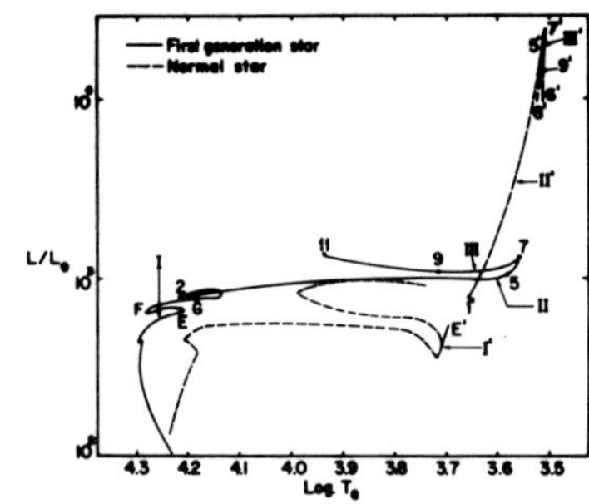

Figure 1. Comparison of the evolutionary track of the
first-generation star of $3M_\Theta$ with that of a
normal star of similar mass, in the H-R diagram.

proton-proton chain until the triple-alpha reactions generate a small
amount of C^{12} toward the end of Hydrogen-burning phase. Core-helium
burning begins in a slightly degenerate, small convetive core while the
star in the blue region of the H-R diagram (Point E). During Core-helium
burning the star first evolves to the blue to some maximum temperature
(Point F), then proceeds to the right direction, but stays in the blue
region until the complete core-helium exhaustion phase (Point G). This
is contrary to the situation of a normal star of $3M_\Theta$, in which helium
burning starts in a larger convective core when the star reaches the tip
of the red-giant branch in the evolutionary track (point E'). Therefore,
the first convective dredge-up process which takes place during the climb
up the red giant branch before core helium ignition, largely effects the
surface abundances. (Becker and Iben, 1979,1980). On the other hand, a
first generation star with the same mass does not experience the first
dredge-up phase, hence no surface abundance changes occur before core-
helium burning phase.

In the evolution of a first-generation star of $3M_\Theta$, following
helium-exhaustion, core contraction is accompanied by envelope contraction. Increased temperature and density at the center of the star favor
the emission of neutrinos at that stage of the evolution. This prevents
the increase of the energy generation rate in the helium-burning shell.
Due to high temperature prevailing inside the star hydrogen-burning shell
quite active and largely responsible for the outflow of the energy up to
the point 2. The evolutionary track starts to move to the red giant region in the H-R diagram as the strength of helium shell grows and hydrogen
burning shell weakens. During the evolution from blue to red, envelope
steadily deepens and covers about 30 % of the outer star's mass, when the
rate of energy generation in helium-burning shell attains its maximum value, at the point 5. Rapid formation of C^{12} during the high rate of energy
generation in the helium burning shell makes the energy production via
CN-cycle reaction in the hydrogen-burning shell to become pronounced. The
evolution proceed on the asymptotic giant branch up to the point 7 at
which the rate of energy generation by the CN-cycle reactions attains its
maximum value. As the helium-burning shell declines in strength the energy
generation in the hydrogen-burning shell becomes the main source of energy
generation in the star, the evolutionary track turns to the left.

On the contrary of the evolution of normal stars, as the hydrogen burning shell becomes the main source of energy for the second time, the evolution does not proceed on the asymptotic giant branch; the star only experiences an abbreviated second dredge-up phase, while hydrogen burning can occur via CN-cycle reactions in the hydrogen-burning shell. However the convective envelope never reaches into the region of variable hydrogen abundance.Therefore, there is no change in surface composition as a consequence of the second dredge-up phase which takes place during short climb up the asymptotic giant branch. Once the hydrogen burning by the p-p chain reaction mostly become responsible for the energy flux in the shell and as the shell moves toward larger mass fractions, the evolutionary track follows horizontal path toward the leftward direction in the H-R diagram until the rate of energy generation by the p-p chain reactionsreaches its maximum value (Point 11). After the point 9, the hydrogen-burning shell is constantly becoming thinner and helium burning rate becomes strong enough to produce a convective region in the upper part of the helium burning shell. The first major thermal pulse appears at the point 9.

In the evolution of normal stars with the same mass, following helium exhaustion, the path followed in the H-R diagram is from blue to red with luminosity L reaching a minimum value at the point 1'. The star, then, begins its ascent of the asymptotic giant branch. The rate of energy generation in the helium-burning shell reaches its maximum value at the point 5'. (Corresponding to the point 5). From there on, helium-burning shell declines in strength, but evolutionary path proceeds on the asymptotic giant branch. The base of the convective envelope starts to move outward in mass at the point 7'. The first major thermal pulse appears at the point 9'. Third dredge-up phase occurs during that part of evolutionary track.It should be noted that the properties of the model at the occurence of the first major thermal pulse are quite different from that of the first-generation star. (See Ezer,1981). The internal structure and location of the first generation star in the H-R diagram during the occurence of the third dredge-up phase suggest that following each helium shell flash any significant amount of He^4 and C^{12} would possibly not convected to the surface.

The evolutionary behaviour of the first generation stars should be properly taken into account for the further evolutionary study of the star. Such a study might effect the occurence of heavy-element enrichment in the helium zones of the stars and the nucleosynthetic yield in the study of the chemical evolution of the Galaxy.

References
Becker,S.A.,Iben,I.Jr.,and Tuggle,R.S.:1977,Ap.J.218,pp.633-653
Becker,S.A.,and Iben,I.Jr.,:1979,Ap.J.232,pp.831-853
Becker,S.A.,and Iben,I.Jr.,:1980,Ap.J.237,pp.111-129
El,Eid,M.F.,Fricke,K.J.,and Ober,W.W.,:1983,Astr.Ap.119,pp.54-60
Ezer,D.,and Cameron,A.G.W.,:1971,Ap.Spa.Sci.10,pp.52-70
Ezer,D.:1972, Ap.Sp.Sci.18,pp.226-245
Ezer-Eryurt,D.:1981, Ap.Sp.Sci.79, pp.265-287
Guenther,D.B., and Demarque,P.:1983,Ast.Ap.118,pp.262-266.

DISCUSSION

Becker: I wonder if the author has an explanation why her model evolves off the asymptotic giant branch shortly after it reaches the double-shell burning phase?

Eryurt-Ezer: Hydrogen-shell burning occurs via p-p chain reactions and CN-cycle reactions. Evolution proceeds on the asymptotic giant branch as the hydrogen-burning shell source is dominating energy source for the star in which the carbon cycle being more dominant over the proton-proton chain. But the operation of CN-cycle depends on the formation of ^{12}C by the 3α-reactions in the evolution of the first giant stars. While the shells move toward the larger mass fractions, the carbon amount formed by the 3α-reactions abruptly decreases, hence the amount of energy generation by the CN-cycle reactions. As the low temperature-sensitive p-p chain reactions becomes main source of energy generation in the shell, the star evolves off the asymptotic giant branch.

CONCLUDING REMARKS : LOOKING AHEAD

E. SCHATZMAN
CHAIRMAN OF THE SCIENTIFIC ORGANIZING COMMITTEE

The absence of concensus on several of the problems which have been
discussed during the meeting is perhaps the best demonstration of the
usefulness of the Symposium .

The history of science is there to testify that consensus does not mean
 that the truth is known . However, the absence of consensus is the proof
that further work has to be carried in order to find out who is right
and who is wrong . And perhaps everyone is wrong !

The conflict between models or rules and new observations has always
been the source of new progresses . It is worth, in this respect, to
quote the story told by H.N.Russell in 1938 :

"... I was visiting my friend and generous benefactor, Professor Edward
C.Pickering . With characteristic kindness, he had volunteered to have
the spectra observed for all the stars - including comparison stars -
which had been observed in the observations for stellar parallax which
Hinks and I made at Cambridge, and I discussed . This piece of apparently
routine work proved very fruitful - it led to the discovery that all the
stars of very faint absolute magnitude were of spectral class M . In
conversation on this subject (as I recall it), I asked Pickering about
certain other faint stars, not on my list, mentionning in particular
o^2 Eridani B. Characteristically, he sent a note to the Observatory
office, and before long the answer came (I think from Mrs.Fleming) that
the spectrum of this star was A . I knew enough about it, even in those
paleozoic days, to realize at once that there was an extreme inconsis-
tency between what we would then have called "possible" values of the
surface brightness and density . I must have shown that I was not only
puzzled, but crestfallen, at this exception to what looked like a very
pretty role of stellar characteristics ; but Pickering smiled upon me,and
said : "It is just these exceptions that lead to an advance in our
knowledge," and so the white dwarfs entered the realm of study ."

I would like then to mention what , I think, are some of the weak points
of our meeting .

A. Maeder and A. Renzini (eds.), Observational Tests of the Stellar Evolution Theory, 583–585.
© *1984 by the IAU.*

Mass loss is an important factor of stellar evolution. From the production
of white dwarfs to the production of WR stars, mass loss has to be introduced in due place, and with the proper intensity . However, it si quite
obvious that we do not understand completely the mechanism of mass loss,
and that we do not have a consistent theory of mass loss. HEAO2, with the
Einstein observatory, has shown in stars the presence of unexpected chromospheres, and it has been shown by Leroy and Lafon (1982) , in the case
of hot stars, that if radiation pressure can achieve the final velocity
of the wind, a chromosphere is necessary, in order to provide the actual
flux of matter . We believe that the formation of a chromosphere is the
key of stellar winds. However, we do not have a full, consistent theory
of the formation of chromospheres. Acoustic waves are important, electro-
magnetic phenomena must probably be included, but their presence and their
effects are Ω - dependant . Are the stellar winds related to the rotation
and how ?And which fracion of the stellar winds is generated by pulsations ?

Mass loss is not a steady process . The rate of mass loss is time dependant . There are activity periods, where $\dot{M}$ is large, follown by quiscient
periods . Is there an explanation of this sort of relaxation process ?
The effect of mass loss on stellar evolution is taken into account by
introducung sort of time average of mass loss . Is that correct ? During
steady mass loss, entropy adjustment of a star takes place easily , but
as mentionned by Shu and Lubow (1981) in the case of binaries, is it
still the case if mass loss takes place through a succession of bursts ?

Finally, what is the Z dependance of mass loss ? The depth of the convective zone, the chromospheric structure, the effect of radiation pressure
are all Z-dependant. We can expect some influence of the chemical composition on mass loss .

Altogether, the only reliable quantitative estimate of mass loss is given
on one side by the estimate of the mass of the progenitors of white
dwarfs (8 to 10 solar masses) , and by the estimate of mass loss from
the WR progenitors ($M \gtrsim 40$ solar masses) compared to the actual mass of
WR stars.

I have just mentionned the connection between chromosphere, rotation and
mass loss . This brings me to a series of unsolved questions related to
rotation . What about the dynamo mechanism ? What is in this connection
the origin and the fate of Ap stars ? What are their descedents ?
We have learned about the conditions of appearance of instabilities .
But what about the conditions of disappearance of the instabilities ?
Is there a sequence of critical Ω's corresponding to various levels of
instability and mixing, up to the Ω of the blue stragglers ? Is it possible to measure the rotation of evolved stars ? and what about Veq of
the blue stragglers ?

Nuclear processing, mixing, dredge-up, winds have been extensively mentionned, but there is a feed back mechanism which remains to be understood . The depth of the convective zone is Z-dependant; the wind production is Z dependant; the way in which chemical elements are expelled

duction is Z dependant; the way in which chemical elements are expelled
into the galaxy is also Z-dependant . We then have also the rate of
mixing in the galaxy, so that the whole loop :

<pre>
                    interstellar matter   ←┐
                    star formation   │
                    stellar evolution   │
                    elements ejection   │
                    galactic mixing   →┘
</pre>

is Z dependant . Is the loop stable or unstable, and at what time scale?

There still question concerning the absorption coefficient which seem
to be yet unsolved, despite the important improvements obtained during
the last years .

High precision astrophysics , with the arrival of high precision parallaxes within a few years (HIPPARCOS) will provide us with new data of
a yet unknown precision : 2% in the luminosities, 3% for the masses ,
for a total number of perhaps100 binaries . In the mean time, high
resolution spectroscopy will allow the detection of subtle and important
variations in the surface abundances . Without writing the full list of
new and high precision measurements which will come, we can remember the
estimate of Harwit (1981) of about one hudred discoveries to come with
the use of better and well adapted resolution in time, frequency, light,
angular size and unlimited range of wave length .

Coming back, as a conclusion to H.N.Russell quotation , we shall conclude
by recalling that the conflict between observation on one side , and
rules , models or theories on the other side is always the source of new
discoveries .

Bibliography .

Harwit M. 1981, Cosmic Discoveries , Basic Books, New York .
Leroy M., Lafon J.-P.J. 1982 , Astron.Astrophys., 106, 266
                                                   106, 345
Shu F.H., Lubow S.H., 1981, Annual rev. Astron. Astrophys., 19,277 .
Russel H.N. 1941, Actualités Scientifiques et industrielles, 903,
   Librairie Hermann, Paris 1941.